AF361857

Advanced Computational Intelligence for Object Detection, Feature Extraction and Recognition in Smart Sensor Environments

Advanced Computational Intelligence for Object Detection, Feature Extraction and Recognition in Smart Sensor Environments

Editor

Marcin Woźniak

MDPI • Basel • Beijing • Wuhan • Barcelona • Belgrade • Manchester • Tokyo • Cluj • Tianjin

Editor
Marcin Woźniak
Silesian University of
Technology
Poland

Editorial Office
MDPI
St. Alban-Anlage 66
4052 Basel, Switzerland

This is a reprint of articles from the Special Issue published online in the open access journal *Sensors* (ISSN 1424-8220) (available at: https://www.mdpi.com/journal/sensors/special_issues/computational_intelligence_object_detection).

For citation purposes, cite each article independently as indicated on the article page online and as indicated below:

LastName, A.A.; LastName, B.B.; LastName, C.C. Article Title. *Journal Name* **Year**, *Volume Number*, Page Range.

ISBN 978-3-0365-1268-6 (Hbk)
ISBN 978-3-0365-1269-3 (PDF)

Contents

About the Editor

Marcin Woźniak received an M.Sc. degree in applied mathematics, a Ph.D. degree in computational intelligence and a D.Sc. degree in computational intelligence, in 2007, 2012, and 2019, respectively. He is currently an associate professor with the Faculty of Applied Mathematics, Silesian University of Technology. He is a scientific supervisor in editions of "The Diamond Grant" and "The Best of the Best" programs for highly talented students from the Polish Ministry of Science and Higher Education. He has participated in various scientific projects (as lead investigator, scientific investigator, manager, or participant) at Polish and Italian universities and in the IT industry. He was a visiting researcher with universities in Italy, Sweden, and Germany. He has authored/coauthored over 150 research papers in international conferences and journals. His current research interests include neural networks with their applications together with various aspects of applied computational intelligence. He was awarded by the Polish Ministry of Science and Higher Education with a scholarship for outstanding young scientists. In 2020, he was nominated among the "Top 2% of Scientists in the World" by Stanford University for his career achievements.

M. Woźniak was an editorial board member or an editor for Sensors, IEEE ACCESS, Measurement, Frontiers in Human Neuroscience, PeerJ Computer Science, International Journal of Distributed Sensor Networks, Computational Intelligence and Neuroscience, Journal of Universal Computer Science, etc., and a session chair at various international conferences and symposiums, including IEEE Symposium Series on Computational Intelligence, IEEE Congress on Evolutionary Computation, etc.

rial

lvanced Computational Intelligence for Object Detection, ature Extraction and Recognition in Smart nsor Environments

cin Woźniak

Faculty of Applied Mathematics, Silesian University of Technology, 44-100 Gliwice, Poland; marcin.wozniak@polsl.pl

check for
updates

on: Woźniak, M. Advanced
utational Intelligence for Object
ion, Feature Extraction and
nition in Smart Sensor
nments. *Sensors* **2021**, *21*, 45.
/dx.doi.org/10.3390/s21010045

ed: 7 December 2020
ted: 22 December 2020
hed: 24 December 2020

her's Note: MDPI stays neu-
th regard to jurisdictional claims
lished maps and institutional
ions.

1. Special Issue

The recent years have seen a vast development in various methodologies for object detection and feature extraction and recognition, both in theory and in practice. When processing images, videos, or other types of multimedia, one needs efficient solutions to perform fast and reliable processing. Computational intelligence is used for medical screening where the detection of disease symptoms is carried out, in prevention monitoring to detect suspicious behavior, in agriculture systems to help with growing plants and animal breeding, in transportation systems for the control of incoming and outgoing transportation, for unmanned vehicles to detect obstacles and avoid collisions, in optics and materials for the detection of surface damage, etc. In many cases, we use developed techniques which help us to recognize some special features. In the context of this innovative research on computational intelligence, contributions to the Special Issue "Advanced Computational Intelligence for Object Detection, Feature Extraction and Recognition in Smart Sensor Environments" present an excellent opportunity for the dissemination of the recent results and achievements for further innovations and development.

Among the total 88 manuscript submissions to this Special Issue, only 24 manuscripts were accepted after a rigorous reviewing process and published in final forms as a separate MDPI *Sensors* volume collection under the link https://www.mdpi.com/journal/sensors/special_issues/computational_intelligence_object_detection. This creates an acceptance rate at the level of 27.2%, which confirms the high level of presented research and the outstanding interest of researchers in contributing their innovative research articles to this venue. The published articles show innovative research results from authors from Europe, Asia, the Americas, and Africa, showing a worldwide research interest in the topic of this Special Issue and the importance of the proposed contributions. The published articles cover important fields of science and technology by showing models and applications for medical image processing, automated drone and vehicle driving systems, marine object detection and recognition, and agriculture and harvesting, with many interesting theoretical aspects of new training models and data augmentation. Additionally, the published articles bring new data sets to the scientific community—i.e., defect detection from optical fabric images and Industrial10 for industrial area image processing.

2. Contributions

The topic of using computer vision for autonomous driving systems, aerial vehicles, and vessel classification has been covered by many innovative ideas. In [1], a model of a system developed for the detection of flying objects for automatic drone protection systems was presented. A proposed solution is composed of a background subtraction model which cooperates with the applied model of the convolutional neural network (CNN). As a result, the system detects flying drones and provides their initial recognition to the operator. In [2], a model was proposed for ship type classification. The proposed complex neural

s **2021**, *21*, 45. https://dx.doi.org/10.3390/s21010045

architecture was based on a time convolutional layer model which helped to compare
extracted ship features. In [3], the authors discuss a model of vehicular traffic conges
with various approaches. As a result of this, a study presented a set of comparative res
for different deep learning models. In [4], a real-time vehicle detection drone system
developed which can detect a car from a bird-view perspective. The model was ba
on an adapted DRFBNet300 structure. In [5], the YOLOv2 model was adapted to
task of multi-scale vehicle detection. The adopted neural network was enhanced v
a proposed foreground–background imbalance estimation. Another interesting m
for non-conventional vessel detection was presented in [6]. An applied system usi
convolutional neural network (CNN) was trained by the Adam algorithm. The aut
compared various architectures and drew conclusions for the best applications in
automatic detection system.

Among interesting propositions for potential industrial applications, we can
applications for various types of images, from object surfaces to whole-scene proces
In [7], a new approach for correlating scene images in industrial areas was discussed
this model, a concept of a regression model of nested markers was used for viewpo
in augmented reality. As a result, the research presented a more efficient image captu
technique for industrial applications, but also a new data set called Industrial10. We ki
encourage the scientific community to adapt this data set in the research on camera p
regression methods. In [8], a model to detect surface regions of interest (ROI) in 3D
presented. As a processing mechanism, a deep convolutional neural network (C
modeling mechanism was adapted with the Adam training algorithm. This combina
was applied in industrial processes to optimal CCD laser image scanning with very g
results. In [9], an idea of composite interpolating feature pyramid (CI-FPN) was app
in a model of fabric defect detection. The result was processed by a cascaded gui
region proposal network (CG-RPN) to classify the detected regions. In addition to
model, this research article also introduced a new data set for defect detection from op
fabric images. In [10], a model of a convolutional neural network (CNN) for indus
application in tool wear identification was presented, where parts of the face milling pro
can be evaluated for potential damage. An application in farming and plant growing
proposed in [11]. A proposed model of a weakly dense connected convolution netv
(WeaklyDenseNet-16) was used to detect plant disease from images. In [12], a sys
model for robotic inspection tasks was proposed. The proposed system enabled dron
detect novelty in inspected areas from a distant viewpoint.

This Special Issue also received interesting research presentations concerning hu
pose detection and recognition. In [13], an innovative video frame analysis mode
surveillance and security applications was presented. The model uses a support ve
machine (SVM) or a convolutional neural network (CNN) as an extractor and detecto
key features from CCTV and operation units. As a result, a faster detection of pote
situations for legal actions was achieved. In [14], a model of active player detection
sport vision system was presented. The solution was based on the idea of a bounding
area, which was associated with motion centroids of the human body pose. As a resu
model of active support for sport transmission to annotate players during the game
developed. In [15], a hand gesture recognition model was proposed. Such a developn
can be very useful for a man–machine interaction system, where the computer should
human intention, i.e., from the hand gesture presented to the camera. The proposed m
was based on EMGNet architecture processing images collected by using electronic ma
devices such as the Myo armband.

Another important category is new models of image processing and feature extrac
and detection by the developed models of computational intelligence. In [16], a
approach to remote sensing image processing was presented, where the image shoul
cleared from radio-frequency interference (RFI) artefacts. The model used a proposed
value conversion from RGB to greyscale as a means to detect such artifacts and rem
them from the adapted neural network. In [17], a semantic segmentation approac

object extraction from images was examined. The model proposed adapted the WASPnet architecture working on the Waterfall Atrous Spatial Pooling (WASP) module. Experiments showed a high efficiency for various types of images. In [18], a comparative review for models of traffic sign detection systems based on various computational intelligence techniques was presented.

The Special Issue received several interesting articles in the domain of medical image processing, where new ideas proposed models of detection and recognition of tissue features. In [19], an applied model of a SegNet convolutional neural network encoder-decoder construction used for more efficient medial image processing was presented. As a result, a processing model for tumor segmentation in CT liver scans in a DICOM format was proposed. In [20], a model for human embryo image generator based on generative adversarial networks (GAN) trained using the Adam algorithm was proposed. The resulting model enables one to manipulate the size, position, and number of artificially generated embryo cells in the composed image. In [21], acute brain hemorrhages on computed tomography scans were detected with the use of an adapted 3-dimensional convolutional neural network. The main goal of such a system is to efficiently reduce the time between diagnosis and treatment.

The Special Issue also received some interesting propositions for various pattern analysis. In [22], a simulation result for vibration signals of high-speed trains for non-stationary object modeling was presented. The research presents the use of intelligent modeling for signal noise reduction. The model proposed in [23] discussed an idea for information retrieval from large-scale text data by using BERT (CLS) representation. To improve this, an efficiency method was based on reasoning paths from a composed cognitive graph structure. In [24], a multi-view approach was discussed for visual question answering (VQA) systems, which are encountered in complex artificial intelligence systems, as operating both in text conversation and image processing and recognition. The proposed approach gave us a chance to boost such systems due to processing several images from one scene, and this therefore enabled the system to consider more aspects on the way to a final decision.

In summary, we should congratulate the authors of the articles accepted in this Special Issue for their outstanding research results and wish them great success in the continuation of their research and projects for future development. The topic of the Special Issue was clearly well accepted in the worldwide scientific community, which gives a sign for the future research and direction of trends for technology and science in the field of computational intelligence for object detection, feature extraction, and recognition in smart sensor environments.

Acknowledgments: The editor would like to thank all the authors who have contributed their innovative works to the Special Issue "Advanced Computational Intelligence for Object Detection, Feature Extraction and Recognition in Smart Sensor Environments". Thanks are also given to all the hardworking reviewers for their detailed comments and suggestions in the revision process. I also acknowledge the great help and involvement of the technical editorial team for managing all the publications.

References

Seidaliyeva, U.; Akhmetov, D.; Ilipbayeva, L.; Matson, E.T. Real-Time and Accurate Drone Detection in a Video with a Static Background. *Sensors* **2020**, *20*, 3856. [CrossRef] [PubMed]

Shen, S.; Yang, H.; Yao, X.; Li, J.; Xu, G.; Sheng, M. Ship Type Classification by Convolutional Neural Networks with Auditory-Like Mechanisms. *Sensors* **2020**, *20*, 253. [CrossRef]

Impedovo, D.; Balducci, F.; Dentamaro, V.; Pirlo, G. Vehicular Traffic Congestion Classification by Visual Features and Deep Learning Approaches: A Comparison. *Sensors* **2019**, *19*, 5213. [CrossRef]

Han, S.; Yoo, J.; Kwon, S. Real-Time Vehicle-Detection Method in Bird-View Unmanned-Aerial-Vehicle Imagery. *Sensors* **2019**, *19*, 3958. [CrossRef]

Wu, Z.; Sang, J.; Zhang, Q.; Xiang, H.; Cai, B.; Xia, X. Multi-Scale Vehicle Detection for Foreground-Background Class Imbalance with Improved YOLOv2. *Sensors* **2019**, *19*, 3336. [CrossRef]

6.   Wlodarczyk-Sielicka, M.; Połap, D.; Sielicka, W. Automatic Classification Using Machine Learning for Non-Conventional Ves on Inland Waters. *Sensors* **2019**, *19*, 3051. [CrossRef]

7.   Yang, C.; Simon, G.; See, J.; Berger, M.-O.; Wang, W. WatchPose: A View-Aware Approach for Camera Pose Data Collectio Industrial Environments. *Sensors* **2020**, *20*, 3045. [CrossRef]

8.   Zhao, L.; Li, F.; Zhang, Y.; Xu, X.; Xiao, H.; Feng, Y. A Deep-Learning-based 3D Defect Quantitative Inspection System in Products Surface. *Sensors* **2020**, *20*, 980. [CrossRef] [PubMed]

9.   Wu, Y.; Zhang, X.; Fang, F. Automatic Fabric Defect Detection Using Cascaded Mixed Feature Pyramid with Guided Localiza *Sensors* **2020**, *20*, 871. [CrossRef] [PubMed]

10.   Wu, X.; Liu, Y.; Zhou, X.; Mou, A. Automatic Identification of Tool Wear Based on Convolutional Neural Network in Face Mil Process. *Sensors* **2019**, *19*, 3817. [CrossRef]

11.   Xing, S.; Lee, M.; Lee, K.-K. Citrus Pests and Diseases Recognition Model Using Weakly Dense Connected Convolution Netw *Sensors* **2019**, *19*, 3195. [CrossRef]

12.   Contreras-Cruz, M.A.; Ramirez-Paredes, J.P.; Hernandez-Belmonte, U.H.; Ayala-Ramirez, V. Vision-Based Novelty Detec Using Deep Features and Evolved Novelty Filters for Specific Robotic Exploration and Inspection Tasks. *Sensors* **2019**, *19*, 2 [CrossRef] [PubMed]

13.   Wilkowski, A.; Stefańczyk, M.; Kasprzak, W. Training Data Extraction and Object Detection in Surveillance Scenario. *Sensors* 2 *20*, 2689. [CrossRef] [PubMed]

14.   Pobar, M.; Ivasic-Kos, M. Active Player Detection in Handball Scenes Based on Activity Measures. *Sensors* **2020**, *20*, 1 [CrossRef] [PubMed]

15.   Chen, L.; Fu, J.; Wu, Y.; Li, H.; Zheng, B. Hand Gesture Recognition Using Compact CNN via Surface Electromyography Sig *Sensors* **2020**, *20*, 672. [CrossRef] [PubMed]

16.   Chojka, A.; Artiemjew, P.; Rapiński, J. RFI Artefacts Detection in Sentinel-1 Level-1 SLC Data Based on Image Proces Techniques. *Sensors* **2020**, *20*, 2919. [CrossRef] [PubMed]

17.   Artacho, B.; Savakis, A. Waterfall Atrous Spatial Pooling Architecture for Efficient Semantic Segmentation. *Sensors* **2019**, *19*, 5 [CrossRef]

18.   Wali, S.B.; Ker, P.J.; Hannan, M.; Hussain, A.; Samad, S.A.; Ker, P.J.; Bin Mansor, M. Vision-Based Traffic Sign Detection Recognition Systems: Current Trends and Challenges. *Sensors* **2019**, *19*, 2093. [CrossRef]

19.   AlMotairi, S.; Kareem, G.; Aouf, M.; Almutairi, B.; Salem, M.A.M. Liver Tumor Segmentation in CT Scans Using Modified Seg *Sensors* **2020**, *20*, 1516. [CrossRef]

20.   Dirvanauskas, D.; Maskeliūnas, R.; Raudonis, V.; Damaševičius, R.; Scherer, R. Hemigen: Human embryo image generator ba on generative adversarial networks. *Sensors* **2019**, *19*, 3578. [CrossRef]

21.   Ker, J.; Singh, S.P.; Bai, Y.; Rao, J.; Lim, C.T.; Wang, L. Image Thresholding Improves 3-Dimensional Convolutional Ne Network Diagnosis of Different Acute Brain Hemorrhages on Computed Tomography Scans. *Sensors* **2019**, *19*, 2167. [Cross [PubMed]

22.   Ning, J.; Fang, M.; Ran, W.; Chen, C.; Li, Y. Rapid Multi-Sensor Feature Fusion Based on Non-Stationary Kernel JADE for Small-Amplitude Hunting Monitoring of High-Speed Trains. *Sensors* **2020**, *20*, 3457. [CrossRef] [PubMed]

23.   Wang, Y.; Yao, B.; Wang, T.; Xia, C.; Zhao, X. A Cognitive Method for Automatically Retrieving Complex Information on a L Scale. *Sensors* **2020**, *20*, 3057. [CrossRef] [PubMed]

24.   Qiu, Y.; Satoh, Y.; Suzuki, R.; Iwata, K.; Kataoka, H. Multi-View Visual Question Answering with Active Viewpoint Select *Sensors* **2020**, *20*, 2281. [CrossRef] [PubMed]

Article

Real-Time and Accurate Drone Detection in a Video with a Static Background

Ulzhalgas Seidaliyeva [1,†,‡], Daryn Akhmetov [2,‡], Lyazzat Ilipbayeva [2,‡] and Eric T. Matson [3,*]

1   Department of Electrical Engineering, Telecommunications and Space Technologies, Satbayev University, Almaty 050000, Kazakhstan; useidali@purdue.edu
2   Department of Radio Engineering, Electronics and Telecommunications, International IT university, Almaty 050000, Kazakhstan; 24168@iitu.kz (D.A.); l.ilipbayeva@edu.iitu.kz (L.I.)
3   Department of Computer and Information Technology, Purdue University, West Lafayette, IN 47907-2021, USA
*   Correspondence: ematson@purdue.edu; Tel.: +1-(765)-494-82-59
†   Current address: 401 North Grant Street, KNOY 255, West Lafayette, IN 47907-2021, USA.
‡   These authors contributed equally to this work.

Received: 6 June 2020; Accepted: 7 July 2020; Published: 10 July 2020

Abstract: With the increasing number of drones, the danger of their illegal use has become relevant. This has necessitated the creation of automatic drone protection systems. One of the important tasks solved by these systems is the reliable detection of drones near guarded objects. This problem can be solved using various methods. From the point of view of the price–quality ratio, the use of video cameras for a drone detection is of great interest. However, drone detection using visual information is hampered by the large similarity of drones to other objects, such as birds or airplanes. In addition, drones can reach very high speeds, so detection should be done in real time. This paper addresses the problem of real-time drone detection with high accuracy. We divided the drone detection task into two separate tasks: the detection of moving objects and the classification of the detected object into drone, bird, and background. The moving object detection is based on background subtraction, while classification is performed using a convolutional neural network (CNN). The experimental results showed that the proposed approach can achieve an accuracy comparable to existing approaches at high processing speed. We also concluded that the main limitation of our detector is the dependence of its performance on the presence of a moving background.

Keywords: unmanned aerial vehicles; object detection; deep learning; computer vision; image processing; drone detection; UAV detection; visual detection

---

1. Introduction

With the constant development of technology, drone companies such as DJI, Parrot, and 3DRobotics are producing different types of unmanned aerial vehicles (UAVs) or systems (UAS). Because of their accessibility and ease of use, UAVs are widely used for commercial purposes, such as the delivery of goods and medicines, surveying, the monitoring of public places, cartography, search and rescue (SAR), first aid, and agriculture. However, the wide and rapid spread of UAVs causes danger when the illegal flight of drones is used for crimes such as smuggling (the illegal transportation of goods at borders, in restricted areas, prisons, etc.), illegal video surveillance, and interference with aircraft flying. In recent years, unmanned aerial vehicles, which are publicly known as drones, have hit the headlines by flying over restricted zones and entering the high-security areas. In January 2015, a drone flown by an intoxicated government officer crashed right in front of the White House's lawn [1]. Another accident happened in 2017 in the Canadian province of Quebec, where during landing, a plane with a light engine crashed into a UAV at an elevation of 450 m [2]. Fortunately, the plane only suffered small damage and was able to land safely. In December 2018, London's Gatwick Airport was shut down for 36 h with reports of drones over the runway, which strangely appeared whenever the airport attempted to reopen [3]. Because of this incident, approximately 1000 flights had to be cancelled, which affected the lives of 140,000 passengers. Due to low visibility of detection, drones can be ideal tools for illegal smuggling. In April 2020, in the state of Georgia, three people were accused of arranging to transport tobacco and phones by means of a drone to a convict at Hays State Prison in Trion [4]. The given examples of drone incidents show the need to monitor the flight of drones. To guarantee security, some drone producer companies have set up no-fly zones by prohibiting drones from flying within a 25 km radius of a few sensitive zones, such as airports, prisons, power plants, and other critical facilities [5]. However, the impact of no-fly zones is exceptionally constrained, and not all drones have those built-in safeguards. Therefore, to solve this problem, the development of anti-drone systems is vigorously developing, and the problem of real-time drone detection is becoming relevant [6]. Drone detection technologies are usually divided into four categories: acoustic, visual, radio-frequency signal-based, and radar [7]. A good balance between price and detection range is achieved using visual drone detection technologies that use images of surveillance areas from cameras. One of the main disadvantages of visual drone detection is the high level of false positives caused by the visual similarity of different objects, especially when they occupy several pixels in an image [8]. As a result, a drone can be mistaken for a bird or background and vice versa. The task becomes even more difficult due to large changes in images caused by varying weather and lighting conditions. At the same time, drones can reach speeds of up to 100 miles per hour, which imposes additional requirements on the speed of detection. To address these problems, the Drone-vs-Bird detection challenge [5] was established. The challenge provides video sequences in which drones are present along with birds. The goal is to detect all drones that appear on video, while birds should not be mistaken for drones. The challenge focuses on the detection accuracy of proposed algorithms, but the execution time of the algorithms is not considered. Our objective was to develop a real-time drone detection algorithm that could achieve a competitive accuracy.

1.1. Drone Detection Modalities

Based on investigations conducted by academia and commercial industries, the primary modalities that can be used for drone detection and classification tasks are radar, radio frequency (RF), acoustic sensors, and camera sensors supported by computer vision algorithms [7].

1.1.1. Radar-Based Drone Detection

Radar is considered to be a traditional sensor that provides the robust detection of flying objects at long-range distances and almost uninfluenced performance in unfavorable light and weather conditions [9,10]. As radar sensors are mostly designed for detecting high velocity ballistic trajectory targets such as military drones, aircrafts, and missiles, they are not suitable to detect small commercial UAVs that fly with relatively lower non-ballistic trajectory velocities [11]. While radar sensors are well-known as reliable solutions for detection, their classification abilities are not optimal [9]. Since UAVs and birds have key characteristics that often make them difficult to distinguish, the above-mentioned drawback of radar sensors makes it an unprofitable solution for the classification task of UAVs and birds. The complexity of installation and high cost of radar sensors are other reasons that necessitate a relatively low-cost anti-drone system.

1.1.2. Acoustic-Based Drone Detection

Relatively low-cost acoustic detection systems use an array of acoustic sensors or microphones to classify specific acoustic patterns of UAV rotors, even in low visible environments [7]. However, the maximum operational range of these systems remains below 200–250 m. Additionally, sensitivity of these systems to environment noise, especially in urban or noisy loud areas and wind conditions, influences detection performance.

1.1.3. RF-Based Drone Detection

RF-based UAV detection system is one of the most popular anti-drone systems in the market, and they detect and classify drones by their RF signatures [12]. An RF sensor is a passive listener between a UAV and its controller, and the sensor does not transfer any signal like in radar-based systems, which makes RF-based detection energy efficient. Unlike acoustic sensors, RF sensors solve the limited detection range problem by utilizing high-gain recipient antennas together with highly sensitive recipient systems to listen to UAV controller signals, and the environmental noise problem is suppressed by using some de-noising methods such as band pass filtering and wavelet decomposition [13]. However, not all drones have RF transmission, and this approach is not suitable for detecting UAVs operating autonomously without communication channels [7].

1.1.4. Camera-Based Drone Detection

The detection of drones that do not have RF transmission can be performed by using low cost camera sensors based on computer vision algorithms. It is well-known that detection and classification abilities are highest when the target is visible, and camera sensors have the advantage of giving a formal vision verification that the detected object is a drone while giving extra visual information such as drone model, dimensions, and payload that other drone detection systems cannot provide [14]. A medium detection range, good localization, affordable price, and easy human interpretation are achieved using visual drone detection technologies that use images of surveillance areas from cameras. However, this modality operates poorly at nighttime and in limited visibility conditions such as in the presence of clouds, fog, and dust. To address some issues of such scenarios, thermal cameras can be utilized in combination. Thermal cameras can solve the detection issue of nighttime surveillance, and sometimes, depending on the used technology, they even can operate better in rain, snow, and fog weather conditions. However, high quality thermal cameras are used for military applications, and low cost commercial thermal cameras might fail in high humidity weather conditions or other unfavorable environmental conditions [9].

1.1.5. Bi- and Multimodal Drone Detection Systems

As we can see, each of these modalities has its specific limitations, and a robust anti-drone system might be complemented by fusing several modalities. In order to develop a cost-efficient drone monitoring system, some researchers [15] considered composing a sensor network with different types of sensors. Depending on the number of sensors used for the detection task, bimodal and multimodal drone detection systems can exist [7]. To improve detection accuracy, a bimodal drone detection system can combine two different modalities such as camera array and audio assistance [16], camera and radar sensors [17], and radar and audio sensors [18]. Meanwhile, a multimodal drone detection system can be performed with the simultaneous use of acoustic arrays; optical and radar sensors [19]; or simple radar, infrared, and visible cameras—as well as an acoustic microphone array [20]. Therefore, a maximal system performance can be achieved by fusing several drone detection modalities. However, our focus is the approach that uses camera images and computer vision algorithms.

1.2. Related Work

1.2.1. UAV Detection and Classification

Drone detection based on visual data (image or video) can be performed using handcrafted feature-based methods [8,21,22] and deep learning-based [6,23–25] algorithms. Handcrafted feature-based methods are based on traditional machine learning algorithms by using traditional descriptors such as scale-invariant feature transform (SIFT), histogram of oriented gradients (HOG), Haar, local binary pattern (LBP), deformable parts model (DPM), and generic Fourier descriptor (GFD) that provide low-level handcrafted features (edges, drops, blobs, and color information) and classical classifiers (support vector machine (SVM), AdaBoost)), whereas the second category relies on the learned features using two-stage (region-based convolutional neural network (R-CNN), Fast R-CNN, Faster R-CNN, and Mask R-CNN) and single-stage (single shot detector (SSD), RetinaNet, and you only look once (YOLO)) deep object detectors.

Drone detection using handcrafted feature-based methods: Unlu et al. [21] developed GFD vision-based features that are invariant to translation and rotation changes to describe the binary forms (such as silhouettes) of drones and birds. In accordance with the system proposed by the authors, the silhouette of a moving object is obtained using a fixed wide-angle camera and a background subtraction algorithm—the region growing algorithm—is used to separate the pixels of the object from the background. In order to avoid the loss of any form information morphological operations are not used after the image segmentation phase, GFD is calculated after the normalization and centering of the silhouette; finally, GFD signals are classified into birds and drones through a neural network consisting approximately 10,000 neurons. To teach their system, the authors created a dataset including 410 drone images and 930 bird images collected from open sources. Training and testing on the custom dataset were performed by using five-fold cross validation. In the test data, CNN classification accuracy was 85.08%, whereas the proposed GFD method showed an accuracy of 93.10%, and the CNN architecture significantly increased the classification efficiency of a small dataset by including the GFD signal vector before classifying the neural network. In [8], the authors proposed two methods of detecting and tracking an unmanned aerial vehicles at a distance of no more than 350 feet during the daytime using image processing and motion detection to control movement and to extract the drone detected by machine learning. The authors made a comparative analysis of the MATLAB, OpenCV, and EmguCV packages currently used in image processing and object detection, and they used OpenCV in their work. According to the proposed system, to reduce the memory, the RGB image captured by a USB camera is converted to grayscale, and the adaptive threshold method is used to adjust the noise level of the image depending on the light condition by setting the threshold value to 60. To eliminate some noise, a dilation morphological operation is used by enlarging the image until it is clearly visible, including a blob tracking algorithm to hold the object and the Dlib technique if the object moves in three frames. To test the proposed system, the authors tested four types of drones (Phantom 4 Pro, Agras MG-1s, Pocket Drone JY019, and Mavic Pro), as well as other objects (birds and balloons). Wang et al. [22] proposed a simple, fast, and efficient detection system for unmanned aerial vehicles based on video images shot with static cameras that covers a large area and is very economical. The method of temporal median background subtraction was used to identify moving objects in a static video camera, and then global Fourier descriptors and local HOG features were obtained from images of moving objects. As a result, the combined Fourier descriptor (FD) and HOG features were sent to the SVM classifier, which performed classification and recognition. To prepare a dataset, the authors converted 10 videos of the unmanned quadcopter Dajiang Phantom 4, taken in various positions, into a series of images; as a result, the drones became a positive class, and other objects such as leaves and buildings were manually annotated as a negative class. For the recognition of "drone" and "non-drone" objects, FD, HOG, and the proposed FD and HOG algorithms were used, the overall accuracy of the proposed recognition method was 98%. The authors also experimentally proved that the proposed FD and HOG algorithm, even with a small dataset, could perform the task of classifying birds and drones with a greater accuracy than the GFD algorithm.

Drone detection using deep learning-based methods: Manja Wu et al. [6] developed a real-time drone detector using the deep learning method. Since training a reliable detector requires a large number of training images, the authors first developed a semi-automatic dataset with a KCF (kernelized correlation filter) tracker instead of manual labeling. The semi-automatic method of labeling datasets based on the KCF tracker accelerated the process of preprocessing the trained images. The authors developed the YOLOv2 deep learning model by changing the resolution structure of input images and adjusting the size parameters of the anchor box. To get the detection network, the authors removed the last convolution layer of Darknet19, which was previously trained on ImageNet dataset, and added three 3×3 convolution layers with 1024 filters and one 1×1 convolution layer with 30 filters at the end of the network. The network was trained using a public-domain USC (University of Southern California) drone dataset and an anti-drone dataset labeled with a KCF tracker. The 2 and 4 GB graphics processing unit (GPU)-random-access memory (RAM) configurations were used to test the detector's operation in real time and at a low cost. With 2 GB of GPU-RAM, the processing speed reached 19 frames per second (FPS), whereas with 4 GB of GPU-RAM, the processing speed reached 33 FPS. Through various experiments, the authors achieved good results in real-time detection using the proposed detector at an affordable price for the system. Yoshihashi et al. [23] proposed a new integrated detection and control system using information on the movement of small flying objects. This system, called the recurrent convolutional network (RCN), consists of four modules, each of which performs a specific task: a convolutional layer, convolutional long short-term memory (ConvLSTM), a cross-correlation layer, and a fully connected layer. The authors used training methods for AlexNet and Visual Geometry Group-16 (VGG-16) tuning systems without training the system from scratch. To evaluate the system, it was first tested on a bird dataset to detect birds around a wind farm, and then the system was tested on a drone dataset of 20 manually captured video components to check whether the system could be applied to other flying objects. The experimental results presented in the form of receiver operating characteristics (ROC) curves showed that the proposed system gave better results than previous solutions. In [24], the authors proposed a drone tracking system that provides the exact location of the drone in the input image based on deep learning. The proposed system consists of detection and tracking modules that complement each other to achieve high performance. While the Faster R-CNN drone detection module detects and localizes the drone from static images, an object tracking module called MDNet (multi-domain network) determines the position of the drone in the next frame based on its position in the current frame, which allows one to identify a drone only in a certain area without looking for the entire frame. To prepare the dataset, the authors used a publicly available drone dataset consisting of 30 YouTube videos shot using different drone models and a USC drone dataset consisting of 30 videos shot using one drone model. Since the number of static drone images for the drone tracking problem was very limited and labeling is a laborious task, the authors developed a method for increasing data based on a model that generates training images and automatically annotates the position of the drone in each frame. The main idea of this technique is to cut out drone images in the background and place them on top of background images. The experiment results showed that, despite training in synthetic data, the proposed system worked well on realistic images of drones against a complex background. Peng et al. [25] used the physical rendering instrumentation tool (PBRT) to solve the problem of limited visual data by creating photorealistic images of UAVs. The authors developed a large-scale training set of 60,480 rendered images, choosing different positions and orientations of UAVs, 3D models, external materials, internal and external camera characteristics, environmental maps, and the post-processing of rendered images. To detect unmanned aerial vehicles, the Faster R-CNN network was precisely tuned to Detectron, recommended by Facebook artificial intelligence (AI) research, using the weights of the basic ResNet-101 model. On the basis of experimental results, the Faster R-CNN network, trained on rendered images, showed an average accuracy of 80.69% in the manually annotated UAV test set, 43.03% in the pre-trained COCO (Common Objects in Context) 2014 dataset, and 43.36% in the PASCAL VOC (Visual Object Classes) 2012 dataset, and it showed an average precision of 56.28% in the rendered training set. According to the results of the experiment, the average precision (AP) of the Faster R-CNN detection network trained on rendered images was relatively higher compared to other methods.

Hu et al. [26] adapted and fine-tuned the YOLOv3 detector [27] to detect drones. The authors collected a dataset consisting of images of drones on which they trained the detector. The video processing speed reached 56.3 frames per second.

The best results in drone detection have been achieved by detectors based on deep learning. This is evidenced by the fact that most studies on the detection of drones [28–31] have partially or totally relied on CNNs for solving the problem.

1.2.2. Drone-vs.-Bird Challenge

The primary related works on UAV detection are the methods proposed specifically for the Drone-vs-Bird detection challenge [5], which was organized in 2017 and 2019. The main goal of the challenge is to detect and distinguish drones from birds in short videos taken from a large distance by a static camera. To perform flying object detection, Saqib et al. [32] evaluated the Faster R-CNN [33] object detector with different CNN backbones. According to the results of the conducted experiment, the Faster R-CNN with VGG-16 backbone network performed better than other networks by reaching 0.66 mAP. The authors concluded that the experiment results might be improved by annotating birds as a separate class, which could reduce false positive factors and enable the trained model to accurately distinguish birds and drones. C. Aker et al. [34] solved the problem of predicting the location of a drone in video and distinguishing drones from birds by adapting and finetuning single-stage YOLOv2 [35] algorithm. An artificial dataset was created by mixing real images of drones and birds subtracted from their backgrounds with frames of coastal area videos. The proposed network was evaluated by using precision–recall (PR) curves, where precision and recall values reached the value of 0.9 at the same time. Nalamati et al. [28] examined the problem of detecting small drones using state-of-the-art deep learning methods such as the Faster R-CNN [33] and SSD [36]. Inception v2 [37] and ResNet-101 [31] were chosen as the backbone networks. The authors fine-tuned backbone networks using the Drone-vs-Bird challenge dataset, which consisted of 8771 frames extracted from 11 Moving Picture Experts Group (MPEG4)-coded videos. For each algorithm, two cases were considered when the drone is close to the camera, i.e., when it is big, and when the drone is far from the camera, that is small. According to the results of the conducted experiment, in the first case, all the algorithms were able to detect a drone, and in the second case, the Faster R-CNN with ResNet-101 backbone network could successfully detect both drones when two drones appeared in the frame simultaneously at large distances, whereas the Faster R-CNN with the Inception v2 backbone network was only able to detect one of two drones. The single-stage SSD detector, on the other hand, could not detect both long-range drones and was ineffective at detecting small objects. According to the conducted evaluation and the challenge results, the Faster R-CNN with ResNet-101 performed best and achieved recall and precision values of 0.15 and 0.1, respectively. The authors did not pay attention to the detection time, but in future work, they will use the detection time as a key indicator to evaluate the effectiveness of the proposed model in real-time drone detection application. David de la Iglesia et al. [38] proposed an approach based on the RetinaNet [39] object detector. To perform drone predictions at different scales, the feature pyramid network (FPN) [40] architecture was used, where lower pyramidal levels are responsible for detecting small objects and upper levels are focused on larger objects. The ResNet-50-D [29] was used as a backbone network that was trained on the Drone-vs-Bird and Purdue UAV [41] datasets. The precision attained on the Drone-vs-Bird challenge was around 0.52, while recall was 0.34. In addition, experimental results in this work showed that the use of motion information can significantly increase the accuracy of detection. According to the challenge results, the F1 score of this approach reached 0.41. In order to improve the accuracy of detection of existing detectors, some approaches used the additional processing of the data. For example, Magoulianitis et al. [30] pre-processed images with deep CNN with skip connection and network in network (DCSCN) super-resolution technique [42] before using the Faster-RCNN detector. As a result, the detector became capable of detecting very distant drones, increasing its recall performance. The results obtained on challenge were 0.59 for recall and 0.79 for precision. In some works [14,43], the solution was divided into two stages. In the first stage, all objects that are highly likely drones are detected. In the second stage, a high-precision classifier is applied to the detections to reduce the number of false positives. For example, Schumann et al. [43] designed a flying object detector using the median background subtraction or deep neural network-based region proposal network (RPN) algorithm. After that, the detected flying objects were classified into drones, birds, and clutter by using the proposed CNN classifier optimized for small targets. In order to train a robust classifier, the authors created their own dataset containing totally 10,386 images of drones, birds, and backgrounds. The proposed framework was evaluated by using five static camera sequences and one moving camera sequence of the Drone-vs-Bird challenge dataset. All appearing birds in the video sequences were manually annotated. The classification of flying objects for different input image sizes such as 16×16, 32×32, and 64×64 was performed separately for the author's own dataset and the Drone-vs-Bird challenge dataset. To participate in the 2017 Drone-vs.-Bird challenge, the authors proposed a VGG-conv5 RPN detector optimized for 16×16 image size, and, based on the challenge metric results, this team took first place in that competition. Celine Craye et al. [14] developed two separate networks—the semantic segmentation network U-net [44] for the detection stage and ResNetV2 network [45] for the classification stage. By achieving a recall value of 0.71 and a precision value of 0.76, their approach won the Drone-vs-Bird detection challenge in 2019.

2. Proposed Approach

In this work, we focused on the real-time detection of drones in the scene with a static background. As illustrated in Figure 1, our approach consists of 2 modules: a moving object detector and a drone–bird-background classifier.

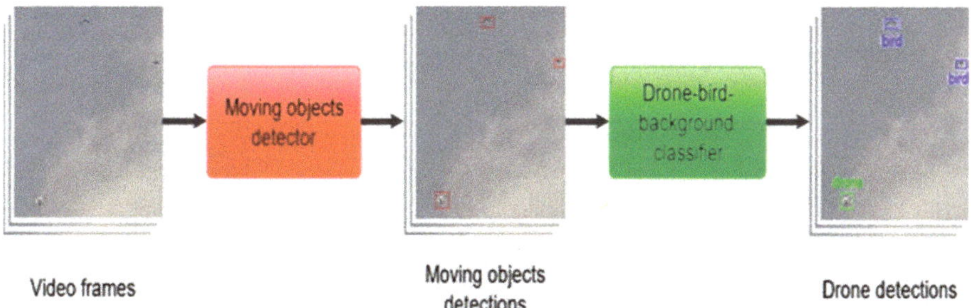

Figure 1. The proposed drone detection pipeline.

The motion detector was based on a background subtraction method. The outputs of this module are all moving objects in the scene. All the detections are fed into a classifier, which differentiates drones from other moving objects. The classifier is a CNN that was trained on the dataset of images of birds, drones, and backgrounds we collected.

2.1. Background Subtraction Method

There are several methods that are used for detecting flying or moving objects from a video sequence such as background subtraction, optical flow method, edge detection, and frame differencing [45]. Optical flow is used for motion estimation in a video and detects moving objects on the basis of objects' relative velocities in the scene. The complicated calculation of the optical flow method makes it inapplicable for real-time detection tasks [45]. By calculating the difference between the current and the previous frames of a video sequence, a frame differencing algorithm extracts the moving objects. Despite its advantages, including quick implementation, flexibility to dynamic changes of the scene, and relatively low computation, frame differencing is generally inefficient for extracting all the relevant pixels of the moving regions [45]. To detect a foreground object from the background of a video sequence background, the subtraction method is used. The background subtraction method is considered one of the widely used detection methods because of its fast and accurate detection, which makes it applicable for real-time detection. Additionally, it is easy to implement. The main drawback of this method is its invalidity for moving cameras, because each frame has different background. In our case, we focused on a video with a static background, and all the short videos of the Drone-vs-Bird detection challenge dataset were taken from a large distance by a static camera. Therefore, our motion detector was based on a background subtraction method.

Moving Objects Detection

The task of a motion detector is to detect all objects moving in a scene. The performance of this module was evaluated by its recall value. We conducted experimental studies of various motion detectors using the Drone-vs-Bird challenge dataset. The greatest recall was achieved by the motion detector based on the two-points background subtraction algorithm [46]. The output of the common background subtraction algorithm is a binary image in which the pixels that change their values in the next frame take the value of 1. The unchanged pixels are set to zero. In addition to moving objects, the output image contains noise in the form of single pixels distributed throughout the image. To remove this noise, the output binary image is filtered. An example of a filtering result is shown in Figure 2.

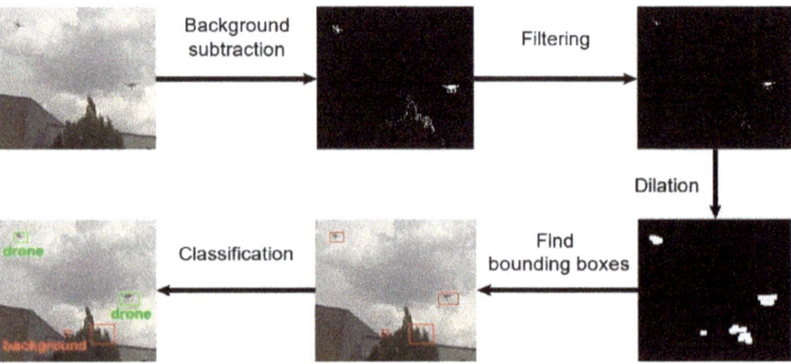

Figure 2. All steps of the proposed drone detection algorithm.

Next, dilation is performed to connect closely spaced pixels. This operation reduces the number of individual regions that are checked by the classifier, therefore increasing the processing speed of the detector. The last step of the moving object detector is to find the bounding boxes covering the regions found in the previous step. All found bounding boxes are sent to the drone–bird-background classifier.

2.2. CNN Image Classification

Audio, image, and text classification tasks are mostly performed using artificial neural networks. For image classification, CNNs are mainly used [45]. Usually, a CNN consists of three primary layers: the convolution, pooling, and fully connected (FC) layers. Convolution layers are the main building blocks of CNN models. Convolution is a mathematical operation for combining two sets of information. Convolution layers consist of filters and feature maps. A convolution operation is performed by sliding the filter along the input. In each place, the element-wise multiplication of the matrices is performed, and the result is summed. This sum is fed into the feature map. That is, trained filters are used to extract important features of the input, and the feature map is the output of the filter applied to the previous layer. The size of the filter that performs the convolution operation is always an odd number size ($1 \times 1, 3 \times 3, 5 \times 5, 7 \times 7, 11 \times 11$, etc.). Deep learning is commonly used to solve non-linear problems. The values obtained from the product of matrices in the convolutional layer are linear. To convert values to non-linear, after each convolutional layer, a non-linear activation function (elu, selu, relu, tanh, sigmoid, etc.) is usually used. The pooling layer is periodically added into a CNN's architecture. Its main function is to reduce the image size and compress each feature map by taking maximum pixel values in the grid. Most CNN architectures use max pooling. Max pooling uses a 2×2 window with stride of 2 and takes the largest elements of the input feature map; as a result, the output of the feature map is half the size. After going through the processes described above, the model is able to understand the features. The fully connected layer comes after the convolution, activation, and pooling layers. The outputs of convolution and pooling layers are always three-dimensional (3D), but a FC layer expects a one-dimensional vector of numbers. Therefore, we flatten the output of a pooling layer to a vector, and it becomes the input of a FC layer [47]. Then, it is inserted into the nodes of the neural network, which performs the classification. Different CNN architectures exist, such as LeNet, AlexNet, VGGNet (VGG-16 and VGG-19), GoogLeNet (Inception v1), ResNet (ResNet-18, ResNet-34, ResNet-50, etc.), and MobileNet (MobileNetV1 and MobileNetV2). They differ in the number of layers and trainable parameters' sizes. These networks have very deep networks and can have thousands or even million parameters. The huge number of parameters lets the network learn more difficult patterns, which improves classification accuracy. On the other hand, the huge number of parameters affects the training speed, required memory for saving the network, and computational complexity. MobileNet [48] is an efficient convolutional neural network architecture that reduces the amount of memory used for computing while maintaining a high predictive accuracy. It was developed by Google and trained on the ImageNet dataset. This network is suitable for mobile devices or any devices with low computational power. MobileNetV1 has two layers, which are the depthwise convolution layer for lightweight filtering using a single convolutional filter for each input channel and a 1×1 convolution (or pointwise) layer for building new features via computing linear combinations of input channels, while MobileNetV2 consists of two blocks, which are a residual block with a stride of 1 and another block with a stride of 2, that are used for downsizing [49]. Each of these blocks has 3 layers: a 1×1 convolution layer with a rectified linear unit (ReLU6), a depthwise convolution, and another 1×1 convolution layer without non-linearity.

Moving Objects Classification

One of the most important part of our approach is the classification of found objects. In real-world scenarios, the detected moving objects are drones, birds, airplanes, insects, and moving parts of scenes. Therefore, we decided to use a classifier that divides all found objects into 3 classes: drones, birds, and background. The MobileNetV2 [50] CNN was chosen as the classifier. The choice of the CNN was due to its low value of inference time and high accuracy. According to [51], the highest detection speed was achieved by a detector with the MobileNet [48] backbone network. MobileNetV2 is an improved version of MobileNet that significantly improves its accuracy [50]. The MobileNetV2 network architecture consists of 19 original basic blocks named bottleneck residual blocks (see Figure 3). These blocks followed by a 1×1 convolution layer with an average pooling layer. The last layer is a classification layer. We used the modified version of the MobileNetV2 network [52]. The author made the network more suitable for tiny images by changing the stride, padding, and filter size. We changed the classification layer so that the number of classes the network classifies became 3.

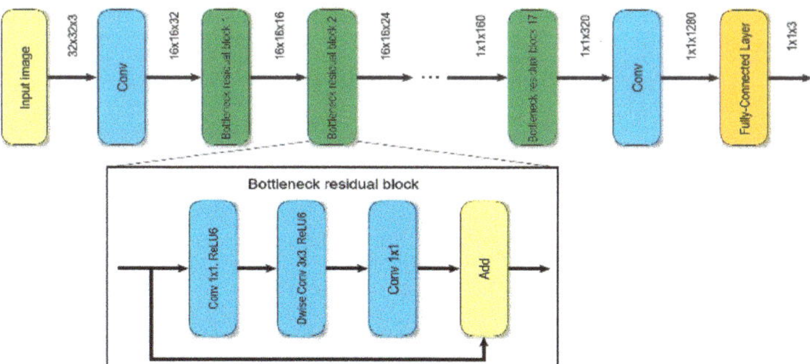

Figure 3. The architecture of the MobileNetv2 network.

3. Experiments and Results

3.1. Data Preparation

The amount of data is crucial for CNN training. Insufficient data affect the generalization ability of the network. As a result, the accuracy of the classification is decreased when the network receives new data. As training data in our work, we used the Drone-vs-Bird challenge dataset, which consisted of 11 videos recorded by a static camera. In addition to a drone, birds and other moving objects may appear in the videos. For each video, annotations of drones appearing in the frames are provided. Annotations are presented in the form of coordinates and sizes of the ground truth bounding boxes. Using the videos and annotations to them, we extracted 10,155 images of drones. In order to extract images of birds and background from the videos, we applied the detector of moving objects described in the previous subsection to the entire dataset. Next, we manually labeled all the images of the detected moving objects. As a result, 1921 images of birds and 9348 images of the background were obtained. Since the number of images of birds was low compared to the other two classes, we additionally used 2651 bird images from Wild Birds in a Wind Farm: Image Dataset for Bird Detection [53] Thus, the total number of images in our dataset was 24,075. The input size of the network was $32 \times 32 \times 3$, so we resized all the images to match the input layer of the network. Some examples of the resized images are shown in Figure 4.

Figure 4. Some of the collected images for training. The first row shows the drones, the second row consists of background images, and the third row is the images of birds.

3.2. Training

We trained the MobileNetV2 CNN from scratch using the dataset described in the previous section. The dataset was divided into a training set and a testing set in a proportion of 80 to 20. To train the network, we used the stochastic gradient descent (SGD) optimization algorithm with a starting learning rate of 0.05, a momentum of 0.9, and a weight decay of 0.001. The training was done on NVIDIA GeForce GT 1030 2 GB GPU with a batch size of 88. We decreased the starting learning rate by a factor of 10 every 50 epochs during the training.

3.3. Evaluation Metrics

For the evaluation of any object detection approach, some statistical and machine learning metrics can be used: ROC curves, precision and recall, F-scores, and false positives per image [54]. Generally, first the results of an object detector are compared to the given list of the ground-truth bounding boxes. To answer the question of when a detection can be considered as correct, most studies related to object detection have used the overlap criterion, which was introduced by Everingham et al. [55] for the Pascal VOC challenge. As noted above, the detections are assigned to ground truth objects, and, by calculating the bounding box overlap, they are judged to be true or false positives. In order to be considered as a correct detection according to [55], the overlap ratio between the predicted and ground truth boxes must be exceed 0.5 (50%). The Pascal VOC overlap criterion is defined as the intersection over union (IoU) and computed as follows:

$$IoU = a_0 = \frac{area\left(B_p \cap B_{gt}\right)}{area\left(B_p \cup B_{gt}\right)} \tag{1}$$

where the IoU is the intersection over union; a_0 is an overlap ratio; B_p and B_{gt} are predicted and ground truth bounding boxes, respectively; $area(B_p \cap B_{gt})$ means the overlap or intersection of predicted and ground truth bounding boxes; and $area(B_p \cup B_{gt})$ means the area of union of these two bounding boxes. Having matched detections to ground truth we can determine the number of correctly classified objects, which are called true positives (TPs), incorrect detections or false positives (FPs), and ground truth objects that are not missed by the detector or false negatives (FNs). Using the total number of TPs, FPs, and FNs, we could compute a wide range of evaluation metrics.

We evaluated our approach using the Drone-vs-Bird challenge's [5] metrics. The challenge provided three test videos for evaluation that were named gopro_001, gopro_004, and gopro_006. The first video contained frames with two drones and a moving background. A feature of the second video was a static background and a very small size of the drone. In the third video, in addition to the drone, several birds were present on the frames. The main evaluation metric used in the challenge is the F1-score:

$$F_1 = 2 * \frac{Precision * Recall}{Precision + Recall} \tag{2}$$

In order to calculate Precision and Recall, we applied our drone detector on the test videos and counted the total number of TPs, FPs, and FNs.

$$Precision = \frac{TP}{TP + FP} \tag{3}$$

$$Recall = \frac{TP}{TP + FN} \tag{4}$$

Precision and recall are the metrics that can be used to evaluate most information extraction algorithms. Sometimes they might be used on their own, or they might be used as a basis for derived metrics such as F-score and precision–recall curves. Precision is the proportion of predicted positive results that are truly true-positive results for all positively predicted objects, whereas recall is the fraction of all true-positively objects to the total number of positively predicted objects—it shows how many samples of all positive examples were classified correctly. Based on these two metrics, we could calculate F1-score metric, which combines information about precision and recall. A detection was counted as a true positive if the value of the IoU between the detected and the ground truth bounding boxes exceeded 0.5.

3.4. Results

As a result of training, the accuracy of the classifier on the entire dataset was 99.83%. The confusion matrix is shown in Figure 5.

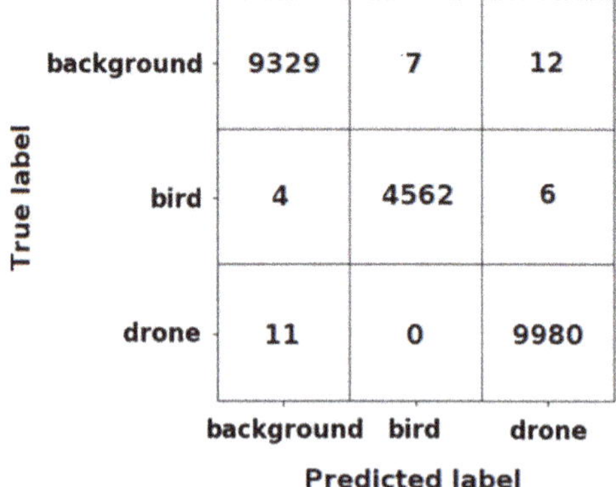

Figure 5. Confusion matrix of the trained convolutional neural network (CNN).

The experiment results obtained by applying our detector to all test videos are shown in Figure 6. True positives and false positives values were counted for an IoU = 0.5. The results were divided into three ranges, depending on the drone size.

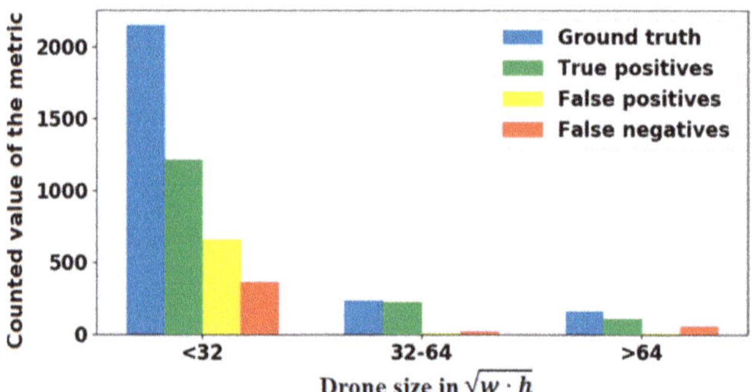

Figure 6. The results of the experiment for various drone sizes.

In Figure 6, w and h are the width and height of the ground truth bounding box in pixels, respectively. The value $\sqrt{w*h}$ reflects the size of the drone in the image. The lower this value, the farther the drone is from the camera. Based on these data, precision and recall values were calculated. Then, we calculated the F1-score by Equation (1) and added it to the last row of Table 1. The same sequence was individually performed for each video.

Table 1. The results of the evaluation for an intersection over union (IoU) = 0.5.

Video Name	Precision	Recall	F1-Score
gopro_001	0.786	0.817	0.801
gopro_004	0.554	0.910	0.689
gopro_006	0.735	0.691	0.712
Overall	0.701	0.788	0.742

For a more detailed analysis of the detector, we conducted experiments for various values of the IoU. The curves were plotted based on the obtained results of recall, precision, and F1-score, as shown in Figure 7.

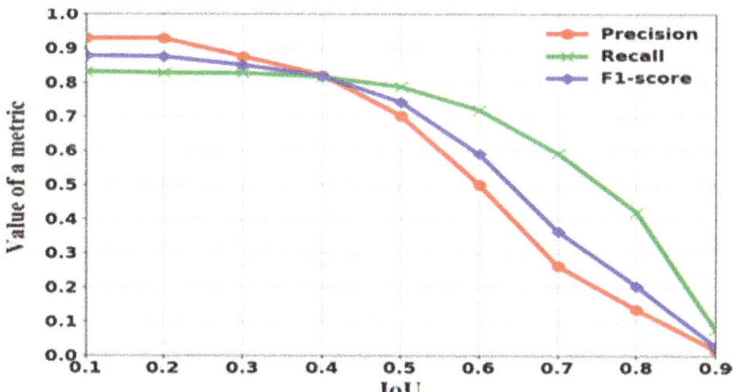

Figure 7. Evaluation metrics values for various IoU values.

Qualitative detection results are depicted in Figure 8.

Figure 8. Qualitative results of our detector. Green bounding boxes are ground truth bounding boxes. Red bounding boxes are the results of applying our detector.

Eighty-five percent of all false positives were caused by inaccurate estimations of the bounding boxes, resulting in a calculated IoU value of less than 0.5. The remaining 15% were classification errors, as a result of which other moving objects were misclassified as drones. Examples of false detections caused by incorrect classification are shown in Figure 9.

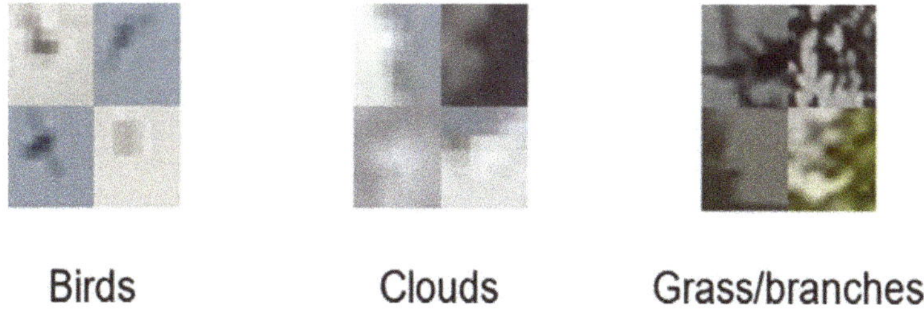

Figure 9. Examples of false detections.

The images shown in Figure 9 were classified by the detector as drones. In most cases, the objects that were misclassified were birds, clouds, swaying tree branches, and grass. On average, the detector processed nine frames of size 1920 × 1080 per second. A third of the processing time was spent on the classification of moving objects, and the rest was spent on their detection. We noticed that the detection speed depended on the background change rate, which increased as the number of bounding boxes fed into the classifier increased. This dependency is shown in more detail in Figure 10.

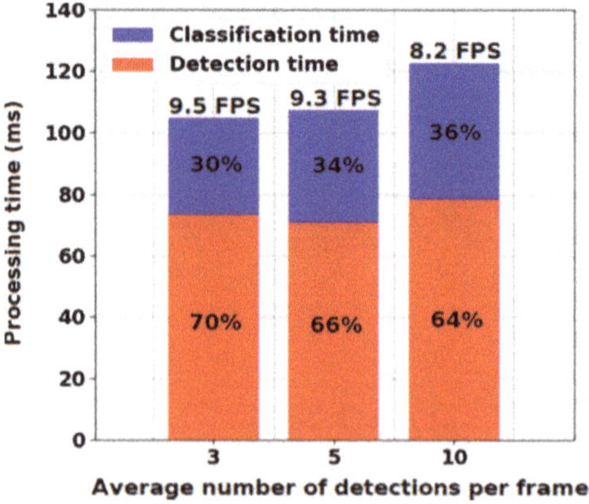

Figure 10. The evaluation results of detection speed.

4. Discussion

Our findings suggest that dividing drone detections task into detecting moving objects and classifying detections can be effective for accurate and fast drone detection. However, the use of motion information for detecting moving objects has several drawbacks. First, as shown in Figure 10, the moving background caused an increase in the number of detected objects, which led to an increase in the classification time and the number of false positives. Secondly, if the drone was flying close to moving objects, then it became impossible to separately distinguish it from other objects, as can be seen in Figure 11.

Figure 11. The result of applying background subtraction to the video segment in which a drone was flying near a swaying grass.

As a result, the drone was not detected, which led to an increase in the number of false negatives. Along with this, the number of false positive results also increased due to the fact that more images were fed to the classifier. Since the accuracy of the classifier was not equal to 100%, this caused a greater number of classification errors. The usage of the metrics of the Drone-vs-Bird detection challenge [5] allowed us to compare our results with the results of the other teams participating in the challenge. According to the experiment results, the accuracy of our approach was comparable with the approaches proposed in [30] and [14]. Compared to [30], in which only the application of the super resolution was performed at a speed of 0.58 FPS, our detector had a significantly higher detection speed. For IoU = 0.3, our approach performed much worse than [14] but still better than [30]. The comparison results for previous works and our approach are shown in Table 2.

Table 2. Comparison results.

Methods	Precision	Recall	F1-Score
[28]	0.103	0.146	0.121
[30]	0.795	0.591	0.678
[14]	0.756	0.713	0.734
[38]	0.524	0.342	0.414
Our approach	0.701	0.788	0.742

5. Conclusions

In this paper, we present a real-time drone detection algorithm, the accuracy of which is comparable to existing algorithms. We provided further evidence that the task of drone detection can be successfully solved by dividing it into the detection and classification stages. The experimental results showed the advantages and disadvantages of this approach. The most important limitation of our detector lies in the fact that its performance is highly dependent on the presence of a moving background. We believe that the accuracy of our detector can be improved by using a larger dataset to train the classifier. For future work, we suggest combining visual information with motion information to detect candidates in the detection stage.

Author Contributions: Conceptualization, U.S., D.A., L.I., and E.T.M.; methodology, U.S. and D.A.; software, D.A.; validation, U.S., D.A., and E.T.M.; formal analysis, U.S. and D.A.; investigation, U.S. and D.A.; resources, U.S. and D.A.; data curation, U.S., D.A., and E.T.M.; writing—original draft preparation, U.S. and D.A.; writing—review and editing, U.S. and E.T.M.; visualization, D.A.; supervision, E.T.M. and L.I. All authors have read and agreed to the published version of the manuscript.

Funding: This research received no external funding.

Acknowledgments: We would like to express our great appreciation to SafeShore project for supporting Drone-vs-Bird challenge dataset. Special thanks to Eric Matson for his patient guidance and useful critiques of this research work.

Conflicts of Interest: The authors declare no conflict of interest.

Abbreviations

AI	Artificial Intelligence
CNN	Convolutional Neural Network
COCO	Common Objects in Context
DPM	Deformable Parts Model
DCSCN	Deep CNN with Skip Connection and Network in Network
FPN	Feature Pyramid Network
FPS	Frames per Second
GFD	Generic Fourier Descriptor
GPU	Graphics Processing Unit
HOG	Histogram of Oriented Gradients
IoU	Intersection over Union
KCF	Kernelized Correlation Filter
LBP	Local Binary Pattern
LSTM	Long short-term memory
MPEG4	Moving Picture Experts Group
RAM	Random-access memory
ReLU	Rectified linear unit
RF	Radio Frequency
RGB	Red, Green and Blue
R-CNN	Region-based Convolutional Neural Network
ROC	Receiver operating characteristic
RPN	Region Proposal Network
SAR	Search and Rescue
SIFT	Scale-Invariant Feature Transform
SGD	Stochastic Gradient Descent
SSD	Single Shot Detector
SVM	Support Vector Machine
UAS	Unmanned Aerial System
UAV	Unmanned Aerial Vehicle
USC	University of Southern California
VGG	Visual Geometry Group
VOC	Visual Object Classes
YOLO	You Only Look Once

References

1. Jansen, B. Drone Crash at White House Reveals Security Risks. *USA Today*. 26 January 2015. Available online: https://www.usatoday.com/story/news/2015/01/26/drone-crash-secret-service-faa/22352857/.
2. Pham, S. Drone Hits Passenger Plane in Canada. CNN. Available online: https://money.cnn.com/2017/10/16/technology/drone-passenger-plane-canada/index.html. (accessed on 16 October 2017).
3. Guardian, T. Gatwick Drone Disruption Cost Airport Just £1.4 m. The Guardian. Available online: https://www.theguardian.com/uk-news/live/2018/dec/21/gatwick-drone-airport-limited-flights-live. (accessed on 21 December 2018).
4. Walker, D. *Report: Trio Planned to Use Drone to Get Tobacco, Phones to Inmate*. Available online: https://www.northwestgeorgianews.com/report-trio-planned-to-use-drone-to-to-get-tobacco-phones-to-inmate/article_44c25a12-7d96-11ea-8c97-73fe94e065d4.html (accessed on 13 April 2020).

5. Coluccia, A.; Saqib, M.; Sharma, N.; Blumenstein, M.; Magoulianitis, V.; Ataloglou, D.; Dimou, A.; Zarpalas, D.; Daras, P.; Craye, C.; et al. Drone-vs-Bird Detection Challenge at IEEE AVSS2019. In Proceedings of the 2019 16th IEEE International Conference on Advanced Video and Signal Based Surveillance (AVSS), Taipei, Taiwan, 18–21 September 2019.

6. Wu, M.; Xie, W.; Shi, X.; Shao, P.; Shi, Z. Real-Time Drone Detection Using Deep Learning Approach. In Proceedings of the 2018 of the 3rd international conference, MLICOM, Hangzhou, China, 6–8 July 2018; pp. 22–32.

7. Taha, B.; Shoufan, A. Machine Learning-Based Drone Detection and Classification: State-of-the-Art in Research. *IEEE Access* **2019**, *7*, 138669–138682. [CrossRef]

8. Hamatapa, R.; Vongchumyen, C. Image Processing for Drones Detection. In Proceedings of the 2019 5th the ICEAST, Luang Prabang, Laos, 2–5 July 2019.

9. Samaras, S.; Diamantidou, E.; Ataloglou, D.; Sakellariou, N.; Vafeiadis, A.; Magoulianitis, V.; Lalas, A.; Dimou, A.; Zarpalas, D.; Votis, K.; et al. Deep Learning on Multi Sensor Data for Counter UAV Applications—A Systematic Review. *Sensors* **2019**, *19*, 4837. [CrossRef] [PubMed]

10. Park, J.; Kim, D.H.; Shin, Y.S.; Lee, S. A Comparison of Convolutional Object Detectors for Real-time Drone Tracking Using a PTZ Camera. In Proceedings of the ICCAS, Jeju, South Korea, 18–21 October 2017.

11. Shi, W.; Arabadjis, G.; Bishop, B.; Hill, P.; Plasse, R.; Yoder, J. *Sensor Fusion—Foundation and Applications*; Thomas, C., Ed.; InTech: Rijeka, Croatia, 2011.

12. Unlu, E.; Zenou, E.; Riviere, N.; Dupouy, P. Deep learning-based strategies for the detection and tracking of drones using several cameras. *IPSJ Trans. Comput. Vis. Appl.* **2019**, *11*. [CrossRef]

13. Ezuma, M.; Erden, F.; Anjinappa, C.K.; Ozdemir, O.; Guvenc, I. Micro-UAV Detection and Classification from RF Fingerprints Using Machine Learning Techniques. In Proceedings of the IEEE AERO, Big Sky, MT, USA, 2–9 March 2019.

14. Craye, C.; Ardjoune, S. Spatio-temporal Semantic Segmentation for Drone Detection. In Proceedings of the AVSS, Taipei, Taiwan, 18–21 September 2019.

15. Shin, S.; Park, S.; Kim, Y.; Matson, E.T. Design and Analysis of Cost-efficient Sensor Deployment for 608 Tracking Small UAS with Agent-based Modeling. Integration of Sensors in Complex, Intelligent Systems Selected Papers from the CHARMS 2015 Workshop. *Sensors* **2016**, *16*, 575. [CrossRef] [PubMed]

16. Liu, H.; Wei, Z.; Chen, Y.; Pan, J.; Lin, L.; Ren, Y. Drone Detection Based on an Audio-Assisted Camera Array. In Proceedings of the IEEE BigMM, Laguna Hills, CA, USA, 19–21 April 2017; pp. 402–406.

17. Caris, M.; Johannes, W.; Stanko, S.; Pohl, N. Millimeter Wave Radar for Perimeter Surveillance and Detection of MAVs (Micro Aerial Vehicles). In Proceedings of the IEEE IRS, Dresden, Germany, 24–26 June 2015.

18. Park, S.; Shin, S.; Kim, Y.; Matson, E.; Lee, K.; Kolodzy, P.J.; Slater, J.C.; Scherreik, M.; Sam, M.; Gallagher, J.C.; et al. Combination of radar and audio sensors for identification of rotor-type unmanned aerial vehicles (UAVs). In Proceedings of the IEEE SENSORS, Busan, South Korea, 1–4 November 2015; pp. 1–4.

19. Hengy, S.; Laurenzis, M.; Schertzer, S.; Hommes, A.; Kloeppel, F.; Shoykhetbrod, A.; Geibig, T.; Johannes, W.; Rassy, O.; Christnacher, F. Multimodal UAV detection: Study of various intrusion scenarios. In Proceedings of the Electro-Optical Remote Sensing XI, Warsaw, Poland, 5 October 2017.

20. Charvat, G.L.; Fenn, A.J.; Perry, B.T. The MIT IAP radar course: Build a small radar system capable of sensing range, Doppler, and synthetic aperture (SAR) imaging. In Proceedings of the IEEE Radar Conference, Atlanta, GA, USA, 7–11 May 2012; pp. 138–144.

21. Unlu, E.; Zenou, E.; Rivière, N. Using Shape Descriptors for UAV Detection. In *Electronic Imaging*; Burlingam, CA, USA; Available online: https://oatao.univ-toulouse.fr/19612/ (accessed on 21 March 2017).

22. Wang, Z.; Qi, L.; Tie, Y.; Ding, Y.; Bai, Y. Drone detection based on FD-HOG descriptor. In Proceedings of the International Conference on Cyber Enabled Distributed Computing and Knowledge Discovery, Zhengzhou, China, 18–20 October 2018.

23. Yoshihashi, R.; Kawakami, R.; Iida, M.; Naemura, T. Construction of a bird image dataset for ecological investigations. In Proceedings of the ICIP, Quebec City, QC, Canada, 27–30 September 2015; pp. 4248–4252.

24. Chen, Y.; Aggarwal, P.; Choi, J.; Kuo, C.J. A Deep Learning Approach to Drone Monitoring. arXiv:1712.00863 [cs.CV]. Available online: https://arxiv.org/pdf/1712.00863.pdf. (accessed on 17 December 2017).

25. Peng, J.; Zheng, C.; Lv, P.; Cui, T.; Cheng, Y.; Si, L. Using Images Rendered by PBRT to Train Faster R-CNN for UAV Detection. *Comput. Sci. Res. Notes* **2018**. [CrossRef]

26. Hu, Y.; Wu, X.; Zheng, G.; Liu, X. Object Detection of UAV for Anti-UAV Based on Improved YOLO v3. In Proceedings of the CCC, Guanzhou, China, 27 July 2019; pp. 8386–8390.

27. Redmon, J.; Farhadi, A. YOLOv3: An Incremental Improvement. ArXiv.org. vol. abs/1804.02767. Available online: https://arxiv.org/pdf/1804.02767.pdf (accessed on 18 April 2018).

28. Nalamati, M.; Kapoor, A.; Saqib, M.; Sharma, N.; Blumenstein, M. Drone Detection in Long-range Surveillance Videos. In Proceedings of the AVSS, Taipei, Taiwan, 18–21 September 2019.

29. He, T.; Zhang, Z.; Zhang, H.; Zhang, Z.; Xie, J.; Li, M. Bag of tricks for image classification with convolutional neural networks. In Proceedings of the IEEE/CVF CVPR, Long Beach, CA, USA, 16–21 June 2019; pp. 558–567.

30. Magoulianitis, V.; Ataloglou, D.; Dimou, A.; Zarpalas, D.; Daras, P. Does Deep Super-Resolution Enhance UAV Detection? In Proceedings of the AVSS, Taipei, Taiwan, 18–21 September 2019.

31. He, K.; Zhang, X.; Ren, S.; Sun, J. Deep Residual Learning for Image Recognition. arXiv:1512.03385 [cs.CV]. Available online: https://arxiv.org/pdf/1512.03385.pdf (accessed on 15 December 2015).

32. Muhammad, S.; Sharma, N.; Khan, S.D.; Blumenstein, M. A study on detecting drones using deep convolutional neural networks. In Proceedings of the AVSS, Lecce, Italy, 29 August–1 September 2017.

33. Ren, S.; He, K.; Girshick, R.; Sun, J. Faster R-CNN: Towards real-time object detection with region proposal networks. In Proceedings of the NIPS, Montreal, QC, Canada, 7–12 December 2015; pp. 91–99.

34. Aker, C.; Kalkan, S. Using deep networks for drone detection. In Proceedings of the AVSS, Lecce, Italy, 29 August–1 September 2017.

35. Redmon, J.; Farhadi, A. YOLO9000: Better, Faster, Stronger. arXiv:1612.08242 [cs.CV]. Available online: https://arxiv.org/pdf/1612.08242.pdf. (accessed on 25 December 2016).

36. Liu, W.; Anguelov, D.; Erhan, D.; Szegedy, C.; Reed, S.; Fu, C.; Berg, A.C. Ssd: Single shot multibox detector. In Proceedings of the European Conference on Computer Vision, Amsterdam, The Netherlands, 8–16 October 2016; pp. 21–37.

37. Szegedy, C.; Vanhoucke, V.; Ioffe, S.; Shlens, J.; Wojna, Z. Rethinking the inception architecture for computer vision. In Proceedings of the CVPR, Las Vegas, NV, USA, 27–30 June 2016; pp. 2818–2826.

38. de la Iglesia, D.; Méndez, M.; Dosil, R.; González, I. Drone detection CNN for close- and long-range surveillance in mobile applications. In Proceedings of the AVSS, Taipei, Taiwan, 18–21 September 2019.

39. Lin, T.; Goyal, P.; Girshick, R.; He, K.; Dollar, P. Focal loss for dense object detection. In Proceedings of the IEEE ICCV, Venice, Italy, 22–29 October 2017; pp. 2999–3007.

40. Lin, T.; Dollar, P.; Girshick, R.; He, K.; Hariharan, B.; Belongie, S. Feature pyramid networks for object detection. In Proceedings of the CVPR, Honolulu, HI, USA, 21–26 July 2017; pp. 2117–2125.

41. Ye, D.H.; Li, J.; Chen, Q.; Wachs, J.P.; Bouman, C.A. Deep Learning for Moving Object Detection and Tracking from a Single Camera in Unmanned Aerial Vehicles (UAVs). *Electron. Imaging* **2018**, *10*, 466-1–466-6. [CrossRef]

42. Yamanaka, S.; Kuwashima, S.; Kurita, T. Fast and Accurate Image Super Resolution by Deep CNN with Skip Connection and Network in Network. Presented at the ICONIP. Available online: https://arxiv.org/ftp/arxiv/papers/1707/1707.05425.pdf (accessed on 18 July 2017).

43. Schumann, A.; Sommer, L.; Klatte, J.; Schuchert, T.; Beyerer, J. Deep cross-domain flying object classification for robust UAV detection. In Proceedings of the AVSS, Lecce, Italy, 29 August–1 September 2017.

44. Ronneberger, O.; Fischer, P.; Brox, T. U-Net: Convolutional Networks for Biomedical Image Segmentation. arXiv:1505.04597 [cs.CV]. Available online: https://arxiv.org/pdf/1505.04597.pdf (accessed on 18 May 2015).

45. Arunachalam, S.T.; Shahana, R.; Vijayasri, R.; Kavitha, T. Flying object detection and classification using deep neural networks. *IJETT* **2019**, *67*.

46. Andrewssobral/Bgslibrary. Available online: https://github.com/andrewssobral/bgslibrary (accessed on June 2013).

47. Dertat, A. Applied Deep Learning—Part 4: Convolutional Neural Networks. Available online: https://towardsdatascience.com/applied-deep-learning-part-4-convolutional-neural-networks-584bc134c1e2 (accessed on 8 November 2017).

48. Howard, A.G.; Zhu, M.; Chen, B.; Kalenichenko, D.; Wang, W.; Weyand, T.; Andreetto, M.; Adam, H. Mobilenets: Efficient Convolutional Neural Networks for Mobile Vision Applications. arXiv:1704.04861 [cs.CV]. Available online: https://arxiv.org/pdf/1704.04861.pdf (accessed on 17 April 2017).

49. Dertat, A. Review: MobileNetV2—Light Weight Model (Image Classification). Available online: https://towardsdatascience.com/review-mobilenetv2-light-weight-model-image-classification-8febb490e61c (accessed on 19 May 2019).

50. Sandler, M.; Howard, A.; Zhu, M.; Zhmoginov, A.; Chen, L. MobileNetV2: Inverted Residuals and Linear Bottlenecks. In Proceedings of the IEEE/CVF CVPR, Salt Lake City, UT, USA, 18–22 June 2018.

51. Park, S.; Kim, Y.; Lee, K.; Smith, A.; Dietz, J.; Matson, E.T. Accessible real-time surveillance radar system for object detection. *Sensors* **2020**, *20*, 2215. [CrossRef]

52. Huyvnphan/PyTorch_CIFAR10. Available online: https://github.com/huyvnphan/PyTorch-CIFAR10 (accessed on 1 June 2020).

53. Yoshihashi, R.; Trinh, T.; Kawakami, R.; You, S.; Iida, M.; Naemura, T. Differentiating Objects by Motion: Joint Detection and Tracking of Small Flying Objects. {arXiv}:1709.04666 [cs.CV]. Available online: https://arxiv.org/pdf/1709.04666.pdf. (accessed on 17 September 2018).

54. Flach, P.A. The geometry of ROC space: Understanding machine learning metrics through ROC isometrics. In Proceedings of the ICML, Washington, DC, USA, 3–8 December 2003; pp. 194–201.

55. Everingham, M.; van Gool, L.; Williams, C.K.I.; Winn, J.; Zisserman, A. The pascal visual object classes (VOC) challenge. *Int. J. Comput. Vis.* **2010**, *88*, 3338. [CrossRef]

Article

Ship Type Classification by Convolutional Neural Networks with Auditory-Like Mechanisms

Sheng Shen, Honghui Yang *, Xiaohui Yao, Junhao Li, Guanghui Xu and Meiping Sheng

School of Marine Science and Technology, Northwestern Polytechnical University, Xi'an 710072, China; shensheng@mail.nwpu.edu.cn (S.S.); yaoxiaohui@mail.nwpu.edu.cn (X.Y.); ljhhjl@mail.nwpu.edu.cn (J.L.); hsugh@mail.nwpu.edu.cn (G.X.); smp@nwpu.edu.cn (M.S.)
* Correspondence: hhyang@nwpu.edu.cn

Received: 1 November 2019; Accepted: 26 December 2019; Published: 1 January 2020

Abstract: Ship type classification with radiated noise helps monitor the noise of shipping around the hydrophone deployment site. This paper introduces a convolutional neural network with several auditory-like mechanisms for ship type classification. The proposed model mainly includes a cochlea model and an auditory center model. In cochlea model, acoustic signal decomposition at basement membrane is implemented by time convolutional layer with auditory filters and dilated convolutions. The transformation of neural patterns at hair cells is modeled by a time frequency conversion layer to extract auditory features. In the auditory center model, auditory features are first selectively emphasized in a supervised manner. Then, spectro-temporal patterns are extracted by deep architecture with multistage auditory mechanisms. The whole model is optimized with an objective function of ship type classification to form the plasticity of the auditory system. The contributions compared with an auditory inspired convolutional neural network include the improvements in dilated convolutions, deep architecture and target layer. The proposed model can extract auditory features from a raw hydrophone signal and identify types of ships under different working conditions. The model achieved a classification accuracy of 87.2% on four ship types and ocean background noise.

Keywords: machine learning; neural network; ship radiated noise; underwater acoustics

---

1. Introduction

The soundscape of the oceans is heavily affected by human activities, especially in coastal waters. The increase of the noise level in oceans is correlated with burgeoning global trade with the expansion of shipping. Automatic recognition of ship type by ship radiated noise is not only affected by the complicated mechanism of noise generation, but is also affected by the complex underwater sound propagation channel. The conventional recognition methods based on machine learning generally include three stages—feature extraction, feature selection and classifier design.

The conventional feature extraction methods for ship radiated noise include waveform features [1,2], auditory features [3,4], wavelet features [5] and so on. These manually designed features are limited in their ability to capture variations in complex ocean environments and ship operative conditions for the use of fixed parameters or filters. Biophysical based models [3,4] are limited to early auditory stages for extracting auditory features. Auditory features designed from perceptual evidence always focus on the properties of signal description rather than the classification purpose [6]. These features do not utilize the plastic mechanism and representation at various auditory stages to improve the recognition performance. Although the noise features or redundant features can be removed by feature selection methods [7], the inherent problem of manually designed features still cannot be solved radically.

Support vector machine (SVM) was used to recognize ship noise using manually designed features [7]. With the increase of data size, hierarchical architectures have been shown to outperform shallow models. Spectrogram [8], probabilistic linear discriminant analysis and i-vectors [9] were used as the input of neural classifiers to detect ship presence and classify ship types. For the application of deep learning, Kamal [10] used a deep belief network and Cao [11] used a stacked autoencoder. A competitive learning mechanism [12,13] was used to increase cluster performance during the training of the deep network. In these works, classifier design and feature extraction were separated from each other. This has a drawback that the designed features may not be appropriate for the classification model.

Deep learning has made it possible to model the original signal as well as to predict targets in a whole model [6,14], to which the auditory system is thought to be adapted. The time convolutional layer in an auditory inspired convolutional neural network (CNN) [6] provided a new way for modeling underwater acoustic signals. However, it did not have enough depth to build an appropriate model to match the expanding acquired dataset. Moreover, the conventional convolutional layer and the fully connected layer led to numerous parameters.

In this paper, we present a deep architecture with the aim of capturing the functionality and robustness of the auditory system to improve recognition performance. The key element in the approach is a deep architecture with time and frequency-tolerant feature detectors, which are inspired by neural cells along the auditory pathway. The early stage auditory model is derived from the auditory inspired CNN, in which the time convolutional layer is improved by dilated convolution. The construction of the deep network refers to inception and residual network [15] in the field of machine vision, in which some ideas are also found in the auditory pathway. Thus, the frequency convolutional layers in Reference [6] are improved to increase the depth of the network. At the final stage, the substitution of the fully connected layer by global average pooling at target layer greatly reduces the parameters. The main findings of this paper are briefly summarized as follows:

- The proposed convolutional neural network could transform the time domain signal into a frequency domain that is similar to gammatone spectrogram.
- Deep architecture of a neural network derived from an auditory pathway improves the classification performance of ship types.
- Auditory filters in convolutional kernals are adaptive in shape during the optimization of the network with the ship type classification task.
- The classification results of the model are robust to ship operative conditions. The increase of distance between ships to hydrophone has a negative effect on recognition results in most cases.

This paper is organized as follows. Section 2 gives an overview of auditory mechanisms and the structure of the proposed model. Section 3 describes details of the model, which include the cochlea model for ship radiated noise modeling and the multistage auditory center model for features learning and targets classification. Section 4 includes experimental data description, experimental setup and results. An overall discussion and directions for future work are concluded in Section 5.

2. Model

2.1. Auditory Mechanisms

Decades of physiological and psychoacoustical studies [16–18] have revealed elegant strategies at various stages of the mammalian auditory system for representation of acoustic signals. Sound is first analyzed in terms of relatively few perceptually significant attributes, followed by higher level integrative processes.

When sound arrives at the ears, the vibration of the eardrum caused by the sound wave is transmitted to the cochlea in the inner ear via ossicular chain in the middle ear. The cochlea performs two fundamental functions. First, through the vibration of different parts of the basement membrane, the cochlea effectively separates the frequency components of sound. The second function of the cochlea is to transform these vibrations into neural patterns with the help of hair cells distributed along the cochlea [18].

The auditory center is one of the longest central pathways in the sensory system. The acoustic spectrum is extracted early in the auditory pathway at the cochlear nucleus, the first stage beyond the auditory nerve. Multiple pathways emerge from the cochlear nucleus up through the midbrain and thalamus to the auditory cortex. Each pathway passes through different neural structures and repeatedly converges onto and diverges from other pathways along the way [18]. The complexity structure extracts rich and varied auditory percepts from the sound to be later interpreted by the brain. Neurons in the primary cortex have been shown to be sensitive to specific spectro-temporal patterns in sounds [19].

As a result of auditory experience, the systematic long-term changes in the responses of neurons to sound are defined as plasticity. Plasticity and learning probably occurs at all stages of the auditory system [20].

By reviewing the process of auditory perception, we can conclude the following four mechanisms of early and higher auditory stages that are useful for establishing an auditory computational model.

- Auditory processing is hierarchical.
- Neurons throughout the auditory pathway are always tuned to frequency.
- Auditory pathways have different neural structures.
- The auditory system has plasticity and learning properties.

2.2. Model Structure

The nature of the auditory computational model is to transform the raw acoustical signal into representations that are useful for auditory tasks. In this paper, mechanisms of auditory system are established mathematically in deep CNN for ship type classification. The model mainly includes the cochlea model and the auditory center model. A complete description of the specifications of the network is given in Figure 1.

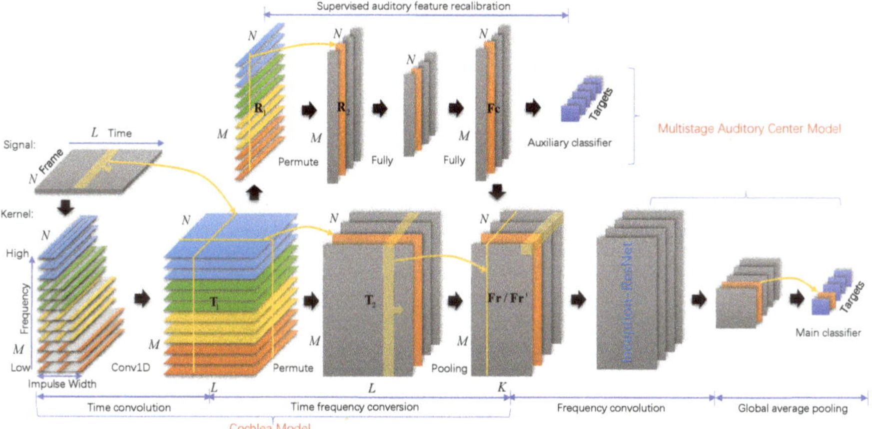

Figure 1. Model structure. The model mainly includes cochlea model and auditory center model. In the time convolutional layer, four colors represent four groups. Dilated convolutions are represented by parallel lines at equal intervals. The time frequency conversion layer includes permute layer and max-pooling layer. At the top of the graph, auditory feature recalibration is implemented by global max-pooling and fully connected layers. Frequency convolutional layers are performed by Inception-ResNet. The final stage includes global average pooling and softmax layer.

Based on the research foundation of the human cochlea, a series of multiscale gammatone auditory filters [21] are used to initial the time convolutional layer with dilated convolutions [22]. Ship noise signals are decomposed by convolution operation with auditory filters. Inspired by the function of hair cells, the time frequency conversion layer transforms these decomposed components into amplitudes of its corresponding frequency components—or its frequency spectrum [18]. We introduce an auxiliary classifier with the goal of enhancing the gradient and recalibrating the learned spectrum in supervised manner. Learned spectra are further extracted by deep architecture with inception structures and residual connections [15] to model the multistage auditory pathway. These layers are defined as frequency convolutional layers. The resulting feature maps of the last frequency convolutional layer are fed into a global average pooling layer [23]. Then ship types are predicted in the softmax layer. During the training of the network, auditory filters and features are subject to classification tasks on the basis of matching human auditory systems.

3. Methodology

3.1. Cochlea Model for Ship Radiated Noise Modeling

The cochlea model is the first stage of the proposed model, it includes the time domain signal decomposition of basement membrane and the time frequency conversion of hair cells. The cochlea model creates a frequency-organized axis known as the tonotopic axis of cochlea.

3.1.1. Time Convolutional Layer with Dilated Auditory Filters

Much is known about the representation of spectral profile in the cochlea [18,21]. The physiologically derived gammatone filter $g(t)$ is shown in (1).

$$g(t) = at^{n-1}e^{-2\pi bt}cos(2\pi ft + \phi), \tag{1}$$

where a is amplitude, t is time in second, n is filter's order, b is bandwidth in Hz, f is center frequency in Hz, and ϕ is phase of the carrier in radians. Center frequency f and bandwidth b are set by an equivalent rectangular bandwidth (ERB) [24] cochlea model in (2) and (3).

$$ERB(f) = 24.7(4.37f/1000 + 1) \tag{2}$$

$$b = 1.019 \times ERB(f). \tag{3}$$

Convolutional kernels of the time convolutional layer represent a population of auditory nerve spikes. A series of gammatone filters of different sizes are used to initialize weight vectors of this layer which forms the primary feature extraction base of the network. This layer performs convolutions over a raw time domain waveform. Suppose the input signal $\mathbf{S}(\mathbf{S} \in \mathbb{R}^{L \times N})$ has N frames, each frame length is L. As shown in (4), signal $\mathbf{S}$ is convolved with kernel $\mathbf{k}_m$ and added to an additive bias b_m. Then, the output puts through activation function f to form the output feature map $\mathbf{t}_m(\mathbf{t}_m \in \mathbb{R}^{L \times N})$.

$$\mathbf{t}_m = f(\mathbf{S} * \mathbf{k}_m + b_m), m \in 1, 2, ..., M. \tag{4}$$

For time convolutional layer that has M kernels, the output $\mathbf{T}_1 = [\mathbf{t}_1, \mathbf{t}_2, ..., \mathbf{t}_M]$ will be obtained. We use 128 gammatone filters with center frequencies ranging from 20 to 8000 Hz. For 16 kHz sampling frequency, the impulse widths range from 50 to 800 points approximately. These filters are divided into 4 groups by quartering according to impulse widths. The convolutional kernel widths of the 4 groups are 100, 200, 400, 800 respectively. Thus, the number of parameters is reduced from 128×800 to $32 \times (100 + 200 + 400 + 800)$.

Bigger kernel size means more parameters, which make the network more prone to overfitting. Dilated convolutions could reduce network parameters, while the receptive fields, center frequencies and band widths remain unchanged. To give the 4 groups an equal number of parameters, dilation rates are 1, 2, 4, and 8 respectively for the 4 groups. The number of parameters in this layer are further reduced to 128×100. Figure 2 illustrates the signal decomposition by using underwater noise radiated from a passenger ship. During the recording period, the ship was 1.95 km away from the hydrophone and its navigational speed was 18.4 kn.

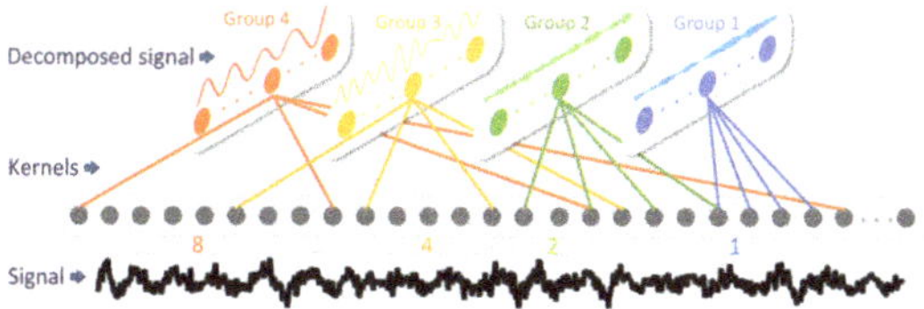

Figure 2. Signal decomposition of time convolutional layer. Black line at the bottom is the input signal. Four colors (orange, yellow, green and blue) at the upper and middle parts represent the four groups. Four capsules at the top represent the outputs of 4 groups. Four colors of straight lines at the middle part represent dilated convolutional kernels with a dilated rate of 8, 4, 2 and 1.

3.1.2. Time Frequency Conversion Layer

Stronger vibrations of basement membrane lead to more vigorous neural responses, which are further transformed by hair cells. The amplitude of the decomposed signal can be regarded as a neural response or frequency spectrum. The proposed time frequency conversion layer transforms the output of the time convolutional layer into a frequency domain. This is accomplished by a permute layer and a max-pooling layer. As shown in (5), $\mathbf{T}_1$ is permuted to $\mathbf{T}_2 = [\tau_1, \tau_2, ..., \tau_N]$, where $\tau_n \in \mathbb{R}^{L \times M}, n = 1, 2, ..., N$.

$$\mathbf{T}_2 = permute(\mathbf{T}_1). \tag{5}$$

As shown in (6), the amplitude of τ_n within regular time bins is calculated as $\mathbf{fr}_n (n = 1, 2, ...N, \mathbf{fr}_n \in \mathbb{R}^{K \times M})$ by max-pooling along time axis. The output of this layer is $\mathbf{Fr} = [\mathbf{fr}_1, \mathbf{fr}_2, ..., \mathbf{fr}_N]$. Thus, the internal representation of sound is calculated as a spectro-temporal excitation pattern which provides a clearer picture of sound.

$$\mathbf{fr}_n = max - pooling(\tau_n). \tag{6}$$

3.2. Multistage Auditory Center Model for Feature Extraction and Classification

The output of the time convolutional layer is divided into two routes. One route is the time frequency conversion layer, then directly through the deep neural network to model the multistage auditory pathway. Another route performs auditory feature recalibration.

3.2.1. Supervised Auditory Feature Recalibration

Given the relatively large depth of the network, the ability to propagate gradients back through all the layers should be enhanced, especially for time convolutional layer at the front of the whole network. Therefore, we propose an auditory feature recalibrate block on the basis of the recalibrate block [25]. This block takes $\mathbf{T}_1$ as the input. As shown in (7), global max pooling is used to aggregate $\mathbf{t}_m$ across frame length.

$$\mathbf{r}_m = max - pooling(\mathbf{t}_m), m = 1, 2, ..., M. \tag{7}$$

The output $\mathbf{r}_m (\mathbf{r}_m \in \mathbb{R}^N)$ is the amplitude of all the frames in $\mathbf{t}_m$. The output of this layer is $\mathbf{R}_1 = [\mathbf{r}_1, \mathbf{r}_2, ..., \mathbf{r}_M]$. It is permuted to $\mathbf{R}_2 = [\gamma_1, \gamma_2, ..., \gamma_N]$, where $\gamma_n \in \mathbb{R}^M, n = 1, 2, ..., N$. Then, it is followed with two fully connected layers to capture the dependencies of frequency components. The activation function of the fully connected layers are Rectified Linear Unit (ReLU) and sigmoid, respectively. The equation of the two layers are shown in (8):

$$\mathbf{fc}_n = sigmoid(v_n ReLU(\omega_n \gamma_n)), n = 1, 2, ..., N. \tag{8}$$

$\mathbf{Fc} = [\mathbf{fc}_1, \mathbf{fc}_2, ..., \mathbf{fc}_N]$ is the output of the fully connected layers, where $\mathbf{fc}_n(\mathbf{fc}_n \in \mathbb{R}^M)$ corresponds to the nth frame. $\mathbf{W} = [\omega_1, \omega_2, ..., \omega_N]$ and $\mathbf{U} = [v_1, v_2, ..., v_N]$ are weight vectors of the two layers. $\mathbf{Fc}$ is also divided into two routes. One route is fed directly into a softmax layer to make an auxiliary classifier. We would expect to increase the gradient signal by adding the loss of the auxiliary classifier to the total loss of the network. Another route is the auditory feature recalibration shown in (9).

$$\mathbf{fr}'_n = \mathbf{fc}_n \times \mathbf{fr}_n, n = 1, 2, ..., N, \tag{9}$$

where $\mathbf{fr}'_n(\mathbf{fr}'_n \in \mathbb{R}^{K \times M})$ is channel-wise multiplication between $\mathbf{fc}_n$ and $\mathbf{fr}_n$. The output of the layer is $\mathbf{Fr}' = [\mathbf{fr}'_1, \mathbf{fr}'_2, ..., \mathbf{fr}'_N](\mathbf{Fr}' \in \mathbb{R}^{K \times M \times N})$. This operation can be interpreted as a means of selecting the most informative frequency components of a signal in supervised manner. The recalibrated auditory features could establish the correlation between features and categories.

3.2.2. Deep Architecture for Feature Learning

Auditory perception depends on the integration of many neurons along the multistage auditory pathway. These neurons likely facilitate frequency topological topographic maps of most hearing region [26]. The proposed frequency convolutional layers perform convolution in both frequency and time axis to extract spectro-temporal patterns embedded in a ship radiated noise signal.

However, a drawback of a deep network constructed with a standard convolutional layer is the dramatically increased use of computational resources. Inception-Resnet [15] has been shown to achieve very good performance in image recognition at a relatively low computational cost. In this paper, a deep neural network with inception structures and residual connections are introduced in frequency convolutional layer to perform the auditory task. Multiscale convolutional kernels in inception block can be interpreted as simulating the different neural structures of auditory pathways. Inception structures have the ability to learn spectro-temporal patterns of different scales with less parameters. Residual connections can be interpreted as simulating the convergence and divergence between different pathways. The architecture allows for increasing the number of layers and units to form the multistage of auditory system. These layers could also preserve locality and reduce spectral variations of the line spectrum in ship radiated noise.

We use an global average pooling layer [23] at the final stage to generate one feature map for each ship type. The resulting vector is fed directly into softmax layer to predict targets. Compared with fully connected layers, global average pooling layer is more native to the convolution structure by enforcing correspondences between feature maps and categories. Moreover, there is no parameter to optimize in global average pooling layer thus overfitting is avoided at this layer.

4. Experiment

4.1. Experimental Dataset

Our experiments were performed on hydrophone acoustic data acquired by Ocean Networks Canada observatory. The data were measured using an Ocean Sonics icListen AF hydrophone placed at 144 m–147 m below sea level. Ship radiated noise data were from ships in a 2 km radius while no other ships were present in a 3 km radius. The duration of the recordings vary from about 5 to 20 min, depending on navigational speed and position. Each recording was sliced into several segments to make up the input of neural network. Each sample was a segment of 3 s duration and was divided into short frames of 1 s. Acoustic data were resampled to a sampling frequency of 16 kHz. Classification experiments were performed on ocean background noise and four ship types (Cargo, Passenger ship, Tanker, and Tug). The four ship type categories were designated by the World Shipping Encyclopedia from Lloyd's Registry of Ships. About 29 months of data were collected, the first 18 months for training and the remaining 11 months for testing. The dataset comprised about 140,000 training samples (771 recordings) and 82,000 validation samples (449 recordings).

4.2. Classification Experiment

To evaluate the classification performance of the proposed model, we report the classification accuracy against the previous proposed auditory inspired CNN and manually designed features. These hand designed features included waveform features [1,2], Mel-frequency Cepstral Coefficients (MFCC) [3], wavelet features [5], auditory features [7] and spectral [11,27]. Waveform features included peak-to-peak amplitude features, zero-crossing wavelength features and zero-crossing wavelength difference features. MFCC features were extracted based on a linear cosine transform of a log power spectrum on a nonlinear Mel scale of frequency. Auditory features were extracted based on the Bark scale and masking properties of the human auditory system. Wavelet features contained a low frequency envelope of wavelet decomposition and entropy of zero-crossing wavelength distribution density of all levels of wavelet signals. For the calculation of spectral, two pass split window (TPSW) was applied subsequently after a short time fast Fourier transform. Signals were windowed into frames of 256 ms before extracting features. The extracted features on frames were stacked or averaged to feed into support vector machine (SVM), back propagation neural network (BPNN) or CNN to classify ship types. The kernel function of SVM was the radial basis function (RBF). The penalty factor and kernel parameter of RBF were selected by grid search. The SVM ensemble was performed by the AdaBoost algorithm. The used BPNN had one hidden layer with 30 hidden units. The structure of CNN from the bottom up was a convolutional layer with 128 feature maps, a max pooling layer, convolutional layer with 64 feature maps, max pooling layer and fully connected layer with 32 units. Kernel size was 5×5 and pooling size was 2×2. The learning rate was 0.1 and momentum was 0.9. When training the proposed network, optimization was performed using RMSprop with learning rate 0.001, momentum 0.9, and a minibatch size of 50. The results are shown in Table 1. Our experiments demonstrate that the proposed method remarkably outperforms manually designed features. Benefiting from the improvements in the network structure, the accuracy has been greatly improved compared with auditory inspired CNN.

The confusion matrix of the proposed model on test data is shown in Table 2. The accuracy is at the bottom right corner. Both the precision and recall of background noise are higher than all ship types. This result indicates that it is easier to detect ship presence than classify ship types. The confusion between Cargo and Tanker are larger than other categories. This may be because the two categories always have similar propulsion systems, gross tonnage and ship size.

Table 1. Comparison of the proposed model and other methods.

Input	Model	Accuracy (%)
Waveform [1,2]	SVM	68.2
MFCC [3]	BPNN	72.1
Wavelet,Waveform,MFCC,Auditory feature [7]	SVM Ensemble	75.1
Wavelet and principal component analysis [5]	BPNN	74.6
Spectral [11]	Stacked Autoencoder	81.4
Spectral [27]	CNN	83.2
Time domain	Auditory inspired CNN [6]	81.5
Time domain	Proposed	87.2

Table 2. Confusion matrix of samples.

Ture \ Predicted	Background	Cargo	Tanker	Passenger	Tug	Recall (%)
Background	15,824	1	202	20	173	97.56
Cargo	16	13,152	2424	560	155	80.65
Tanker	120	1479	13,283	881	610	81.13
Passenger	133	356	233	14,908	748	91.02
Tug	334	317	590	1098	14,083	85.76
Precision (%)	96.33	85.93	79.39	85.35	89.31	87.2

We evaluated the performance of the proposed model on recordings by majority voting. One recording would be classified to a category to which the most samples in the recording are classified. The confusion matrix is shown in Table 3. The obtained accuracy is 94.75% at the bottom right corner. The recall and precision of all categories are improved compared with Table 2. Although individual samples could be misidentified, we can still make a correct recognition results of the whole signal by majority voting.

Table 3. Confusion matrix of recordings.

Ture \ Predicted	Background	Cargo	Tanker	Passenger	Tug	Recall (%)
Background	50	0	0	0	0	100
Cargo	0	107	9	2	0	90.68
Tanker	0	3	76	2	1	92.68
Passenger	0	0	1	137	3	97.16
Tug	0	0	0	2	56	94.92
Precision (%)	98.04	97.27	88.37	95.80	93.33	94.75

4.3. Operative Conditions Analysis

We analyzed the recognition results and operative conditions together in order to observe how the accuracy varies with ship operative conditions. Speed over ground (SOG), course over ground (COG) and distance to hydrophone were analyzed. Different ship types usually had different speeds and routes. Most ships were northbound, whereas only some passenger ships and tugs went in other directions. Because of these differences between ship types, it is necessary to analyse each ship type separately. From Figure 3, we can see that, with the increase of distance between ships and hydrophone, the recall rate of Cargo, Passenger ships and Tanker decreased. It is hard to find obvious laws about the influence of SOG and COG. The results indicate that the proposed model is robust to ship operative conditions. The detection results of passenger ships in each operative condition were obviously better than for the other ship types. This may be because the samples of a passenger ship are uniformly distributed in operative conditions and the classification model can fit it better.

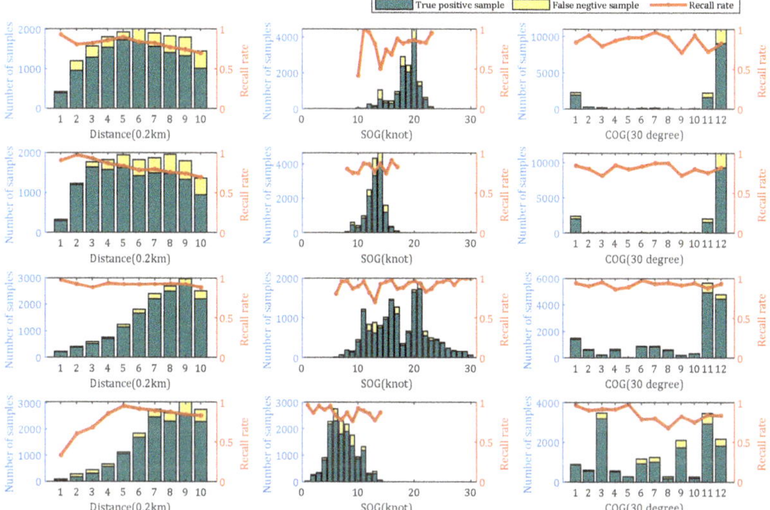

Figure 3. Operative condition and classification results analysis. Four rows from top to bottom represent Cargo, Tanker, Passenger ship and Tug, respectively. Histograms are the number of samples under different operative conditions. Yellow and green histograms represent the false negative samples and true positive samples, respectively. Orange lines are recall rates.

4.4. Visualization

4.4.1. Learned Auditory Filter Visualization

To observe learned auditory filters in the time convolutional layer, we selected one convolutional kernel (learned filters) from each of the 4 groups. Output feature maps corresponding to the 4 selected kernels were also extracted. As shown in Figure 4, the training of the network modified the shapes of these filters. The use of dilated convolution enables the 4 groups to have the same kernel width. The output feature maps are decomposed signals whose center frequencies are consistent with the original auditory filters. The dilated convolution can reduce parameters as well as preserve the center frequency of the auditory filter.

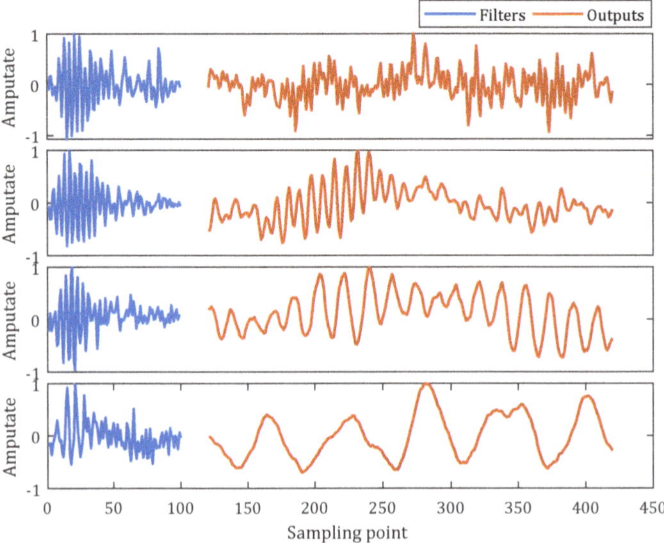

Figure 4. Visualization of learned auditory filters and their outputs in the time convolutional layer. Blue lines are learned auditory filters and orange lines are the corresponding output feature maps. The 4 panels from top to bottom represent the dilation rate of 1, 2, 4, and 8, respectively. For each panel, vertical axis represents the amplitude of filter and decomposed signal and horizontal axis represents the time domain sampling points of the filter and decomposed signal.

4.4.2. Learned Spectrogram Visualization

Outputs of the time frequency conversion layer were extracted as a learned spectrogram. It was compared with the gammatone spectrogram in Figure 5. The frame length and hop time for the gammatone spectrogram were the same as the kernel size and strides in the time convolutional layer. Thus, the dimension of the gammatone spectrogram was the same as the learned spectrogram. The learned spectrogram generated by the network is similar to the gammatone spectrogram. The network reserved low frequency components in signal and smoothed noises in high frequency components.

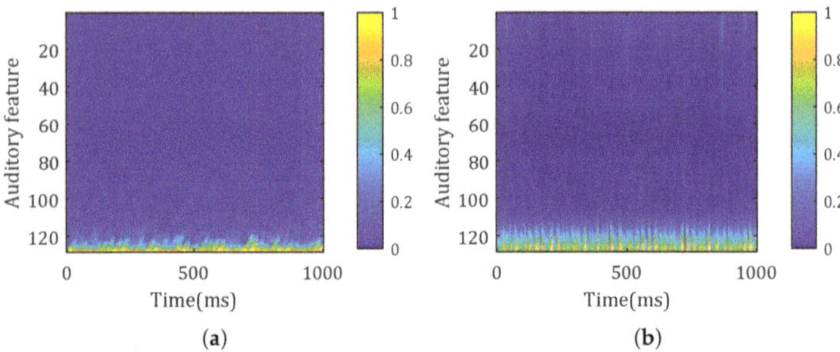

Figure 5. Comparison of learned spectrogram and gammatone spectrogram. (**a**) Learned spectrogram. (**b**) Gammatone spectrogram.

The data visualization method t-distributed stochastic neighbor embedding (t-SNE) [28] was used to visualize extracted features by giving each sample a location in a two dimensional map.

In Figure 6, the output of the whole network, learned spectrogram in time frequency conversion layer and gammatone spectrogram were visualized. As shown in Figure 6a, the last layer constructs a map in which most classes are separate from other classes. Figure 6b,c are the samples distribution of learned spectrograms and gammatone spectrograms, respectively. There are larger overlaps between the classes compared with Figure 6a. The samples distribution in Figure 6b is slightly better than that in Figure 6c. The results indicate that the proposed model could provide better insight into class structure of ship radiated noise data. Features extracted by the deeper layer are more discriminative than those from the shallow layer.

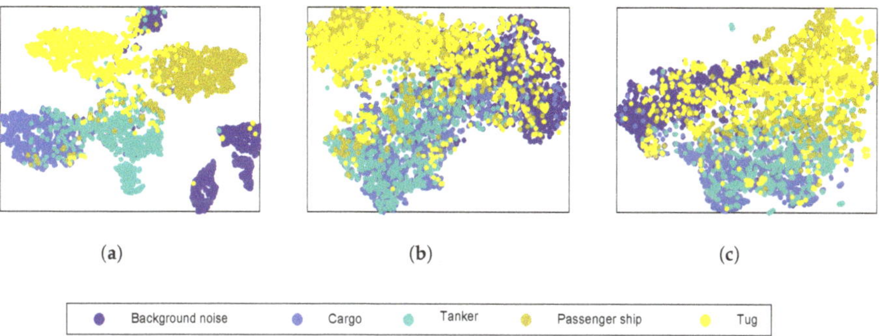

Figure 6. Feature visualization by t-distributed stochastic neighbor embedding (t-SNE). Five thousand samples selected randomly from test data were used to perform the experiments. Dots of different colors represent different types of ships. (**a**) Output of the whole network. (**b**) Learned spectrogram. (**c**) Gammatone spectrogram.

5. Conclusions

A deep convolutional neural network with auditory-like mechanisms is proposed to simulate the processing procedure of an auditory system for ship type classification. The integrated auditory mechanisms from early to higher auditory stages include auditory filters at basement membrane, neural pattern transformation by hair cells, spectro-temporal patterns along hierarchical structure, multiple auditory pathways and plasticity.

The classification experiments demonstrate that the proposed method outperforms manually designed features and classifiers. This study analyzes the recognition results in a way that is closer to the real-world scenario. The accuracy of recordings obtained by majority voting is much higher than the accuracy of segments. The increase of distance between ships to hydrophone has a negative effect on the recognition results in most cases. The proposed method has robustness to ship operative conditions. The network could generate a spectrogram that is similar to gammatone spectrogram, but smooth noises of high frequency components. The auditory filter banks in the network are adaptive in shape to ship radiated noise.

The proposed method facilitates the development of a smart hydrophone that could not only measure underwater acoustic signals, but also send alerts if it detects a specific underwater acoustic event. It will make it easier for researchers to listen to the ocean.

Author Contributions: Conceptualization, S.S. and H.Y.; Data curation, S.S., X.Y., J.L. and G.X.; Investigation, H.Y.; Methodology, S.S.; Project administration, H.Y.; Writing–original draft, S.S.; Writing–review and editing, S.S., H.Y., J.L. and M.S. All authors have read and agreed to the published version of the manuscript.

Funding: This work was supported by a grant from the Science and Technology on Blind Signals Processing Laboratory.

Acknowledgments: Data used in this work were provided by Ocean Networks Canada. The authors would like to gratefully acknowledge the support from Ocean Networks Canada.

Conflicts of Interest: The authors declare no conflict of interest.

References

1. Meng, Q.; Yang, S.; Piao, S. The classification of underwater acoustic target signals based on wave structure and support vector machine. *J. Acoust. Soc. Am.* **2014**, *136*, 2265. [CrossRef]
2. Meng, Q.; Yang, S. A wave structure based method for recognition of marine acoustic target signals. *J. Acoust. Soc. Am.* **2015**, *137*, 2242. [CrossRef]
3. Zhang, L.; Wu, D.; Han, X.; Zhu, Z. Feature Extraction of Underwater Target Signal Using Mel Frequency Cepstrum Coefficients Based on Acoustic Vector Sensor. *J. Sens.* **2016**, *2016*, 1–11. [CrossRef]
4. Mohankumar, K.; Supriya, M.; Pillai, P.S. Bispectral gammatone cepstral coefficient based neural network classifier. In Proceedings of the 2015 IEEE Underwater Technology (UT), Chennai, India, 23–25 February 2015; pp. 1–5.
5. Wei, X.; Gang-Hu, L.I.; Wang, Z.Q. Underwater Target Recognition Based on Wavelet Packet and Principal Component Analysis. *Comput. Simul.* **2011**, *28*, 8–290.
6. Shen, S.; Yang, H.; Li, J.; Xu, G.; Sheng, M. Auditory Inspired Convolutional Neural Networks for Ship Type Classification with Raw Hydrophone Data. *Entropy* **2018**, *20*, 990. [CrossRef]
7. Yang, H.; Gan, A.; Chen, H.; Pan, Y. Underwater acoustic target recognition using SVM ensemble via weighted sample and feature selection. In Proceedings of the International Bhurban Conference on Applied Sciences and Technology, Islamabad, Pakistan, 12–16 January 2016; pp. 522–527.
8. Filho, W.S.; de Seixas, J.M.; de Moura, N.N. Preprocessing passive sonar signals for neural classification. *IET Radar Sonar Navig.* **2011**, *5*, 605. [CrossRef]
9. Karakos, D.; Silovsky, J.; Schwartz, R.; Hartmann, W.; Makhoul, J. Individual Ship Detection Using Underwater Acoustics. In Proceedings of the 2018 IEEE International Conference on Acoustics, Speech and Signal Processing (ICASSP), Calgary, AB, Canada, 15–20 April 2018; pp. 2121–2125.
10. Kamal, S.; Mohammed, S.K.; Pillai, P.R.S.; Supriya, M.H. Deep learning architectures for underwater target recognition. In Proceedings of the Ocean Electronics, Kochi, India, 23–25 October 2013; pp. 48–54.
11. Cao, X.; Zhang, X.; Yu, Y.; Niu, L. Deep learning-based recognition of underwater target. In Proceedings of the IEEE International Conference on Digital Signal Processing, Beijing, China, 16–18 October 2016, pp. 89–93. [CrossRef]
12. Yang, H.; Shen, S.; Yao, X.; Sheng, M.; Wang, C. Competitive Deep-Belief Networks for Underwater Acoustic Target Recognition. *Sensors* **2018**, *18*, 952. [CrossRef] [PubMed]
13. Shen, S.; Yang, H.H.; Sheng, M.P. Compression of a Deep Competitive Network Based on Mutual Information for Underwater Acoustic Targets Recognition. *Entropy* **2018**, *20*, 243. [CrossRef]
14. Hu, G.; Wang, K.; Peng, Y.; Qiu, M.; Shi, J.; Liu, L. Deep Learning Methods for Underwater Target Feature Extraction and Recognition. *Comput. Intell. Neurosci.* **2018**, *2018*, 1214301. [CrossRef] [PubMed]
15. Szegedy, C.; Ioffe, S.; Vanhoucke, V.; Alemi, A.A. Inception-v4, inception-resnet and the impact of residual connections on learning. In Proceedings of the Thirty-First AAAI Conference on Artificial Intelligence, San Francisco, CA, USA, 4–9 February 2017.
16. Smith, E.C.; Lewicki, M.S. Efficient auditory coding. *Nature* **2006**, *439*, 978. [CrossRef] [PubMed]
17. Moore, B.C.J. Temporal integration and context effects in hearing. *J. Phon.* **2003**, *31*, 563–574. [CrossRef]
18. Shamma, S. Encoding Sound Timbre in the Auditory System. *IETE J. Res.* **2015**, *49*, 145–156. [CrossRef]
19. Chechik, G.; Nelken, I. Auditory abstraction from spectro-temporal features to coding auditory entities. *Proc. Natl. Acad. Sci. USA* **2012**, *109*, 18968–18973. [CrossRef] [PubMed]
20. Weinberger, N.M. Experience-dependent response plasticity in the auditory cortex: Issues, characteristics, mechanisms, and functions. In *Plasticity of the Auditory System*; Springer: New York, NY, USA, 2004; pp. 173–227.
21. Slaney, M. *An Efficient Implementation of the Patterson-Holdsworth Auditory Filter Bank*; Apple Computer: Cupertino, CA, USA, 1993.
22. Den Oord, A.V.; Dieleman, S.; Zen, H.; Simonyan, K.; Vinyals, O.; Graves, A.; Kalchbrenner, N.; Senior, A.W.; Kavukcuoglu, K. WaveNet: A Generative Model for Raw Audio. *arXiv* **2016**, arXiv:1609.03499 .
23. Lin, M.; Chen, Q.; Yan, S. Network In Network. In Proceedings of the International Conference on Learning Representations, Banff, AB, Canada, 14–16 April 2014.

24. Glasberg, B.R.; Moore, B.C. Derivation of auditory filter shapes from notched-noise data. *Hear. Res.* **1990**, *47*, 103–38. [CrossRef]
25. Hu, J.; Shen, L.; Sun, G. Squeeze-and-excitation networks. In Proceedings of the IEEE Conference on Computer Vision and Pattern Recognition, Long Beach, CA, USA, 16–20 June 2018; pp. 7132–7141.
26. Gazzaniga, M.; Ivry, R.B. *Cognitive Neuroscience: The Biology of the Mind: Fourth International Student Edition*; WW Norton: New York, NY, USA, 2013.
27. Yue, H.; Zhang, L.; Wang, D.; Wang, Y.; Lu, Z. *The Classification of Underwater Acoustic Targets Based on Deep Learning Methods*; CAAI 2017; Atlantis Press: Paris, France, 2017.
28. Hinton, G.E. Visualizing High-Dimensional Data Using t-SNE. *Vigiliae Christ.* **2008**, *9*, 2579–2605.

 *sensors* 

Article

Vehicular Traffic Congestion Classification by Visual Features and Deep Learning Approaches: A Comparison

Donato Impedovo *, Fabrizio Balducci, Vincenzo Dentamaroand Giuseppe Pirlo

Dipartimento di Informatica, Università degli Studi di Bari Aldo Moro, 70125 Bari, Italy;
fabrizio.balducci@uniba.it (F.B.); vincenzo@gatech.edu (V.D.); giuseppe.pirlo@uniba.it (G.P.)
* Correspondence: donato.impedovo@uniba.it; Tel.: +39-080-544-2280

Received: 17 October 2019; Accepted: 26 November 2019; Published: 28 November 2019

Abstract: Automatic traffic flow classification is useful to reveal road congestions and accidents. Nowadays, roads and highways are equipped with a huge amount of surveillance cameras, which can be used for real-time vehicle identification, and thus providing traffic flow estimation. This research provides a comparative analysis of state-of-the-art object detectors, visual features, and classification models useful to implement traffic state estimations. More specifically, three different object detectors are compared to identify vehicles. Four machine learning techniques are successively employed to explore five visual features for classification aims. These classic machine learning approaches are compared with the deep learning techniques. This research demonstrates that, when methods and resources are properly implemented and tested, results are very encouraging for both methods, but the deep learning method is the most accurately performing one reaching an accuracy of 99.9% for binary traffic state classification and 98.6% for multiclass classification.

Keywords: vehicular traffic flow detection; vehicular traffic flow classification; vehicular traffic congestion; deep learning; video classification; deep learning; benchmark

---

1. Introduction

Urban roads and highways have, nowadays, plenty of surveillance cameras initially installed for various security reasons. Traffic videos coming from these cameras can be used to estimate the traffic state, to automatically identify congestions, accidents, and infractions, and thus helping the transport management to face critical aspects of the mobility. At the same time, this information can also be used to plan the mid and long-term roads mobility strategy. This is a clear example of a smart city application having also a strong impact on citizens' security [1].

Studies dealing with traffic state estimation by videos adopt a common processing pipeline, which includes the following:

- The pre-processing of the video frames to highlight the useful elements (vehicles) and hide the unnecessary ones (background, etc.);
- The extraction of visual features able to describe the traffic state (e.g., number of vehicles, speed, etc.);
- One or more methods to classify the traffic state.

It is difficult to identify and select the best algorithms to be adopted because, often, systems reported in many studies are different at many stages, as well as adopt different datasets and testing conditions. This research provides a brief review of the most used techniques and reports an extended and systematic experimental comparison under common set-up conditions, thus highlighting strengths and weaknesses of each approach and supporting interested readers in the most profitable choices. More specifically:

- Haar Cascade, You Only Look Once (YOLO), the Single Shot MultiBox Detector (SSD), and Mask R-Convolutional Neural Networks (R-CNN) are adopted and compared for vehicle detection;
- Results provided by the two most accurately performing detectors, among the aforementioned, are used in a comparative schema to evaluate a set of visual features able to characterize the traffic state. These features were fed to four different classic classifiers, thus highlighting the most accurately performing one;
- Results obtained with classic approaches (previous point of this list) were compared to the use of deep learning techniques.

The article is organized as follows: Section 2 describes related studies, Section 3 presents the object detectors candidate to identify vehicles, the visual features useful to characterize a traffic video frame, and classifiers. The video datasets, evaluation metrics, along with experimental results are shown in Section 4. Section 5 presents conclusions and future researches.

2. Related Studies

Among the different methods that can be adopted to provide traffic flow (congestion) estimation, surveillance cameras play a crucial role. These systems can be installed without interfering with road infrastructures. Moreover, a large plethora of retrofit solutions are available in many cases, or systems have been already installed for some initial different aim. In any case, surveillance cameras can supply real-time information. The estimation of the traffic can be provided to users and police patrols to help in departures planning and congestion avoiding. Road panels or integrated vehicular monitors can also be used to reach the aim [2].

One of the first steps within the pipeline is vehicle identification [3]. Vehicles can be identified using features such as colors, pixels, and edges along with some machine learning algorithms. More specifically, detectors able to locate and classify objects in video frames exploiting visual features must be considered. State-of-the-art features are SURF (Speeded Up Robust Features) and bag-of-features [4], Haar Features [5], Edge Features [6], Shape Features [7], and Oriented Gradients Histograms [8]. Approaches based on visual features and machine learning models greatly benefited in efficiency with the introduction Convolutional Neural Networks (CNN) [9].

Choudhury et al. [10] proposed a system based on Dynamic Bayesian Network (DBN). It was tested on three videos after extracting visual features such as the number of moving vehicles [11]. Other examples of vehicle detector are in [12] and in [13] where features to characterize the traffic density are extracted from video footage. Li et al. [14] exploited the texture difference between congestion and unobstructed images, while similar approaches were used to recognize tow-away road signs from videos [15].

Vehicular traffic behavior can be also revealed by observable motion [16], in fact, it can be used to determine the number of vehicles performing the same action (e.g., by exploiting trajectory clustering for scene description). Drones and radio-controlled model aircrafts are exploited in the works of Liu et al. [17] and Gao et al. [18] to shoot live traffic flow videos and to study roads conditions. Shen and He [19] analyzed the vehicular behaviors at traffic bottlenecks and their movement to verify the decision-making process. In the research of Avinash et al. [20], Multiple Linear Regression (MLR) Technique was adopted to comprehend the factors influencing pedestrian safety. Koganti et al. [21] adopted Lane Distribution Factor (LDF) to describe the distribution of vehicular traffic across the roadway. Thomas et al. [22] presented a perceptual video summarization technique on a stack of videos to solve accident detection.

Real-time traffic videos have been used to observe temporal state of vehicles on a pedestrian crossing lane by using several image processing algorithms, connected components, and ray-casting technique [23]. Lin and Wang [24] implement a Vehicular Digital Video Recorder System able to support an online real-time navigator and an offline data viewer: The system was able to consider data security about vehicular parameters and to depict the instant status of vehicles.

3. Methods and Materials

Two approaches are compared in this research: The first one relies on visual features evaluated from traffic videos through computer vision algorithms using state-of-the-art object detectors and classifiers, the latter considers deep learning models able to automatically extract features from videos needed for the final classification.

3.1. Object Detectors

Four different object detectors have been explored: Haar Cascade, You Only Look Once (YOLO), Single Shot MultiBox Detector (SSD), and Mask R-CNN.

The Haar Cascade object detector, originally developed by Viola and Jones [25], relies on a set of visual features able to exploit rectangular regions at a specific location identifying pixel intensities and differences between the regions. The term "Cascade" refers to the decision sequencing, in fact the algorithm constructs a "strong" classifier as a combination of weighted weak classifiers using a boosting technique. A search window moves through the whole image to cover all the pieces and, at the same time, it is resized to different scales to find objects of different sizes. The detector requires a set of positive and negative samples. Haar features are extracted in the test phase and then compared with those used in the training phase, in order to obtain a positive or negative identification. Haar Cascade has been successfully used for Vehicle Detection [26], also to evaluate a traffic congestion index [27].

The YOLO (You Only Look Once) object detector consists of a CNN called Darknet [28] with an architecture made by 24 convolutional layers working as feature extractors and 2 dense layers for the prediction. Successively, YOLO v2 introduced anchors as a set of boxes used to predict the bounding boxes, while in YOLO v3 the prediction is performed at different scales. The algorithm looks at the image only once and splits it into an NxN grid where each cell predicts a fixed number of bounding boxes to associate an object to the supposed class providing a confidence score. Many of the bounding boxes have a very low confidence. Therefore, they can be filtered by applying a threshold. YOLO is a fast algorithm because it requires only one image processing, at the same time accuracy decreases when two different objects are very close to each other. YOLO has been successfully used for Vehicle Detection in [29] where the focal loss is exploited and validated on the BIT-Vehicle dataset and in [30] where, on the same dataset, a mean Average Precision (mAP) of 94.78% is reported.

Convolutional Neural Network (CNN) is also at the basis of the *Single Shot MultiBox Detector* (SSD), producing a set of fixed-size bounding, a confidence score is provided representing the probability that the object in the box belongs to a specific class. The CNN is constituted by several layers that progressively decrease in size so that objects can be predicted on several scales. SSD has been successfully used for Vehicle Detection in [12] while Sreekumar et al. [31] developed a multi-algorithm method for the real-time traffic pattern generation and Asvadi et al. [32] exploited a Dense Reflection Map (DRM) inputted to a Deep Convolutional Neural Network for the vehicle detection.

Mask R-CNN is the evolution of R-CNNs [33]. The original R-CNN detector used selective search to extract a huge number of candidate (proposal) regions (2000) from the input image. The algorithm first generates many candidate regions by performing initial sub-segmentation. Then, a greedy approach combines similar (adjacent) regions in larger ones. The selective search and the greedy approach result in very low computing [34]. R-CNN developed by Microsoft Research addresses this speed computation problem. One single model is used to extract features from each region, predict the belonging class, and compute the box coordinates. This is performed on a filtered image, which uses the low rank approximation with Singular Value Decomposition (SVD) of the original image. Fast R-CNN is 10× faster than R-CNN. Faster R-CNN [34] improves Fast R-CNN by using an additional network, called Region Proposal Network, in place of the selective search (originally derived from R-CNN) used for the generation of the regions of interest. This increases the prediction speed by about

10x [35]. Mask R-CNN is built on top of Faster R-CNN and, in addition to Faster R-CNN, it provides also the object segmentation. The mask of the segmented object could be used for inferring the valid shape of the classified object.

It is important to note that, while SSD used pre-trained weights on the Pascal Visual Object Classes (VOC) dataset [36], YOLO v3 and Mask R-CNN used pre-trained weights on the Coco dataset [37].

3.2. Visual Features

The output produced by a specific detector can be exploited to evaluate the traffic state and its density. In this research, five visual descriptors were considered: Total Vehicles, Traffic Velocity, Traffic Flow, Road Occupancy, and the Texture Feature.

The *Total Vehicles* feature is the number of bounding boxes provided by the vehicle detector and evaluated for each video frame [10,13,20,21,24,38–43].

The Traffic Velocity (average speed of all vehicles in the frame) is evaluated by tracking the vehicular bounding boxes centers and calculating the Euclidean distance with the corresponding position in the next frame [31]. The distance of each vehicle has been normalized according to the fps (frame per second) rate thus finding the individual speed and the global average speed (the sum of all vehicles velocities divided by their number). The last parameter can be considered as an estimation of the global traffic velocity [12,13,42,44,45].

The third visual feature is *Traffic Flow*, it is calculated considering the difference between the incoming and the outgoing traffic, respectively evaluated as the number of vehicles entering the camera field of view and those leaving it [15,39,40,42,45].

Background suppression and morphological transformations (i.e., opening and closing) can be used to isolate vehicle shapes: This process returns the *Road Occupancy* providing a relationship between the flat road surface (white pixels in Figure 1) and vehicles (black pixels in Figure 1). To gain this result, frames are converted to grayscale. Successively, pixel values are subtracted from the next frame highlighting non-modified areas (background) and modified ones (foreground): Pixels belonging to vehicles shape are those related to changes in the scene. A threshold operator (i.e., Regional Minimum) is adopted to distinguish flat areas from areas occupied by vehicles which result from the ratio between black and white pixels dynamically changing according to the number of vehicles [9,12,16,41,42,44,45].

Figure 1. The morphological operator applied to the frame of the vehicular traffic video.

The *Texture Feature* was calculated according to the Gray Level Co-occurrence Matrix (GLCM) method [14,45–47]. This parameter is typically used to estimate vehicles' density by exploiting the corresponding 'energy' and 'entropy' values. More specifically, energy reveals whether the texture is thick (high value) or thin, while the entropy expresses how the elements are distributed (uniformly featuring a high value or not) [48]. The value of energy is inversely proportional with vehicle density, and the value of entropy is proportional with vehicle density. In other words, the gray histogram of image should be distributed uniformly and the texture of the image should be thin in case there are many vehicles in a frame. Therefore, the energy feature value should be small, and the entropy feature value should be big.

3.3. Machine Learning Classifiers

In this research three classifiers were considered and compared: k-Nearest Neighbors, Support Vector Machine, and Random Forest.

The K-Nearest Neighbors (KNN) is used for classification and regression tasks exploiting sample vectors in a multi-dimensional feature space. K is a user-defined parameter referred to the number of class labels: the unknown input vector is classified assigning it the "nearest class" among those already known. Many distance measures can be considered as, for example, the Euclidean or the Manhattan one. K-NN has been applied to this specific field in [38] and [49].

Support Vector Machine (SVM) maps feature vectors of two different classes within a hyperspace and searches for the best separating hyperplane taking into account only a reduced set of the initial amount of examples (called support vectors), which are those difficult to be classified. According to data distribution, different separating hyperplanes (kernel) can be considered. In this research the linear kernel and *a Gaussian Radial Basis* (rbf) function were considered as in [50] where the traffic congestion is classified through a comparison between AlexNet+SVM and VGGNet+SVM.

The Random Forest (RF) classifier relies on a bagging method to build several "base learners", usually decision trees. The base learners are successively combined to provide the final classification. RF repeatedly selects a bootstrap sample from the training set. It selects a random subset of features, and then fits a decision tree to this sample. Due to this randomness, the bias of the forest increases, but, due to averaging, its variance also decreases. In extremely randomized trees (ET), randomness is taken a step further by also completely randomizing the cut-point choice while splitting each node of a tree. This allows the variance to be reduced a little more at the expense of a slight increase in bias. In this research, 100 decision trees were used as base learners for RF and ET.

3.4. Deep Learning Classification Models

He et al. [51] proposed a residual deep learning framework for image classification, where layers were redrafted in order to learn residual functions with respect to the input layer. The proposed ResNET had 34 layers that follow the same pattern while performing 3×3 convolutions with a fixed feature map dimension, the input is bypassed every two convolutions. Moreover, the width and height dimensions remain constant for the entire layer thus reducing the complexity per layer of the network. The output is a binary classification (congested/not congested). The ResNET has been also re-trained in [52] on the Shaanxi Province dataset.

Kurniawan et al. [53] used two convolutional layers, a max pooling layer, and a fully connected layer where the first two layers are convolute with 3×3 filters and 32 feature maps, the third one is a 2×2 max pooling layer used for down-sampling, and the last one is a fully connected layer with 128 neurons. Rectified Linear Units (ReLU) activation function has been exploited in both the convolutional and fully connected layers while a sigmoid activation function has been used for the output layer [45,50].

4. Experiments and Discussion

4.1. Video Datasets

Different datasets, used for different aims, were adopted in this research. The GRAM Road Traffic Monitoring (RTM) is generally used for vehicle detection and it was adopted here to evaluate and compare performance of object (vehicles) detectors [54]. Trafficdb contains annotations related to the state of the traffic and it is here used to compare classification techniques [55,56].

4.1.1. GRAM RTM

The Road-Traffic Monitoring [54] is a dataset specifically used for vehicular detection in a traffic environment. It consists of three video sequences from which individual frames were labeled with bounding box around vehicles. The "M-30" video includes 7520 frames recorded on a sunny day with a Nikon Coolpix L20 camera having a resolution of 640 × 480 pixels at 30 fps. The second video "M-30-HD" includes 9390 frames recorded in the same place of the previous video but on a cloudy day at a higher resolution (1280 × 720 pixels at 30 fps using a Nikon DX3100 camera). The last "Urban1" video contains 23,435 frames in low resolution (480 × 320 pixels at 25 fps). This dataset offers the possibility to evaluate vehicles detectors under different working conditions.

From each video of the dataset, the ground-truth consists in bounding boxes around all vehicles per each frame. Information about the acquisition properties are provided, together with pixel masks useful to extract region of interests (ROI) and decrease the computational load of subsequent processing phases (Figure 2).

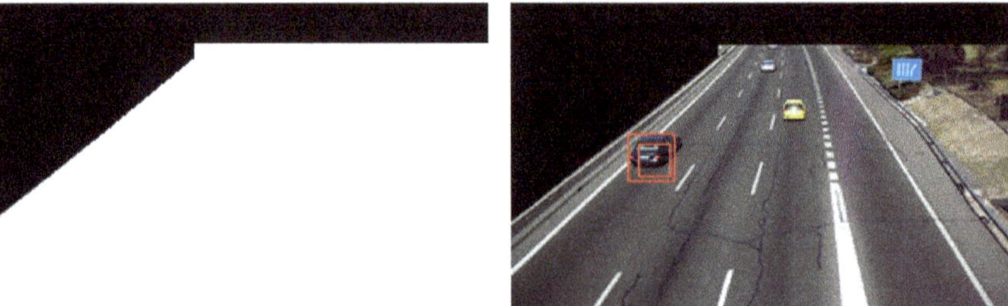

Figure 2. GRAM dataset: (**left**) a pixel mask that highlights the Region of Interest about vehicles present, annotated (and subsequently trackable) in each video frame (**right**).

4.1.2. Trafficdb

Trafficdb dataset is a state-of-the-art dataset used for vehicular traffic state classification since it is provided with specific annotations [55,56]. It is constituted by 254 videos acquired between 8 April and 8 June 2004 on Seattle (USA) highway segments. The ground-truth includes three classes (Figure 3): 'Heavy'—very congested traffic, 'Medium'—low vehicular flow, and 'Light'—normal travel velocity.

Figure 3. Three images from Trafficdb that depict a traffic state classified as Light, Medium, and Heavy.

4.2. Evaluation Metrics

Vehicle bounding boxes provided by the object detector can be compared to the real vehicle annotation highlighting the most accurately performing one. It is quite clear that the result of this phase has a strong impact on all the subsequent stages within the processing pipeline.

The metric *Correct detections* refers to correctly detected vehicles for each frame and it is measured as the pixel intersection between the original ground-truth bounding box area (named G) and the predicted one (named P). The *Jaccard Index* or *Intersection over Union* (1) was considered.

$$J(G, P) = \frac{|G \cap P|}{|G \cup P|}.$$ (1)

The second performance metric here considered is the *Computing time* evaluated by adding the processing time for each video frame at each iteration. It is useful to select the most suitable detector for a specific problem (real-time, on site, off-line, etc.).

The accuracy was considered and evaluated as follow:

$$accuracy = (TP + TN)/(TP + TN + FP + FN),$$ (2)

where:
- TP: the true positive samples, example of class X and classified by the system as X;
- TN: the true negative samples, object not of class X and not classified by the system as X;
- FP: the false positive samples, object not of class X but classified by the system as X;
- FN: the false negative samples, object of class X, but not classified by the system as X;

Experiments were performed on a System featuring Ubuntu 18.04 as Operating System, AMD Ryzhen threadripper 1920x with 12 cores, Nvidia Titan RTX 24 GB RAM, 64 GB RAM DDR4.

4.3. Vehicle Detectors Evaluation

Table 1 reports results of the object detectors on the Road-Traffic Monitoring GRAM dataset.

Table 1. Performance comparison according to the processing time and vehicle detection accuracy of the four object detectors on the three videos in the GRAM dataset.

		M-30	M-30-HD	Urban1
Haar Cascade	Time [s]	0.08–0.13	0.3–0.44	0.02–0.06
	Accuracy	43%	75%	40%
SSD	Time [s]	4–7	11–14	2.6–5.6
	Accuracy	22%	70%	69%
YOLO v3	Time [s]	1.0–1.8	1.0–1.8	1.0–1.8
	Accuracy	82%	86%	91%
Mask R-CNN	Time [s]	2.4–3.0	2.4–3.0	2.4–3.0
	Accuracy	89%	91%	46%

The experimental phase pointed out that the Haar Cascade detector is the fastest one on each dataset. However, it provides good accuracy only on M-30-HD (75%).

The lowest processing time is achieved on the 'Urban1' video due to the low frame resolution that also impacts on the correctly identified vehicles, with 40% of accuracy also due to incorrect multiple detections.

The Single Shot MultiBox Detector (SSD) is the slowest object detector. The experiment on the "M-30" video reported an accuracy of 22%, which is the lowest between the four solutions. On the "M-30-HD" video, the computational time is much heavier (11 to 14 s) with frequent peaks between 14 and 17 s, the reported accuracy reaches 70%.

YOLO detector offers the best compromise with acceptable execution times and very good performances on all datasets.

Mask R-CNN exhibits very discordant results in terms of accuracy. In particular, the poor performance on Urban1 dataset is probably due to the poor image quality: JPEG compression kneaded colors, and thus cheated the Region Proposal Network.

The most accurately performing detector, considering the processing time, is Haar Cascade, while YOLO represents the compromise between time resources and detection accuracy.

4.4. Traffic State Classification: Visual Features and Machine Learning Classification

Results obtained in the previous section clearly report YOLO and Haar Cascade as the most accurately performing vehicles detectors in terms of accuracy and processing time. For this reason, they have been chosen to support the visual features extraction to build the input vector for vehicle classifiers.

The first experimental session involves the five visual features calculated using the best two selected object detectors with the different machine learning classifiers seen in Section 3.3. A 10-fold cross validation setup was adopted to minimize the effect of variance when choosing the training and testing examples. Classification results on the Trafficdb dataset are reported in Table 2. Visual features were extracted on a sampling rate of 30 frames.

Table 2. Traffic state classification accuracy on the Trafficdb dataset.

	KNN	SVM (Linear)	SVM (rbf)	Random Forest
YOLO v3	0.81 ± 0.10	0.78 ± 0.12	0.79 ± 0.16	**0.84 ± 0.13**
Haar Cascade	0.66 ± 0.21	0.64 ± 0.11	0.64 ± 0.08	0.68 ± 0.21

Table 2 shows that YOLO detector combined with the Random Forest is the most accurately performing solution with an accuracy of 84%. The confusion matrix provided by this solution is reported in Table 3. Due to the unbalance of the classes in Trafficdb (the 'Heavy' class instances are about four times the other two), the classification results are provided in a normalized form.

Table 3. Normalized confusion matrix of the traffic state classification reached by the Random Forest classifier on the Trafficdb video dataset.

Random Forest	Light (Pred)	Medium (Pred)	Heavy (Pred)
Light	**0.94**	0.02	0.04
Medium	0.42	**0.47**	0.11
Heavy	0.20	0.11	**0.68**

4.5. Traffic State Classification: Deep Learning

The deep learning models described in Section 3.4 were implemented and re-trained on the Trafficdb video dataset in a 10-fold cross validation setup. The ResNet [51] and the deep network architecture proposed in [52] were originally tested by respective authors on a two class (Heavy vs. Light) traffic state classification. To perform similar tests, at a first stage, samples labeled as 'Medium' were removed from the Trafficdb: results are shown in Table 4.

Table 4. Normalized confusion matrix about the binary traffic state classification performed by the deep neural network of Kurniawan et al. [53] on the Trafficdb video dataset.

Deep Learning Architecture [44]	Light (Pred)	Heavy (Pred)
Light	**0.995**	0.004
Heavy	0	**1**

Finally, to compare results to the cases if the previous section, the two deep learning architectures were extended to perform traffic state classification on three classes. In this case, the best performance was reached by the ResNet [51] with an accuracy of 98.61% (results are in Table 5).

Table 5. Normalized confusion matrix about the multiclass traffic state classification performed by the deep neural network of Kurniawan et al. [53] on the Trafficdb video dataset.

Deep Learning Architecture [43]	Light (Pred)	Medium (Pred)	Heavy (Pred)
Light	**0.997**	0.003	0.
Medium	0.004	**0.972**	0.025
Heavy	0.	0.040	**0.959**

5. Conclusions

A pipeline to develop state-of-the-art traffic state classification systems from videos has been presented in this research. The pipeline is made up of three main steps: vehicle detection, feature extraction, and classification. Several state-of-the-art approaches have been considered and compared. A preliminary comparison between object detectors, performed on the GRAM Road Traffic Monitoring video dataset, has pointed out that YOLO v3 can be used for real-time vehicle detector exhibiting a detection accuracy of over 80%.

For the traffic state classification, two different approaches have been studied and tested on the Trafficdb video dataset. The first approach relies on visual features calculated through computer vision techniques and machine learning classifiers while the second one exploits deep learning able to embed the features extraction when training the model on the annotated dataset. In the classic approach (visual features and machine learning classifiers), the Random Forest has gained 84% of accuracy while the deep learning approach has reached an accuracy of over 98% with the same experimental setup, and thus showing a noticeable increase of +14% in the results.

The problem here considered is obviously complex, and the provided results need further improvement as for example: refinements of the object detection algorithms, validation on more traffic datasets. The last aspect is non-trivial because different road settings (country road, city road, road junction and crossroad, double lane roads, etc.) and weather conditions (rainy nights, fog, snow, gusts of wind, and so on) could have significant impact on systems.

Author Contributions: Conceptualization, D.I. and G.P.; methodology, D.I.; software, F.B. and V.D.; validation, D.I. and G.P.; investigation, D.I. and G.P.; resources, D.I.; data curation, F.B. and V.D.; writing—original draft preparation, F.B. and D.I.; writing—review and editing, F.B., D.I., and V.D.; supervision, D.I.; project administration, D.I.; funding acquisition, D.I.

Funding: This research is within the "Metrominuto Advanced Social Games" project funded by POR Puglia FESR-FSE 2014–2020—Fondo Europeo Sviluppo Regionale—Asse I—Azione 1.4—Sub-Azione 1.4.b—Avviso pubblico "Innolabs".

Conflicts of Interest: The authors declare no conflict of interest.

Code: https://gitlab.com/islabuniba/traffic_benchmark.

References

1. Ministero Delle Infrastrutture e dei Trasporti, "I Sistemi di Trasporto Intelligenti (ITS)". 2016. Available online: mit.gov.it/documentazione/sistemi-trasporto-intelligenti (accessed on 22 April 2019).
2. Wan, J.; Yuan, Y.; Wang, Q. Traffic congestion analysis: A new Perspective. In Proceedings of the IEEE International Conference on Acoustics, Speech and Signal Processing (ICASSP), New Orleans, LA, USA, 5–9 March 2017; pp. 1398–1402.
3. Shi, X.; Shan, Z.; Zhao, N. Learning for an aesthetic model for estimating the traffic state in the traffic video. *Neurocomputing* **2016**, *181*, 29–37. [CrossRef]

4. Nguyen, H.N.; Krishnakumari, P.; Vu, H.L.; Van Lint, H. Traffic COngestion pattern classification using multi-class SVM. In Proceedings of the IEEE 19th International Conference on Intelligent Transportation Systems (ITSC), Rio de Janeiro, Brazil, 1–4 November 2016; pp. 1059–1064.

5. Viola, P.A.; Jones, M.J.; Snow, D. Detecting pedestrians using patterns of motion and appearance. *Int. J. Comput. Vis.* **2005**, *63*, 153–161. [CrossRef]

6. Wu, B.; Nevatia, R. Detection of multiple, partially occluded humans in a single image by bayesian combination of edgelet part detectors. In Proceedings of the 10th IEEE International Conference on Computer Vision (ICCV 2005), Beijing, China, 17–21 October 2005; pp. 90–97.

7. Sabzmeydani, P.; Mori, G. Detecting pedestrians by learning shapelet features. In Proceedings of the IEEE Conference on Computer Vision and Pattern Recognition (CVPR), Minneapolis, MN, USA, 17–22 June 2007; pp. 1–8.

8. Dalal, N.; Triggs, B. Histograms of oriented gradients for human detection. In Proceedings of the IEEE Computer Society Conference on Computer Vision and Pattern Recognition (CVPR'05), San Diego, CA, USA, 20–25 June 2005; Volume 1, pp. 886–893.

9. Wali, S.B.; Abdullah, M.A.; Hannan, M.A.; Hussain, A.; Samad, S.A.; Ker, P.J.; Mansor, M.B. Vision-Based Traffic Sign Detection and Recognition Systems: Current Trends and Challenges. *Sensors* **2019**, *19*, 2093. [CrossRef] [PubMed]

10. Chaudhary, S.; Indu, S.; Chaudhury, S. Video-based road traffic monitoring and prediction using dynamic Bayesian networks. *IET Intell. Transp. Syst.* **2018**, *12*, 169–176. [CrossRef]

11. Choudhury, S.; Chattopadhyay, S.P.; Hazra, T.K. Vehicle detection and counting using haar feature based classifier. In Proceedings of the 8th Annual Industrial Automation and Electromechanical Engineering Conference (IEMECON), Bangkok, Thailand, 16–18 August 2017; pp. 106–109.

12. Qiong, W.U.; Sheng-bin, L. Single Shot MultiBox Detector for Vehicles and Pedestrians Detection and Classification. In *DEStech Transactions on Engineering and Technology Research*; DEStech Publications, Inc.: Lancaster, PA, USA, 2018.

13. Dailey, X.D.; Pumin, S.; Cathey, F.W. An algorithm to estimate mean traffic speed using uncalibrated cameras. *IEEE Trans. Intell. Transp. Syst.* **2000**, *1*, 98–107. [CrossRef]

14. Balducci, F.; Impedovo, D.; Pirlo, G. Detection and validation of tow-away road sign licenses through deep learning methods. *Sensors* **2018**, *18*, 4147. [CrossRef] [PubMed]

15. Morris, B.; Trivedi, M. Understanding vehicular traffic behavior from video: A survey of unsupervised approaches. *J. Electron. Imaging* **2013**, *22*, 041113. [CrossRef]

16. Liu, S.; Zheng, Y.; Luo, H.; Duan, S.; Wang, H. Vehicle trajectory observation based on traffic video provided by radio-controlled model aircraft. In Proceedings of the 4th International Conference on Transportation Engineering, Chengdu, China, 19–20 October 2013; pp. 122–128.

17. Gao, H.; Kong, S.; Zhou, S.; Lv, F.; Chen, Q. Automatic extraction of multi-vehicle trajectory based on traffic videotaping from quadcopter model. *Appl. Mech. Mater.* **2014**, *552*, 232–239. [CrossRef]

18. Shen, X.; He, Z. Analysis of vehicular behavior at bottlenecks considering lateral separation. *Smart Innov. Syst. Technol.* **2017**, *53*, 169–185.

19. Avinash, C.; Jiten, S.; Shriniwas, A.; Gaurang, J.; Manoranjan, P. Evaluation of pedestrian safety margin at mid-block crosswalks in India. *Saf. Sci.* **2018**, *119*, 188–198. [CrossRef]

20. Koganti, S.; Raja, K.; Sajja, S.; Narendra, M.S. A study on volume, speed and lane distribution of mixed traffic flow by using video graphic technique. *Int. J. Eng. Technol. (UAE)* **2018**, *7*, 59–62. [CrossRef]

21. Thomas, S.; Gupta, S.; Subramanian, V. Event detection on roads using perceptual video summarization. *IEEE Trans. Intell. Transp. Syst.* **2018**, *19*, 2944–2954. [CrossRef]

22. Goma, J.C.d.; Ammuyutan, L.A.B.; Capulong, H.L.S.; Naranjo, K.P.; Devaraj, M. Vehicular obstruction detection in the zebra lane using computer vision. In Proceedings of the IEEE 6th International Conference on Industrial Engineering and Applications (ICIEA), Tokyo, Japan, 12–15 April 2019; pp. 362–366.

23. Lin, C.; Wang, M. An implementation of a vehicular digital video recorder system. In Proceedings of the IEEE/ACM Int'l Conference on Green Computing and Communications Int'l Conference on Cyber, Physical and Social Computing, Hangzhou, China, 18–20 December 2010; pp. 907–911.

24. Viola, P.; Jones, M. Robust Real-Time Object Detection. *Int. J. Comput. Vis.* **2001**, *57*, 137–154. [CrossRef]

25. Dubey, A.; Rane, S. Implementation of an intelligent traffic control system and real time traffic statistics broadcasting. In Proceedings of the International conference of Electronics, Communication and Aerospace Technology (ICECA), Coimbatore, India, 20–22 April 2017; pp. 33–37.

26. Lam, C.; Gao, H.; Ng, B. A real-time traffic congestion detection system using on-line images. In Proceedings of the IEEE 17th International Conference on Communication Technology (ICCT), Chengdu, Chian, 27–30 October 2017; pp. 1548–1552.

27. Redmon, J.; Divvala, S.; Girshick, R.; Farhadi, A. You only look once: Unified, real-time object detection. In Proceedings of the IEEE Conference on Computer Vision and Pattern Recognition, Las Vegas, NV, USA, 27–30 June 2016; pp. 779–788.

28. Wu, Z.; Sang, J.; Zhang, Q.; Xiang, H.; Cai, B.; Xia, X. Multi-Scale Vehicle Detection for Foreground-Background Class Imbalance with Improved YOLOv2. *Sensors* **2019**, *19*, 3336. [CrossRef] [PubMed]

29. Sang, J.; Wu, Z.; Guo, P.; Hu, H.; Xiang, H.; Zhang, Q.; Cai, B. An Improved YOLOv2 for Vehicle Detection. *Sensors* **2018**, *18*, 4272. [CrossRef]

30. Sreekumar, U.K.; Devaraj, R.; Li, Q.; Liu, K. TPCAM: Real-time traffic pattern collection and analysis model based on deep learning. In Proceedings of the IEEE SmartWorld, Ubiquitous Intelligence Computing, San Francisco, CA, USA, 4–8 August 2017; pp. 1–4.

31. Asvadi, A.; Garrote, L.; Premebida, C.; Peixoto, P.; Nunes, U.J. Real-time deep convnet-based vehicle detection using 3d-lidar reflection intensity data. In *ROBOT 2017: Third Iberian Robotics Conference*; Springer International Publishing: Cham, Switzerland, 2017; pp. 475–486.

32. Girshick, R.; Donahue, J.; Darrell, T.; Malik, J. Rich feature hierarchies for accurate object detection and semantic segmentation. In Proceedings of the IEEE Conference on Computer Vision and Pattern Recognition, Columbus, OH, USA, 23–28 June 2014; pp. 580–587.

33. Girshick, R. Fast R-CNN Object Detection with Caffe; Microsoft Research. *arXiv* **2015**, arXiv:1504.08083v2.

34. Ren, S.; He, K.; Girshick, R.; Sun, J. Faster R-CNN: Towards real-time object detection with region proposal networks. *IEEE Trans. Pattern Anal. Mach. Intell.* **2017**, *39*, 1137–1149. [CrossRef]

35. Everingham, M.; Van Gool, L.; Williams, C.K.I.; Winn, J.; Zisserman, A. The Pascal Visual Object Classes (VOC) Challenge. *Int. J. Comput. Vis.* **2009**, *88*, 303–338. [CrossRef]

36. Lin, T.Y.; Maire, M.; Belongie, S.; Hays, J.; Perona, P.; Ramanan, D.; Zitnick, C.L. Microsoft coco: Common objects in context. In *European Conference on Computer Vision*; Springer: Cham, Switzerland, 2014; pp. 740–755.

37. Asmaa, O.; Mokhtar, K.; Abdelaziz, O. Road traffic density estimation using microscopic and macroscopic parameters. *Image Vis. Comput.* **2013**, *31*, 887–894. [CrossRef]

38. Zhu, F. A video-based traffic congestion monitoring system using adaptive background subtraction. In Proceedings of the Second International Symposium on Electronic Commerce and Security, Nanchang, China, 22–24 May 2009; pp. 73–77.

39. Zhu, F.; Li, L. An optimized video-based traffic congestion monitoring system. In Proceedings of the Third International Conference on Knowledge Discovery and Data Mining, Phuket, Thailand, 9–10 January 2010; pp. 150–153.

40. Kanungo, A.; Sharma, A.; Singla, C. Smart traffic lights switching and traffic density calculation using video processing. In Proceedings of the 2014 Recent Advances in Engineering and Computational Sciences (RAECS), Chandigarh, India, 6–8 March 2014; pp. 1–6.

41. Perkasa, O.; Widyantoro, D.H. Video-based system development for automatic traffic monitoring. In Proceedings of the International Conference on Electrical Engineering and Computer Science (ICEECS), Kuta, Indonesia, 24–25 November 2014; pp. 240–244.

42. Eamthanakul, B.; Ketcham, M.; Chumuang, N. The traffic congestion investigating system by image processing from cctv camera. In Proceedings of the International Conference on Digital Arts, Media and Technology (ICDAMT), Kuta, Indonesia, 24–25 November 2017; pp. 240–245.

43. Xun, F.; Yang, X.; Xie, Y.; Wang, L. Congestion detection of urban intersections based on surveillance video. In Proceedings of the 18th International Symposium on Communications and Information Technologies (ISCIT), Bangkok, Thailand, 26–28 September 2018; pp. 495–498.

44. Ke, X.; Shi, L.; Guo, W.; Chen, D. Multi-dimensional traffic congestion detection based on fusion of visual features and convolutional neural network. *IEEE Trans. Intell. Transp. Syst.* **2019**, *20*, 2157–2170. [CrossRef]

45. Li, W.; Dai, H.Y. Real-time road congestion detection based on image texture analysis. *Procedia Eng.* **2016**, *137*, 196–201.

46. Haralick, R.M.; Shanmugnm, K.; Dinstein, I. Textural features for image classification. *IEEE Trans. Syst. Man. Cybern.* **1973**, *3*, 610–621. [CrossRef]

47. Xu, Y. Crowd density estimation using texture analysis and learning. In Proceedings of the IEEE International Conference on Robotics and Biomimetics, Kunming, China, 17–20 December 2006; pp. 214–219.

48. Lozano, A.; Manfredi, G.; Nieddu, L. An algorithm for the recognition of levels of congestion in road traffic problems. *Math. Comput. Simul.* **2009**, *79*, 1926–1934. [CrossRef]

49. Wang, P.; Li, L.; Jin, Y.; Wang, G. Detection of unwanted traffic congestion based on existing surveillance system using in freeway via a CNN-architecture trafficnet. In Proceedings of the 13th IEEE Conference on Industrial Electronics and Applications (ICIEA), Wuhan, China, 31 May–2 June 2018; pp. 1134–1139.

50. He, K.; Zhang, X.; Ren, S.; Sun, J. Deep residual learning for image recognition. In Proceedings of the IEEE Conference on Computer Vision and Pattern Recognition, Las Vegas, NV, USA, 27–30 June 2016; pp. 770–778.

51. Wang, P.; Hao, W.; Sun, Z.; Wang, S.; Tan, E.; Li, L.; Jin, Y. Regional detection of traffic congestion using in a large-scale surveillance system via deep residual TrafficNet. *IEEE Access* **2018**, *6*, 68910–68919. [CrossRef]

52. Kurniawan, J.; Syahra, S.G.; Dewa, C.K. Traffic Congestion Detection: Learning from CCTV Monitoring Images using Convolutional Neural Network. *Procedia Comput. Sci.* **2018**, *144*, 291–297. [CrossRef]

53. Guerrero-Gomez-Olmedo, R.; Lopez-Sastre, R.J.; Maldonado-Bascon, S.; Fernandez-Caballero, A. Vehicle tracking by simultaneous detection and viewpoint estimation. In Proceedings of the IWINAC 2013, Part II, LNCS 7931, Mallorca, Spain, 10–14 June 2013; pp. 306–316.

54. Chan, B.; Vasconcelos, N. Probabilistic kernels for the classification of auto-regressive visual processes. In Proceedings of the IEEE Computer Society Conference on Computer Vision and Pattern Recognition (CVPR'05), San Diego, CA, USA, 20–25 June 2005; pp. 846–851.

55. Chan, V.N.; Antoni, B. Classification and retrieval of traffic video using auto-regressive stochastic processes. In Proceedings of the IEEE Intelligent Vehicles Symposium, Las Vegas, NV, USA, 6–8 June 2005; pp. 771–776.

56. Snijders, C.; Matzat, U.; Reips, U.D. Big Data: Big gaps of knowledge in the field of Internet. *Int. J. Internet Sci.* **2012**, *7*, 1–5.

Article

Real-Time Vehicle-Detection Method in Bird-View Unmanned-Aerial-Vehicle Imagery

Seongkyun Han [1], Jisang Yoo [1] and Soonchul Kwon [2,*]

[1]   Department of Electrical Engineering, Kwangwoon University, 20 Kwangwoon-ro, Nowon-gu, Seoul 01897, Korea; skhan3410@naver.com (S.H.); jsyoo@kw.ac.kr (J.Y.)
[2]   Department of Smart Convergence, Kwangwoon University, 20 Kwangwoon-ro, Nowon-gu, Seoul 01897, Korea
*   Correspondence: ksc0226@kw.ac.kr; Tel.: +82-2-940-8637

Received: 31 July 2019; Accepted: 10 September 2019; Published: 13 September 2019

Abstract: Vehicle detection is an important research area that provides background information for the diversity of unmanned-aerial-vehicle (UAV) applications. In this paper, we propose a vehicle-detection method using a convolutional-neural-network (CNN)-based object detector. We design our method, DRFBNet300, with a Deeper Receptive Field Block (DRFB) module that enhances the expressiveness of feature maps to detect small objects in the UAV imagery. We also propose the UAV-cars dataset that includes the composition and angular distortion of vehicles in UAV imagery to train our DRFBNet300. Lastly, we propose a Split Image Processing (SIP) method to improve the accuracy of the detection model. Our DRFBNet300 achieves 21 mAP with 45 FPS in the MS COCO metric, which is the highest score compared to other lightweight single-stage methods running in real time. In addition, DRFBNet300, trained on the UAV-cars dataset, obtains the highest AP score at altitudes of 20–50 m. The gap of accuracy improvement by applying the SIP method became larger when the altitude increases. The DRFBNet300 trained on the UAV-cars dataset with SIP method operates at 33 FPS, enabling real-time vehicle detection.

Keywords: vehicle detection; object detection; UAV imagery; convolutional neural network

1. Introduction

In recent years, studies have been conducted to apply a large amount of information obtained from Unmanned Aerial Vehicles (UAVs) to various systems. Representative UAV applications exist in social-safety, surveillance, military, and traffic systems, and the field is increasingly expanding [1–7]. In this application, vehicle detection, which is detecting the position and size of vehicles in UAV imagery, is very important as background information. Zhu et al. [5] and Ke et al. [6] proposed a vehicle-flow- and density-calculating system in UAV imagery using the vehicle-detection model. Yang et al. [7] proposed an Intelligent Transportation System (ITS).

Traditional methods are less accurate because of poor generalization performance, and only vehicles on asphalt roads are detectable in top-view images, where only relatively standardized vehicle shapes are shown [8–10]. However, AlexNet [11] won the 2012 ImageNet Large Scale Visual Recognition Competition (ILSVRC) [12] and showed excellent generalization performance on a Convolutional Neural Network (CNN). As a result, in 2015, a CNN was able to classify more accurately than humans [12]. Various CNN-based object-detection models have been proposed, such as the Single Shot MultiBox Detector (SSD) series [13–15] and Region proposals with CNNs (RCNN) series [16–18], which utilize CNN. A diversity of vehicle-detection methods has been proposed using various CNN-based object detectors, but these UAV imagery vehicle-detection methods fail to find small objects and operate at low altitudes [19] using YOLOv2 [20]. There is also a real-time problem using complex models to improve accuracy [4,5,21,22], like Faster-RCNN [18], deep and complex SSD [13], and YOLOs [15,20].

The MS COCO [23] and PASCAL VOC [24] used in the training of a general object-detection model consists of front-view images. In addition, each dataset has labels that are not needed for UAV imagery vehicle detection, such as suitcase, fork, wine glass or bottle, potted plant, and sofa, respectively. Vehicles in UAV imagery captured at a high altitude have different composition and distortion peculiarities than general front-view images. For those reasons, vehicle detection using a model trained with a general object detection dataset is not accurate in UAV imagery.

Therefore, in this paper, we propose a real-time vehicle-detection method in bird-view UAV imagery using a lightweight single-stage CNN-based object detector. First, we designed a DRFB module and DRFBNet300, which is a light and fast detection model that uses the MobileNet v1 backbone [25]. Our DRFB module has multi-Receptive Field-size branches to improve the expressive power of feature maps, using dilated convolution [26] to minimize increments of computational complexity. We propose the UAV-cars dataset, which includes distortion peculiarities of UAV imagery to train object-detection models. We also propose a Split Image Processing (SIP) method to improve the accuracy of the detection model. Our SIP method improves accuracy by using divided input frames, different from existing CNN-based object-detection methods. Thus, using the DRFBNet300 trained on the UAV-cars dataset and by using the SIP method, we propose a real-time bird-view UAV imagery vehicle-detection method.

In Section 3.1, we describe DRFBNet300 using the DRFB module, which is optimized for small-object detection. Section 3.2 outlines the SIP method, which improved the accuracy of the object-detection model. In Section 3.3, we describe the UAV-cars dataset that contains the distortion peculiarities of vehicles in UAV imagery. Section 3.4 describes the overall flow of our vehicle-detection method. In Section 4, we lay out the environment of the experiments. Section 5.1 outlines a performance comparison between each model using the MS COCO dataset. In Section 5.2, we compare the performance of the models trained on our UAV-cars dataset with and without the SIP method.

All models used MobileNet v1 [25] with one *Width Multiplier* (α) and *Resolution Multiplier* (ρ) as a backbone network for fast detection. In the MS COCO experiment, SSD300 [13], RFBNet300 [14], YOLOv3 320 [27], FSSD300 [28], and our DRFBNet300, which are single-stage object detectors, were used for the performance comparison. In the UAV-cars dataset experiment, SSD300, RFBNet300, and our DRFBNet300, which are lightweight single-stage methods, were used for the comparison. At each model, the number means input size.

2. Related Work

Vehicle-detection algorithms require a high-end computing model that needs high amounts of power. It is difficult to mount these high-end computing models directly on small battery-powered UAVs. Therefore, most of them use precaptured or transmitted images from the UAV and run detection algorithms on the server PC [1–10,19,21,22,29–34].

Vehicle detection has been widely studied as a background research area for various applications, like surveillance systems and traffic systems [1–7]. Various studies using traditional handcrafted methods have been carried out. Zhao et al. [8] detected vehicles using the edge-contour information of vehicles on the road in top-view low-resolution images. Breckon et al. [29] detected vehicles using a cascaded Haar classifier [35] in bird-view UAV imagery. Shao et al. [31] found vehicles using various algorithms, such as Histogram of Oriented Gradients (HOG) [36], Local Binary Pattern (LBP) [37], and exhaustive search in top-view high-resolution images. Yang et al. [9] tracked vehicles with Scale Invariant Feature Transform (SIFT) [38] and the Kanade–Lucas–Tomasi (KLT) feature tracker [39] after finding a vehicle using blob information in top-view images. Xu et al. [10] improved the original Viola–Jones object-detection method [35] to enhance the accuracy of UAV imagery vehicle-detection models at low altitudes. However, these handcrafted methods are not robust, and are only accurate in certain environments, such as on roads with one direction. They are also only optimized for top-view images, where vehicles are seen as a formulaic square shape, or for low-altitude UAV imagery.

To overcome the low generalization performance of traditional handcrafted methods, various UAV imagery vehicle-detection methods using CNN-based object detectors have been proposed. Yang et al. [6] proposed the Enhanced-SSD, which modified the SSD structure [13], and detect vehicles in ultrahigh-resolution UAV imagery. Tang et al. [33] proposed a UAV imagery vehicle-detection model using the original structure of YOLOv2 [20]. Radovic et al. [19] used the structure of YOLO [15] to detect vehicles in UAV imagery. Xu et al. [22] proposed a deeper YOLO model, DOLO, using the structure of YOLOv2. Xie et al. [34] proposed a UAV imagery vehicle-detection model by modifying the structure of RefineDet [40]. Fan et al. [21] proposed a vehicle-detection model using Faster-RCNN [18], which is a two-stage method. However, most of these methods use simple top-view images. They also designed the models to operate at low altitudes [19] or to use ultrahigh-resolution images [6], and heavyweight models are used to achieve high accuracy, which has high computational complexity [6,21,22,33,34].

3. Proposed Method

Figure 1 shows imagery capturing the schematic concept of the proposed vehicle-detection method in bird-view UAV imagery. The angle of the camera was 30 degrees from the ground, and the video was taken at various altitudes while maintaining the camera angle. The altitude of the UAV was 10–50 m above ground. The vehicle-detection model used a prerecorded bird-view UAV image to infer the location and size of the vehicle on the server PC. Imaging was done using a built-in UAV camera; its detailed specifications are covered in Section 4.1, and those of the server PC used in the experiment are covered in Section 4.2.

Figure 1. Unmanned-aerial-vehicle (UAV) imagery-capturing schematic concept.

Figure 2 shows the overall flowchart of the proposed UAV imagery vehicle-detection method. First, the input image was separated left and right through the Image Separation part of the SIP method. The two separated images were inputted, respectively, to the DRFBNet300 trained on the UAV-cars dataset. Next, using the coordinates indicating the position and size of objects found in the overlapped area at each separated image, the overlapping results were combined at the Box Merging part. Finally, result boxes were drawn on the input image using the generated coordinate values. In Section 3.1, we explain the DRFB module and DRFBNet300, used for vehicle detection. Section 3.2 describes the SIP method that separates the input image of DRFBNet300 and combines or removes duplicated coordinates from the inference result. Section 3.3 describes the UAV-cars dataset, which is used to train and validate the proposed vehicle model. In Section 3.4, we discuss the overall framework.

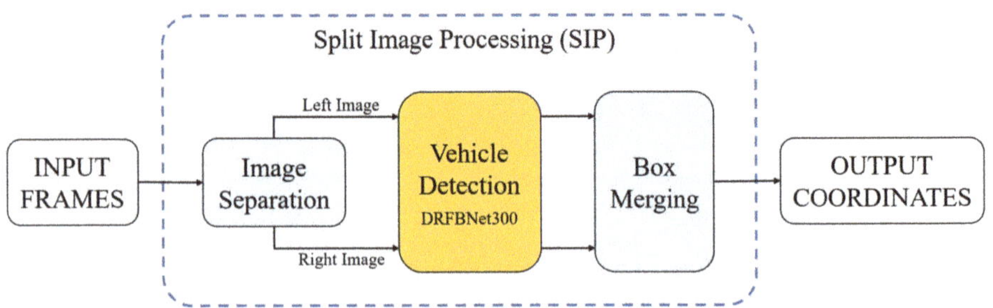

Figure 2. Overall flowchart of our UAV imagery vehicle-detection method.

3.1. DRFBNet300

Speed and precision, the main performance indices of the object detector are directly related to the structure of the backbone network and the meta-architecture of the detector. Therefore, there are performance differences according to the meta-architecture even for the same backbone network [41]. To improve the accuracy of the object-detection model, several studies use a heavyweight backbone network such as ResNet [42] or VGGs [43], or meta-architecture like the RCNN series [16–18]. Such a heavyweight structure is computationally complex and expensive, resulting in the real-time problem of the object-detection model. This problem can be solved using a lightweight backbone like MobileNet v1 [25] or single-stage meta-architecture such as SSD [13]. However, a lightweight structure results in low accuracy because of the lack of network capacity.

In this paper, we designed a DRFB module to improve the accuracy of MobileNet v1 backbone SSD300 [13], which is a light and fast detector. The values of *Width Multiplier* (α) and *Resolution Multiplier* (ρ) of MobileNet v1 were 1. The proposed DRFB module was designed based on the human population Receptive Field (pRF) [44], Inception series [45–47], and RFBNet [14]. The DRFB module improved the quality of feature maps with weak expressive power. DRFBNet300 is an object-detection model using our DRFB module and RFB basic module [14] on the MobileNet v1 backbone SSD300 structure.

DRFB module. Using a multisized Receptive Field (RF) branch structure rather than a fixed-sized RF layer in CNN increases scale invariance and produces better-quality feature maps [48,49]. In addition, if the Inception family-based module [45–47] that concatenates the feature maps generated by the multisized RF convolution is applied to the CNN, the expressiveness of the feature maps and the accuracy of the model are improved, with faster training speed [46,47]. These Inception module-based approaches have been verified in classification, semantic-segmentation, and object-detection tasks [14,47–49].

The proposed DRFB (Deeper Receptive Field Block) module was connected to feature maps for detecting small objects and consists of branches with variously sized RFs. The left-hand side of Figure 3 shows the structure of our DRFB module. Each branch uses dilated convolution [26] to generate good-quality feature maps using large RF. The module has a shortcut branch of ResNet [42] and Inception-ResNet V2 [46], and follows the multibranch structure of inception [45,46]. This makes it possible to enhance the expressiveness of the feature maps and speed up model training while minimizing parameter increase.

Our DRFB module used 1×1 convolution to increase nonlinearity and depth. This minimizes the amount of computation increases and improves the capacity of the structure [50]. Instead of using 3×3 convolutions, 1×3 and 3×1 convolutions were used to reduce computational complexity with nonlinearity increments. The depth of the 5×5-dilated convolution branch was deeper than other branches. The SSD series object-detection model deduces the position, size, and label of multiple objects in a single-input image at once. Therefore, we used a deeper structure to increase the capacity of the large RF branch by adding nonlinearity in order to extract better features from objects that were scattered in the image. We also used a cascaded structure to enhance the expressiveness of the feature maps. The deeper branch had a bottleneck structure based on Szegedy et al. [45,46] to increase efficiency while minimizing the number of parameter increasing.

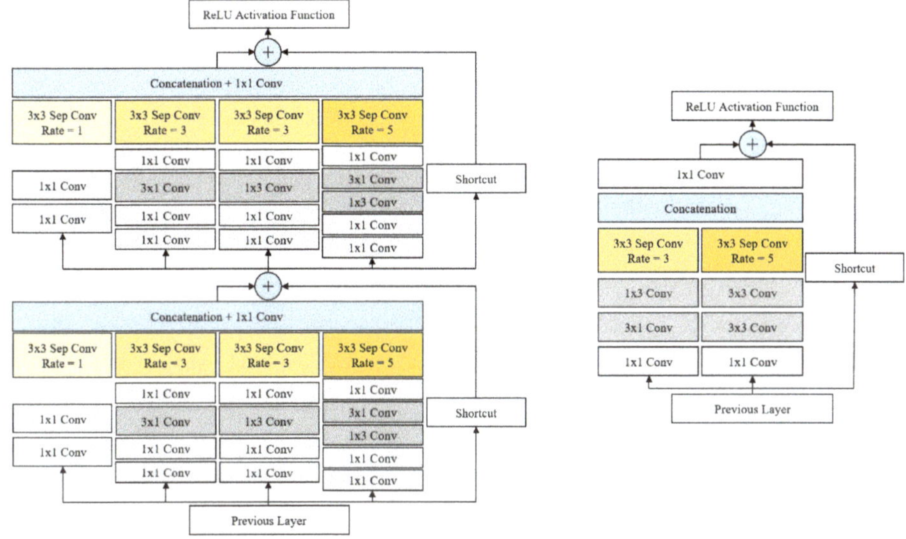

Figure 3. Structure of the (**left**) DRFB module and (**right**) RFB basic module.

In each structure in Figure 3, each layer in every branch includes batch normalization and ReLU activation after the convolution layer (**Conv**). However, a separable convolution (**Sep Conv**), shortcut and the concatenation layer did not include an activation function. Table 1 shows the number of channels in each layer before the DRFB module was cascaded. In Table 1, the top and bottom row mean input and output, respectively, and each number sequentially indicates the number of input/output channels, the application of batch normalization (**BN**), and the ReLU activation function (**ReLU**). The DRFB module was composed of the structure cascade in Table 1. The spatial size of all inputs and outputs equaled 19×19. The shortcut branch, shown in Figure 3, was multiplied by a scale factor (0.1 [14]) and added to each feature map. The structure of the RFB basic module was the same as the one of Liu et al. [14].

Table 1. Input and output channels before the cascaded structure of the DRFB module.

Branch 0	Branch 1	Branch 2	Branch 3
Input ($19 \times 19 \times 512$)			
512/128, BN, ReLU	512/128, BN, ReLU	512/128, BN, ReLU	512/64, BN, ReLU
128/128, BN, ReLU	128/128, BN, ReLU	128/128, BN, ReLU	64/64, BN, ReLU
128/128, -, ReLU	128/128, BN, ReLU	128/128, BN, ReLU	64/96, BN, ReLU
-	128/128, BN, ReLU	128/128, BN, ReLU	96/128, BN, ReLU
-	128/128, -, ReLU	128/128, -, ReLU	128/128, BN, ReLU
-	-	-	128/128, -, ReLU
Concatnation + Conv (BN, $19 \times 19 \times 512$)			

DRFBNet300. The SSD object-detection model has various combined backbone versions. Among them, the MobileNet v1 backbone SSD300 uses depthwise convolution [25], which reduces the number of parameters and computational complexity, preserving its accuracy. However, the SSD

object detector was trained to detect small-sized objects using feature maps from the front side of the feature extractor. Accordingly, feature maps for small-sized object detection have relatively low expressive power. Therefore, the SSD could quickly detect objects, but overall accuracy is low.

In this paper, we propose our DRFB module-applied MobileNet v1 backbone SSD300 with RFB basic module [14] and define it as DRFBNet300. The right-hand side of Figure 3 shows the structure of the RFB basic module. Figure 4 shows the structure of the proposed DRFBNet300. For the backbone network, we used ImageNet [51] pretrained MobileNet v1. All of the structures in Figure 4 were identical to MobileNet v1 backbone SSD300 except the RFB basic and DRFB modules. The feature extractor consisted of the MobileNet v1 backbone, DRFB module, RFB basic module, and six additional convolution layers. The quality of the feature maps for small-object detection, $19 \times 19 \times 512$ shapes, was enhanced through the DRFB module. The RFB basic module was connected to the front side of the extra layers. As a result, the expressiveness of the feature maps for large-object detection was enhanced, improving the overall accuracy of the detection model.

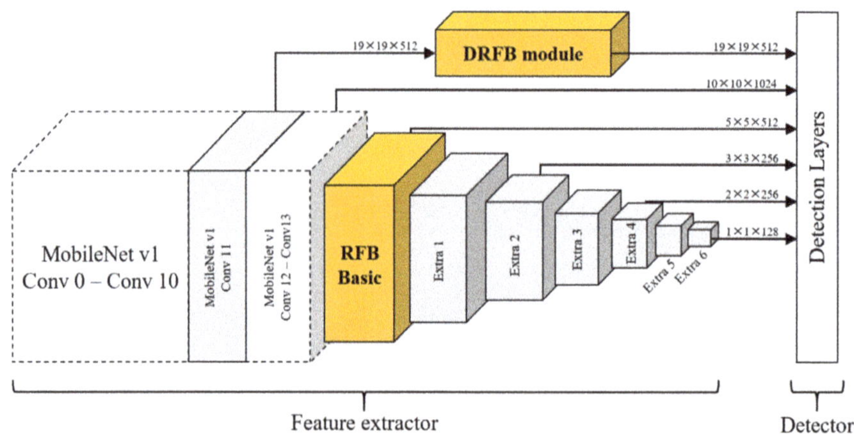

Figure 4. DRFBNet300 structure.

3.2. Split Image Processing

In general object-detection methods, the input image of a single-stage detector is resized to a specific size. An SSD is divided into SSD300 and SSD512 according to the resized input image [13]. Single-stage object-detection models deduce the position, size, and label of the object in the input image with only one network forward pass. Therefore, the SSD512 uses high-resolution input detect objects relatively well, but the SSD300 does not. On the other hand, the SSD300 using small-sized inputs deduces results using only 9.7% (90,000 pixels) of the input image when the input size is 720P (921,600 pixels). This makes SSD300 relatively fast, but low accuracy is inevitable.

Therefore, in this paper, we propose a SIP method that reduces information loss at the input-image-resizing process of the network. The bottom side of Figure 5 shows the schematic concept of the SIP method. Unlike the conventional method shown in the upper part of Figure 5, the detection method with SIP inputs separated images into two segments at the Image Separation part and outputs the final result through the Box Merging part. Overall flow is shown in Figure 2.

Image Separation. A single input image is separated into two images so that 12.5% of the original width is overlapped at the center. If the input image is 720P (1280×720 pixels), then 160×720 pixels are overlapped at the center. A single 720P image is separated into two 720×720 pixel images. The separated images are inputted in object-detection model DRFBNet300 through normalization. The network infers the positions of the objects in each left and right image. The Box Merging part of the SIP method merges the coordinates of the objects in the overlapped area to generate the final result.

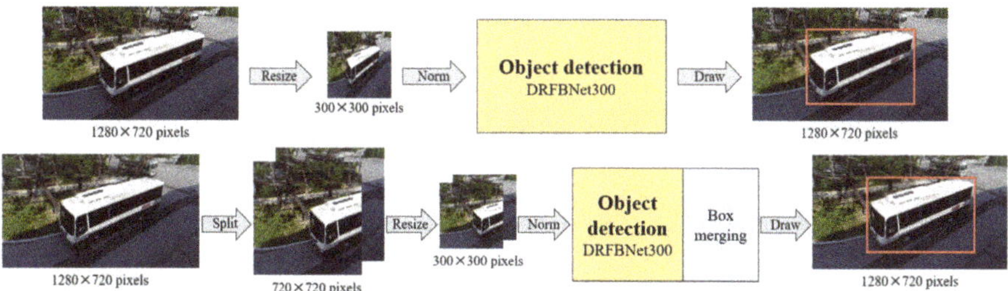

Figure 5. Schematic concept of (**top**) existing object-detection method and (**bottom**) proposed Split Image Processing (SIP) method.

Box Merging. Figure 6 shows a flowchart of the Box Merging part. All thresholds were 720P image referenced values, and optimal values were obtained through experiments. The object-detection model outputs a result in a coordinate format. Values used in the Box Merging part are the coordinate values of objects in the overlapped area. If the detector locates the same object in the overlapped area at each left and right image, the box is truncated or overlapped, as shown in the left image in Figure 7. This happens when objects in the overlapped area are simultaneously detected in the left and right images.

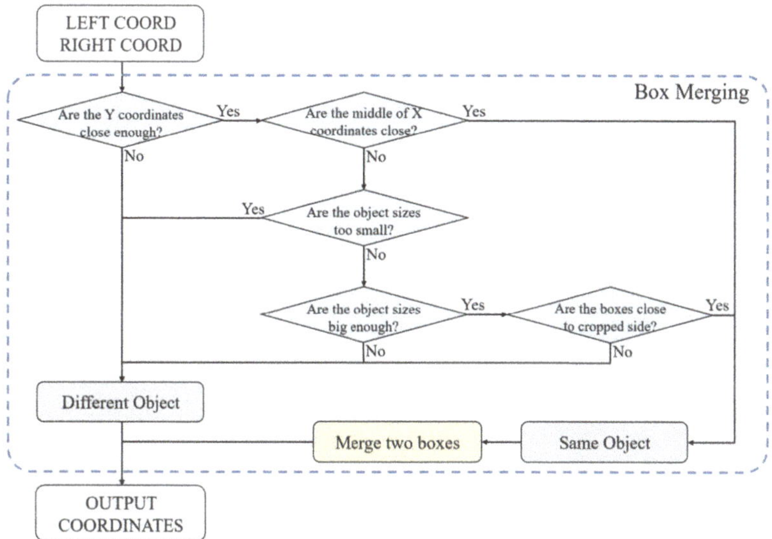

Figure 6. Box merging flowchart.

(**a**)   (**b**)

Figure 7. Experiment results (**a**) before and (**b**) after applying the Box Merging part.

In general, when the same object is detected in each of the left and right UAV imagery, the difference of the Y coordinates is not large. Using this, considering the difference between minimum (top) and maximum (bottom) Y values between each left and right box, if the difference is larger than 20 pixels, it is decided as another object. In the comparison of Y coordinates, the top and bottom values are separately compared in the overlapping boxes. The final result can be true when both respective values satisfy the condition. When the Y coordinate-value condition is satisfied, the Box Merging part uses the center point between the same objects of the result. To do this, the X-coordinate center point of each box was calculated, and, if the distance between them was less than 40 pixels, it was decided as the same object. If the size of the bounding box was smaller than 30×30 pixels even if all of the previous conditions were satisfied, it was decided as another box. This is a condition that considers when the size of the vehicles is very small at a high altitude. If the box size was larger than 160×160 pixels, the maximum X value of the left image box was in the range of 710–720 pixels, and if the minimum X value of the right image box was in the range of 560–570 pixels, it was decided to the same object. This is a condition that considers when the size of the vehicles is very large at low altitude. Figure 7 shows the results before and after applying the Box Merging part.

3.3. UAV-Cars Dataset

To train the general object-detection model, most studies used datasets such as MS COCO [23] or PASCAL VOC [13–18,20,24,40,50]. Each dataset has 81 and 21 labels, including backgrounds, and labels such as frisbee, hot dog, and potted plant. These labels are very insignificant in UAV imagery vehicle detection. Furthermore, UAV imagery is captured using a wide-angle camera, resulting in object composition, ratio, and angle distortion. Most general object-detection datasets consist of front-view images, and even equally labeled objects have different characteristics. Figure 8 shows feature differences of the vehicle between MS COCO and UAV imagery. If the general object-detection dataset is used for UAV imagery vehicle-detection model training, detection accuracy deteriorates because it does not have the peculiarities of UAV imagery.

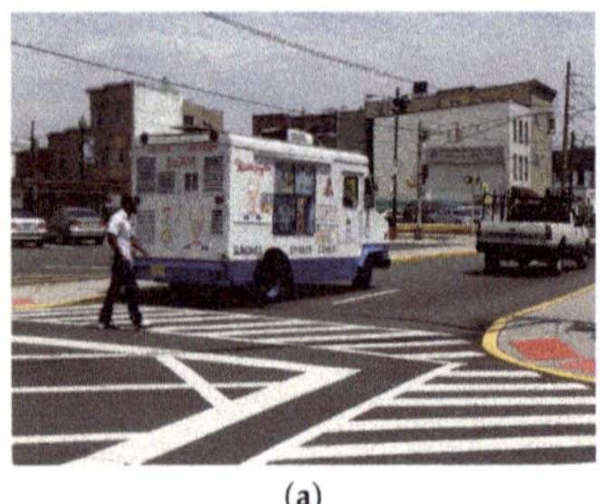

(a)   (b)

Figure 8. Vehicle-feature difference between (**a**) MS COCO and (**b**) UAV imagery.

In this paper, we propose a dataset for vehicle detection in bird-view UAV imagery and UAV-cars. The UAV-cars dataset includes a training and a validation set. The training set consists of 4407 images containing 18,945 objects, and the validation set consists of 628 images containing 2637 objects. To generate the dataset, the vehicles on the road were directly captured using the built-in UAV camera, and then the video was sampled to make images at a constant frame rate. A total of 5035 images were used to generate ground truth (GT) using LabelImg [52]. We used LabelImg to create the GT boxes and save coordinates in the form of XML files. In addition, 628 images that were not included in the training set were used as the validation set. The validation set included the altitude condition, which was 10 m intervals from 10 to 50 m.

We used a camera equipped with a wide-angle lens and UAV to capture the UAV imageries. Detailed specifications of UAV and camera will be covered in Section 4.1. All images were taken in various vehicle compositions with altitude changes between 10 and 50 m. As a result, the UAV-cars dataset contained all various distortion peculiarities of UAV imagery.

3.4. Vehicle Detection in UAV Imagery

Network Training. We used GPU-enabled Pytorch 1.0.1, which is the deep learning library, to implement DRFBNet300. Our training strategies were similar to SSD, including data augmentation and the aspect ratios of the default box. The size of the default box was modified to detect small objects well. For weight-parameter initialization, the weight values of ImageNet [51]-pretrained MobileNet v1 were used for the backbone network. All remaining layers were initialized using the MSRA method [53]. The loss function used in the training phase was multibox loss [13]. The Stochastic Gradient Descent (SGD) momentum optimizer was used to optimize the loss function. DRFBNet300 was trained on the UAV-cars training set for 150 epochs. Further details are covered in Section 4.3.

Vehicle detection. The proposed vehicle-detection model in bird-view UAV imagery is implemented by applying the SIP method to DRFBNet300 trained on the UAV-cars dataset. Figure 2 shows a flowchart of the entire vehicle-detection method. We use precaptured bird-view UAV imagery as input. The video input to the program was divided into frames and conveyanced to our vehicle-detection model. The input frame was separated into left and right images through the Image Separation part. The separated images were fed to DRFBNet300 pretrained on the UAV-cars dataset to infer the coordinates and scores. The Box Merging part used the coordinates of the bounding boxes inferred from DRFBNet300 to merge redundant detection results when objects were in the overlapped area. Finally, the completed coordinate values were drawn in the bounding-box shape on the input image, and results were displayed on the screen and saved.

4. Experimental Environment

4.1. UAV Specification

The experiment used images taken with the DJI Phantom 4 Advanced model (Shenzhen, China). The weight of the fuselage was 1368 g and the size was diagonally 350 mm except for the propellers. The fuselage was equipped with four front- and bottom-side cameras, a GPS, and a gyroscope for the autonomous flight system. The UAV used these sensors to fly at a vertical error of ± 10 cm. The built-in camera used a 20M pixel one-inch CMOS sensor and it was equipped with an 8.8/24 mm lens of 84 FOV. The gimbal that connects the camera to the fuselage has three axes to compensate for yaw, pitch and roll motion. All images were shot at 720P (1280 × 720 pixels) with 30 FPS. Figure 9 shows the UAV and its built-in camera used in the experiment.

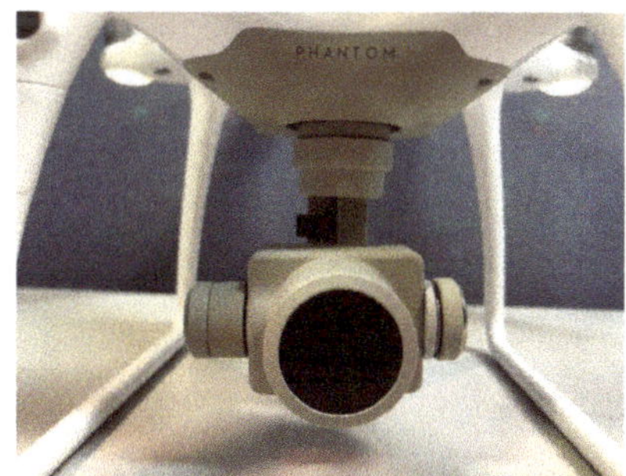

Figure 9. DJI Phantom 4 Advanced UAV (**left**) and its built-in camera (**right**).

4.2. Experiment Environment

During the implementation of the proposed method, we used GPU-enabled Pytorch 1.0.1. Pytorch uses CUDA 9.0 and the cuDNN v 7.5 GPU library. Table 2 shows the specifications of the server PC used for model training and operating our vehicle-detection model.

Table 2. Server PC specification table.

CPU	**Inter Core I7-7700K**
RAM	DDR4 16GB
GPU	Nvidia GeForce GTX Titan X (Maxwell)
O/S	Ubuntu 16.04 LTS
GPU Library	CUDA 9.0 with cuDNN v7.5
Toolkit	Pytorch-GPU 1.0.1

4.3. Training Strategies

The same training strategies were applied to training models using MS COCO [23] and the UAV-cars dataset. Data augmentation, including distortions such as cropping, expanding, and mirroring, was applied to the training phase. Data normalization was applied for fast training and global-minima optimization using mean RGB values (104,117,124) of MS COCO [14]. The models were trained with 32 batch sizes during 150 epochs. The learning rate started at 2×10^{-3} and was reduced by 1/10 at 90, 120, and 140 epochs, respectively. We applied a warm-up epoch [54] that helped global-minima convergence during the initial five epochs, linearly increasing the learning rate from 1×10^{-6} to 2×10^{-3} during the very first five epochs. The SGD momentum using a 0.9 momentum coefficient and 5×10^{-4} weight decay coefficient was used as an optimizer.

Different methods were applied to the backbone and the remaining layers to initialize the weight parameters of the network. The initial weight parameter of the backbone network used ImageNet [51] pretrained MobileNet v1 [25], and all other layers were initialized using the MSRA method [53].

5. Experimental Results

5.1. MS COCO

In this experiment, we used the MobileNet v1 [25] backbone SSD300 [13], RFBNet300 [14], YOLOv3 320 [27], FSSD300 [28] and our DRFBNet300, which are single-stage object-detection methods. We trained each model using MS COCO *trainval35k* [23]. The training strategies in Section 4.3 were used for each model. All models in this experiment were evaluated using MS COCO *minival2014* [23].

Table 3 shows the speed and mean Average Precision (mAP) of each model trained on MS COCO. The experiment result showed that the proposed DRFBNet300 achieves 21 mAP. This is the highest mAP score compared to the SSD300 and RFBNet300, which are lightweight single-stage object-detection models running in real time. The network inference of DRFBNet300 also only took 22.3 ms, meaning real-time detection is possible at about 45 FPS. The FSSD300 and YOLOv3 320 were accurate, but the number of parameters was 19.1M and 24.4M, respectively. In addition, operation speed was 60.9 and 40.1 ms, meaning real-time object detection is impossible. Figure 10 shows the detection results of DRFBNet300 of MS COCO val2017 [23].

Figure 11 shows the results of person detection in bird-view UAV imagery using MS COCO-trained SSD300, RFBNet300, and our DRFBNet300. Comparing the experiment results of each model, the detection results of DFRBNet300 were better than the other models. This is because the generalization performance of DRFBNet300 is the best and it was designed to detect small objects well.

Table 3. Experiment results of the MS COCO dataset.

Method	Backbone	# of Params	Time (ms)	Avg. Precision, IoU			Avg. Precision, Area		
				0.5:0.95	0.5	0.75	Small	Medium	Large
SSD300		7.8 M	19.8	0.181	0.318	0.181	0.014	0.173	0.369
FSSD300		19.1 M	60.9	0.229	0.402	0.236	0.055	0.258	0.386
YOLOv3 320	MobileNet v1	24.4 M	40.1	0.236	0.407	0.241	0.082	0.240	0.386
RFBNet300		6.8 M	21.3	0.206	0.358	0.209	0.018	0.210	0.381
DRFBNet300		7.6 M	22.3	0.210	0.368	0.212	0.018	0.212	0.387

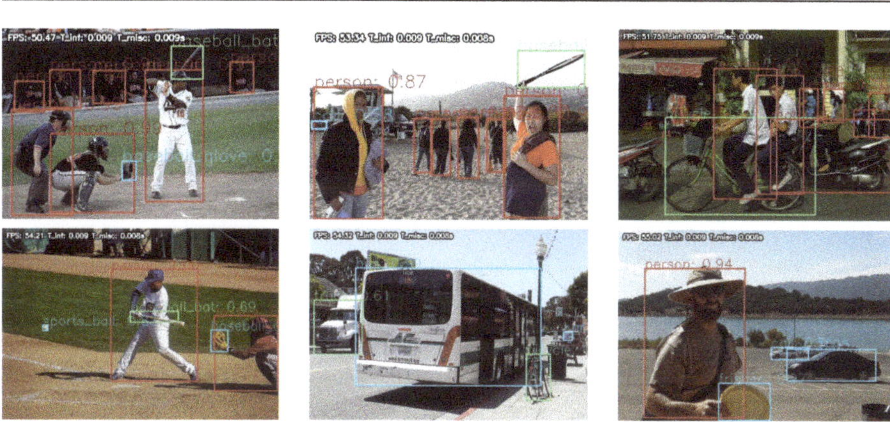

Figure 10. Object-detection results of DRFBNet300 on MS COCO val2017.

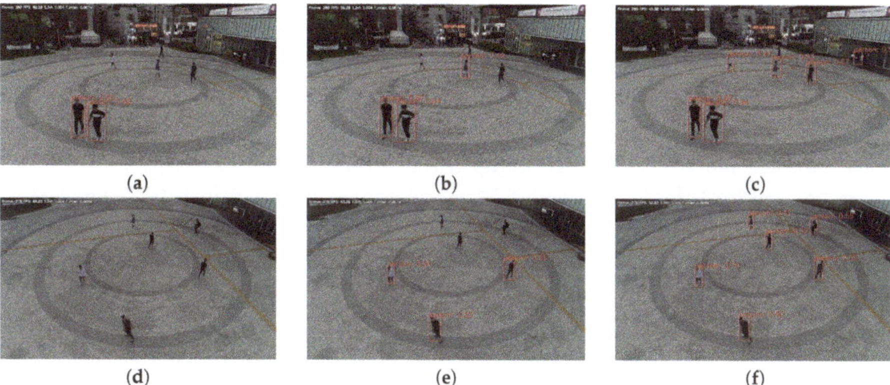

Figure 11. Person detection in bird-view UAV imagery of each model trained on MS COCO. (**a**,**d**) SSD300 results; (**b**,**e**) RFBNet300 results; and (**c**,**f**) DRFNet300 results.

Figure 12 shows the experiment results of applying the SIP method to DRFBNet300 trained on MS COCO. The model with the SIP method slowed down because the amount of computation increased. However, unlike undetected or misdetected objects when the SIP method is not applied, the accuracy of the applied model was greatly improved.

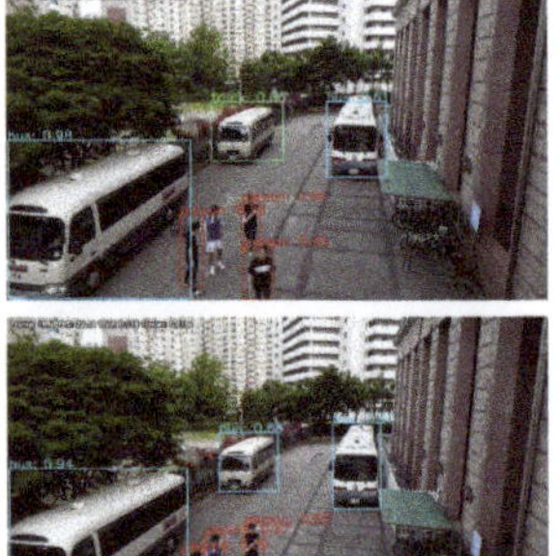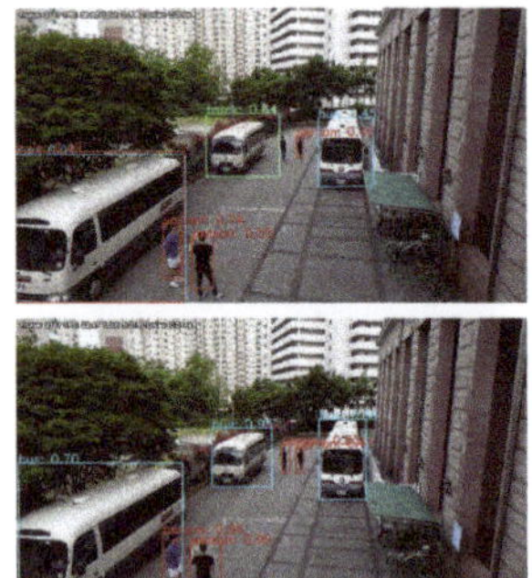

Figure 12. Experiment results of adjacent-frame object-detection (**top**) before and (**bottom**) after applying SIP of MS COCO-trained DRFBNet300.

5.2. UAV-Cars Dataset

In this experiment, we trained SSD300, RFBNet300, and our DRFBNet300, which are lightweight single-stage object detectors running in real time, using the training strategies described in Section 4.3. A training set consisting of 4407 images, including 18,945 objects, in the UAV-cars dataset was used for each model's training. The trained models were evaluated using the UAV-cars validation set consisting of 628 images containing 2637 objects. The validation set was divided into five cases at altitude intervals of 10 m from 10 to 50 m. The True Positive criterion was set to a 0.5 Intersection over Union (IoU) threshold, which was the same value of the PASCAL VOC [24].

Table 4 and Figure 13 show the AP and detection results of models trained on the UAV-cars training set. In Table 4, we can see that DRFBNet300 achieved the highest AP score at all altitudes except for at 10 m. In addition, since inference time was only 17.5 ms, real-time vehicle detection was possible at 57 FPS. In Figure 13, we can see that DRFBNet300 detected small-sized vehicles better than the other models.

In Table 4, accuracy at all altitudes except for at 10 m was greatly improved when the SIP method was applied. Especially as altitude increased, the AP score also further increased. Even at 50 m altitude, the AP score of DRFBNet300 with the SIP method was 57.28, which is a 30.07 AP increase at 27.21 AP when not applied. This is more than double the AP score when it was not applied.

Figure 14 shows UAV imagery vehicle-detection results in the practical case of DRFBNet300 according to whether SIP was applied or not. Figure 14 shows DRFBNet300 with SIP detected vehicles that cannot be detected by normal DRFBNet300 in bird-view UAV imagery at high altitudes. In addition, it ran in real time at 33 FPS even when SIP was applied.

Table 4. Experiment results of UAV-cars dataset.

Method	Meta Architecture	Backbone	Time (ms)	AP by Altitude (%)				
				10 m	20 m	30 m	40 m	50 m
W/O SIP method	SSD300		13.3	98.59	72.19	36.07	26.73	5.01
	RFBNet300		14.9	99.98	82.81	64.35	54.75	24.35
	DRFBNet300	MobileNet v1	17.5	99.54	90.19	71.38	55.22	27.21
W/ SIP method	SSD300		21.6	95.58	81.99	58.65	45.54	18.23
	RFBNet300		26.2	94.28	84.73	71.11	64.65	47.09
	DRFBNet300		30.2	94.82	91.13	76.85	68.44	57.28

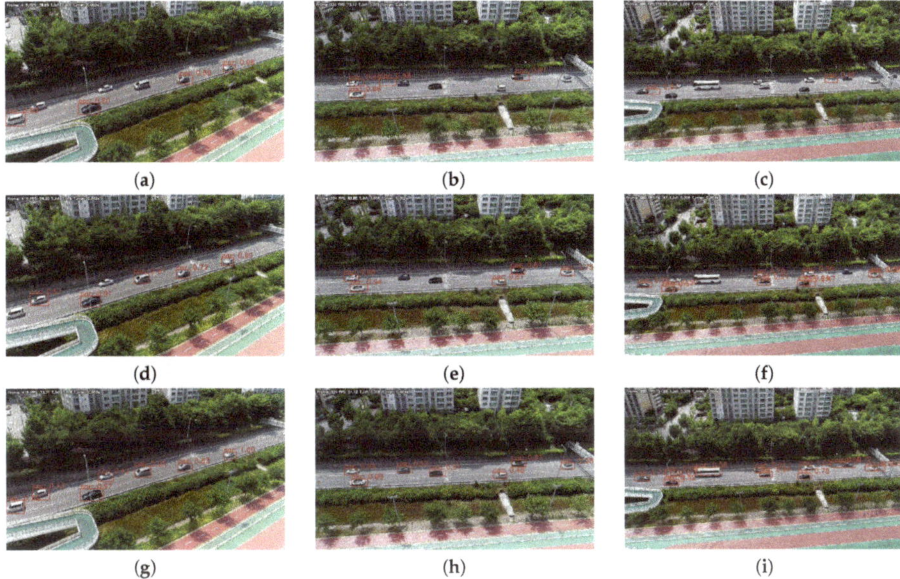

Figure 13. Experiment results of the UAV-cars validation set without applying SIP method. (left to right) Altitudes of 30, 40, and 50 m. Results of (**a–c**) SSD300; (**d–f**) RFB300; and (**g–i**) our DRFBNet300.

Figure 14. Experiment results of DRFBNet300 trained on UAV-cars dataset (**top**) before and (**bottom**) after applying SIP.

6. Conclusions

In this paper, we proposed the use of DRFBNet300 with a DRFB module for bird-view UAV imagery vehicle detection, the UAV-cars dataset to train DRFBNet300, and the SIP method to improve the accuracy of our vehicle detector. The single-stage object-detection model, SSD, has low computational complexity and is fast, but does not detect small objects well. Accuracy is also lower when using a lightweight backbone network for speeding up. DRFBNet300 is a DRFB and RFB basic module attached to the MobileNet v1 backbone SSD300, which is a lightweight object-detection model. The DRFB module was designed to have a multisized RF branch, and dilated convolution was implemented to minimize the increase of computation amount. The multibranched and cascaded structure of our DRFB module improved the quality of feature maps, which improved the accuracy of the vehicle-detection model. We also proposed a UAV-cars dataset consisting of 5035 images containing 21,582 objects, including distortion peculiarities of vehicles in bird-view UAV imagery. Lastly, we proposed the SIP method to improve DRFBNet300 accuracy. DRFBNet300 with the DRFB module achieved the highest score among other lightweight single-stage methods running in real time with 21 mAP at 45 FPS on the MS COCO metric. In the experiment on the UAV-cars dataset, DRFBNet300 also obtained the highest AP score, regardless of whether the SIP method was applied or not at 20–50 m altitudes. The DRFBNet300 increased the accuracy improvement by the SIP method as the UAV altitude increased, and accuracy was improved by more than two times at an altitude of 50 m. Because of DRFBNet300 and the SIP method, the proposed method can more accurately detect vehicles in real-time in UAV imagery at 33 FPS.

Author Contributions: Conceptualization, S.H.; methodology, S.K.; software, S.H.; investigation, S.H.; writing—original-draft preparation, S.H.; writing—review and editing, J.Y., and S.K.; supervision, S.K.; project administration, S.K.

Funding: This research received no external funding.

Acknowledgments: This research was supported by the Ministry of Science and ICT (MSIT), Korea, under the Information Technology Research Center (ITRC) support program (IITP-2019-2016-0-00288), supervised by the Institute for Information and Communications Technology Planning and Evaluation (IITP).

Conflicts of Interest: The authors declare no conflict of interest.

References

1. Zacharie, M.; Fuji, S.; Minori, S. Rapid Human Body Detection in Disaster Sites Using Image Processing from Unmanned Aerial Vehicle (UAV) Cameras. In Proceedings of the 2018 International Conference on Intelligent Informatics and Biomedical Sciences (ICIIBMS), Bangkok, Thailand, 21–24 October 2018; Volume 3, pp. 230–235.
2. Doherty, P.; Rudol, P. A uav search and rescue scenario with human body detection and geolocalization. In Proceedings of the Australasian Joint Conference on Artificial Intelligence, Gold Coast, Australia, 2–6 December 2007; pp. 1–13.
3. Ma'sum, M.A.; Arrofi, M.K.; Jati, G.; Arifin, F.; Kurniawan, M.N.; Mursanto, P.; Jatmiko, W. Simulation of intelligent Unmanned Aerial Vehicle (UAV) For military surveillance. In Proceedings of the International Conference on Advanced Computer Science and Information Systems (ICACSIS), Bali, Indonesia, 28–29 September 2013; pp. 161–166.
4. Xiaozhu, X.; Cheng, H. Object detection of armored vehicles based on deep learning in battlefield environment. In Proceedings of the 2017 4th International Conference on Information Science and Control Engineering (ICISCE), Changsha, China, 21–23 July 2017; pp. 1568–1570.
5. Zhu, J.S.; Sun, K.; Jia, S.; Li, Q.Q.; Hou, X.X.; Lin, W.D.; Liu, B.Z.; Qiu, G.P. Urban Traffic Density Estimation Based on Ultrahigh-Resolution UAV Video and Deep Neural Network. *IEEE J. Sel. Top. Appl. Earth Obs. Remote Sens.* **2018**, *12*, 4968–4981. [CrossRef]
6. Ke, R.; Li, Z.; Tang, J.; Pan, Z.; Wang, Y. Real-time traffic flow parameter estimation from UAV video based on ensemble classifier and optical flow. *IEEE Trans. Intell. Transp. Syst.* **2018**, *20*, 54–64. [CrossRef]
7. Yang, Z.; Pun-Cheng, L.S. Vehicle Detection in Intelligent Transportation Systems and its Applications under Varying Environments: A Review. *Image Vis. Comput.* **2017**, *69*, 143–154. [CrossRef]
8. Zhao, T.; Nevatia, R. Car detection in low resolution aerial images. *Image Vis. Comput.* **2003**, *21*, 693–703. [CrossRef]
9. Yang, Y.; Liu, F.; Wang, P.; Luo, P.; Liu, X. Vehicle detection methods from an unmanned aerial vehicle platform. In Proceedings of 2012 IEEE International Conference on Vehicular Electronics and Safety (ICVES 2012), Istanbul, Turkey, 24–27 July 2012; pp. 411–415.
10. Xu, Y.; Yu, G.; Wu, X.; Wang, Y.; Ma, Y. An enhanced Viola-Jones vehicle detection method from unmanned aerial vehicles imagery. *IEEE Trans. Intell. Transp. Syst.* **2017**, *18*, 1845–1856. [CrossRef]
11. Krizhevsky, A.; Sutskever, I.; Hinton, G.E. ImageNet classification with deep convolutional neural networks. In Proceedings of the Advances in Neural Information Processing Systems, Lake Tahoe, NV, USA, 3–8 December 2012; pp. 1097–1105.

12. Russakovsky, O.; Deng, J.; Su, H.; Krause, J.; Satheesh, S.; Ma, S.; Huang, Z.; Karpathy, A.; Khosla, A.; Bernstein, M.; et al. Imagenet Large Scale Visual Recognition Challenge. *Int. J. Comput. Vis. (IJCV)* **2015**, *115*, 211–252. [CrossRef]

13. Liu, W.; Anguelov, D.; Erhan, D.; Szegedy, C.; Reed, S.; Fu, C.Y.; Berg, A.C. SSD: Single Shot MultiBox Detector. In Proceedings of the European Conference on Computer Vision, Amsterdam, The Netherlands, 8–16 October 2016; Springer: Cham, Switzerland, 2016; pp. 21–37.

14. Liu, S.; Huang, D. Receptive field block net for accurate and fast object detection. In Proceedings of the European Conference on Computer Vision (ECCV), Munich, Germany, 8–14 September 2018; pp. 385–400.

15. Redmon, J.; Divvala, S.; Girshick, R.; Farhadi, A. You only look once: Unified, real-time object detection. In Proceedings of the IEEE Conference on Computer Vision and Pattern Recognition, Las Vegas, NV, USA, 27–30 June 2016.

16. Girshick, R.; Donahue, J.; Darrell, T.; Malik, J. Rich feature hierarchies for accurate object detection and semantic segmentation. In Proceedings of the IEEE Conference on Computer Vision and Pattern Recognition, Columbus, OH, USA, 23–28 June 2014; pp. 580–587.

17. Girshick, R. Fast R-CNN. In Proceedings of the IEEE International Conference on Computer Vision, Santiago, Chile, 7–13 December 2015; pp. 1440–1448.

18. Ren, S.; He, K.; Girshick, R.; Sun, J. Faster R-CNN: Towards real-time object detection with region proposal networks. In Proceedings of the International Conference on Neural Information Processing Systems, Montreal, QC, Canada, 7–12 December 2015; MIT Press: Cambridge, MA, USA, 2015; pp. 91–99.

19. Radovic, M.; Adarkwa, O.; Wang, Q. Object recognition in aerial images using convolutional neural networks. *J. Imaging* **2017**, *3*, 21. [CrossRef]

20. Redmon, J.; Farhadi, A. YOLO9000: Better, faster, stronger. In Proceedings of the IEEE Conference on Computer Vision and Pattern Recognition, Honolulu, HI, USA, 21–26 July 2017; pp. 7263–7271.

21. Fan, Q.; Brown, L.; Smith, J. A closer look at Faster R-CNN for vehicle detection. In Proceedings of the 2016 Intelligent Vehicles Symposium (IV), Gothenburg, Sweden, 19–22 June 2016; pp. 124–129.

22. Xu, Z.; Shi, H.; Li, N.; Xiang, C.; Zhou, H. Vehicle Detection Under UAV Based on Optimal Dense YOLO Method. In Proceedings of the 2018 5th International Conference on Systems and Informatics (ICSAI), Nanjing, China, 10–12 November 2018; pp. 407–411.

23. Lin, T.Y.; Maire, M.; Belongie, S.; Hays, J.; Perona, P.; Ramanan, D.; Dollár, P.; Zitnick, C.L. Microsoft coco: Common objects in context. In Proceedings of the European Conference on Computer Vision, Zurich, Switzerland, 6–12 September 2014; pp. 740–755.

24. Everingham, M.; Van Gool, L.; Williams, C.K.I.; Winn, J.; Zisserman, A. The PASCAL visual object classes (VOC) challenge. *Int. J. Comput. Vis. IJCV* **2010**, *88*, 303–338. [CrossRef]

25. Howard, A.G.; Zhu, M.; Chen, B.; Kalenichenko, D.; Wang, W.; Weyand, T.; Andreetto, M.; Adam, H. MobileNets: Efficient Convolutional Neural Networks for Mobile Vision Applications. *ar*Xiv **2017**, arXiv:1704.04861.

26. Yu, F.; Koltun, V. Multi-scale context aggregation by dilated convolutions. In Proceedings of the International Conference on Learning Representations (ICLR 2016), San Juan, PR, USA, 2–4 May 2016.

27. Redmon, J.; Farhadi, A. YOLOv3: An Incremental Improvement. *ar*Xiv **2018**, arXiv:1804.02767.

28. Li, Z.; Zhou, F. FSSD: Feature Fusion Single Shot Multibox Detector. *ar*Xiv **2017**, arXiv:1712.00960.

29. Breckon, T. P.; Barnes, S. E.; Eichner, M. L.; Wahren, K. Autonomous real-time vehicle detection from a medium-level UAV. In Proceedings of the 24th International Conference on Unmanned Air Vehicle Systems, Bristol, UK, 30 March–1 April 2009.

30. Gaszczak, A.; Breckon, T.P.; Han, J. Real-time people and vehicle detection from UAV imagery. In Proceedings of the Intelligent Robots and Computer Vision XXVIII: Algorithms and Techniques, San Francisco, CA, USA, 24 January 2011; pp. 536–547.

31. Shao, W.; Yang, W.; Liu, G.; Liu, J. Car detection from high-resolution aerial imagery using multiple features. *IEEE Int. Geosci. Remote Sens. Symp.* **2012**, *53*, 4379–4382.

32. Moranduzzo, T.; Melgani, F. Detecting Cars in UAV Images With a Catalog-Based Approach. *IEEE Trans. Geosci. Remote Sens.* **2014**, *52*, 6356–6367. [CrossRef]

33. Tang, T.; Deng, Z.; Zhou, S.; Lei, L.; Zou, H. Fast vehicle detection in UAV images. In Proceedings of the 2017 International Workshop on Remote Sensing with Intelligent Processing (RSIP), Shanghai, China, 18–21 May 2017; pp. 1–5.
34. Xie, X.; Yang, W.; Cao, G.; Yang, J.; Zhao, Z.; Chen, S.; Liao, Q.; Shi, G. Real-time Vehicle Detection from UAV Imagery. In Proceedings of the Fourth IEEE International Conference on Multimedia Big Data (BigMM), Xi'an, China, 13–16 September 2018.
35. Viola, P.; Jones, M. Rapid object detection using a boosted cascade of simple features. In Proceedings of the 2001 IEEE Computer Society Conference on Computer Vision and Pattern Recognition, Kauai, HI, USA, 8–14 December 2001.
36. Dala, N.; Triggs, B. Histograms of Oriented Gradients for Human Detection. In Proceedings of the 2005 IEEE Computer Society Conference on Computer Vision and Pattern Recognition, San Diego, CA, USA, 20–26 June 2005.
37. He, D.-C.; Wang, L. Texture Unit, Texture Spectrum, In addition, Texture Analysis. *IEEE Trans. Geosci. Remote Sens.* **1990**, *28*, 509–512.
38. Lowe, D.G. Object recognition from local scale-invariant features. In Proceedings of the Seventh IEEE International Conference on Computer Vision, Kerkyra, Greek, 20–27 September 1999.
39. Suhr, J.K. Kanade-lucas-tomasi (klt) feature tracker. In *Computer Vision (EEE6503)*; Yonsei University: Seoul, Korea, 2009; pp. 9–18.
40. Zhang, S.; Wen, L.; Bian, X.; Lei, Z.; Li, S.Z. Single-Shot Refinement Neural Network for Object Detection. In Proceedings of the IEEE/CVF Conference on Computer Vision and Pattern Recognition, Salt Lake City, UT, USA, 18–22 June 2018; pp. 4203–4212.
41. Huang, J.; Rathod, V.; Sun, C.; Zhu, M.; Korattikara, A.; Fathi, A.; Fischer, I.; Wojna, Z.; Song, Y.; Murphy, K.; et al. Speed/Accuracy Trade-Offs for Modern Convolutional Object Detectors. In Proceedings of the IEEE 2017 Conference on Computer Vision and Pattern Recognition (CVPR) 2017, Honolulu, HI, USA, 21–26 July 2017; pp. 7310–7311.
42. He, K.; Zhang, X.; Ren, S.; Sun, J. Deep residual learning for image recognition. In Proceedings of the IEEE Conference on Computer Vision and Pattern Recognition, Washington, DC, USA, 27–30 June 2016; pp. 770–778.
43. Simonyan, K.; Zisserman, A. Very Deep Convolutional Networks for Large-Scale Image Recognition. *arXiv* **2014**, arXiv:1409.1556.
44. Wandell, B. A.; Winawer, J. Computational neuroimaging and population receptive fields. *Trends Cognit. Sci.* **2015**, *19*, 349–357. [CrossRef] [PubMed]
45. Szegedy, C.; Liu, W.; Jia, Y.; Sermanet, P.; Reed, S.; Anguelov, D.; Erhan, D.; Vanhoucke, V.; Rabinovich, A. Going deeper with convolutions. In Proceedings of the IEEE Conference on Computer Vision and Pattern Recognition, Boston, MA, USA, 7–12 June 2015.
46. Szegedy, C.; Vanhoucke, V.; Ioffe, S.; Shlens, J.; Wojna, Z. Rethinking the Inception Architecture for Computer Vision. In Proceedings of the IEEE Conference on Computer Vision and Pattern Recognition, Boston, MA, USA, 7–12 June 2015.
47. Szegedy, C.; Ioffe, S.; Vanhoucke, V.; Alemi, A. Inception-v4, inception-resnet and the impact of residual connections on learning. In Proceedings of the Thirty-First, AAAI Conference on Artificial Intelligence, San Francisco, CA, USA, 4–9 February 2017; pp. 4278–4284.
48. Chen, L.C.; Papandreou, G.; Schroff, F.; Adam, H. Rethinking Atrous Convolution for Semantic Image Segmentation. *arXiv* **2017**, arXiv:1706.05587.
49. Dai, J.; Qi, H.; Xiong, Y.; Li, Y.; Zhang, G.; Hu, H.; Wei, Y. Deformable Convolutional Networks. In Proceedings of the International Conference on Computer Vision, Venice, Italy, 22–29 October 2017; pp. 764–773.
50. Mehta, R.; Ozturk, C. Object detection at 200 frames per second. In Proceedings of the European Conference on Computer Vision (ECCV), Munich, Germany, 8–14 September.
51. Deng, J.; Dong, W.; Socher, R.; Li, L.-J.; Li, K.; Li, F.-F. Imagenet: A large-scale hierarchical image database. In Proceedings of the IEEE Conference on Computer Vision and Pattern Recognition, Miami, FL, USA, 20–25 June 2009; pp. 248–255.
52. Tzutalin. LabelImg. Available online: https://github.com/tzutalin/labelImg (accessed on 30 July 2019)

53. He, K.; Zhang, X.; Ren, S.; Sun, J. Delving deep into rectifiers: Surpassing human-level performance on imagenet classification. In Proceedings of the IEEE International Conference on Computer Vision, Santiago, Chile, 7–13 December 2015.

54. Goyal, P.; Dollár, P.; Girshick, R.; Noordhuis, P.; Wesolowski, L.; Kyrola, A.; Tulloch, A.; Jia, Y.; He, K. Accurate, Large Minibatch SGD: Training ImageNet in 1 Hour. *arXiv* **2017**, arXiv:1706.02677.

Multi-Scale Vehicle Detection for Foreground-Background Class Imbalance with Improved YOLOv2

Zhongyuan Wu [1,2], Jun Sang [1,2,*], Qian Zhang [1,2], Hong Xiang [1,2], Bin Cai [1,2] and Xiaofeng Xia [1,2,*]

[1] Key Laboratory of Dependable Service Computing in Cyber Physical Society of Ministry of Education, Chongqing University, Chongqing 400044, China

[2] School of Big Data & Software Engineering, Chongqing University, Chongqing 401331, China

* Correspondence: jsang@cqu.edu.cn (J.S.); xiaxiaofeng@cqu.edu.cn (X.X.); Tel.: +86-139-8369-7592 (J.S.); +86-138-8352-3657 (X.X.)

Received: 22 June 2019; Accepted: 26 July 2019; Published: 30 July 2019

Abstract: Vehicle detection is a challenging task in computer vision. In recent years, numerous vehicle detection methods have been proposed. Since the vehicles may have varying sizes in a scene, while the vehicles and the background in a scene may be with imbalanced sizes, the performance of vehicle detection is influenced. To obtain better performance on vehicle detection, a multi-scale vehicle detection method was proposed in this paper by improving YOLOv2. The main contributions of this paper include: (1) a new anchor box generation method Rk-means++ was proposed to enhance the adaptation of varying sizes of vehicles and achieve multi-scale detection; (2) Focal Loss was introduced into YOLOv2 for vehicle detection to reduce the negative influence on training resulting from imbalance between vehicles and background. The experimental results upon the Beijing Institute of Technology (BIT)-Vehicle public dataset demonstrated that the proposed method can obtain better performance on vehicle localization and recognition than that of other existing methods.

Keywords: vehicle detection; YOLOv2; focal loss; anchor box; multi-scale

---

1. Introduction

Vehicle detection is one of the essential parts in computer vision, which aims to locate the vehicles and recognize the vehicle types. In recent years, vehicle detection has been applied in numerous fields, such as traffic surveillance, unmanned vehicle, gate monitoring and so on. However, due to the complicated background, the greatly varying illumination intensity, the occlusion problem and the small variations of each vehicle types, vehicle detection is still a challenging task and a hot research field of Artificial Intelligence (AI).

Numerous vehicle detection methods have been proposed, which can be divided into two categories: traditional machine learning methods and deep learning-based methods. For the traditional methods, Tsai et al. [1] used a new color transformation model to obtain the key vehicle colors and locate the candidate objects. Then, a multichannel classifier was adopted to recognize the candidate objects. For vehicle detection in video, Jazayeri et al. [2] thought that it should combine the temporal information of the features and the vehicle motion behaviors, which can compensate the complexity in recognizing vehicle colors and shapes. In Refs. [3–5], the Histogram of Oriented Gradient (HOG) method was applied to extract the vehicle features in the image, which have a lower false positive rate. In recent years, with the continuous improvement of computing power, deep learning [6] gradually becomes the main way for vehicle detection. The methods based on deep learning have surpassed the methods based on traditional methods [7–10] on detection and recognition performance. Different from designing feature by human beings, Ref. [11–13] proposed to learn features automatically with Convolutional Neural Network (CNN), which only needs large labeled vehicle images to train the network with supervision, while it does not need to design feature manually. The first systematic framework for object detection is R-CNN [14]. In R-CNN, the selective search algorithm [15] was used to generate the regions of interest, and then CNN was applied to recognize whether the region of interest is object or background. After R-CNN, numerous methods [16–19] adopted the same framework with slight changes for object detection. In Ref. [20] and Ref. [21], the Faster R-CNN was applied to detect the vehicles, which surpassed the previous methods. Ref. [22] developed an accurate vehicle proposal network to extract vehicle-like targets. Then, a coupled R-CNN method was proposed to extract the vehicle's location and attributes simultaneously. To improve the detection speed, Redmon et al. [23] proposed YOLO, which was the first one to convert detecting object to regression and achieved end-to-end detection. Then, some other methods, such as SSD [24], YOLOv2 [25], YOLOv3 [26], etc., were proposed, which reduce the detection time greatly. Ref. [27] proposed an improved YOLOv2 for vehicle detection. k-means++ [28] was used to generate anchor boxes, instead of k-means [29], and the loss function was improved with normalization. In Ref. [30], by modifying the net resolution and depth of YOLOv2, the proposed model gave near Faster R-CNN performance at more than4 times speed. In object detection, usually the objects only occupy a small part of the image, while the majority in the image is background. The imbalance between objects and background can hinder object detection models from converging in a correct direction in the training stage. Therefore, to reduce the impact of the imbalance, Lin et al. [31] proposed Focal Loss to focus training on a sparse set of hard examples. Their experiments validated that the method can improve the detection result on end-to-end methods.

To improve the accuracy of localization and recognition simultaneously, an improved vehicle detection method based on YOLOv2 was proposed in this paper. A new anchor box generation method was applied to enhance the network localization ability. In addition, Focal Loss was introduced in YOLOv2 to improve the network recognition ability.

2. Brief Introduction on YOLOv2

In YOLOv2 [25], the input image is divided into S × S grids. Each grid predicts the location information, the class probability and the object confidence for each anchor box. The location information includes x, y, w and h of the bounding box, where x, y represent the abscissa and ordinate of the center pixel of the bounding box, and w, h represent the length and height of the bounding box. The class probability indicates which class the object in the current grid is most likely to belong to. The object confidence represents the confidence that there exists objects in the current grid. Then, the Non-Maximum Suppression (NMS) is applied to remove the bounding boxes with low object confidence. Finally, the rest bounding boxes are decoded to obtain the detection boxes. In this section, YOLOv2 is introduced in brief, mainly including the generation of anchor boxes, the network structure and the loss function.

2.1. The Generation of Anchor Boxes

Anchor boxes were first proposed in Faster R-CNN [18], which aims to generate bounding boxes with a certain ratio instead of predicting the sizes of bounding boxes directly. The authors of YOLOv2 indicated that generating anchor boxes with manual design was absurd. Instead, they applied k-means cluster on training set to obtain better anchor boxes.

When implementing k-means, instead of the traditional Euclidean distance, YOLOv2 adopted the Intersection over Union (IoU) distance to measure the closeness degree of two bounding boxes. The main reason is that, if the Euclidean distance is adopted, the anchor boxes with big size will produce more errors than those with small size. Consequently, by adopting the IoU distance, the errors will be irrelevant to the sizes of the anchor boxes. The distance of k-means in YOLOv2 can be expressed as Equation (1).

$$d(box, centroid) = 1 - IOU(box, centroid) \tag{1}$$

2.2. The Network Structure

In recent years, most of the detection methods take VGG [7] or ResNet [8] as the base feature extractor. The authors of YOLOv2 argued that these networks were accurate and powerful, but they were needlessly complex. Therefore, Darknet19 was adopted as the backbone of YOLOv2, which has less parameters and may obtain better performance than VGG and ResNet. The network of YOLOv2 is shown in Figure 1.

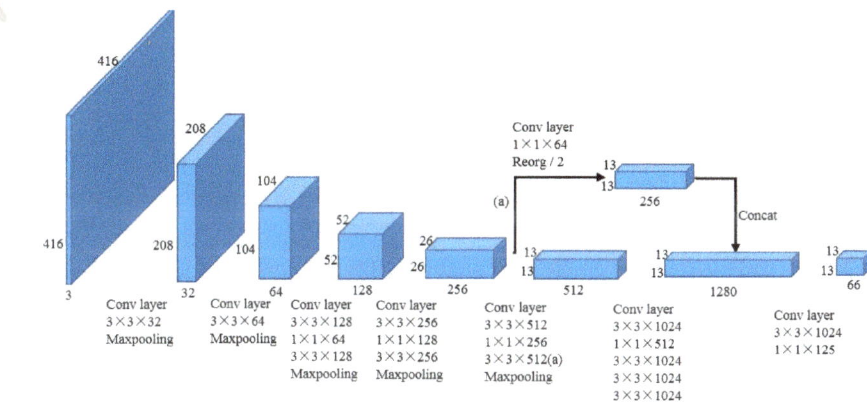

Figure 1. The network of YOLOv2.

As shown in Figure 1, the network includes 32 layers, which have 19 convolutional layers and five maxpooling layers. Similar to VGG, 3×3 convolutional layer is used to double the channel of feature maps after each pooling layers, and 1×1 convolutional layer is used to halve the channels and fuse the features. Feature fusion is applied on the feature maps from the direct path and path (a) in Figure 1, which can retain the features from the shallower layer to improve the ability of detecting the small objects.

2.3. The Loss Function

In YOLOv2, multi-part loss function is adopted, which includes the bounding box loss, the confidence loss and the class loss. The bounding box loss includes the coordinate loss and the size loss of the bounding box. The confidence loss includes the confidence loss of bounding box with objects and without objects. The class loss is calculated by softmax to obtain the class probability. The loss function in YOLOv2 can be expressed as Equation (2).

$$
\begin{aligned}
&\lambda_{coord} \sum_{i=0}^{S^2} \sum_{j=0}^{B} \mathbb{1}_{ij}^{obj} [(x_i - \hat{x}_i)^2 + (y_i - \hat{y}_i)^2] \\
&+\lambda_{coord} \sum_{i=0}^{S^2} \sum_{j=0}^{B} \mathbb{1}_{ij}^{obj} [(\sqrt{w_i} - \sqrt{\hat{w}_i})^2 + (\sqrt{h_i} - \sqrt{\hat{h}_i})^2] \\
&+\sum_{i=0}^{S^2} \sum_{j=0}^{B} \mathbb{1}_{ij}^{obj} (C_i - \hat{C}_i)^2 \\
&+\lambda_{noobj} \sum_{i=0}^{S^2} \sum_{j=0}^{B} \mathbb{1}_{ij}^{noobj} (C_i - \hat{C}_i)^2 \\
&+\sum_{i=0}^{S^2} \mathbb{1}_{i}^{obj} \sum_{c \in classes} (p_i(c) - \hat{p}_i(c))^2
\end{aligned}
\tag{2}
$$

As shown in Equation (2), x_i and y_i denote the center coordinates of the box relative to the current grid bounds in the i-th grid. w_i and h_i denote width and height of the bounding box relative to the whole image in the i-th grid. C_i denotes the confidence of the bounding box in the i-th grid. $p_i(c)$ denotes the class probability of the bounding box in the i-th grid. $\hat{x}_i, \hat{y}_i, \hat{w}_i, \hat{h}_i, \hat{C}_i, \hat{p}_i(c)$ denote the corresponding predicted values of $x_i, y_i, w_i, h_i, C_i, p_i(c)$. S^2 denotes the S × S grids. B denotes the bounding boxes. λ_{coord} denotes the weight of the coordinate loss and λ_{noobj} denotes the weight of the loss of bounding boxes without objects. $\mathbb{I}_i^{obj}$ denotes whether the object is on the i-th grid or not and $\mathbb{I}_{ij}^{obj}$ denotes whether the j-th box predictor in the i-th grid is "responsible" for that prediction or not.

In the loss function, the first line is to compute the coordinate loss, the second line is to compute the bounding box size loss, the third line is to compute the bounding box confidence loss containing objects and the last line is to compute the bounding box confidence loss not containing objects. To prevent the sizes of the bounding boxes from making a significant impact on the loss, the square roots of width and length of bounding boxes are applied to decrease their magnitudes. Since usually only a few bounding boxes with object exist in the real pictures, the confidence loss of bounding box with object is much smaller than the other losses. Consequently, the weighted method is applied to balance the different kinds of losses. Usually, λ_{coord} is set as 5 and λ_{noobj} is set as 0.5 to balance each loss. Otherwise, each loss may result in different contributions to the total loss, which can cause some losses ineffective for network training.

3. The Proposed Method

The proposed vehicle detection method was based on YOLOv2, in which a new anchor box generation method was proposed and Focal Loss was introduced in YOLOv2. In this section, these two improved points will be introduced in detail.

3.1. The Generation of Anchor Boxes

The quality of anchor boxes is important for the end-to-end detection methods. It is efficient by using k-means or k-means++ to generate anchor boxes. However, the anchor boxes generated by such methods are usually suitable to the common sizes of objects, while the sizes of the generated anchor boxes may be far away from those of the objects with usual sizes. By analyzing the detection boxes of YOLOv2, we found that the common sizes of the vehicles can be predicted well, while some unusual sizes of vehicles, such as the size of a truck which is usually much bigger than that of the common vehicles, were predicted terribly. As shown in Figure 2, the black boxes denote the ground truths and the other color boxes denote the detection boxes predicted by YOLOv2. It is obvious that the sedan was detected well and the minivan was detected terribly, since the size of minivan is unusual and the size of sedan is common.

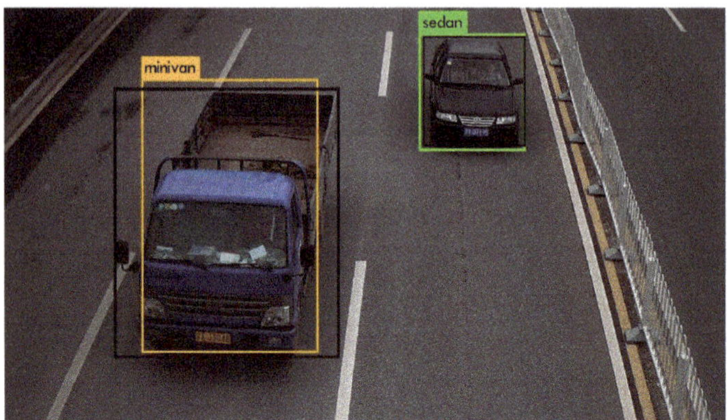

Figure 2. The detection boxes generated by YOLOv2.

To improve the localization accuracy, it is better to generate the anchor boxes to match most sizes of the ground truths, instead of only matching the common sizes of the ground truths. Therefore, a clustering method called Rk-means++ was proposed in this paper. As shown in Figure 3, in Rk-means++, 2 ratios regarding the width and the length of the ground truths were obtained by applying k-means++. Then, we applied a certain proportion to two ratios and generate anchor boxes with different scale. Compared with the methods of clustering anchor boxes directly, such as k-means and k-means++, Rk-means++ generates anchor boxes on different hierarchies, which may match most sizes of the ground truths much better. In our experiments, the proportion was set as 1:2:4:6:8:10 and six anchor boxes were obtained.

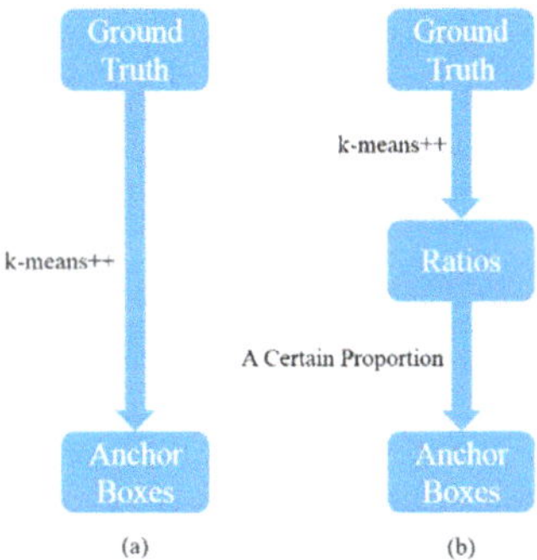

Figure 3. The computing procedures of k-means++ and Rk-means++. (**a**) k-means++; (**b**) Rk-means++. The anchor boxes obtained by Rk-means++ ensures that each size of the vehicle with different scale can be matched with one of the anchor boxes. In other words, the proposed anchor box generation method can enhance the robustness of YOLOv2 for different scales of vehicles and improve the localization accuracy.

3.2. Focal Loss

For vehicle detection, usually the vehicles are only a small part of the whole image, while the majority in the image is background. Consequently, thousands of candidate bounding boxes will be generated in one image with YOLOv2, while only a few of them include vehicles. Obviously, the problem of imbalance between the positives (i.e., vehicles) and the negatives (i.e., backgrounds) is serious. In the training stage of YOLOv2, because of the large quantity of candidate bounding boxes not containing object, the loss will be overwhelmed by them and a few quantity of candidate bounding boxes containing object cannot influence the loss effectively. The imbalance of quantity will make training inefficient since most candidate bounding boxes tend to be easy negative, which is useless for CNN learning. The numerous easy negatives will overwhelm some important training examples, which leads the model to converge in a wrong direction. Therefore, in order to decrease the imbalance between the positives and negatives, Focal Loss [31] was introduced in YOLOv2 to detect the vehicles.

Before Focal Loss was introduced, the cross entropy loss is commonly employed to calculate the classification loss, which is shown in Equation (3) for binary classification.

$$
L_{ce} = -y \log y' - (1-y) \log(1-y') = \begin{cases} -\log y', & y = 1 \\ -\log(1-y'), & y = 0 \end{cases} \tag{3}
$$

In Equation (3), y denotes the ground truth, which is 1 for positive and 0 for negative. y' denotes the predicted value ranging from 0 to 1. For the positives, the higher the predicted probability is, the smaller the cross entropy loss is. For the negatives, the lower the predicted probability is, the smaller the cross entropy loss is. Consequently, it is inefficient to train with the iteration of numerous easy examples. More seriously, the model may not be optimized to a good state.

Focal Loss is based on cross entropy loss, which aims to reduce the weights of easy examples in loss and make the model focus on distinguishing the hard examples. Focal Loss can be expressed as Equation (4).

$$L_{fl} = \begin{cases} -\alpha(1-y')^{\gamma}\log y', & y = 1 \\ -(1-\alpha)y'^{\gamma}\log(1-y'), & y = 0 \end{cases} \tag{4}$$

As shown in Equation (4), corresponding to cross entropy loss, 2 factors, i.e., γ and α, are added. γ is used to reduce the loss of easy examples. For instance, by setting γ as 2, the loss will be smaller than the cross entropy loss if an easy positive with predicted value 0.95 or an easy negative with predicted value 0.05. The factor γ makes the training more focus on the hard examples and reduce the impact of easy examples. α is used to balance the weights of positives and negatives, which can control the decline rate for the weights of examples. Within the limited numbers of experiments, we found that setting γ as 2 and setting α as 0.25 can obtain the best accuracy.

4. Experiments

The implementation of our experiments was based on Darknet, which is a light deep learning framework and provided by the author of YOLOv2. The experiments were conducted on the GPU server, which includes 8 pieces of GPUs, 24 G video memory and 64 G memory. The experimental platform was equipped with 64-bit ubuntu14.04, Opencv2.7.0 and CUDA Toolkit8.0.

4.1. Dataset

The experiments were conducted on BIT-Vehicle [32] public dataset provided by Beijing Institute of Technology. The BIT-Vehicle dataset contains 9580 images, including 6 vehicle types, i.e., bus, microbus, minivan, sedan, truck and suv. The numbers of each vehicle type are 558, 883, 476, 5922, 822 and 1392. As shown in Figure 4, the BIT-Vehicle dataset includes day scene and night scene, and the images are seriously influenced by the variation of illumination intensity.

Figure 4. Some images in BIT-Vehicle dataset.

In our experiments, the BIT-Vehicle dataset was divided into training set and test set with the ration of 8:2, namely, 7880 images were used for training and 1970 images were used for testing.

4.2. Implementation

In our experiments, the size of input image was 416×416. The initial learning rate was set to be 0.001. The model was trained by 120 epochs and the learning rate was decreased 10 times when the iteration reached 60 and 90 epochs, respectively. The size of batch was set to be 8 and the momentum was set as 0.9. In order to make the model detect well for the images with different sizes, the multi-scale training trick was applied in our experiments. A new size of input image was selected randomly for training every 10 epochs. Since the factor of down sampling is 32 in YOLOv2, the sizes of the randomly selected input images all were multiple of 32, i.e., the maximum size was 608×608 and the minimum size was 352×352. Other hyperparameters followed as those in YOLOv2.

4.3. Experimental Results and Analysis

The proposed method was compared with YOLOv2_Vehicle [27] and the other methods for vehicle detection. In addition, all of these methods were implemented on our platform, so the results of these methods were obtained under the same experimental environment. The experimental results are shown in Table 1.

Table 1. Experimental results for different methods.

Method	The Class AP (%)						mAP (%)	IoU (%)	Speed (s)
	Bus	Microbus	Minivan	Sedan	SUV	Truck			
YOLOv2 [25]	98.34	95.03	91.11	97.42	93.62	98.41	95.65	90.44	0.0496
YOLOv2_Vehicle [27]	98.42	97.04	95.02	97.37	93.73	97.80	96.56	91.06	0.0486
YOLOv3 [26]	98.65	96.98	94.04	97.65	94.36	98.17	96.64	88.50	0.1100
SSD300 VGG16 [24]	97.97	**97.98**	90.28	97.15	91.25	97.75	93.75	91.60	**0.0440**
Faster R-CNN VGG16 [18]	**99.05**	93.75	91.38	98.14	94.75	98.17	95.87	92.19	0.4257
Our Method	98.86	96.63	**95.90**	**98.23**	**94.86**	**99.30**	**97.30**	**92.97**	0.0522

As shown in Table 1, Class Average Precision (AP) denotes the recognition performance for each vehicle class, mean Average Precision (mAP) denotes the average recognition performance and IoU denotes the localization performance. It can be seen that mAP and IoU obtained with our proposed method are 97.30% and 92.97%, respectively. Specifically, the mAP of our method is higher than those of other methods by nearly 1 percentage point and the IoU is higher than those of other methods by nearly 1 percentage points. Significantly, except the speed and AP of bus, microbus, all the metrics obtained by our proposed method are better than those of the others. For the AP of bus and microbus, our method is not the best one. The reason may be that the different models will focus on different classes due to the random initial parameters, the network architecture or the anchor boxes being different. Consequently, it is acceptable that there are two class APs slightly smaller than others. For the speed, our method is not the best one, which may result from the fact that the more candidate bounding boxes were proposed before the NMS, which slow down the speed lightly. In general, our proposed method showed better performance than others on the accuracy of localization and recognition, while maintaining the speed.

Figure 5 showed the detection results of our proposed method. Whether it was daytime or night, the abilities of localization and recognition were not affected. Whether the vehicle is with small size or with large size, our method can locate and recognize them as well. In addition, it can be seen from the second picture and last picture of the second column in Figure 5, though the vehicle was incomplete, they can be detected accurately. Consequently, regardless of the variation of illumination intensity, the variation of vehicle sizes or the vehicle completeness, our method can recognize and locate the vehicles accurately. It demonstrates the strong robustness of our proposed method.

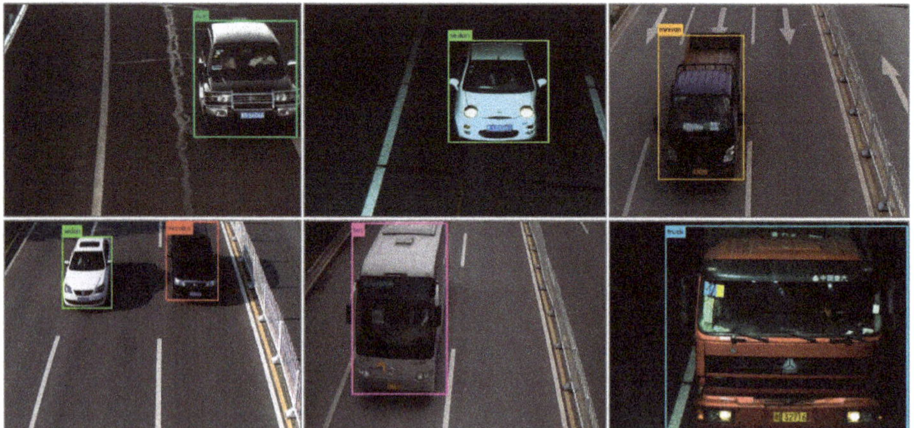

Figure 5. Examples of detection results.

4.4. The Validation of Rk-means++

In order to validate the effect of Rk-means++ for YOLOv2 in vehicle detection, numerous experiments were conducted to compare the performance of different methods with different anchor box generation methods. In the experiments, YOLOv2 and YOLOv3 were adopted as the basic model. The experimental results are shown in Table 2.

Table 2. Experimental results on comparing different anchor box generation methods.

Method		Bus	Microbus	Minivan	Sedan	SUV	Truck	mAP (%)	IoU (%)
				The Class AP (%)					
YOLOv2	k-means	98.34	95.03	91.11	97.42	93.62	98.41	95.65	90.44
	k-means++	98.60	96.29	93.16	97.47	93.72	98.15	**96.23**	91.05
	Rk-means++	98.68	96.66	91.50	97.48	93.59	97.33	95.88	**92.18**
YOLOv3	k-means	98.65	96.98	94.04	97.65	94.36	98.17	**96.64**	88.50
	k-means++	98.70	96.44	93.80	97.66	94.76	97.78	96.52	88.86
	Rk-means++	98.32	97.08	92.65	97.70	94.27	97.96	96.33	**90.49**

As shown in Table 2, it is obvious that the IoU with Rk-means++ are the best with YOLOv2 and YOLOv3. The reason could be that each size of vehicle with different scales can be matched with one of the anchor boxes, since the anchor boxes generated by Rk-means++ are on different hierarchies. However, the mAP with Rk-means++ is slightly lower than that of others. This may result from the fact that more candidate bounding boxes were proposed with Rk-means++, which makes the detection precision drop slightly. In general, with Rk-means++, the model can obtain better localization performance while the precision is slightly lower.

4.5. The Validation of Focal Loss

In order to validate the effect of Focal loss for YOLOv2 in vehicle detection, YOLOv2 was adopted as the basic model with different anchor box generation methods. The experimental results are shown in Table 3, where 'Wo' denotes 'Without' and 'FL' denotes 'Focal Loss'.

Table 3. Experimental result on comparing YOLOv2 with Focal Loss (FL).

YOLOv2		The Class AP (%)						mAP (%)	IoU (%)
		Bus	Microbus	Minivan	Sedan	SUV	Truck		
k-means	Wo FL	98.34	95.03	91.11	97.42	93.62	98.41	95.65	90.44
	With FL	98.34	96.80	94.81	98.20	95.34	99.32	**97.13**	**92.06**
k-means++	Wo FL	98.60	96.29	93.16	97.47	93.72	98.15	96.23	91.05
	With FL	98.48	97.44	96.03	98.24	95.68	99.17	**97.51**	**92.30**
Rk-means++	Wo FL	98.68	96.66	91.50	97.48	93.59	97.33	95.88	92.18
	With FL	98.86	96.63	95.90	98.23	94.86	99.30	**97.30**	**92.97**

As shown in Table 3, k-means, k-means++ and Rk-means++ were applied to compare the performance without or with Focal Loss. It is obvious that all the YOLOv2 models with Focal Loss will surpass those without Focal Loss regardless of the different anchor box generation methods, which indicates that Focal Loss can reduce the influence on imbalance of positives and negatives. It also validates that Focal Loss is useful for improving both mAP and IoU.

5. Conclusions

In this paper, a multi-scale vehicle detection method based on YOLOv2 was proposed to improve the accuracy of localization and recognition simultaneously. The main contributions of this paper include:

(1) By analyzing the detection boxes of YOLOv2, it was found that the common sizes of the vehicles can be predicted well, but some unusual sizes of vehicles were predicted terribly. Therefore, to adapt most sizes of vehicles with different scales, instead of clustering the anchor boxes directly, a new anchor box generation method, Rk-means++, was proposed.
(2) By analyzing the loss of the training stage, it was found that most of the loss was from the irrelevant background, which is due to the imbalance between the vehicles and background in each image. Therefore, the Focal Loss was introduced in YOLOv2 for vehicle detection to reduce the impact of easy negatives (background) on loss and make the loss focus on the hard examples.
(3) The experiments conducted on the BIT-Vehicle public dataset demonstrated that our improved YOLOv2 obtained superior performance than others on vehicle detection.
(4) By conducting numerous comparison experiments, the effectiveness of Rk-means++ and Focal Loss was validated.

Although the method proposed in this paper shows good performance on vehicle detection, the scales of data and the types of vehicle are relatively few, which cannot show the distinctions of different methods well. In the future work, we will collect more vehicle data and try to improve the accuracy on the larger vehicle dataset.

Author Contributions: Conceptualization, Z.W. and J.S.; Data curation, Q.Z. and B.C.; Funding acquisition, H.X.; Investigation, Q.Z.; Methodology, Z.W. and Q.Z.; Supervision, J.S.; Validation, B.C.; Visualization, X.X.; Writing—original draft, Z.W. and X.X.; Writing—review & editing, J.S. and H.X.

Funding: This research was funded by National Key R&D Program of China (No. 2017YFB0802400).

Conflicts of Interest: The authors declare no conflict of interest.

References

1. Tsai, L.W.; Hsieh, J.W.; Fan, K.C. Vehicle Detection Using Normalized Color and Edge Map. *IEEE Trans. Image Process.* **2007**, *16*, 850–864. [CrossRef] [PubMed]
2. Jazayeri, A.; Cai, H.; Zheng, J.Y.; Tuceryan, M. Vehicle Detection and Tracking in Car Video Based on Motion Model. *IEEE Trans. Intell. Transp. Syst.* **2011**, *12*, 583–595. [CrossRef]

3. Cao, X.; Wu, C.; Yan, P.; Li, X. Linear SVM classification using boosting HOG features for vehicle detection in low-altitude airborne videos. In Proceedings of the 2011 IEEE International Conference Image Processing (ICIP), Brussels, Belgium, 11–14 September 2011; pp. 2421–2424.

4. Guo, E.; Bai, L.; Zhang, Y.; Han, J. Vehicle Detection Based on Superpixel and Improved HOG in Aerial Images. In Proceedings of the International Conference on Image and Graphics, Shanghai, China, 13–15 September 2017; pp. 362–373.

5. Laopracha, N.; Sunat, K. Comparative Study of Computational Time that HOG-Based Features Used for Vehicle Detection. In Proceedings of the International Conference on Computing and Information Technology, Helsinki, Finland, 21–23 August 2017; pp. 275–284.

6. LeCun, Y.; Bengio, Y.; Hinton, G. Deep learning. *Nature* **2015**, *521*, 436–444. [CrossRef] [PubMed]

7. Simonyan, K.; Zisserman, A. Very Deep Convolutional Networks for Large-Scale Image Recognition. *arXiv* **2014**, arXiv:1409.1556.

8. He, K.; Zhang, X.; Ren, S.; Sun, J. Deep Residual Learning for Image Recognition. In Proceedings of the IEEE Conference on Computer Vision and Pattern Recognition (CVPR), Las Vegas, NV, USA, 27–30 June 2016; pp. 770–778.

9. Huang, G.; Liu, Z.; Van Der Maaten, L.; Weinberger, K.Q. Densely Connected Convolutional Networks. In Proceedings of the IEEE Conference on Computer Vision and Pattern Recognition, Honolulu, HI, USA, 22–25 July 2017; pp. 2261–2269.

10. Pyo, J.; Bang, J.; Jeong, Y. Front collision warning based on vehicle detection using CNN. In Proceedings of the International SoC Design Conference (ISOCC), Jeju, Korea, 23–26 October 2016; pp. 163–164.

11. Tang, Y.; Zhang, C.; Gu, R.; Li, P.; Yang, B. Vehicle Detection and Recognition for Intelligent Traffic Surveillance System. *Multimed. Tools Appl.* **2017**, *76*, 5817–5832. [CrossRef]

12. Gao, Y.; Guo, S.; Huang, K.; Chen, J.; Gong, Q.; Zou, Y.; Bai, T.; Overett, G. Scale optimization for full-image-CNN vehicle detection. In Proceedings of the IEEE Intelligent Vehicles Symposium (IV), Los Angeles, CA, USA, 11–14 June 2017; pp. 785–791.

13. Huttunen, H.; Yancheshmeh, F.S.; Chen, K. Car type recognition with deep neural networks. In Proceedings of the IEEE Intelligent Vehicles Symposium (IV), Gothenburg, Sweden, 19–22 June 2016; pp. 1115–1120.

14. Girshick, R.; Donahue, J.; Darrell, T.; Malik, J. Rich Feature Hierarchies for Accurate Object Detection and Semantic Segmentation. In Proceedings of the 2014 IEEE Conference on Computer Vision and Pattern Recognition, Columbus, OH, USA, 24–27 June 2014; pp. 580–587.

15. Uijlings, J.R.; Van De Sande, K.E.; Gevers, T.; Smeulders, A.W. Selective Search for Object Recognition. *Int. J. Comput. Vis.* **2013**, *104*, 154–171. [CrossRef]

16. He, K.; Zhang, X.; Ren, S.; Sun, J. Spatial Pyramid Pooling in Deep Convolutional Networks for Visual Recognition. In Proceedings of the 2014 IEEE International Conference of European Conference on Computer Vision, Zurich, Switzerland, 6–12 September 2014; pp. 346–361.

17. Girshick, R. Fast R-CNN. In Proceedings of the 2015 IEEE International Conference on Computer Vision (ICCV), Santiago, Chile, 7–13 December 2015; pp. 1440–1448.

18. Ren, S.; He, K.; Girshick, R.; Sun, J. Faster R-CNN: Towards Real-Time Object Detection with Region Proposal Networks. *arXiv* **2016**, arXiv:1506.01497v3. [CrossRef] [PubMed]

19. Dai, J.; Li, Y.; He, K.; Sun, J. R-FCN: Object Detection via Region-Based Fully Convolutional Networks. In Proceedings of the 2016 IEEE International Conference of Advances in Neural Information Processing Systems, Barcelona, Spain, 5–8 December 2016; pp. 379–387.

20. Azam, S.; Rafique, A.; Jeon, M. Vehicle Pose Detection Using Region Based Convolutional Neural Network. In Proceedings of the International Conference on Control, Automation and Information Sciences (ICCAIS), Ansan, Korea, 27–29 October 2016; pp. 194–198.

21. Tang, T.; Zhou, S.; Deng, Z.; Zou, H.; Lei, L. Vehicle Detection in Aerial Images Based on Region Convolutional Neural Networks and Hard Negative Example Mining. *Sensors* **2017**, *17*, 336. [CrossRef] [PubMed]

22. Deng, Z.; Sun, H.; Zhou, S. Toward Fast and Accurate Vehicle Detection in Aerial Images Using Coupled Region-Based Convolutional Neural Networks. *IEEE J.-Stars.* **2017**, *10*, 3652–3664. [CrossRef]

23. Redmon, J.; Divvala, S.; Girshick, R.; Farhadi, A. You only look once: Unified, real-time object detection. In Proceedings of the IEEE Conference on Computer Vision and Pattern Recognition (CVPR), Las Vegas, NV, USA, 27–30 June 2016; pp. 779–788.

24. Liu, W.; Anguelov, D.; Erhan, D.; Szegedy, C.; Reed, S.; Fu, C.Y.; Berg, A.C. SSD: Single Shot Multibox Detector. In Proceedings of the European Conference on Computer Vision (ECCV), Amsterdam, The Netherlands, 8–16 October 2016; pp. 21–37.
25. Redmon, J.; Farhadi, A. YOLO9000: Better, Faster, Stronger. In Proceedings of the IEEE Conference on Computer Vision and Pattern Recognition (CVPR), Honolulu, HI, USA, 21–26 July 2017; pp. 6517–6525.
26. Redmon, J.; Farhadi, A. Yolov3: An Incremental Improvement. *arXiv* **2018**, arXiv:1804.02767.
27. Sang, J.; Wu, Z.; Guo, P.; Hu, H.; Xiang, H.; Zhang, Q.; Cai, B. An Improved YOLOv2 for Vehicle Detection. *Sensors* **2018**, *18*, 4272. [CrossRef] [PubMed]
28. Arthur, D.; Vassilvitskii, S. k-means plus plus: The Advantages of Careful Seeding. In Proceedings of the Eighteenth Annual ACM-SIAM Symposium on Discrete Algorithms, New Orleans, LA, USA, 7–9 January 2007; pp. 1027–1035.
29. Hartigan, J.A.; Wong, M.A. Algorithm AS 136: A k-means Clustering Algorithm. *J. R. Stat. Soc.* **1979**, *28*, 100–108. [CrossRef]
30. Carlet, J.; Abayowa, B. Fast vehicle detection in aerial imagery. *arXiv* **2018**, arXiv:1709.08666.
31. Lin, T.Y.; Goyal, P.; Girshick, R.; He, K.; Dollar, P. Focal Loss for Dense Object Detection. In Proceedings of the IEEE International Conference on Computer Vision (ICCV), Venice, Italy, 22–29 October 2017; pp. 2980–2988.
32. Dong, Z.; Wu, Y.; Pei, M.; Jia, Y. Vehicle type classification using a semisupervised convolutional neural network. *IEEE Trans. Intell. Transp. Syst.* **2015**, *16*, 2247–2256. [CrossRef]

Article

Automatic Classification Using Machine Learning for Non-Conventional Vessels on Inland Waters

Marta Wlodarczyk-Sielicka [1,*] **and Dawid Polap** [2]

[1] Institute of Geoinformatics, Department of Navigation, Maritime University of Szczecin, Waly Chrobrego 1-2, 70-500 Szczecin, Poland
[2] Marine Technology Ltd., ul. Roszczynialskiego 4/6, 81-521 Gdynia, Poland
* Correspondence: m.wlodarczyk@am.szczecin.pl; Tel.: +48-513-846-391

Received: 23 May 2019; Accepted: 8 July 2019; Published: 10 July 2019

Abstract: The prevalent methods for monitoring ships are based on automatic identification and radar systems. This applies mainly to large vessels. Additional sensors that are used include video cameras with different resolutions. Such systems feature cameras that capture images and software that analyze the selected video frames. The analysis involves the detection of a ship and the extraction of features to identify it. This article proposes a technique to detect and categorize ships through image processing methods that use convolutional neural networks. Tests to verify the proposed method were carried out on a database containing 200 images of four classes of ships. The advantages and disadvantages of implementing the proposed method are also discussed in light of the results. The system is designed to use multiple existing video streams to identify passing ships on inland waters, especially non-conventional vessels.

Keywords: machine learning; image analysis; feature extraction; ship classification; marine systems

---

1. Introduction

Monitoring marine vessels is crucial to navigational safety, particularly in limited waters. At sea, vessel traffic services (VTS) are the systems responsible for supervision and traffic management [1]. In the case of inland waters, the support system used to track ships is called river information services (RIS) [2]. These systems use a variety of technologies to assist in observation, e.g., the automatic identification system (AIS), radars, and cameras [3]. The AIS is an automated tracking system that uses information concerning the position of the ship and additional data entered by the navigator, provided that the ship has an appropriate transmitter [4,5]. The international convention for the safety of life at sea (SOLAS convention) [6] developed by the International Maritime Organization requires that AIS transmitters be installed on vessels with a gross tonnage of 300 and more engaged in international voyages, vessels with a gross tonnage of 500 and more not engaged in international voyages, and passenger vessels. All other vessels navigating at sea and in inland waters cannot be unambiguously identified by traffic monitoring systems (for example, non-commercial craft, recreation craft, yachts, and other small boats).

No system or method is available at present for the automatic identification of small vessels. The only way to recognize them is to observe markings on their sides. RIS and VTS information systems have video monitoring, where cameras are mounted mainly on river bridges and in ports. It provides a useful opportunity to observe ships from different perspectives. However, the entire process is not automated. These systems are operated round the clock or during the day (in the case of daily navigation) by human workers. Therefore, to minimize cost, the automation of system operations is desirable. This research problem is a part of the SHREC project, the main purpose of which is to develop a prototype system for the automatic identification of vessels in limited areas [7,8]. At this stage, the authors propose a method to detect and classify vessels belonging to four classes based on an image database. The proposed method is a component of the system and forms part of the detection and recognition layer.

The recognition process is one of the most popular tasks due to the number of problems. Feature extraction is particularly important, thanks to which it is possible to further assign these features to a given object. However, quite often, two-dimensional images contain various kinds of additional elements such as sky, trees, or houses. In the case when an image is taken with a good camera, further objects can be blurred, which is a certain distortion causing a better extraction of the features of the closer elements. Unfortunately, this situation is not always available. Hence, scientists from around the world are modeling different solutions to get the best results. First of all, artificial intelligence techniques have developed quite dynamically in recent years. In [9], the authors presented the recent advances in convolutional neural networks, which are a popular tool for image classification. The authors noticed the growing popularity trend as well the modeling of more hybrids and modifications. In [10], the idea of using this mathematical model of neuron nets was used in medical diagnosis. Again in [11], the idea of long short-term memory was introduced as an improvement of the classic approach, which can be used for obtaining better efficiency. The problem of classification of water facilities was analyzed in [12,13], where deep representation for different vessels was examined. The problem of the amount of data in the classification process was analyzed in [14] by the use of data augmentation. Existing classification solutions apply to marine waters for large ships. It should be emphasized that, the designed system will be using multiple existing video streams to identify passing ships on inland waters, especially non-conventional vessels.

The remainder of this article is organized as follows: Section 2 presents the problem of detecting the features of ships in images using different filters, and Section 3 contains a description of the convolutional neural network used for classification. Section 4 details the proposed ship classification system model, and Section 5 describes our experiments and the conclusions of this study. The main goal of this paper is to propose a detection algorithm based on parallel image processing that allows the creation of two different samples from one image and the architecture of a deep convolutional neural network. Our tests are the basis during the construction of the SHREC system architecture.

2. Detecting Features of Ships

The proposed technique of detecting features of ships in images is based on the parallel processing of a given image using different filters. This processing allows the distortion of images in such a way that only important elements giving features are mapped. Images obtained from the parallel filtering process are used to find features that appear in both images with different filters. The Hessian matrix was used for comparison and detection. The proposed technique relies on classical image processing to reduce the area used for further classification. The features of vessels were detected in order to minimize the amount of data in images such as background or surroundings. We used the idea of key points to detect only significant elements. An additional advantage of the proposed detection method is extension of the number of samples in each class. The proposed detection technique returns two rectangles, where both are interpreted as different images. Before being used in a subsequent classification process, they are scaled down in size, resulting in additional distortion.

The first of the images is blurred using a fast box algorithm [15]. To formulate the expression for this action, a convolution needs to be defined. The convolution of two functions is called an operation, the result of which is a new function $h(\cdot)$. Thus, image convolution consists of a moving window (matrix) with function values $g(\cdot)$ called a filter along function value $f(\cdot)$ that calculates the sum of products of the obtained values. This can be formally described as follows (in a discrete way):

$$h[i, j] = (f * g)[i, j] = \sum_{y=i-r}^{i+r} \sum_{x=j-r}^{j+r} f[y, x]\, g[y, x] \tag{1}$$

The discreteness implies that the functions $f(\cdot)$ and $g(\cdot)$ are matrices. The first one stores the pixel values of the processed image, and the second one is much smaller (its dimensions are 3×3 or 4×4 and are marked as $r \times r$). Its coefficients are constant and predetermined for the corresponding filter. For Gaussian blur, the following matrix is obtained:

$$g = [\cdot, \cdot] = \begin{pmatrix} \frac{1}{16} & \frac{1}{8} & \frac{1}{16} \\ \frac{1}{8} & \frac{1}{4} & \frac{1}{8} \\ \frac{1}{16} & \frac{1}{8} & \frac{1}{16} \end{pmatrix} \tag{2}$$

Box blur involves blurring certain areas (the size of the filter) and allows for multiple instances of blurring in a given area (approximated) compared with the transition of the complete image several times. This solution reduces the number of calculations. The blur distorts the image, but to enhance the features, the image is converted into grayscale. This is done by replacing each pixel, *pxl*, (composed of the three components of the RGB model (Red–Green–Blue)) according to the following formula:

$$R(pxl) = G(pxl) = B(pxl) = \frac{R(pxl) + G(pxl) + B(pxl)}{3} \tag{3}$$

Then, the contrast and brightness are increased. Changing the contrast allows for the darkening of the dark pixels and the brightening of the bright ones, and can be achieved using

$$C(pxl) = [C[pxl]\ 128]\cdot\eta + 128 \tag{4}$$

where $\eta \leq 0$ is a threshold, and $C(pxl)$ is a specific function defined as a component of the RGB model. The brightness of this image is increased, and all colors are approximated to white. The effect is the elimination of pixels close to white in color:

$$C\ (pxl) = C(pxl) + \sigma \tag{5}$$

where σ is a threshold. The colors of the pixels are then inverted, which allows for better visualization of the other elements. To this end, the process involves converting each component of the pixel as

$$C(pxl) = 255 - C(pxl) \tag{6}$$

It is easy to see that the inverted color highlights small elements that would otherwise be invisible. They can be removed using two operations: Improving the gamma value and binarization. Gamma correction with a coefficient γ allows for the scaling of the brightness of the image (that is, the other elements):

$$C(pxl) = 255\cdot\frac{C(pxl)^{\gamma}}{255} \tag{7}$$

The last step in processing the first image is its binarization, i.e., the conversion of all pixels into white or black. The conversion is carried out by calculating the average value of all components of a given pixel, *pxl*, and checking if it is within the range $\langle 0, \alpha \rangle$ or $(\alpha, 255\rangle$ as

$$\begin{cases} \frac{\sum_{C \in \{R,G,B\}} C\{pxl}{3} \leq \alpha \ then\ C\{pxl = 0 \\ \frac{\sum_{C \in \{R,G,B\}} C\{pxl}{3} > \propto\ then\ C\{pxl = 255 \end{cases} \tag{8}$$

The second image is processed in almost the same way using other coefficients, such as the threshold values. The most significant difference is the absence of binarization on the second image. Images created in this way can be used to detect features that can help locate the ship. For this purpose, the SURF algorithm (Speeded-Up Robust Features) is used [16]. In a given image, it searches for features by analyzing the neighborhood based on the determinant of the Hessian matrix defined as follows:

$$H(p, \omega) = \begin{bmatrix} L_{xx}[p, \omega] & L_{xy}[p, \omega] \\ L_{xy}[p, \omega] & L_{yy}[p, \omega] \end{bmatrix} \tag{9}$$

where $L_{ij}(p, \omega)$ represents the convolution of the integrated image I at point $p = (x, y)$ with smoothing using the Gaussian kernel through parameter ω. The coefficients of this matrix can be represented by the following formulae:

$$L_{xx}(p, \omega) = I(p) \frac{\partial^2}{\partial x^2} g(\omega) \tag{10}$$

$$L_{yy}(p, \omega) = I(p) \frac{\partial^2}{\partial y^2} g(\omega) \tag{11}$$

$$L_{xy}(p, \omega) = I(p) \frac{\partial^2}{\partial xy} g(\omega) \tag{12}$$

The aforementioned integrated image is an indirect representation called rectangle features. For a given position $p = (x, y)$, the sum of all pixels in the original image $I'(\cdot)$ is calculated as

$$I(p) = I((x, y)) = \sum_{i=0}^{i \leq x} \sum_{j=0}^{j \leq y} I'(i, j) \tag{13}$$

The algorithm chooses points that form the trace of the matrix shown in Equation (9) and the local extreme. For this purpose, the values of the Hessian determinant and the trace are determined:

$$det(H(p, \omega)) = \omega^2 \left(L_{xx}(p) L_{yy}(p) - L_{xy}^2(p) \right) \tag{14}$$

$$trace(H(p, \omega)) = \omega \left(L_{xx}(p) + L_{yy}(p) \right) \tag{15}$$

In this way, the key points are determined, i.e., the second image is processed without binarization. Then, for each key point found, its color is checked on the second image (after binarization). As a result, two sets of points are obtained belonging to one of two colors—black or white. However, the number of points can be very large, and therefore it is worth minimizing them using the average distance for the entire set.

It is assumed that the set of key points for one of the colors is marked as $\{p^{\{0\}}, p^{\{1\}}, \ldots, p^{\{n-1\}}\}$. The average distance ξ for all points in the set is calculated using the Euclidean metric defined as

$$\xi = \frac{\sum_{i=0}^{n-1} \sum_{j=0}^{n-1} \sqrt{\left(p_x^{(i)} - p_x^{(j)}\right)^2 + \left(p_y^{(i)} - p_y^{(j)}\right)^2}}{(n-1)^2 - (n-1)} \tag{16}$$

The calculated distance is used to analyze all points in the set. If, for any point, there is a neighbor at a distance smaller than ξ, the given point remains in the set. Otherwise, it is removed.

Finally, the method returns two sets of points that can be called features of the processed image. These points are used to find the maximum number of dimensions of the found object by finding the largest and smallest values of the coordinates x and y. The image between these extremes is cut out and used as a sample in the ship classification process. The model of image processing is shown in Figure 1.

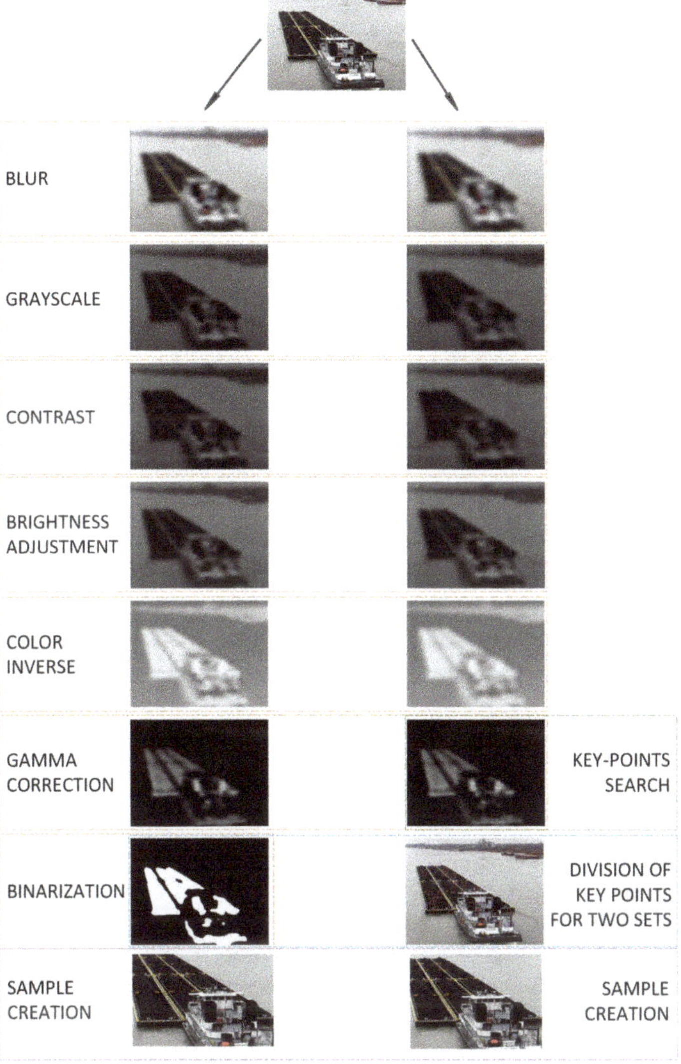

Figure 1. Graphical hierarchy of image processing.

3. Convolutional Neural Network for Classification

The image in computing is a tensor, i.e., a mathematical object described by three components $w \times h \times d$. The first two components form the dimensions of the image, i.e., width and height, and the third is its depth. In the classic approach to this subject, the depth of the image is the number of components. Thus, for the RGB model, there are three values and the depth is three. As a result, three matrices are created in which the numerical values of a given color are stored. Such data can be used in the recognition process using artificial neural networks called convolutional neural networks. Such structures are inspired by the action and mechanisms of the cerebral cortex [10,17].

This input structure accepts a graphic file and processes it to classify it. It is composed of several types of layers, where the first is called the convolutional layer. It is interpreted as a filter of a given size $k \times k$. This filter moves on the image and modifies it. However, it is worth noting that by dividing the image according to filters of different sizes, a grid of rectangles can be obtained. Each rectangle represents a part of the image to which a weight w is assigned, that is, a numerical value used for further training. Each part is interpreted as a neuron and returns the following value:

$$f\left(\sum_{i=0}^{k}\sum_{j=0}^{k} w_{i,j}\cdot a_{x+i,y+j}\right) \tag{17}$$

where $f(\cdot)$ is the activation function, and $a_{x,y}$ is the activation value for the output from position (x, y).

The convolutional layer is responsible for the extraction of features and eliminating redundancy in the data. It is easy to see that the grids can overlap, which may result in multiple appearances of the same fragment. To prevent this, a pooling layer is used to reduce the size of the image by extracting only the most appropriate features. For this selection, the maximum value is typically used. Both layers may occur several times depending on the problem and its size (understood as image size). A third type of layer, a fully connected layer, is also used. It is a classical neural network, the input to which consists of data from the previous layer (for each pixel, there are three values). The operation of the classical neural network can be described as the processing of numerical values. Neurons are arranged in columns called layers, where neurons between neighboring layers are related to one another by weighted connections. For each neuron, the output value x_i and weight w from the previous neurons are sent. The value of each neuron is calculated as in Equation (17), that is, as the sum of the products of weights and values of neurons from the previous layer rescaled by the activation function.

As the output of the entire network, values are obtained that are not yet a probability distribution. They are normalized using the softmax function. The approximate values should indicate the assignment to the class predicted by the classifier.

However, all weights in the neural network are assigned in a random manner, which causes the classifier to return meaningless information. To remedy this, an important step is to train the classifier. The training process consists of checking the assignment returned by the network and comparing it with the correct one. The comparison is made using a loss function. The idea of training is to minimize the loss function by modifying weights in the network. An example of an optimization algorithm used for training is adaptive moment estimation, often called the Adam optimizer [18]. The algorithm is based on calculating the variance v and mean m for each iteration t. If a single weight is $w^{(t)}$, the values have the following form:

$$m_t = \beta_1 m_{t-1} + (1-\beta_1)g_t \tag{18}$$

$$v_t = \beta_2 v_{t-1} + (1-\beta_2)g_t^2 \tag{19}$$

where β_1 and β_2 are distribution values. Using them, the following correlations can be defined:

$$\hat{m}_t = \frac{m_t}{1-\beta_1^t} \tag{20}$$

$$\hat{v}_t = \frac{v_t}{1-\beta_2^t} \tag{21}$$

The above correlation formulae are used to update weights in the network using the following equation:

$$\theta_{t+1} = \theta_t - \frac{\eta}{\sqrt{\hat{v}_t}+\epsilon}\hat{m}_t \tag{22}$$

where η is the training rate, and ϵ is a very small value preventing an indeterminate form.

4. Model of Ship Classification System

The idea of the proposed system is to create software that can process real-time video samples and categorize ships appearing in them. The camera recording the image can be placed in any position, such as a water buoy or a bridge. The video being recorded is composed of many frames, which may not necessarily feature a vessel. Moreover, it makes no sense to process each frame—one second of a recorded video usually contains 24 frames, it is not possible to process this amount of data in real time.

Hence, image processing can be minimized to one frame every one or two seconds. This prevents the task queue from overloading and blocking. However, keeping the camera in one position leads it to record the same elements in space. Therefore, to avoid searching for ships in nearly identical images, a pattern was created to avoid the need for continuous processing. The idea of this solution is to find key points at different times of the day and in different weather (for this purpose, the SURF algorithm was used). If a given frame has only points that coincide with the previous pattern, the frames do not need to be analyzed in terms of searching for a floating vehicle. If there are greater deviations from the standard (for example, a greater number of points), image processing should be performed as described in Section 2, and classification should then be performed using the convolutional neural network described in Section 3.

The proposed data processing technique allows reducing the area of the incoming image. An additional advantage is the creation of two images, which are created by parallel image processing. In the classification process, the images are reduced in order to normalize the size of all samples. The effect of this action is to obtain two images that are distorted by size reduction.

5. Experiments

The previous sections described the image processing model to extract only fragments and the model of the classifier that categorizes incoming images into selected classes. Depending on the number of classes, this is the output of the network. For n classes, there are n neurons in the last layer, and the classes are saved as n-element vectors consisting of $n - 1$ zeros and a single one. For each class, the "1" is in a different position in the vector. The experimental part is divided into two parts: Tests on public images and tests on real data collected by means of cameras mounted on bridges and placed on the bank in the port of Szczecin.

5.1. Experiments on Test Data

To check the operation of the proposed method, the authors used an image database divided into four classes, each of them composed of 200 samples. The classes represented four types of ships—a boat, a motor yacht, a navy ship, and a historic ship. The image processing method presented in Section 2 was tested on different values of parameters in the range {0, 15}. The adjustment of these values for the system led to the setting of an average number of key points. All coefficients were used in different combinations, and the authors selected one that returned approximately 15–20 key points (after processing). The selected configuration for image processing is presented in Table 1. This number is not accidental because the authors noticed that if the number of key points was small, the worse area was returned. On the contrary, when the number of key points was large, the original image was returned. This is why the configuration was chosen in an empirical manner.

In this way, the best configuration was selected and the files were processed. It is worth noting that two images were always generated from one sample, as presented in Figure 2. Before these samples were used for training, images among them that did not contain vessels were removed. The database of each class (composed of 200 samples) was thus extended to an average value of 320 images (first class, 343; second class, 316; third class, 324; and fourth class, 299). The obtained samples did not always contain the entire vessel, but only fragments or other objects in the image. This was important for the training process because it could not be guaranteed that an incoming vessel would be oriented in a particular manner. In addition, using the same method of extraction yielded a similar sample.

Table 1. The selected configuration for image processing.

Filter	First Image	Second Image
Blur	8	8
Grayscale	yes	yes
Contrast	5	5
Brightness	15	15
Inverse	yes	yes
Gamma	6	7
Binarization	yes	no

a.

b.

c.

d.

Figure 2. Sample examples of images, where the rectangles indicate the extracted images: (**a**) boat; (**b**) motor yacht; (**c**) navy ship; and (**d**) historic ship.

The proposed technique for processing and extracting features allowed for an average of 320 samples, whereas the ideal solution would return exactly 400 (double the amount from the entry database). However, the different conditions in which the images were taken (in terms of quality, distance, and exposure) and the background posed problems. For ships in high seas, the extraction was perfect. However, a large part of these samples was set against the background of forests or cities. Thus, the recorded average extraction efficiency of 80% is a good result. It might have been affected by incorrect filter configuration. An empirical approach to the evaluation of the obtained data indicated the best adjustment of these values.

Table 2 presents the structure of the classification network. The layers were selected in order to obtain the highest effectiveness. It was noted that the proposed technique for the division of samples in the ratio 80:20 (training:verification) and the learning rate of $\eta = 0.001$ gives the best results. We trained the classifier during 100 epochs. The resulting efficiency was 89%. A graph of the training is shown in Figure 3.

Table 2. Architecture of convolutional neural network.

Type of Layer	Shape
Convolutional 3×3	(None, 96, 96, 32)
Activation—ReLU function	(None, 96, 96, 32)
Batch normalization	(None, 96, 96, 32)
Max pooling 3×3	(None, 32, 32, 32)
Dropout 0.25	(None, 32, 32, 32)
Convolutional 3×3	(None, 32, 32, 64)
Activation—ReLU function	(None, 32, 32, 64)
Batch normalization	(None, 32, 32, 64)
Convolutional 3×3	(None, 32, 32, 64)
Activation—ReLU function	(None, 32, 32, 64)
Batch normalization	(None, 32, 32, 64)
Max pooling 2×2	(None, 16, 16, 64)
Dropout 0.25	(None, 16, 16, 64)
Convolutional 3×3	(None, 16, 16, 128)
Activation–ReLU function	(None, 16, 16, 128)
Batch normalization	(None, 16, 16, 128)
Convolutional 3×3	(None, 16, 16, 128)
Activation—ReLU function	(None, 16, 16, 128)
Batch normalization	(None, 16, 16, 128)
Max pooling 2×2	(None, 8, 8, 128)
Dropout 0.25	(None, 8, 8, 128)
Flatten	(None, 8192)
Dense 1024	(None, 1024)
Activation—ReLU function	(None, 1024)
Batch normalization	(None, 1024)
Dropout 0.5	(None, 1024)
Dense 4	(None, 4)
Activation—Softmax function	(None, 4)

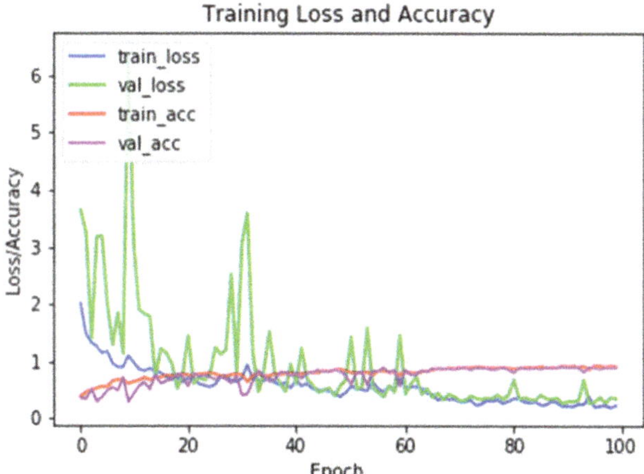

Figure 3. Learning history of the classifier obtained using the created database.

To more accurately verify the operation of the classifier, 118 false samples were added to the database. In our experiments, different sizes of images were analyzed. For a given architecture, we tested the classifier in order to achieve the best possible accuracy. The obtained measurements are presented in Figure 4.

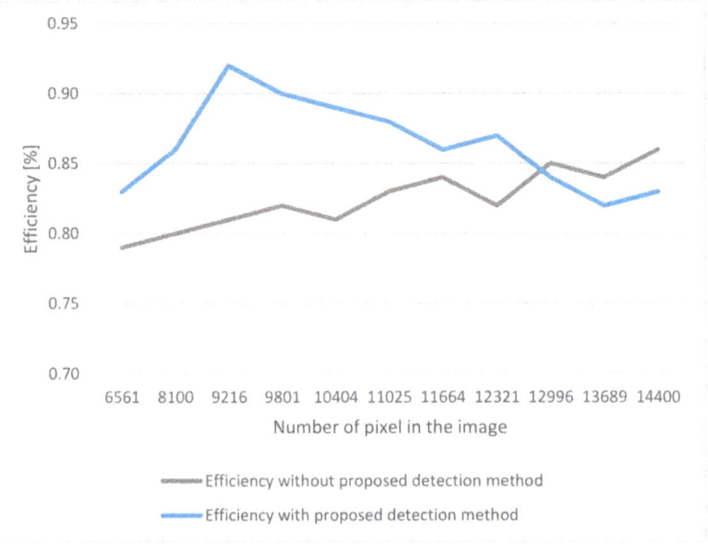

Figure 4. Measurement of the effectiveness of the techniques relative to the size of the incoming image.

The classification results were then checked and are presented as a confusion matrix in Figure 5. In addition, matrices for individual classes were made, as shown in Figure 6. The proposed classifier categorized every wrong sample into one of the classes of vessels. A condition was thus added (to the extraction stage) that if no key point was found in an image, the image was rejected.

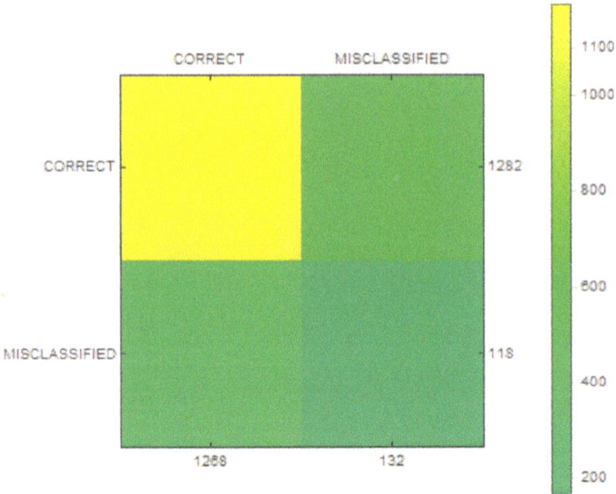

Figure 5. The confusion matrix of the classification model.

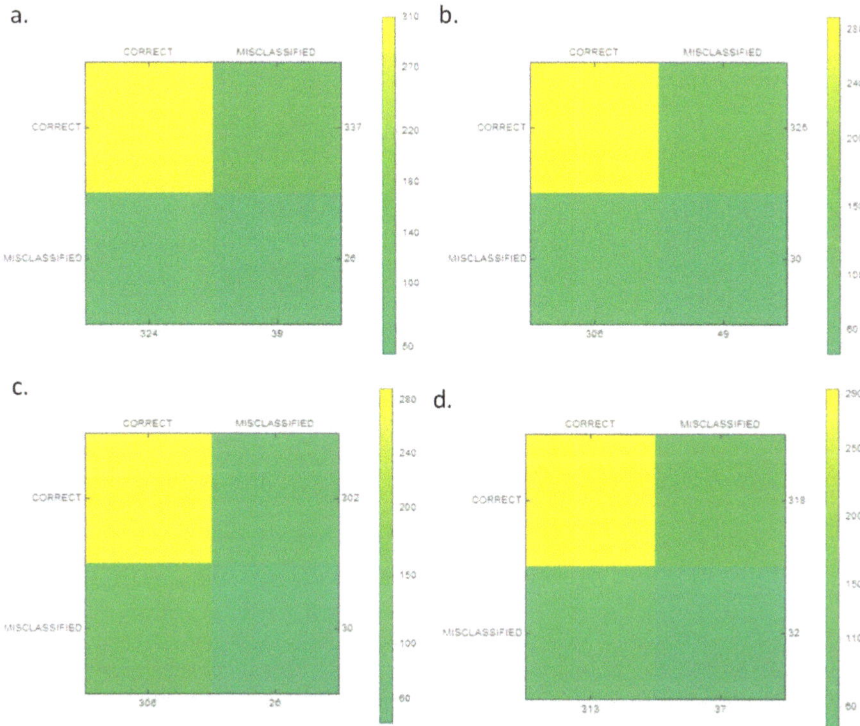

Figure 6. The confusion matrices of the classification model for each class: (**a**) boat; (**b**) motor yacht; (**c**) navy ship; and (**d**) historic ship.

Based on the results, the authors determined the accuracy, overlap, sensitivity, specificity, and the F-measure. The results are presented in Table 3.

Table 3. Mean quality measures for the trained classifier.

Measure	Boat	Motor Yacht	Navy Ship	Historic Ship	Mean
Accuracy	0.915	0.909	0.928	0.91	0.916
Sensitivity	0.957	0.958	0.962	0.96	0.959
Specificity	0.679	0.558	0.659	0.438	0.583
Precision	0.944	0.939	0.957	0.941	0.945
Negative prediction rate	0.735	0.649	0.692	0.538	0.654
Miss rate	0.043	0.042	0.037	0.04	0.041
False discovery rate	0.056	0.061	0.043	0.059	0.055
False omission rate	0.265	0.351	0.308	0.462	0.346
F1 score	0.951	0.948	0.96	0.95	0.952
Recall	0.944	0.939	0.957	0.941	0.945

The resulting effectiveness was nearly 92%, which is high considering that a small number of samples were used for training. A higher value was obtained for sensitivity, which implies accurate prediction of true samples. The F1 score was also high, which indicates that the classifier was efficient. Considerably worse results were obtained for specificity, nearly 0.541, implying that the classification of true negatives was substandard. The same result was obtained for the rate of false omission, indicating that a large portion (almost 30%) of the false samples had been rejected.

As part of the conducted research, classification tests on a given database were performed with and without the proposed detection method and the results are presented in Figure 7.

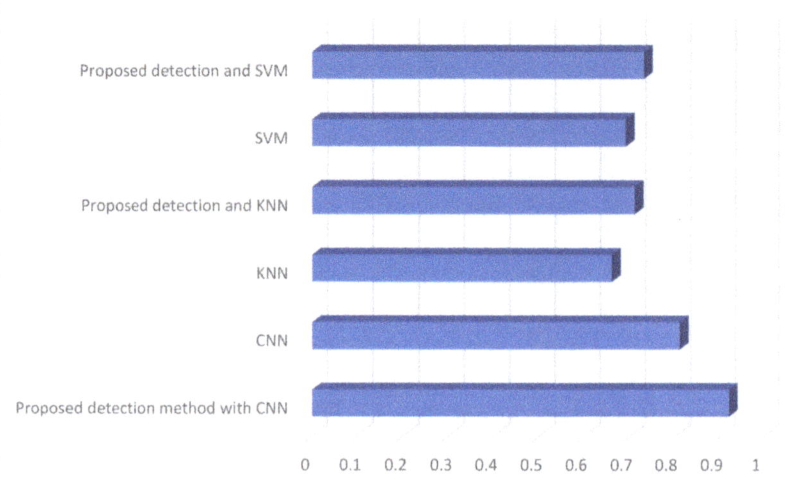

Figure 7. Comparison of different classification techniques on a selected database. SVM—support vector machine; KNN—k-nearest neighbors; CNN— convolutional neural network.

First of all, we made a test that uses a convolutional neural network without detection. The efficiency was about 80%. Again, for classic classifiers such as k-nearest neighbors (KNN) and support vector machine (SVM), the results were similar. In both cases, the proposed detection technique increased the value of effectiveness. However, the highest results were obtained for the proposed technique along with a convolutional neural network.

5.2. Experiments on Real Data

In the next step, we analyzed the proposed solution on real data. To test the proposed solution in real conditions, we manage to create a database contained 18,361 image files that stored six classes of boats—barge (1703 images), yacht (1489 images), marine service (2362 images), motorboat (6980 images), passenger (3629 images), and other kinds (2198 images). The data were obtained by the setting of cameras mounted on bridges and placed on the bank in Szczecin in Poland. The resulting video data were processed with detection to get video frames containing the passing ship on inland waters. Within one second, a maximum of one frame was collected. For the needs of the experiments, a maximum of 20 photos of one ship crossing the bridge were left in the database. The examples of this images are presented in Figure 8.

Figure 8. Sample examples of images: (**a**) barge; (**b**) yacht; (**c**) marine service; (**d**) motorboat; (**e**) passenger; and (**f**) other.

Images obtained at the stage of detection differ from each other in size and quality. As part of the research, the same classification architecture was used, which is presented in Table 2, with one difference being that there were six neurons in the output layer because of the number of classes. For the purposes of conducting experiments, the images within each class were divided in the ratio 80:20 (training:verification), with a training coefficient\eta = 0.001 and 100 iterations. After the training process, the database was enlarged by an additional 300 pictures that did not fit into any of the six classes. For each of the classes, plus the previously mentioned fake images, the efficiency of the trained classifier was checked. The results in the form of confusion matrices are presented in Figure 9 for the classifier and for each class in Figure 10.

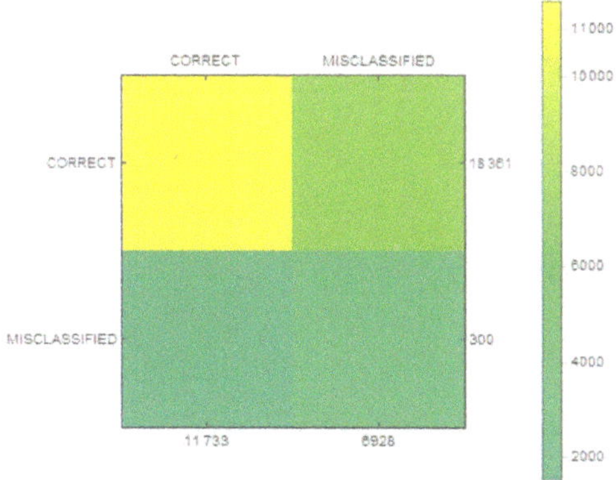

Figure 9. The confusion matrix of the classification model for real data.

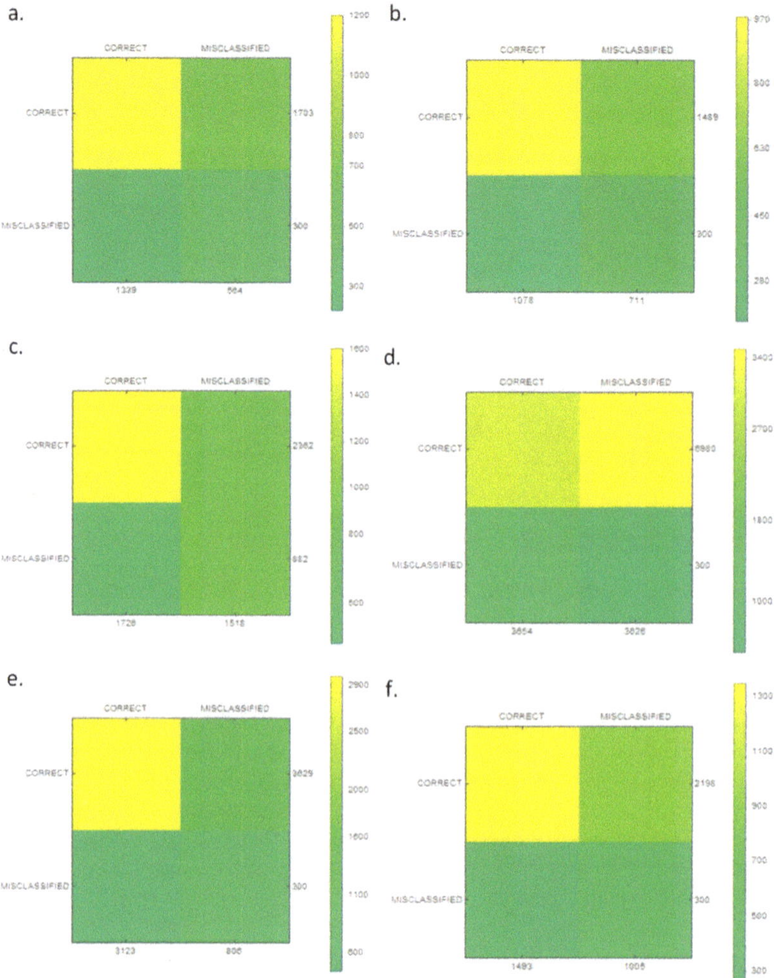

Figure 10. The confusion matrices of the classification model for each class: (**a**) barge; (**b**) yacht; (**c**) marine service; (**d**) motorboat; (**e**) passenger; and (**f**) other.

Statistical tests for each class are presented in Table 4. In the case of average performance of the proposed classification, accuracy reached 63%. As in the case of test data, the higher value was obtained for sensitivity. In addition, the recall is on the same level. The F1 score was also high, which indicates that the classifier is efficient. Here, we also see that worse results were obtained for specificity and miss rate.

Table 4. Mean quality measures for the trained classifier.

Measure	Barge	Yacht	Marine Services	Motorboat	Passenger	Other
Accuracy	0.668	0.603	0.648	0.502	0.795	0.598
Sensitivity	0.882	0.83	0.901	0.963	0.95	0.897
Specificity	0.217	0.164	0.139	0.046	0.18	0.146
Precision	0.705	0.657	0.679	0.5	0.821	0.613
Negative prediction rate	0.463	0.333	0.41	0.557	0.473	0.483
Miss rate	0.118	0.17	0.099	0.037	0.05	0.103
False discovery rate	0.295	0.343	0.321	0.5	0.179	0.387
False omission rate	0.537	0.667	0.59	0.443	0.527	0.517
F1 score	0.783	0.733	0.774	0.658	0.881	0.728
Recall	0.882	0.83	0.901	0.963	0.95	0.897

Next, for the database thus created, the operation of the proposed method was verified with other classic architectures of the convolution network as shown in Figure 11.

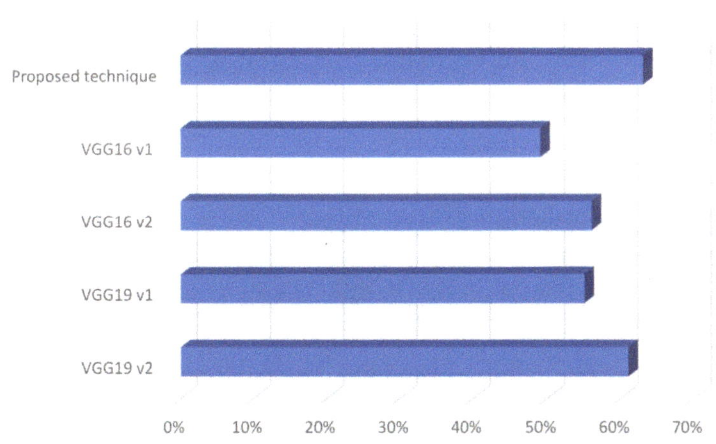

Figure 11. Comparison of different convolutional neural networks on a selected database.

In our experiments, we used a strategy to train some layers and leave others frozen in existing models of CNN such as VGG16 and VGG19 [19]. Such a strategy was used to transfer knowledge from previously trained classifiers so as not to train classifiers from scratch. While choosing to leave specific layers, we paid attention to possible overfitting. We have chosen a few examples of architectures to illustrate the proposed solution with existing ones. In the case of the first version (on the chart v1), the last five layers were frozen, and in the second case (on the charts v2), it was the last six layers. In both cases, two layers were added to the architecture as fully connected ones—in both cases, a first layer was composed of 512 neurons and a ReLu function, and the last one had six neurons. This architecture yielded less effectiveness than the proposed technique. Summarizing, the proposed solution can be used in the system being built. Although the classes are similar, the network was able to classify the images. However, the number of samples in the database should be increased.

6. Conclusions

The identification and categorization of ships in images can be very useful for creating a register of passing ships, observing boundaries, and even recording information on a catastrophic recorder. This paper proposed a model to extract features to identify marine vessels from a single frame of an image and a deep learning-based method to categorize them. The authors analyzed only four categories of vessels, each of which was initially composed of 200 sample images. Features of ships were detected in images through the parallel processing of an image using different filters. The aim was to distort images so that only important elements yielded feature maps. Both images obtained from the parallel filtering process were used to find features common to them using different filters. The Hessian matrix was used for this. For feature extraction, the database was extended to include additional samples used to train the classifier. During classification, an image was assumed to be a tensor with its number of components as its depth (RGB model). The convolutional neural network was used for the recognition process. Its structure was inspired by the action and mechanisms of the cerebral cortex. Following tests and the reconfiguration of the network, the obtained efficiency for test data was nearly 90%, which is a very good result considering the small number of samples. In the case of a real database with six classes the average accuracy reached 63%. The proposed solution will be implemented in the identification system being built. However, the database should be increased.

In the SHREC system, the camera stream is analyzed using image processing methods for detection, classification, and text recognition to properly identify a vessel. The method for ship classification presented in this article will form part of ongoing research in the SHREC project. Future work will focus on improving the proposed classification method. The authors intend to modify the parameters of the method to be able to classify more types of ships. For this purpose, a more extensive database of images of ships is being gathered using video cameras at varying resolutions.

Author Contributions: Conceptualization, M.W.-S. and D.P.; methodology, D.P.; software, D.P.; data curation, M.W.-S. and D.P.; writing—original draft preparation, M.W.-S. and D.P.; writing—review and editing, M.W.-S. and D.P.

Funding: This work was supported by the National Centre for Research and Development (NCBR) of Poland under grant no. LIDER/17/0098/L-8/16/NCBR/2017 and grant no. 1/S/IG/16, financed by a subsidy from the Ministry of Science and Higher Education for statutory activities.

Conflicts of Interest: The authors have no conflict of interest to declare.

References

1. IALA Recommendation. *V–128: On Operational and Technical Performance Requirements for VTS Equipment*, 2nd ed.; IALA: Saint-Germain-en-Laye, France, 2005.
2. Stateczny, A. Sensors in River Information Services of the Odra River in Poland: Current State and Planned Extension. In Proceedings of the Baltic Geodesy Congress (BGC), Gdansk, Poland, 22–25 June 2017. [CrossRef]
3. Bloisi, D.; Iocchi, L.; Fiorini, M.; Graziano, G. Camera-based Target Recognition for Maritime Awareness. In Proceedings of the 15th International Conference on Information Fusion (FUSION), Singapore, 9–12 July 2012; pp. 1982–1987.
4. Chaturvedi, S.; Yang, C.-S.; Ouchi, K.; Shanmugam, P. Ship recognition by integration of SAR and AIS. *J. Navig.* **2012**, *65*, 323–337. [CrossRef]
5. Kazimierski, W.; Stateczny, A. Radar and automatic identification system track fusion in an electronic chart display and information system. *J. Navig.* **2015**, *68*, 1141–1154. [CrossRef]
6. International Maritime Organisation. *SOLAS International Convention for the Safety of Life at Sea*; International Maritime Organisation: London, UK, 1974.
7. Wawrzyniak, N.; Stateczny, A. Automatic watercraft recognition and identification on water areas covered by video monitoring as extension for sea and river traffic supervision systems. *Pol. Marit. Res.* **2018**, *25*, 5–13. [CrossRef]
8. Wawrzyniak, N.; Hyla, T. Automatic Ship Identification Approach for Video Surveillance Systems. In Proceedings of the 14th International Conference on Systems, ICONS 2019, Valencia, Spain, 24–28 March 2019.
9. Gu, J.; Wang, Z.; Kuen, J.; Ma, L.; Shahroudy, A.; Shuai, B.; Liu, T.; Wang, X.; Wang, G.; Cai, J.; et al. Recent advances in convolutional neural networks. *Pattern Recognit.* **2018**, *77*, 354–377. [CrossRef]
10. Acharya, U.R.; Oh, S.L.; Hagiwara, Y.; Tan, J.H.; Adeli, H. Deep convolutional neural network for the automated detection and diagnosis of seizure using EEG signals. *Comput. Biol. Med.* **2018**, *100*, 270–278. [CrossRef] [PubMed]

11. Hanson, J.; Paliwal, K.; Litfin, T.; Yang, Y.; Zhou, Y. Accurate prediction of protein contact maps by coupling residual two-dimensional bidirectional long short-term memory with convolutional neural networks. *Bioinformatics* **2018**, *34*, 4039–4045. [CrossRef] [PubMed]

12. Solmaz, B.; Gundogdu, E.; Yucesoy, V.; Koc, A. Generic and attribute-specific deep representations for maritime vessels. *IPSJ Trans. Comput. Vis. Appl.* **2017**, *9*, 22. [CrossRef]

13. Leclerc, M.; Tharmarasa, R.; Florea, M.C.; Boury-Brisset, A.C.; Kirubarajan, T.; Duclos-Hindié, N. Ship Classification Using Deep Learning Techniques for Maritime Target Tracking. In Proceedings of the 2018 21st International Conference on Information Fusion (FUSION), Cambridge, UK, 10–13 July 2018; pp. 737–744.

14. Milicevic, M.; Obradovic, I.; Zubrinic, K.; Sjekavica, T. Data Augmentation and Transfer Learning for Limited Dataset Ship Classification. *Wseas Trans. Syst. Control* **2018**, *13*, 460–465.

15. Koves, P. Fast Almost-Gaussian Filtering. In Proceedings of the International Conference on Digital Image Computing: Techniques and Applications (DICTA), Sydney, NSW, Australia, 1–3 December 2010; pp. 121–125.

16. Bay, H.; Tuytelaars, T.; Van Gool, L. SURF: Speeded-Up Robust Features. In *Computer Vision–ECCV 2006: Lecture Notes in Computer Science*; Leonardis, A., Bischof, H., Pinz, A., Eds.; Springer: Berlin/Heidelberg, Germany, 2006; Volume 3951.

17. Zhang, C.; Sun, G.; Fang, Z.; Zhou, P.; Pan, P.; Cong, J. Caffeine: Towards uniformed representation and acceleration for deep convolutional neural networks. *IEEE Trans. Comput. Aided Des. Integr. Circuits Syst.* **2018**. [CrossRef]

18. Kingma, D.; Ba, J. Adam: A method for stochastic optimization. In Proceedings of the 3rd International Conference on Learning Representations, San Diego, CA, USA, 12 November 2014.

19. Liu, S.; Deng, W. Very deep convolutional neural network based image classification using small training sample size. In Proceedings of the 3rd IAPR Asian Conference on Pattern Recognition (ACPR), Kuala Lumpur, Malaysia, 3–6 November 2015; pp. 730–734.

Article

WatchPose: A View-Aware Approach for Camera Pose Data Collection in Industrial Environments

Cong Yang [1], Gilles Simon [1], John See [2], Marie-Odile Berger [1] and Wenyong Wang [3],*

[1] MAGRIT Team, INRIA/LORIA, 54600 Nancy, France; cong.yang@loria.fr, cong.yang@uni-siegen.de (C.Y.); gilles.simon@loria.fr (G.S.); marie-odile.berger@loria.fr (M.-O.B.)
[2] Faculty of Computing and Informatics, Multimedia University, Cyberjaya 63100, Selangor, Malaysia; johnsee@mmu.edu.my
[3] School of Information Science and Technology, Northeast Normal University, Changchun 130000, China
* Correspondence: wangwy546@nenu.edu.cn; Tel.: +86-(0431)84536228

Received: 29 April 2020; Accepted: 25 May 2020; Published: 27 May 2020

Abstract: Collecting correlated scene images and camera poses is an essential step towards learning absolute camera pose regression models. While the acquisition of such data in living environments is relatively easy by following regular roads and paths, it is still a challenging task in constricted industrial environments. This is because industrial objects have varied sizes and inspections are usually carried out with non-constant motions. As a result, regression models are more sensitive to scene images with respect to viewpoints and distances. Motivated by this, we present a simple but efficient camera pose data collection method, WatchPose, to improve the generalization and robustness of camera pose regression models. Specifically, WatchPose tracks nested markers and visualizes viewpoints in an Augmented Reality- (AR) based manner to properly guide users to collect training data from broader camera-object distances and more diverse views around the objects. Experiments show that WatchPose can effectively improve the accuracy of existing camera pose regression models compared to the traditional data acquisition method. We also introduce a new dataset, Industrial10, to encourage the community to adapt camera pose regression methods for more complex environments.

Keywords: data acquisition; augmented reality; pose estimation; deep learning; industrial environments

---

1. Introduction

Camera pose (location and orientation) estimation is a fundamental task in Simultaneous Localization and Mapping (SLAM) and Augmented Reality (AR) applications [1–3]. Recently, end-to-end approaches based on convolutional neural networks have become popular [4–13]. Instead of using machine learning for only specific parts of the estimation pipeline [14–17], these methods aim to learn the full pipeline with a set of training images and their corresponding poses. Building on that, the trained models directly regress the camera pose from an input image. Several works [4,7] in the literature report that those methods are plausible in regular living environments (e.g., along the street or path), achieving around a 9~25 m and 4~17° accuracy in localization and orientation, respectively.

However, the utility of such methodology is limited in industrial environments. This is because people in such environments usually exhibit two kinds of typical motions: (1) They inspect industrial objects with non-constant motions and views and (2) they inspect industrial objects from different distances. For the first case, traditional data collection methods cannot cover enough viewpoints to properly train a generalized pose regression model. For example, in [5] a smartphone was used by a pedestrian to take videos around each scene while in [7,18], robot cars were designed to take pictures along the street. Meanwhile, viewpoints in industrial environments may be restricted to only random collection and limited moving directions. Figure 1a visualizes the viewpoints of the video collection method and it is obvious that most regions around the industrial object are uncovered. As a result, the trained models easily over-fit to such limited training data and may not generalize well to uncovered scenes. For the second case, a simple idea is to apply data augmentation techniques such as image zooming to imitate different camera-object distances (hereinafter is referred to as "camera distance"). However, over-zooming could reduce the quality of training data and decay the regression accuracy [19]. To solve these problems, one possible way is to reconstruct a 3D model of the target object and then generate training data via rendering [20–22]. Nevertheless, the problem we tackle here involves more complex scenarios: 3D object models are not always available in industrial environments and in many cases, they are hard to construct because of the presence of textureless and specular surfaces under sharp artificial lights [1]. Figure 1b shows an example of an industrial object and its reconstructed 3D model using Kinect Fusion [23,24]. We can easily observe the missing parts (the black holes) in the model. In fact, as shown in Figure 2, industrial environments are typically inundated with such textureless and occluded objects and specular surfaces, etc.

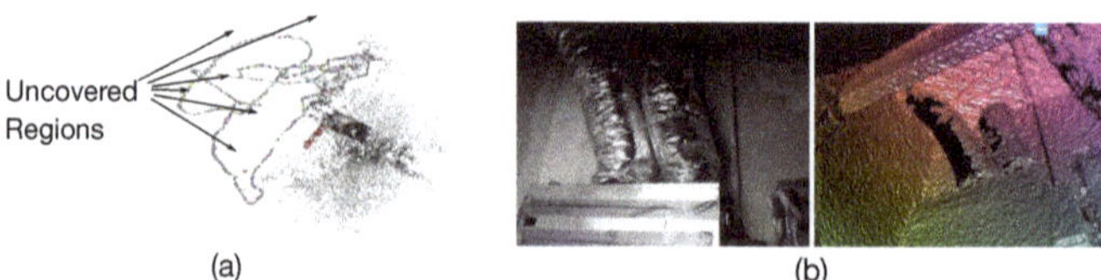

Figure 1. Challenges of collecting training data in industrial environments. (**a**) There are some uncovered regions within the training data using traditional video collection [6]. (**b**) There are some missing parts within the reconstructed 3D model on specular surfaces using Kinect Fusion [23,24].

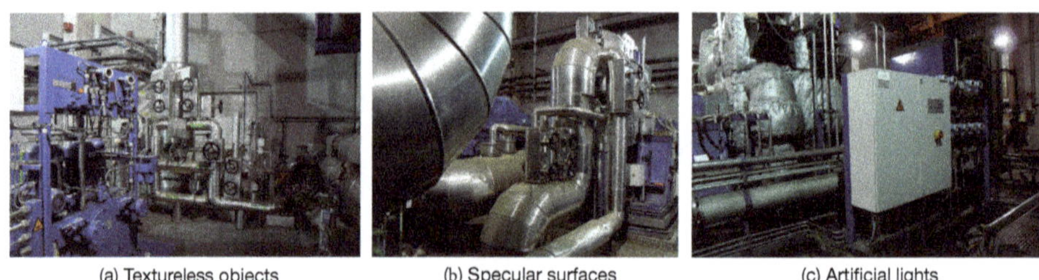

Figure 2. Sample images in industrial environments.

Thus, we seek to find an efficient approach that can collect pose training data from sufficient viewpoints and camera distances. Towards this goal, as shown in Figure 3, a view-aware approach, WatchPose, is introduced. The basic idea of WatchPose is to place a marker close to the target object and the training data is collected through marker tracking [25,26] with dynamic viewpoint control around the object. Here, we introduce three strategies to improve the efficiency of this process: (1) We propose Nestmarker, a combination, and nesting of traditional markers, for marker tracking. The Nestmarker performs detection and tracking from different camera distances since it contains two markers with fixed relative positions and different sizes. Thus, we can flexibly collect training data from a range of larger and smaller camera distances as compared to traditional markers. (2) During the data collection process, virtual imagery is drawn on the marker for checking the correctness of marker tracking (the blue box). (3) A virtual ball is drawn for visualizing the captured viewpoints (the red points) and navigating the uncovered regions. With these strategies, our data collection approach is applied in a real-time setting and it is robust towards variations in objects, surfaces, and lights since the camera poses are directly generated from marker tracking. Moreover, it is easier for us to visually control the camera distances, viewpoint locations, and densities since this approach can navigate the camera controller (e.g., people, robot, etc.) to move the camera dynamically so that most regions around the object can be covered. Once the training data are collected, we apply a set of post-processing steps such as calibrating the camera poses from the marker to the target object, data augmentation, etc.

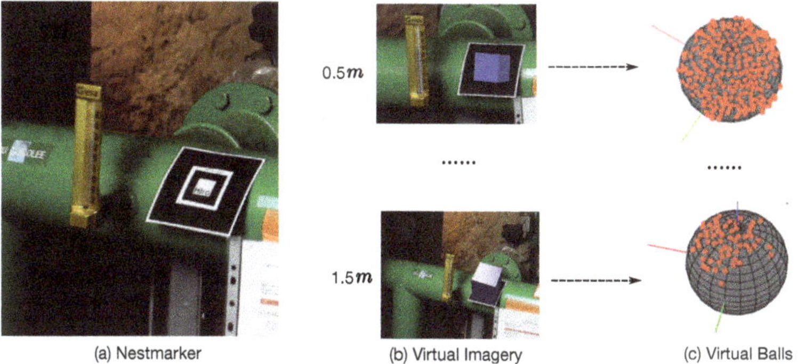

Figure 3. Main idea of WatchPose. (**a**) The Nestmarker is placed close to the target object. (**b**) A virtual imagery is drawn for checking the correctness of marker tracking (the blue box) from different camera distances. (**c**) For each camera distance interval, a virtual ball is drawn and automatically switched for visualizing the captured viewpoints (the red points) and navigating the uncovered regions.

To encourage the community to adapt camera pose regression methods towards more complex domains, we collected a new benchmark dataset, Industrial10, to reflect the challenges of industrial environments. Industrial10 comprises of training and testing data of 10 industrial objects. To assess the efficiency of the proposed method, five actively used pose regression methods are employed in our experiments. Evaluations show that the proposed method can effectively improve the camera pose regression accuracy compared to the traditional data collection method.

The contributions of this work are summarized as follows: (1) We propose a novel training data collection method called WatchPose to improve the performance of camera pose regression in industrial environments. The proposed method is sufficiently general and can be extended to other scenarios. (2) We introduce a new dataset, Industrial10, to spur the computer vision systems community towards innovating and adapting existing camera pose regression approaches to cope with more complex environments.

2. Related Works

Here, we briefly glanced through several existing camera pose data collection strategies followed by a review of camera pose regression methods. For a more detailed treatment of this topic in general, the recent compilation by Shavit and Ferens [27] offers a sufficiently good review.

2.1. Camera Pose Data Collection

In general, camera pose collection methods can be classified into two ways: Direct and indirect approaches. With direct approaches, camera poses are acquired from markers or physical sensors. For example, Brachmann et al. [28] collected the ground-truth camera poses via integrating a set of traditional markers densely surrounding the target object. Tiebe et al. [29] collected the camera poses via a finely-controlled robot arm. In most cases, camera poses are collected indirectly. For instance, the actively used dataset 7Scenes [30] was first recorded from a handheld Kinect RGB-D camera [31]. After that, the ground-truth camera poses were extracted by Kinect Fusion technique [24]. To generate the Cambridge Landmarks dataset, Kendall et al. [5] first captured high definition videos from around each scene using a smartphone and then proceeded to extract ground-truth camera poses using Structure from Motion (SfM) techniques [32]. Alternatively, Wohlhart et al. [20] first reconstructed 3D models of the target objects and then extracted training data (images and their pose annotations) by rendering the 3D models. More recently, the WoodScape [18] dataset were collected from a car mounted with a set of sensors (e.g., Inertial Measurement Unit,GPS, LiDAR, multiple cameras, etc.) along selected streets in USA, Europe, and China. Camera poses of this dataset could be extracted by fusion of LiDAR, IMU, and camera sensors. However, these indirect collection methods cannot be properly applied in industrial environments since industrial objects are normally textureless and may possess specular surfaces under strong artificial lights. For direct methods, it is also not feasible to use traditional markers or robot arms as well. This is because in addition to high costs, there are normally limited spaces around industrial objects and people often inspect them from varying distances. Therefore, we introduce WatchPose, which can better imitate watchers' motion and avoid the aforementioned problems. WatchPose extends the idea of traditional markers [28] to support data collection from both close and far distances. Moreover, it is robust to textureless and specular surfaces since the camera poses are directly collected from Nestmarker. Finally, as more viewpoints are covered in the data collection of WatchPose, the trained models can generalize more robustly.

2.2. Camera Pose Estimation

Leveraging on the idea of transfer learning, more and more researchers attempt to learn models for pose estimation tasks. Generally, these methods function by training descriptors, classifiers, and regressors using a variety of ways. For descriptors, Gu et al. [33] built discriminative models by training a mixture of HOG templates, while Aubry et al. [34] employed them for 3D object detection and pose estimation. In contrast to mixed descriptors, Masci et al. [35] trained a single-layer neural network with different hashing approaches to compute discriminative descriptors for omnidirectional image matching. Wohlhart et al. [20] extend this idea further and train a LeNet [36] to compute features for rendered object views that capture both the object identity and camera poses. Though these approaches can efficiently handle a large number of objects under a large range of poses, they highly rely on handcrafted representations [33,34] and 3D object models [20–22]. For classifiers, Schwarz et al. [22] computed image features by a pre-trained Convolutional Neural Network (CNN) and then fed them to a Support Vector Machine (SVM) to determine object class, instance, and pose. Brachmann et al. [28,37] employed image features from [30] to train a random forest for object detection, tracking, and pose estimation. Those approaches achieved promising performance in cluttered scenes, but they normally require additional reconstruction steps to generate dense scenes and object models. For regressions, Shotton et al. [30] introduced a regression forest that is capable of inferring an estimate of each pixel's correspondence from a given image to 3D points in the scene's world coordinate frame. With this, the computation of feature descriptors are not required. Unlike their approach, Gupta et al. [38] proposed a 3-layer convolutional neural network to regress coarse poses using detected and segmented object instances. However, these algorithms are constrained by the use of RGB-D images to generate the scene coordinate label, which in practice limits its use in industrial environments. To improve it, Kendall et al. [5,6] proposed PoseNet that directly regresses the camera pose from RGB images. However, it easily overfitted with its training data while its localization error on indoor and outdoor datasets was an order of magnitude larger. Such limitations motivated a surge of absolute pose estimation methods such as Bayesian PoseNet [39], MapNet [7], LSTM-Pose [40], Hourglass-Pose [8], SVS-Pose [9], BranchNet [10], and NNnet [11] to involve deeper encoders, better loss functions, and more sensor data. However, as discussed in [4], those pose regression models is more closely related to pose approximation via image retrieval. In other words, the training data should cover viewpoints and camera-distances as much as possible. Thus, we try to improve the performance of pose regression models by feeding better data. Experiments in Section 4 showed that performance improvement with better training data is more obvious, and particularly suitable for industrial type of objects.

3. WatchPose

WatchPose employs a Nestmarker-based strategy to generate training images and their camera pose labels built on AR frameworks [1]. The reason is that AR frameworks have powerful libraries for robust and real-time marker tracking [25]. In other words, we can directly collect accurate camera poses during the marker tracking process. The proposed pose generation strategy is general and can be built on top of most existing pose training and estimation measures.

3.1. Nestmarker

Inspired by [41], Nestmarker is a combination and nesting of traditional markers (Figure 4). In particular, the inner 50% of the Nestmarker is a traditional 40 mm square marker plus a 10 mm white square gap. We call this inside marker as the small marker. Using it as a pattern, we add a continuous border to make a 100 mm square marker (big marker). In other words, a Nestmarker contains two square markers: A small marker and a big marker. Hence, the proposed Nestmarker has three distinct features: (1) Since patterns of small and big markers are rotationally asymmetric, both markers can be trained and tracked independently. (2) By marker tracking, the collected camera poses from small and big markers are correlated since their relative positions are fixed and their sizes are known. (3) The big marker can be tracked from larger camera distances than the small marker. Based on our preliminary experiment using a 1080P HD Webcam and 1024×768 frame resolution, the small and big markers could be detected between 5 cm to 80 cm and 13 cm to 300 cm camera distances, respectively. With these features, we developed a Nestmarker tracking system as illustrated in Figure 4c. Particularly, the big marker is triggered and tracked (e.g., red box) when the small marker cannot be detected due to big camera distances. Otherwise, the small marker is triggered and tracked (e.g., blue box). In both cases, the square marker, from the black continuous border to the inner part, is tracked.

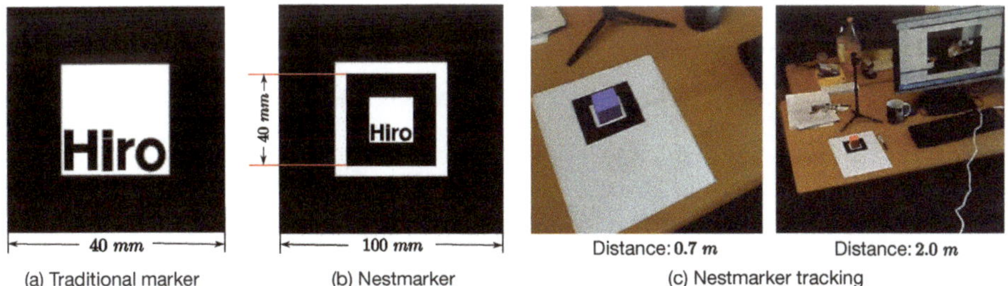

Figure 4. The Nestmarker and its tracking system.

3.2. Data Collection

In our work, the ARToolKit [25] library is employed for Nestmarker learning, tracking, and camera pose collection. Figure 5 shows an overview of our strategy. Firstly, the small and big marker patterns (Figure 5a,b) are independently learned by our system thereby they can be detected in the next steps. After that, the Nestmarker is placed close to an object (Figure 5c) and is tracked by a camera around the object. During tracking (the second row in Figure 5), a virtual image (blue box) is synchronously drawn above the marker so that we can visually check whether the marker has been detected. If the marker is correctly detected from the camera, a viewpoint (shown with red point) is simultaneously plotted on the virtual ball and the current image and the real camera pose are saved. In some cases, the collected image could be badly blurred due to fast camera motion or out-of-focus condition. For this, we quantitatively evaluate its blur effect based on the method in [42] and discard the image and its camera pose if its blur value exceeds a predefined threshold ϱ, which is set to 0.5 in our experiments.

During the capturing process, the virtual ball is automatically switched and rolled depending on the camera position and orientation. This strategy can effectively navigate the camera controller by determining the direction that the camera should be moved in order to capture training data from uncovered regions and distances. The pose computed using our algorithm is expressed in the Marker Coordinate System (MCS). The labels and/or added graphics must be expressed in this frame. To assist in this placement of the virtual scene in the MCS, reference points representing the object may be added in that frame. To do this, two views of the object with the marker are all that is needed. Using the marker, the poses for both views can be calculated and the 3D reference points can be reconstructed in the MCS by triangulation [43]. Pairing the points between the two views can be done manually if a few reference points are sufficient, or with the help of a Scale-invariant Feature Transform (SIFT)-like descriptor [44] if a denser point cloud is desired.

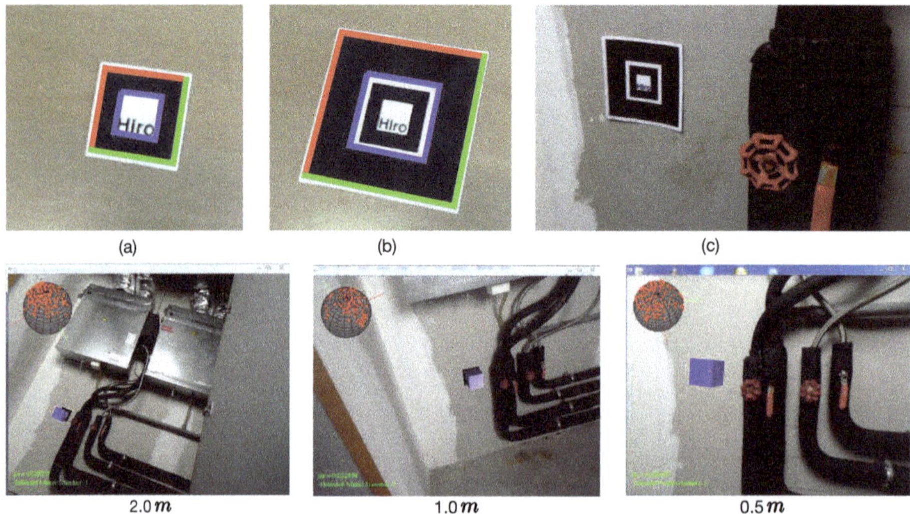

(a)   (b)   (c)

2.0 *m*   1.0 *m*   0.5 *m*

Figure 5. Camera pose training data collection using WatchPose.

In order to properly plot viewpoints on the virtual ball, as shown in Figure 6a, a viewpoint (red point) is mapped to the virtual ball surface (arc) from a real camera position (triangle) with respect to the ball center point (black point). In our case, the real 3D camera position $\mathbf{x}$ is calculated by $\mathbf{x} = -R^{\mathsf{T}} * T$, where R^{T} is a 3×3 camera rotation matrix and T is a 1×3 camera translation vector that are both directly generated in our system based on the ARToolKit library.

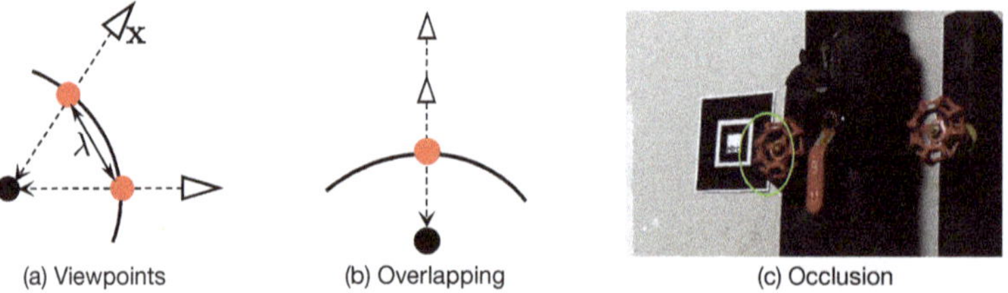

(a) Viewpoints   (b) Overlapping   (c) Occlusion

Figure 6. Viewpoint calculation and density control.

As shown in Figure 6a, we introduce a threshold λ to control the minimal viewpoint distance. With this, we can dynamically control the viewpoint distribution density by changing the λ value. In other words, if the current viewpoint is too close to a plotted one based on the threshold λ (i.e., $< \lambda$), the system will discard the current camera pose. It should be noted that there is a high probability that viewpoints from different camera positions could overlap on the virtual ball (Figure 6b) if they have the same orientation and their locations are on the same line with respect to the object. As such, we use multiple virtual balls to visualize and control the viewpoint distribution based on real camera distance intervals (Figure 5). In this work, we predefine 3 camera distances: 0.5 m, 1.0 m, and 2.0 m to control and visualize viewpoint distributions separately. For the overlapping and uncovered viewpoints within other distances (e.g., 0.3 m, 1.3 m, etc.), we introduce a data augmentation strategy, described in Section 3.3, to enrich the collected data. There is another possibility that the Nestmarker may be occluded by the object under certain viewpoints thereby it cannot be properly detected (Figure 6c). In this case, the small marker will be triggered if the big marker is occluded. If both markers are occluded, we can place multiple markers around the object in advance depending on scenarios. In our case, we only place another small marker close to the object to avoid occlusion since this phenomenon normally appears when the camera is close to an object.

In practise, some industrial objects may have limited space (e.g., in the corner or occluded) for data collection. In such a case, as an example shown in Figure 7a, a flexible Nestmarker is proposed in which the black continuous border and the nested small marker are detachable. In particular, we first capture training data from large camera distances using the large marker (Figure 7b). After that, the black continuous border is removed and only the small marker is kept for capturing training data from smaller camera distances (such as 5 cm to 80 cm in our case) (Figure 7c). Using this strategy, the proportion of the marker in an image can remain small.

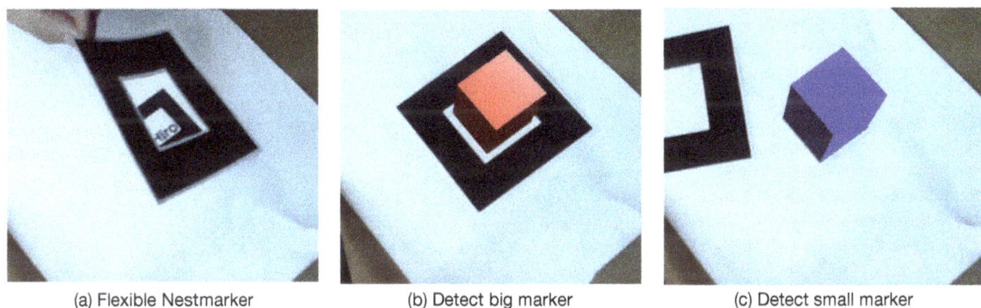

(a) Flexible Nestmarker (b) Detect big marker (c) Detect small marker

Figure 7. Flexible use of Nestmarker in narrow environments.

3.3. Parametrization and Augmentation

For a collected camera pose $\mathbf{p}$, we parameterize it with a 7-dimensional pose vector $\mathbf{p} = [\mathbf{x}, \mathbf{q}]$, where $\mathbf{x}$ contains 3 values representing the camera position while quaternion $\mathbf{q}$ contains 4 values representing the camera orientation. Here, the quaternion $\mathbf{q}$ can be directly converted from the camera rotation matrix R. Consequently, for each collected image I, a 7-dimensional $\mathbf{p}$ is constructed to represent the camera pose of I. We also apply an inpainting process using Coherency Sensitive Hashing (CSH) [45] to remove the marker from each image. The main reason is that in practice there is a time distance between learning and testing ages. It is quite normal to have a marker during the learning age and to remove it during the application time. CSH relies on hashing to seed the initial local matching and then on image coherence to propagate good matches. As a marker location is automatically detected and saved by ARToolKit, CSH can quickly find matched parts in the marker neighbourhood in the image plane and thereby replace the marker region. Moreover, to reduce the inpainting error, markers can be placed in the place with a homogeneous color or texture. This is not challenging in industrial environments since most walls or surfaces are painted with monotonous colors.

After inpainting, we apply a data augmentation process to imitate different camera locations and orientations (Figure 8). We enrich each collected image by employing rotation and slight zooming (i.e., scaling along the camera axis). It should be noted that the number of times zooming is applied within a particular camera distance is highly dependent on the scenario. For instance, our experiment in Section 4.3.2 suggests that augmenting four times within a camera distance is enough to achieve promising pose estimation accuracy. Thus, if the training data are densely captured (with small distance intervals), the number of times augmentation is needed can be reduced or even cancelled altogether. To properly incorporate viewpoint density control and augmentation, we suggest that the viewpoint density λ (Figure 6a) should be reduced when the camera is closer to the object so as to avoid redundant images resulting from augmentation.

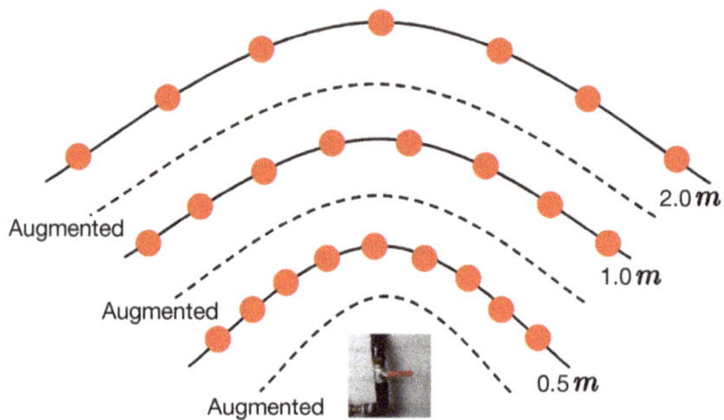

Figure 8. Camera pose augmentation.

4. Experiments

In this section, we first introduce the proposed Industrial10 dataset and its properties. Built on that, we evaluate the WatchPose and widely used pose regression approaches.

4.1. Industrial10

To the best of our knowledge, there is no existing dataset specially designed for evaluating camera pose estimation in industrial environments. For this, we collected a dataset containing 10 industrial objects using our proposed WatchPose approach. As shown in Figure 9, the main purpose of selecting these objects is to reflect the real challenges from industrial environments. For example, Object1 and Object10 have similar appearances under certain viewpoints that may confuse object detection algorithms. Object6 has limited training data since it is located in a narrow environment. Since Object9 is relatively big, only part of its appearance was captured with small camera distances. Object7 is occluded by a green pipe so its appearance could look different depending on camera pose. For each object, we collected the training data from three camera distances: 0.5, 1.0, and 2.0 m (Figure 8) using the proposed WatchPose approach. Figure 10 presents an example depicting the viewpoint distributions of Object1 with different camera distances. We can observe that most regions around Object1 are covered. The number of original training data for the 10 objects is also shown in Figure 9 after their respective names. For testing, each object has a set of 200 testing images that are randomly collected within 0.5 to 2.0 m camera distances. The Industrial10 dataset is publicly available to the community (Please check congyang.de for more details).

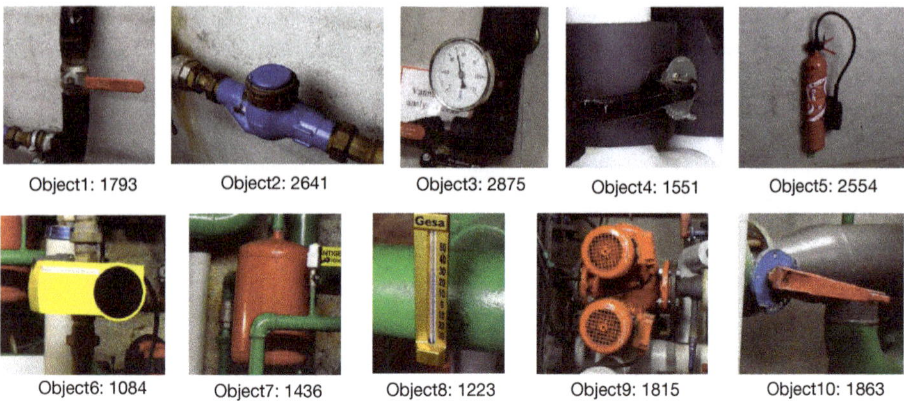

Figure 9. Sample images of 10 objects in industrial environments.

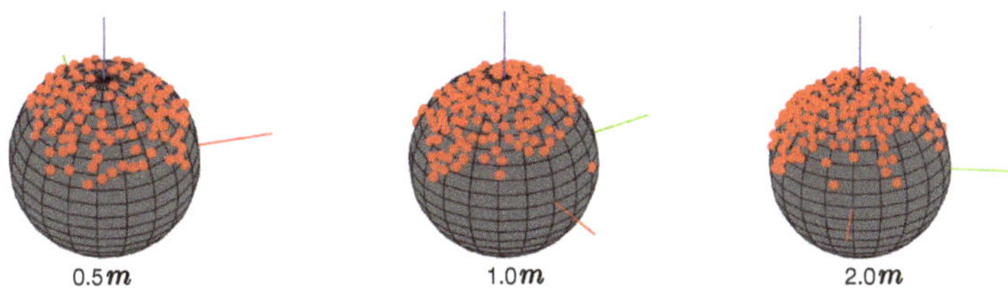

Figure 10. Viewpoints of Object1 in different camera distances.

In addition to the Industrial10 training data collected with WatchPose (named Industrial10-WatchPose), we also extracted another set of training data, named Industrial10-Traditional, using the traditional data acquisition approach. Specifically, Industrial10-Traditional was collected by capturing a high definition video around each object from camera distances 0.5, 1.0, and 2.0 m. In total, 3×10 videos were recorded and each video was then sub-sampled at 2Hz to generate its frames. These images are then inputted to the SfM pipeline to extract the camera poses. For fair comparison, two post-processing steps were followed: (1) We converted the camera poses from SfM to the same format as of Industrial10-WatchPose based on the pre-measured datum line. (2) We uniformly selected the same number of training images according to the training data distribution in Industrial10-WatchPose. For example, Object1 has 1793 training images in both Industrial10-WatchPose and Industrial10-Traditional datasets.

4.2. Target Object Detection

Built on the target object detector, the pre-trained pose regression models can be easily switched based on the detected object. For this, we train a Faster R-CNN [46] using the additional collected and annotated images. For each object, there are 1000 and 300 images used for training and verification, respectively. For target object annotation, the popular annotation tool LabelMe [47] is used. We did not train the detector from scratch since the industrial objects are mostly rigid and textureless, and they resemble some objects from ImageNet. For this, the pretrained ImageNet model with the lightweight ResNet18 [48] backbone network was selected, while the other parameter settings introduced in [46] were kept. After 20 epochs, some detection results using the trained model and testing data are presented in Figure 11. We evaluated the detector's Average Precision (AP) on the testing images of each object and the results are presented in Table 1. We also calculated the mean Average Precision (mAP) by averaging the detection APs across the 10 objects. The mAP was around 99.98%, which is comparatively higher than the reported results in [46]. This is because most objects' surroundings in industrial environments are much simpler than images taken from natural environments such as in PASCAL VOC [49] and MS COCO [50]. Moreover, objects are more distinguishable from their backgrounds in industrial environments since they normally have different paint surfaces and specular effects. Considering the number of training images and the performance we achieved, the training and testing images for pose regression could also be annotated for training the object detector in practice.

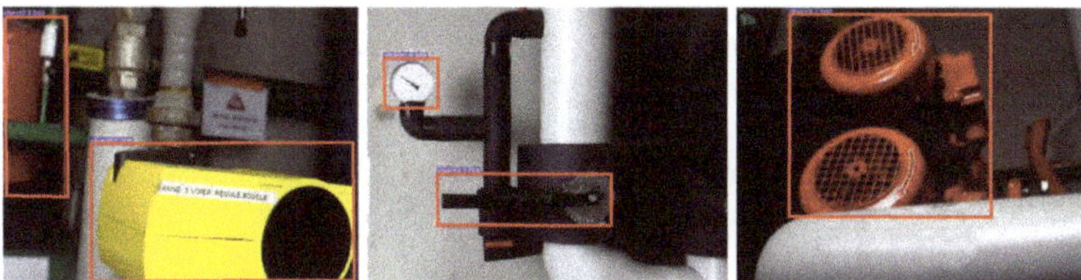

Figure 11. Object detection results (shown in red rectangles) using the fine-tuned Faster R-CNN [46].

Table 1. Average precision of detecting object-of-interest on the testing images.

Object	Average Precision	Object	Average Precision
Object1	95.0%	Object6	97%
Object2	99.2%	Object7	99.6%
Object3	100%	Object8	100%
Object4	99.5%	Object9	100%
Object5	100%	Object10	99.5%

4.3. Ablation Studies of WatchPose

To fully assess the proposed Watchpose scheme and its properties, we performed a set of experiments by fixing the pose regressor to PoseNet [5]. Specifically, we first evaluate and compare the trained PoseNet using original images from both Industrial10-Traditional and Industrial10-WatchPose. The main purpose is to show the performance improvement using our proposed method against the traditional method of collection. We also evaluated the usability of our proposed data augmentation strategy on both datasets. It should be noted that the β value in PoseNet's loss function should be calibrated based on the scenario. As reported in [6], different β brings considerable performance gaps. Based on the preliminary experiments and discussion in [5], β is mainly influenced by the camera location unit and the scene type (i.e., indoor and outdoor). Since our locations are labeled in millimeter (mm) scale, as compared to the original PoseNet experiments in [5], the computed Euclidean loss values are normally much bigger than the one from orientation. Thus, we systematically searched for appropriate β values by ascending order. Figure 12 illustrates some pose estimation results with different β values using a subset of Industrial10-WatchPose training and testing data (for ease of experimentation). As mentioned in [6], the balanced choice of β must be struck between the orientation and translational errors, both of which are highly coupled as they are regressed from the same model weights. Therefore, we considered both location and orientation estimations and finally selected $\beta = 40000$ for the experiments below.

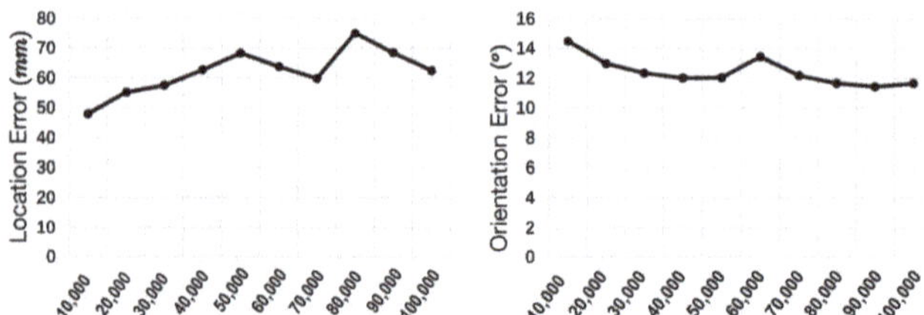

Figure 12. Estimated camera location (mm) and orientation (°) errors using different β values (x axis). $\beta = 40,000$ was finally selected for our experiments.

4.3.1. Original Images

Here, we compare the pose regression performance using the original training data from Industrial10-WatchPose and Industrial10-Traditional. In other words, the images were not augmented. The mean results of each object are detailed in Table 2. We can clearly observe that our proposed WatchPose scheme significantly improved the pose regression performance in this challenging dataset by around 4.7 times for location error and 3.9 times for orientation error, compared to the traditional data collection approaches [5,7].

Based on Industrial10-WatchPose, we also experimentally verified the necessity and usability of image inpainting for pose regression in our scenario. Sample images before and after inpainting are shown in Figure 13a,c and b,d, respectively. Particularly, we applied experiments with cross combinations of with-inpainting and without-inpainting for training and testing. Table 3 presents the mean pose estimation results among 10 objects. We can clearly find that with-inpainting training and testing as well as without-inpainting training and testing were similar to each other (around 60~70 mm location error and 20° orientation error), but both apparently outperformed the other combinations (more than 150 mm location error and 5° orientation error). As in practice the markers were removed after pose generation as it is necessary to apply the inpainting process on training images to meet a promising performance. The followed experiments were all performed based on marker inpainted training and testing images.

Table 2. Estimated camera location (mm) and orientation (°) errors of objects using traditional and WatchPose data collection methods.

Objects	Traditional [5,7]		WatchPose		Objects	Traditional [5,7]		WatchPose	
Object1	368.2857 mm	77.0220°	79.1773 mm	19.7309°	Object6	391.6417 mm	79.5715°	82.9850 mm	17.1254°
Object2	350.5098 mm	63.8454°	81.3936 mm	20.2622°	Object7	291.6453 mm	81.2755°	61.2231 mm	23.2668°
Object3	381.9421 mm	80.2673°	84.8384 mm	21.4031°	Object8	237.5793 mm	59.0616°	50.4712 mm	16.5201°
Object4	341.1610 mm	83.6164°	69.9175 mm	20.8298°	Object9	280.1659 mm	81.4738°	56.7338 mm	18.3799°
Object5	355.6238 mm	86.3258°	73.7141 mm	22.0913°	Object10	278.5611 mm	77.3197°	58.4822 mm	19.9535°

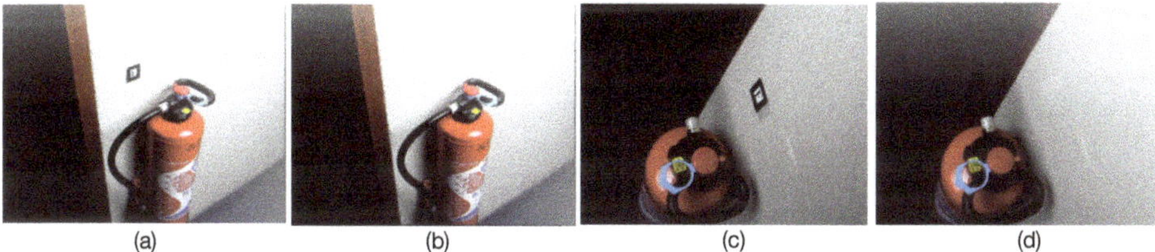

(a)   (b)   (c)   (d)

Figure 13. Sample images before (**a**,**c**) and after inpainting (**b**,**d**).

Table 3. Pose estimation using with-marker/no-marker data for training/testing.

Train Types	Testing Set: With-Inpainting		Testing Set: Without-Inpainting	
Training set: with-inpainting	69.8936 mm	19.9563°	210.5763 mm	25.1174°
Training set: without-inpainting	360.1666 mm	31.7290°	60.6378 mm	20.5549°

4.3.2. Augmentation

In this section, we experimentally explore the efficacy of the data augmentation strategy introduced in Section 3.3 on the Industrial10-WatchPose and Industrial10-Traditional datasets. Following the rotation augmentation introduced in existing reports [6,39], all images were first rotated 4 times. After that, we empirically enriched each training data 4 times within each camera distance interval. For instance, augmenting 4 times within the 0.5 m~1.0 m camera distance interval meant that the camera distances 0.6, 0.7, 0.8, and 0.9 m were imitated by performing zooming-in and center cropping from the 1.0 m image. Evaluations were done on the same testing data as in Section 4.1 and the result are detailed in Table 4.

We can observe that the improvement of pose regression accuracy using augmented Industrial10-Traditional was limited, around 4%. In contrast, the pose regression accuracy improved around 5.3 times in location and 4.2 times in orientation using the augmented Industrial10-WatchPose. The main reason was that the uncovered regions by the traditional approach were still not properly covered after augmentation. As a result, the performance gap between the traditional and proposed WatchPose approaches was enlarged after data augmentation. Specifically, we achieved errors of around 13.3 mm for location and 4.7° for orientation using Industrial10-WatchPose after augmentation, which is around 19 times better than Industrial10-Traditional.

In Figure 14, we plot the pose estimation results in each iteration during the training process using augmented Industrial10-WatchPose. Horizontal axis represents the iteration numbers and vertical axis represents the estimated median performance with respect to camera location and orientation. To optimise the visualisation effect, results from 5 objects are plotted together in each sub-figure. We can clearly observe that both location and orientation errors dramatically dropped after 10,000 training iterations and then stabilized after around 15,000 training iterations.

Table 4. Estimated camera location (mm) and orientation (°) errors of each object using models trained on the augmented Industrial10-WatchPose and Industrial10-Traditional.

Objects	Traditional		WatchPose		Objects	Traditional		WatchPose	
Object1	321.4866 mm	69.1536°	16.6907 mm	4.6411°	Object6	389.0706 mm	76.0915°	7.3524 mm	3.9631°
Object2	311.8624 mm	60.1966°	16.5327 mm	5.6037°	Object7	280.2132 mm	79.5342°	12.4772 mm	5.5193°
Object3	360.3298 mm	76.3290°	19.7768 mm	5.5316°	Object8	229.0576 mm	51.5531°	7.9837 mm	3.0772°
Object4	339.3160 mm	80.3916°	15.6313 mm	5.0524°	Object9	266.6300 mm	75.0026°	10.7575 mm	4.3059°
Object5	354.3400 mm	81.3303°	15.3658 mm	5.3537°	Object10	271.8297 mm	72.1473°	10.4555 mm	4.6450°

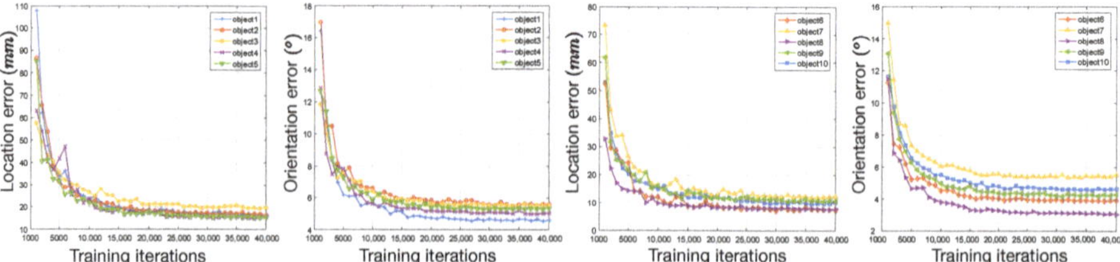

Figure 14. Estimated camera location and orientation errors of 10 objects in each training iteration.

In addition to the quantitative comparison, we also compare the pose estimation results based on the marker-reprojection approach (some samples shown in Figure 15) so that we can visually observe the differences in performance. For better visualization, we employed the original images before marker inpaining and transferred the relative camera pose of the object back to that of the Nestmarker. If an estimated pose is closer to the ground truth (blue box), the reprojected marker (green box) covers the physical location more accurately. Promising reprojections shown in the lower row of images in Figure 15 demonstrate the robustness of WatchPose. As shown in the upper row, we find that the estimated poses trained with Industrial10-Traditional performed less satisfactorily. The main reason is that the Industrial10-Traditional did not cover enough regions around the target objects thereby the trained PoseNet model could not generalize well on the testing data from uncovered regions.

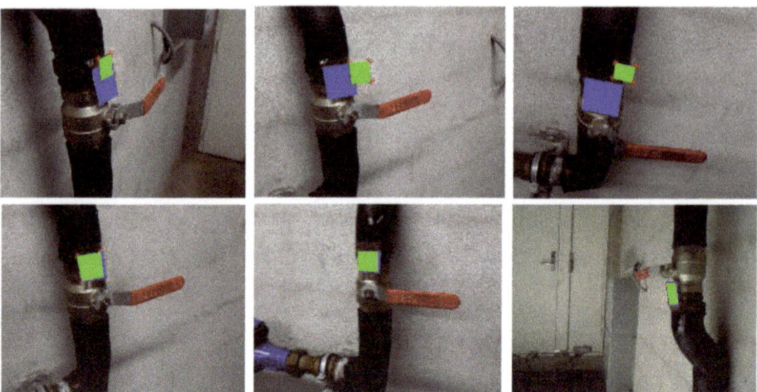

Figure 15. Comparison of marker reprojections (green) between the Industrial10-Traditional (upper) and Industrial10-WatchPose (lower) trained PoseNet. The ground truth is marked in blue.

4.3.3. Dense Control

In this section, we quantitatively evaluate the influence of viewpoint dense control parameter λ in Section 3.3 to the final pose regression performance. To this end, we employ Object1 ($\lambda = 0.2$) in the Industrial10-WatchPose dataset for the experiment. We also fix other factors such as β in PoseNet, the augmentation strategy, and the testing data. With the increase of λ value (from 0 to 1), original viewpoints of Object1 were proportionally and randomly selected from the virtual ball to imitate the data collection from dense to sparse. For $\lambda = 0$, 20% of the original viewpoints were randomly selected and duplicated to imitate the overlapped views. The corresponding viewpoints of original images were selected (or duplicated) to generate different training sets. Table 5 presents the pose estimation results with different λ values (the original number of training images before augmentation is also provided for reference). We observed that the trained model performed less and less satisfactorily when λ increased. This is expected since more and more regions were not covered by the training data. We also found that both orientation and location errors were surprisingly slightly higher at $\lambda = 0$, compared to the performance at $\lambda = 0.2$. The main reason is that in some regions the viewpoints were densely distributed and redundant with $\lambda = 0$. As a result, the trained model was slightly overfitted to these regions.

Table 5. Estimated camera location (mm) and orientation (°) errors of Object1 in using training data with different viewpoint density controlled by λ.

λ	Original Images	Orientation Error (mm)	Location Error (°)
0	2367	16.7402	4.6613
0.2	1793	16.6907	4.6411
0.4	1434	29.989	7.2499
0.6	1075	103.663	26.8137
0.8	717	212.5046	47.2115
1	358	325.8071	72.7677

Based on the evaluation in Table 5, we can observe that λ is important to balance the coverage and redundancy of viewpoints. In practice, λ is configured empirically depending on the target object size and the collection conditions. If none of them can be determined in advance, $\lambda = 0$ is a reasonably acceptable choice of value to ensure the sufficient coverage of different viewpoints, though redundant training images could be generated with this value.

4.4. Deep Absolute Pose Estimators

Using the augmented Industrial10-Traditional and Industrial10-WatchPose datasets, we compare the pose regression performance of five existing approaches. Specifically, Bayesian-PoseNet [39] was released by the same authors of the PoseNet approach. Bayesian-PoseNet first generates some samples by dropping out the activation units of convolutional layers of PoseNet based on a probability value. The final pose is then computed by averaging over the individual samples' predictions. Meanwhile, PoseNet+ [6] introduced a loss with learned uncertainty parameters (learnable weights pose loss) for optimizing PoseNet. Hourglass-Pose [8] focused on optimizing the architecture of PoseNet by suggesting an encoder-decoder architecture implemented with a ResNet34 [48] encoder (removing the average pooling and softmax layers). In contrast to the aforementioned approaches, a more recent method called MapNet [7] suggested to include additional data sources in order to constrain the loss. It is trained with both absolute and relative ground truth data.

On the Industrial10-Traditional dataset, we analyzed the estimation performance of each object category so that deeper observations could be made. The final results are detailed in Table 6. We found that the performances varied among different objects. We also observed that the improved optimization of loss in PoseNet+ and the architecture design in Hourglass-Pose both improved PoseNet's accuracy by around 2 times, particularly in the location errors which were dramatically reduced. Overall, MapNet achieved the best performances, around 3 times better in location and 2 times better in orientation, compared to the poses estimated by the original PoseNet.

Table 6. Estimated camera location (mm) and orientation (°) errors of different pose regression approaches using the post-processed Industrial10-WatchPose dataset.

Traditional	PoseNet [5]		Bayesian-PoseNet [39]		Hourglass-Pose [8]		PoseNet+ [6]		MapNet [7]	
Object1	16.691 mm	4.641°	17.717 mm	3.223°	7.132 mm	2.952°	6.551 mm	2.338°	5.259 mm	2.100°
Object2	16.533 mm	5.604°	15.505 mm	4.213°	11.040 mm	4.041°	11.608 mm	4.084°	9.646 mm	3.568°
Object3	19.777 mm	5.532°	18.740 mm	4.102°	12.183 mm	4.095°	11.953 mm	3.843°	10.676 mm	3.288°
Object4	15.631 mm	5.052°	17.946 mm	3.986°	9.635 mm	3.541°	8.325 mm	3.476°	9.466 mm	3.096°
Object5	15.366 mm	5.354°	14.364 mm	3.585°	10.865 mm	3.193°	12.225 mm	2.778°	9.105 mm	2.405°
Object6	7.352 mm	3.963°	7.311 mm	2.475°	5.803 mm	2.087°	4.285 mm	2.058°	4.882 mm	1.988°
Object7	12.477 mm	5.519°	13.307 mm	3.985°	7.523 mm	3.371°	6.950 mm	3.112°	5.973 mm	3.020°
Object8	7.984 mm	3.077°	8.154 mm	2.603°	5.419 mm	2.677°	5.727 mm	2.539°	3.785 mm	2.395°
Object9	10.758 mm	4.306°	10.587 mm	3.369°	6.276 mm	2.923°	5.785 mm	2.716°	4.308 mm	2.467°
Object10	10.456 mm	4.645°	11.988 mm	3.555°	6.405 mm	3.114°	6.620 mm	2.996°	4.685 mm	2.411°
Mean	13.303 mm	4.769°	13.562 mm	3.510°	8.228 mm	3.199°	8.003 mm	2.99°	6.779 mm	2.674°

In contrast to Table 6, the estimation errors in Table 7 are smaller among 5 approaches using the Industrial10-WatchPose dataset. This is because the WatchPose data covers more camera poses thereby leading to better generalization in pose regression models. In particular, while the Bayesian-PoseNet only marginally improved the pose regression accuracy (over PoseNet), the Hourglass-Pose and PoseNet+ approaches achieved around 14% improvement in location and 4% in orientation estimation. Once again, MapNet achieved the best performance, with around 226.187 mm in location error and 57.551°in orientation error. It also shows that the proposed WatchPose method of collecting data enabled much higher in both location and orientation performance improvements on MapNet compared to the traditional data collection method. This shows that the proposed protocol is particularly suitable for complex industrial applications as the efficacy of the pose estimation methods are wholly improved.

Table 7. Estimated camera location (mm) and orientation (°) errors of different pose regression approaches using the augmented Industrial10-Traditional dataset.

WatchPose	PoseNet [5]		Bayesian-PoseNet [39]		Hourglass-Pose [8]		PoseNet+ [6]		MapNet [7]	
Object1	321.487 mm	69.154°	315.742 mm	66.544°	295.021 mm	62.792°	284.158 mm	57.688°	260.900 mm	55.758°
Object2	311.862 mm	60.197°	312.967 mm	55.300°	253.023 mm	56.655°	241.538 mm	51.312°	230.533 mm	48.319°
Object3	360.330 mm	76.329°	355.475 mm	71.164°	290.545 mm	70.936°	279.047 mm	72.325°	233.519 mm	64.396°
Object4	339.316 mm	80.392°	341.601 mm	76.356°	295.942 mm	74.763°	280.991 mm	71.527°	239.109 mm	63.830°
Object5	354.340 mm	81.330°	352.981 mm	74.372°	299.807 mm	79.449°	286.456 mm	72.869°	256.954 mm	57.061°
Object6	389.071 mm	76.092°	265.340 mm	72.558°	322.610 mm	73.795°	310.644 mm	69.460°	269.646 mm	56.306°
Object7	280.213 mm	79.534°	282.665 mm	77.425°	261.754 mm	79.773°	250.111 mm	77.497°	233.322 mm	67.701°
Object8	229.058 mm	51.553°	221.250 mm	56.541°	210.119 mm	49.858°	190.307 mm	47.544°	155.271 mm	40.131°
Object9	266.630 mm	75.003°	275.934 mm	75.422°	215.861 mm	71.598°	203.967 mm	69.300°	181.857 mm	57.695°
Object10	271.830 mm	72.147°	268.961 mm	71.184°	234.627 mm	72.644°	222.443 mm	70.313°	200.762 mm	64.317°
Mean	312.414 mm	72.173°	299.292 mm	69.687°	267.931 mm	69.226°	254.966 mm	65.984°	226.187 mm	57.551°

4.5. Discussion on Restrictions

Theoretically, WatchPose is applicable to a scenario when it fulfills the following conditions: (1) Inspections are carried out from within 0.2 to 2 meters of the target object. Otherwise, the Nestmarker cannot be properly tracked. (2) There should be enough space around the target object for data acquisition via a handheld camera. (3) There should be a homogeneous place close to the target object for pasting Nestmarkers. However, there could be more factors that influence the efficiency of WatchPose and the performance of pose regression in the application phase. To explore these factors, we employed the marker reprojection approach (similar to Figure 15) so that the pose estimation performance of MapNet could be visually observed, as shown in Figure 16. We can see that most reprojected markers (in blue) on most objects could cover the ground truth (the original marker) quite accurately. However, there were still some testing images with poor reprojection results, as shown in Figure 16b. These images were mainly affected by specific types of challenging conditions: Under- and over-exposed images, partially blurred regions, incorrectly detected objects, and also lesser training data. This is partly attributed to the lighting conditions of the industrial environment. Other factors such as camera configurations and moving speed, shaking, and camera resolution could also impact the image quality and the regression performance. In practice, a camera with high resolution and fixed exposure time is recommended. During the capturing process, the camera should also be moved as slow as possible. Moreover, some tools such as handheld gimbals [51] could be used to stabilize the acquisition. In Section 5, we also introduce two future works to deal with these challenging problems.

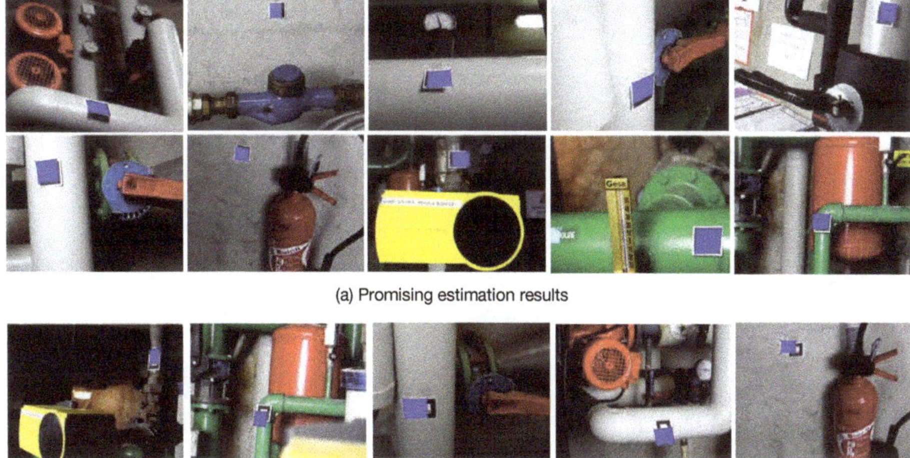

(a) Promising estimation results

(b) Poor estimation results

Figure 16. Marker reprojection results (blue) using the estimated camera poses from MapNet [7]. For better visualization, the original marker is attached in each image.

Our experiments were performed on two platforms: A laptop and a desktop machine. Training images and camera pose labels were processed on a laptop with Intel Core i7 2.9 GHz CPU, 16 GB installed memory and 64-bit Windows 7 OS. An ELP 1080P HD Webcam was connected to the laptop via a 2 m USB-cable for data collection. Model training, object detection, and pose estimation experiments were accomplished on the desktop machine with 6 Intel Xeon Core 3.5 GHz CPUs, 64 GB installed memory, a Quadro M4000 GPU (8 GB global memory and 1664 CUDA Cores), and Ubuntu 14.04 LTS. The pose generation was implemented with C++. The object detection, pose training, and estimation tasks were implemented with Python. For object detection, the entire training process took around 26 hours and the detection (at inference time) took about 0.28 seconds per testing image. For camera pose estimation, each camera pose could be regressed in about 5 to 100 ms depending on the model, which puts the system at real-time speeds.

5. Conclusions

In this paper, we introduced a simple but efficient data collection method for complex industrial environments named WatchPose so as to learn effective absolute camera pose regression models. The proposed method integrated the advantageous properties of marker tracking and viewpoint visualization approaches. The features of WatchPose could properly handle the challenges of industrial-type objects: Textureless and specular surfaces under strong artificial lights, and highly variable distances and views angles. We also proposed two post-processing steps (inpainting and augmentation) to improve the robustness and stability of the trained models. To advance pose estimation research in industrial environments, we introduced a new challenging dataset, Industrial10, to represent the aforementioned challenges of industrial-type objects and environments. Experiments showed that the proposed WatchPose method could effectively improve the pose regression performance of five widely-used approaches.

In the future, we propose two further directions. Firstly, we will collect more data to cover other kinds of challenges in industrial environments. For instance, we can enrich the training data by varying an image brightness [52] to imitate different lighting conditions. Secondly, we will compare the motion tracking performance between ARToolKit and ARCore [53], which can be applied with and without markers, respectively. Finally, we will release further baselines for the Industrial10 dataset using a few other camera pose estimation methods such as SfM [32], 3D Scene [30], etc.

Author Contributions: Conceptualization, C.Y.; Data curation, C.Y.; Funding acquisition, M.-O.B. and W.W.; Methodology, C.Y.; Project administration, G.S.; Software, C.Y.; Supervision, G.S., M.-O.B. and W.W.; Validation, G.S. and J.S.; Writing—original draft, C.Y.; Writing—review & editing, G.S., J.S. and M.-O.B. All authors have read and agreed to the published version of the manuscript.

Funding: This work was funded by the Region Lorraine and Université de Lorraine. Grant Number: 1.87.02.99.216.331/09

Acknowledgments: The authors would like to thank Pierre Rolin, Antoine Fond, Vincent Gaudilliere, and Philippe Vincent from INRIA/LORIA for their invaluable help.

Conflicts of Interest: The authors declare no conflict of interest.

References

1. Fernando, L.; Mariano, F.; Zorzal, E. A survey of industrial augmented reality. *Comput. Ind. Eng.* **2020**, *139*, 106–159.

2. Williams, B.; Klein, G.; Reid, I. Real-time SLAM relocalisation. In Proceedings of the 2007 IEEE 11th International Conference on Computer Vision, Rio de Janeiro, Brazil, 14–21 October 2007; pp. 1–8.

3. Billinghurst, M.; Clark, A.; Lee, G. A survey of augmented reality. *Found. Trends Hum.-Comput. Interact.* **2015**, *8*, 73–272.

4. Sattler, T.; Zhou, Q.; Pollefeys, M.; Leal-Taixe, L. Understanding the Limitations of CNN-based Absolute Camera Pose Regression. In Proceedings of the IEEE Conference on Computer Vision and Pattern Recognition, Long Beach, CA, USA, 16–20 June 2019; pp. 1–11.

5. Kendall, A.; Grimes, M.; Cipolla, R. PoseNet: A convolutional network for real-time 6-dof camera relocalization. In Proceedings of the IEEE International Conference on Computer Vision, Santiago, Chile, 7–13 December 2015; pp. 2938–2946.

6. Kendall, A.; Cipolla, R. Geometric loss functions for camera pose regression with deep learning. In Proceedings of the IEEE Conference on Computer Vision and Pattern Recognition, Honolulu, HI, USA, 21–26 July 2017; pp. 5974–5983.

7. Brahmbhatt, S.; Gu, J.; Kim, K.; Hays, J.; Kautz, J. Geometry-Aware Learning of Maps for Camera Localization. In Proceedings of the IEEE Conference on Computer Vision and Pattern Recognition, Salt Lake City, UT, USA, 18–22 June 2018; pp. 2616–2625.

8. Melekhov, I.; Ylioinas, J.; Kannala, J.; Rahtu, E. Image-based localization using hourglass networks. In Proceedings of the IEEE International Conference on Computer Vision, Venice, Italy, 22–29 October 2017; pp. 879–886.

9. Naseer, T.; Burgard, W. Deep regression for monocular camera-based 6-dof global localization in outdoor environments. In Proceedings of the International Conference on Intelligent Robots and Systems, Vancouver, BC, Canada, 24–28 September 2017; pp. 1525–1530.

10. Wu, J.; Ma, L.; Hu, X. Delving deeper into convolutional neural networks for camera relocalization. In Proceedings of the IEEE International Conference on Robotics and Automation, Singapore, 29 May–3 June 2017; pp. 5644–5651.

11. Laskar, Z.; Melekhov, I.; Kalia, S.; Kannala, J. Camera relocalization by computing pairwise relative poses using convolutional neural network. In Proceedings of the IEEE International Conference on Computer Vision, Venice, Italy, 22–29 October 2017; pp. 929–938.

12. Brachmann, E.; Krull, A.; Nowozin, S.; Shotton, J.; Michel, F.; Gumhold, S.; Rother, C. DSAC-differentiable RANSAC for camera localization. In Proceedings of the IEEE Conference on Computer Vision and Pattern Recognition, Honolulu, HI, USA, 21–26 July 2017; pp. 6684–6692.

13. Brachmann, E.; Rother, C. Learning less is more-6d camera localization via 3d surface regression. In Proceedings of the IEEE Conference on Computer Vision and Pattern Recognition, Salt Lake City, UT, USA, 18–23 June 2018; pp. 4654–4662.

14. He, Y.; Sun, W.; Huang, H.; Liu, J.; Fan, H.; Sun, J. PVN3D: A Deep Point-wise 3D Keypoints Voting Network for 6DoF Pose Estimation. In Proceedings of the IEEE Conference on Computer Vision and Pattern Recognition, Seattle, WA, USA, 13–19 June 2020; pp. 1–10.

15. Moo, Y.K.; Ono, E.T.Y.; Lepetit, V.; Salzmann, M.; Fua, P. Learning to Find Good Correspondences. In Proceedings of the IEEE Conference on Computer Vision and Pattern Recognition, Salt Lake City, UT, USA, 18–23 June 2018; pp. 1–11.

16. Toft, C.; Stenborg, E.; Hammarstrand, L.; Brynte, L.; Pollefeys, M.; Sattler, T.; Kahl, F. Semantic Match Consistency for Long-Term Visual Localization. In Proceedings of the European Conference on Computer Vision, Munich, Germany, 8–14 September 2018; pp. 383–399.

17. Schonberger, J.L.; Pollefeys, M.; Geiger, A.; Sattler, T. Semantic Visual Localization. In Proceedings of the IEEE Conference on Computer Vision and Pattern Recognition, Salt Lake City, UT, USA, 18–23 June 2018; pp. 1–11.

18. Yogamani, S.; Hughes, C.; Horgan, J.; Sistu, G.; Varley, P.; ODea, D.; Uricar, M.; Milz, S.; Simon, M.; Amende, K.; et al. WoodScape: A multi-task, multi-camera fisheye dataset for autonomous driving. In Proceedings of the IEEE International Conference on Computer Vision, Seoul, Korea, 27–28 October 2019; pp. 1–9.

19. Shorten, C.; Khoshgoftaar, T.M. A survey on Image Data Augmentation for Deep Learning. *J. Big Data* **2019**, *6*, 1–60.

20. Wohlhart, P.; Lepetit, V. Learning descriptors for object recognition and 3d pose estimation. In Proceedings of the IEEE Conference on Computer Vision and Pattern Recognition, Boston, MA, USA, 7–12 June 2015; pp. 3109–3118.

21. Cao, Z.; Sheikh, Y.; Banerjee, N.K. Real-time scalable 6DOF pose estimation for textureless objects. In Proceedings of the IEEE International Conference on Robotics and Automation, Stockholm, Sweden, 16–21 May 2016; pp. 2441–2448.

22. Schwarz, M.; Schulz, H.; Behnke, S. RGB-D object recognition and pose estimation based on pre-trained convolutional neural network features. In Proceedings of the IEEE International Conference on Robotics and Automation, Seattle, WA, USA, 26–30 May 2015; pp. 1329–1335.

23. Zollhöfer, M.; Stotko, P.; Görlitz, A.; Theobalt, C.; Nießner, M.; Klein, R.; Kolb, A. State of the Art on 3D Reconstruction with RGB-D Cameras. *Comput. Graph. Forum* **2018**, *37*, 625–652. [CrossRef]

24. Newcombe, R.A.; Izadi, S.; Hilliges, O.; Molyneaux, D.; Kim, D.; Davison, A.J.; Kohli, P.; Shotton, J.; Hodges, S.; Fitzgibbon, A.W. KinectFusion: Real-time dense surface mapping and tracking. In Proceedings of the IEEE/ACM International Symposium on Mixed and Augmented Reality, Basel, Switzerland, 26–29 October 2011; pp. 127–136.

25. Rabbi, I.; Ullah, S.; Rahman, S.U.; Alam, A. Extending the Functionality of ARToolKit to Semi Controlled/Uncontrolled Environment. *Information* **2014**, *17*, 2823–2832.

26. Azuma, R.; Baillot, Y.; Behringer, R.; Feiner, S.; Julier, S.; MacIntyre, B. Recent advances in augmented reality. *IEEE Comput. Graph. Appl.* **2001**, *21*, 34–47. [CrossRef]

27. Shavit, Y.; Ferens, R. Introduction to Camera Pose Estimation with Deep Learning. *arXiv* **2019**, arXiv:1907.05272v3 .

28. Brachmann, E.; Krull, A.; Michel, F.; Gumhold, S.; Shotton, J.; Rother, C. Learning 6d object pose estimation using 3d object coordinates. In Proceedings of the European Conference on Computer Vision, Zurich, Switzerland, 6–12 September 2014; pp. 536–551.

29. Tiebe, O.; Yang, C.; Khan, M.H.; Grzegorzek, M.; Scarpin, D. Stripes-Based Object Matching. In *Computer and Information Science*; Springer: Cham, Switzerland, 2016; pp. 59–72.

30. Shotton, J.; Glocker, B. Scene coordinate regression forests for camera relocalization in RGB-D images. In Proceedings of the IEEE Conference on Computer Vision and Pattern Recognition, Portland, OR, USA, 23–28 June 2013; pp. 2930–2937.

31. Bui, M.; Baur, C.; Navab, N.; Ilic, S.; Albarqouni, S. Adversarial Networks for Camera Pose Regression and Refinement. In Proceedings of the IEEE International Conference on Computer Vision Workshops, Seoul, Korea, 27–28 October 2019; pp. 1–9.

32. Wu, C. Towards Linear-Time Incremental Structure from Motion. In Proceedings of the International Conference on 3D Vision, Seattle, WA, USA, 29 June–1 July 2013; pp. 127–134.

33. Gu, C.; Ren, X. Discriminative Mixture-of-templates for Viewpoint Classification. In Proceedings of the European Conference on Computer Vision, Crete, Greece, 5–11 September 2010; pp. 408–421.

34. Aubry, M.; Maturana, D.; Efros, A.A.; Russell, B.C.; Sivic, J. Seeing 3d chairs: Exemplar part-based 2d-3d alignment using a large dataset of cad models. In Proceedings of the IEEE Conference on Computer Vision and Pattern Recognition, Columbus, OH, USA, 23–28 June 2014; pp. 3762–3769.

35. Masci, J.; Migliore, D.; Bronstein, M.M.; Schmidhuber, J. Descriptor Learning for Omnidirectional Image Matching. In *Registration and Recognition in Images and Videos*; Springer: Berlin/Heidelberg, Germany, 2014; pp. 49–62.

36. LeCun, Y.; Bottou, L.; Bengio, Y.; Haffner, P. Gradient-based learning applied to document recognition. *Proc. IEEE* **1998**, *86*, 2278–2324. [CrossRef]

37. Gall, J.; Yao, A.; Razavi, N.; Van Gool, L.; Lempitsky, V. Hough forests for object detection, tracking, and action recognition. *IEEE Trans. Pattern Anal. Mach. Intell.* **2011**, *33*, 2188–2202. [CrossRef] [PubMed]

38. Gupta, S.; Arbeláez, P.; Girshick, R.; Malik, J. Aligning 3D models to RGB-D images of cluttered scenes. In Proceedings of the IEEE Conference on Computer Vision and Pattern Recognition, Boston, MA, USA, 7–12 June 2015; pp. 4731–4740.

39. Kendall, A.; Cipolla, R. Modelling uncertainty in deep learning for camera relocalization. In Proceedings of the IEEE International Conference on Robotics and Automation, Stockholm, Sweden, 16–21 May 2016; pp. 4762–4769.

40. Walch, F.; Hazirbas, C.; Leal-Taixe, L.; Sattler, T.; Hilsenbeck, S.; Cremers, D. Image-based localization using lstms for structured feature correlation. In Proceedings of the IEEE International Conference on Computer Vision, Venice, Italy, 22–29 October 2017; pp. 627–637.

41. Tateno, K.; Kitahara, I.; Ohta, Y. A Nested Marker for Augmented Reality. In Proceedings of the IEEE Virtual Reality Conference, Charlotte, NC, USA, 10–14 March 2007; pp. 259–262.

42. Crete, F.; Dolmiere, T.; Ladret, P.; Nicolas, M. The blur effect: Perception and estimation with a new no-reference perceptual blur metric. In Proceedings of the Human Vision and Electronic Imaging, San Jose, CA, USA, 29 January–1 February 2007; pp. 1–11.

43. Hartley, R.; Zisserman, A. *Multiple View Geometry in Computer Vision*; Cambridge University Press: Cambridge, UK, 2003.

44. Zhong, B.; Li, Y. Image Feature Point Matching Based on Improved SIFT Algorithm. In Proceedings of the International Conference on Image, Vision and Computing, Xiamen, China, 5–7 July 2019; pp. 489–493.

45. Korman, S.; Avidan, S. Coherency Sensitive Hashing. *IEEE Trans. Pattern Anal. Mach. Intell.* **2016**, *38*, 1099–1112. [CrossRef] [PubMed]

46. Ren, S.; He, K.; Girshick, R.; Sun, J. Faster r-cnn: Towards real-time object detection with region proposal networks. In Proceedings of the Advances in Neural Information Processing Systems, Montreal, QC, Canada, 7–12 December 2015; pp. 91–99.

47. Russell, B.; Torralba, A.; Murphy, K.; Freeman, W. LabelMe: A database and web-based tool for image annotation. *Int. J. Comput. Vis.* **2008**, *77*, 157–173. [CrossRef]

48. He, K.; Zhang, X.; Ren, S.; Sun, J. Deep Residual Learning for Image Recognition. In Proceedings of the IEEE Conference on Computer Vision and Pattern Recognition, Las Vegas, NV, USA, 26 June–1 July 2016; pp. 770–778.

49. Everingham, M.; Gool, L.V.; Williams, C.K.I.; Winn, J.; Zisserman, A. The PASCAL Visual Object Classes Challenge Results. 2007. Available online: http://host.robots.ox.ac.uk/pascal/VOC/voc2007/ (accessed on 26 May 2020).

50. Lin, T.Y.; Maire, M.; Belongie, S.; Hays, J.; Perona, P. Microsoft coco: Common objects in context. In Proceedings of the European Conference on Computer Vision, Zurich, Switzerland, 6–12 September 2014; pp. 740–755.

51. Tian, Y.; Jiang, W. Gimbal Handheld Holder. U.S. Patent 10,208,887. February 19, 2019.

52. Kim, Y.T. Contrast enhancement using brightness preserving bi-histogram equalization. *IEEE Trans. Consum. Electron.* **1997**, *43*, 1–8.

53. ARCore. Google ARCore. 2017. Available online: https://developers.google.com/ar/ (accessed on 26 May 2020).

Article

A Deep-Learning-based 3D Defect Quantitative Inspection System in CC Products Surface

Liming Zhao, Fangfang Li, Yi Zhang *, Xiaodong Xu, Hong Xiao and Yang Feng

Research Center of Intelligent System and Robotics, Chongqing University of Posts and Telecommunications, Chongqing 400065, China; zhaolm@cqupt.edu.cn (L.Z.); liffang123@gmail.com (F.L.); xxd@cqupt.edu.cn (X.X.); xiaohong@cqupt.edu.cn (X.H.); fy1222111@163.com (Y.F.)
* Correspondence: zhangyi@cqupt.edu.cn; Tel.: +01-267-366-4048 or +86-188-8376-9122

Received: 27 December 2019; Accepted: 7 February 2020; Published: 12 February 2020

Abstract: To create an intelligent surface region of interests (ROI) 3D quantitative inspection strategy a reality in the continuous casting (CC) production line, an improved 3D laser image scanning system (3D-LDS) was established based on binocular imaging and deep-learning techniques. In 3D-LDS, firstly, to meet the requirements of the industrial application, the CCD laser image scanning method was optimized in high-temperature experiments and secondly, we proposed a novel region proposal method based on 3D ROI initial depth location for effectively suppressing redundant candidate bounding boxes generated by pseudo-defects in a real-time inspection process. Thirdly, a novel two-step defects inspection strategy was presented by devising a fusion deep CNN model which combined fully connected networks (for defects classification/recognition) and fully convolutional networks (for defects delineation). The 3D-LDS' dichotomous inspection method of defects classification and delineation processes are helpful in understanding and addressing challenges for defects inspection in CC product surfaces. The applicability of the presented methods is mainly tied to the surface quality inspection for slab, strip and billet products.

Keywords: continuous casting; surface defects; 3D imaging; neural network; deep learning; defect detection

---

1. Introduction

In recent years, with the advent of the industrial 4.0 enterprises undergoing transformation and upgrading manufacturing processes, continuous casting (CC) as a main solidification process for molten steel has been widely popularized to produce metal semi-finished products [1]. In the iron and steel industry with the maturity of CC technology, hot charging and direct rolling (HC-DR) as an energy-efficient production pattern is currently experiencing rapid development [2,3]. Technically, to implement HC-DR, the defect-free CC products will undoubtedly be an essential prerequisite [4,5]. Although the technical objectives to be improved have been identified, no manufacturer in the world has reported one-hundred percent defect-free CC semi-products manufacturing technology in such a complex and systematic setting [6]. Therefore, complementary technologies such as automatic nondestructive examination (NDE) for CC products surface quality evaluation have become essential in the promotion of HC-DR [7,8]. This is an advisable method to eliminate flaw segments according to accurate NDE evaluation results [9]. Machine vision (MV) in NDE combined with AI algorithms is becoming a burgeoning method which can perform with a fast response, a high signal-to-noise ratio and a strong anti-jamming capability [10,11] compared with ultrasonic, eddy current and other contact methods. The MV merits make it more competitive in harsh environment application like CC manufacturing field [12,13]. On the other hand, MV-based 3D optical metrology has gradually demonstrated superiority, such as [14–16] stereoscopy triangulation (mm), interferometry (nm), con-focal vertical scanning, and fringe projection (um). ArcelorMittal Corp. developed a conoscopic holography rangefinders system tested in ACERALIA Crop. (Spain). The Cognex Corp. in the US developed a SmartView detection system that applied to a wide variety of surface defects inspection tasks. Elkem Corp. in Norway and Honeywell Corp. in the United States conducted infrared and visible-light MV detection methods [17]. Xu et al. [18], based on MV technology, carried out extensive research on CC slab and rolled strip surface defects inspection. To obtain effective 3D defects shapes, Zhao et al. [19] combined line array CCD and area array CCD imaging methods and devised the informative image scanning method. As a fast developing subfield of machine learning, multilayer perceptron convolutional neural networks combined with deep learning(CNN-DL) strategies in MV inspection field have shown state-of-the-art performance [20]. CNN-DL methods do not require laborious hand-craft features for classifier design [21] and as a branch of ANN, they make the complex function approximation feasible by learning a deep nonlinear network. He Di et al. [22] trained a classifier for strip defects recognition based on convolutional auto-encoder (CAE) and a devised semi-supervised Generative Adversarial Networks. To overcome the trivial image pre-processing and feature extraction process, Wangzhe Du et al. [23] presented an X-ray defect detection system based on the Feature Pyramid Network and a data augmentation method for model generalization training. Veitch-Michaelis et al. [24] studied the 3D cracks recognition method through the combination of morphological detection and SVM classifier. Hongwen Dong in Northeastern University proposed a pyramid feature fusion and a global context attention network for pixel-wise detection of surface defect in the industrial production process [25]. Fatima A. Saiz et al. [26] reported a deep-learning based automatic defects recognition system in which CNN was utilized in the model design, which achieved an outstanding classification rate. CNN-DL strategies need to make full use of training datasets and learning algorithms to make the detection results relatively stable. Therefore, they generally require a large number of training samples as input. In high-noise environments, MV based-intelligent inspection methods as mainstream schemes have been successfully applied in the CC products line, although the accuracy and mechanism of the AI algorithm require in-depth research with the improvement of application requirements.

In the CC production line, with the improvement of quality requirements, the defect depth has become a significant factor, which, especially for the CC slab, sometimes may cause potential security problems. In other words, some defects can be ignored or repaired by the follow-up finishing process if the depth of the defects does not exceed a certain value. Furthermore, conventional optical imaging 2D inspection methods are susceptible to high-temperature radiation interference. In this work, we refer to the entire defects inspection process as two separate steps: recognition and delineation, and based on our previous work in [6], a novel two-step defects inspection strategy was presented by devising a fusion deep CNN model (fully connected CNN with fully convolution CNN). The entire scheme, as shown in Figure 1, was implemented by the devised flexible binocular 3D quantitative inspection deep-learning system (3D-LDS). In this system, unlike traditional inspection methods the 3D depth point cloud mapping images will be feed into 3D-LDS. Furthermore, a region proposal method was designed using 3D-LDS ROI location that can effectively suppress redundant candidate bounding boxes in a real-time defects recognition process. Systematically a 3D-LDS-based CNN-DL strategy was attempted for CC products surface defects inspection that allows a feasible method of AI algorithms and powerful ROI recognition and delineation strategies to be further studied in industrial applications.

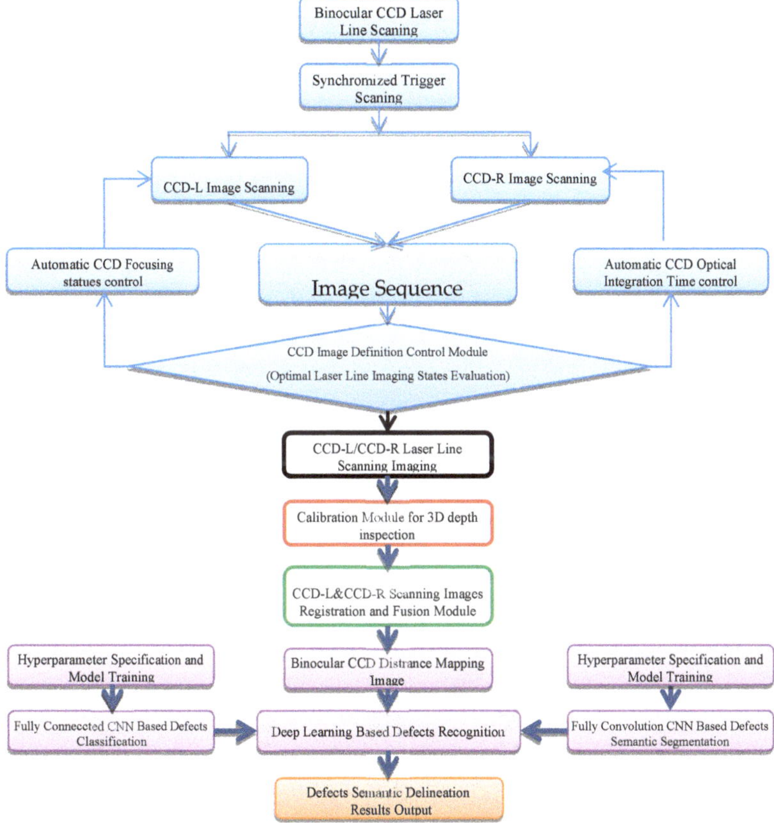

Figure 1. The scheme of binocular CCD-based 3D image deep-learning CC products surface defects inspection.

2. An Improved 3D Image Scanning System

2.1. Optimal Image Laser Scanning Method

In image-based ROI inspection methods, it is a prerequisite for the imaging sensor to be able to capture objects informatively and adjust imaging parameters adaptively as the peripheral environment changes. Therefore, the 3D-LDS as a structured light assistant active imaging system needs a laser stripe with a maximal color contrast and the most homogeneous gray-level. Namely, the imaging sensor should be set to an appropriate optical integral time (OIT) and focus status. When it comes to a rigid system architecture, the focus status can be fixed as the imaging distance and imaging depth of field (DOF) are constants. However, the automatic OIT controlling method needs to be focused on if imaging sensor works in an unstable high-temperature radiation environment. According to the Planck theorem [27], we took the CC production as an blackbody and assumed that its surface emissivity is equal to 1. While T > 500 °C (like the CC slab roughly varied between 600 °C to 900 °C when it comes out of the second cooling area), the visible red-light radiation can be sensed by unaided eyes. We tested the optical spectrum radiation interference in different temperatures in hot CC slab surface from 720 °C to 1021 °C, as in shown in Figure 2. We can observe the regular patterns of light strength distribution with different OIT and object surface temperatures. The experiments present a quantitative guidance for determining laser luminous wavelength and controlling the imaging sensor's parameters. In 3D-LDS, to minimize radiation interference, we selected a 532 nm green laser emitter. On the one hand, it can ensure that the CCD sensor is in the imaging spectral sensitive range and on the other hand, it can avoid high-temperature radiation interference as much as possible. We can observe that the radiation intensity of the laser stripe at 3 ms is easily distinguishable from the hot slab surface (1000°C) at the integral time of 10 ms.

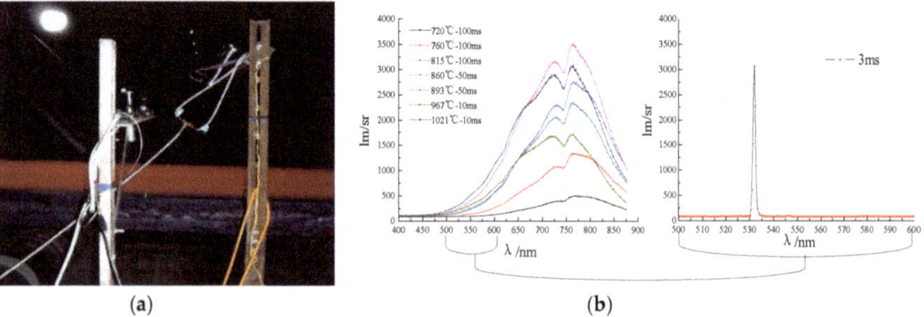

Figure 2. The spectral radiation intensity at different temperatures in slab surface and light intensity distribution of the laser stripe. (**a**) High temperature spectrum measurement; (**b**) Spectral intensity comparison.

On the basis of the light radiation principle, we presented an improved method to determine threshold T_L in 3D-LDS, which allows the CCD cameras to scan the laser stripe precisely without being interfered by a high temperatures radiation. Based on the CCD imaging principle, theoretically, the objects luminance can be formulated as follows [28]:

$$E_0 = \left(\frac{n'}{n}\right)^2 K\pi L \sin^2 U',$$
(1)

where n and n' denote, respectively, the refractive index in object space and image space, K is the optical system transmittance, L is the light luminance, and U' represents the image aperture angle. Supposing that the laser reflected luminance can be expressed by $L' = \rho E$, where E represents laser transmitter luminance and ρ is the reflectivity ($0 < \rho < 1$), then, the diffuse reflection of the laser stripe on the object surface can be formulated by

$$E_0' = \left(\frac{n'}{n}\right)^2 K\pi\rho E \sin^2 U',$$
(2)

Apparently, as shown in Figure 3, quantitatively determining the best color distance between the slab surface and the laser stripe depends on the threshold at the optimal light integration time [29]. It also shows that in the figure, the laser stripe shape is easily extracted when the light intensity is concentrated.

Therefore, T_L can be found by the following method. Firstly, we convert the 24-bit color image into gray level directly by assigning $R = G$ and $B = G$. While the CCD sensor's images have pixel levels of $[1, ..T.., L]$, let n_i and N denote the number of pixels at level i and the total number in one frame, then the T_L should be between the μ_b and μ_f [30]:

$$\begin{cases} \mu_b = \sum\limits_{i=1}^{T} ip_i/\omega_b = \mu(T)/\omega(T) \\ \mu_f = \sum\limits_{i=T+1}^{L} ip_i/\omega_f = \dfrac{\mu_t-\mu(T)}{1-\omega(T)} \end{cases},$$
(3)

where $p_i = n_i/N$, $\mu(T) = \sum\limits_{i=1}^{T} ip_i$, $\mu_t = \sum\limits_{i=1}^{L} ip_i$, $\omega_t = \sum\limits_{i=1}^{T} p_i = \omega(T)$, $\omega_f = \sum\limits_{i=T+1}^{L} p_i = 1 - \omega(T)$. The variances of the foreground of the laser stripe and the background are formulated as follows:

$$
\begin{cases}
\sigma_b^2 = \sum\limits_{i=1}^{T} (i - \mu_b)^2 p_b / \omega_b \\
\sigma_f^2 = \sum\limits_{i=T+1}^{L} (i - \mu_f)^2 p_f / \omega_f
\end{cases}, \tag{4}
$$

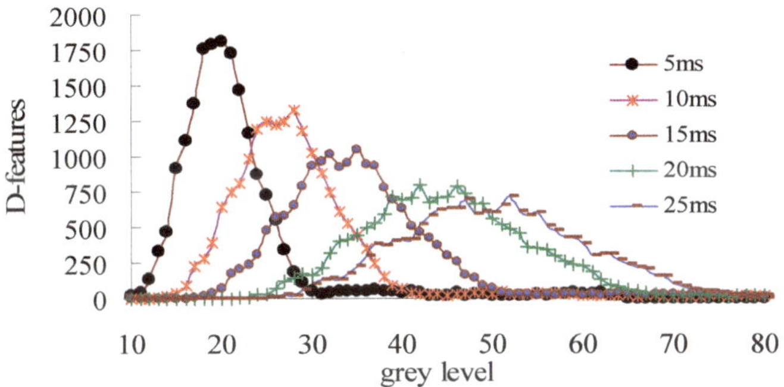

Figure 3. Laser stripe shape and intensity distribution counts at different CCD integral times.

Based on the Otsu and CCD imaging definition variance evaluation function, the optimal scanning threshold T can be determined by the following discriminate criterion measure:

$$
f(T) = \sigma_B^2(T)/\sigma_t^2, \tag{5}
$$

where $\sigma_B^2(T) = \omega_b(\mu_b - \mu_t)^2 + \omega_f(\mu_f - \mu_t)^2$ denotes the classes variance and $\sigma_t^2 = \sum\limits_{i=1}^{L} (i - u_t)^2 p_i$ represents the current frame total variance. In fact, optimal T_o can be computed by searching the threshold interval $[1, ..T.., L]$ to meet the requirement:

$$
T_L = \max_{1 \leq T \leq L} \sigma_B^2(T), \tag{6}
$$

Figure 4 displays the laser imaging results that Figure 4b is the most convenient shape for data processing through experiments under optimal imaging states.

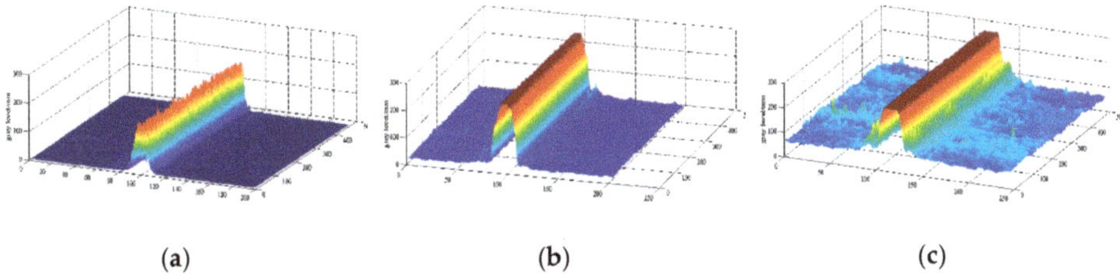

(a)   (b)   (c)

Figure 4. Laser stripe imaging features at different CCD imaging states. (**a**) Short CCD integral time; (**b**) Optimal laser imaging; (**c**) Long CCD integral time.

2.2. System Construction

To implement the deep learning 3D inspection method and create a reliable detection system to meet the special requirements, we devised an improved experimental system based on our previous research. Figure 5a is the schematic principle of the devised binocular CCD laser image scanning system. Figure 5b is the corresponding experimental system devised that we updated from our previous multi-source CCD imaging system in the literature [6]. The previous system mainly utilized the traditional inspection methods, and the 3D laser scanning system just played auxiliary role in defect location. In the new 3D-LDS system, the integrity of defects can be captured properly without the line scanning CCD. In this system, we employed two MERCURY CCD cameras (model: MER-500-14GC-P) and the lens model selected was M0814-MP2. Here, the deep learning defects recognition process was conducted on the fusion image from the two imaging sensors. In this system, the two laser scanning images were overlaid informatively by a registration method and this process is a rigid transformation of rotation and translation. Once the system calibration was completed, the imaging parameters between the two CCD cameras were settled. Notice that the applicability of the proposed experimental system is not tied to CC products surface defects inspection exclusively.

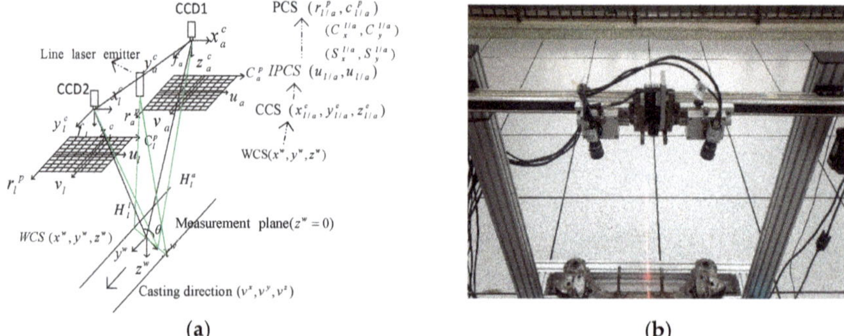

Figure 5. The schematic illustration of the binocular CCD laser image scanning system. (**a**) The schematic working principle diagram; (**b**) The corresponding devised experimental system.

In the system, the 3D images pixels (12-bit) are indirectly mapped from the calibrated laser triangulation strategies (the metric is millimeter). Therefore, the image ROI was reconstructed by converting the 3D distance point cloud of the object surface. From the experiments in Figure 6, we can visually observe that the system can change its detection accuracy and sensitivity for depth information by finely adjusting θ according to the detection requirements. Generally, CNN-DL model training requires a large number of labeled examples. We utilized the angular fine adjustment to acquire different scanning images for the testing samples as an auxiliary data augmentation method. The depth of variation was explicitly added to the training samples. Based on this method, we also used the typical variation, including changes in contrast, rotations and translations. Deep-learning is extremely data-hungry and performance grows only logarithmically with the amount of data used. This is one of main limitations that the field is currently facing.

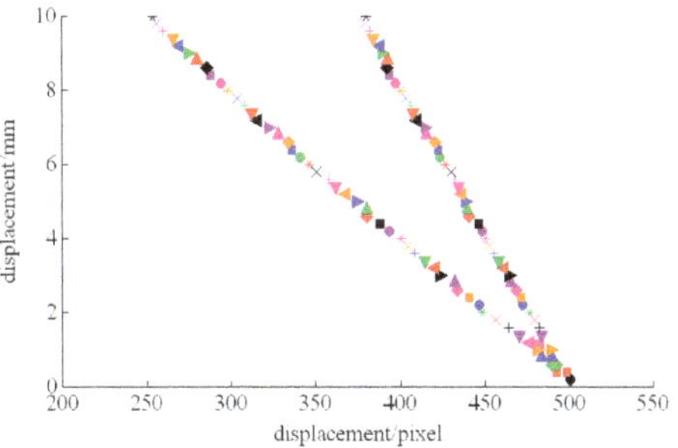

Figure 6. The influence of different oblique angles on detection accuracy and sensitivity (left:$\theta/2 = 60°$ and right: $\theta/2 = 45°$).

3. CNN-DL Inspection Method Design in 3D-LDS

3.1. CNN Networks Design in 3D-LDS

In neural networks, a neuron is the fundamental unit that takes a bias w_0 and a weight vector $\omega = (w_0, \ldots w_n)$ as parameters to a decision model: $f(x) = h(\omega^T x + w_0)$ where $h(x)$ is a non-linear activation function. More complex nonlinear mapping is usually based on the combination of lots of neurons that are arranged in layers. Commonly, a single layer network can be expressed as a linear combination of N individual neurons [31]:

$$\tilde{f}(x) = \sum_{i=0}^{N-1} v_i h(w^T x + w_{0,i}),\tag{7}$$

where the trainable parameters for this network can be summarized as (v_0, $w_{0,0}$, w_0,..., v_N, $w_{0,N}$, w_N). Appropriate parameters can decrease the ideal function and its approximation: $|f(x) - \tilde{f}(x)|$. Theoretically, any function can be approximated using a single layer network only if we give a large number of neurons and have the proper parameters within the same compact set that the network can be trained. The more layers (deeper networks) the network creates, the stronger the networks' modeling capacity. However, the deeper the number of layers, the more challenging it is to train the network parameters. In recent years, deep learning technology has been widely used in many fields, especially the proposed convolutional and pooling payers make the model have a robust ability to extract local and macro characteristics. In Figure 7, the convolutional and pooling process in DL networks achieved locality perception and parameter-sharing mechanism, which dramatically reduce the amount of model training parameters. In addition, the End-to-End training strategy makes the feature extraction-selection and classifier design integrated in a streamlined process. The hand-crafting features are no longer required while everything is learned by the network model based on a data-driven mode.

Based on the end-to-end training mechanism, we built a complete deep neural network model in 3D-LDS. As shown in Figure 8, we devised a dichotomous defects inspection strategy that includes two steps and a two-branch deep neural network for defects types classification (recognition) and ROI delineation. In the overall inspection process, the input images mapped from the laser triangulation were finally converted to a predication map and a classification label. The proposed methodology is helpful in understanding and addressing challenges for CC production surface inspection. In the recognition process, 3D point cloud images in 3D-LDS was utilized to locate the defect positions accurately according to the depth detection results. Through the initial location of the possible ROI(defects) the candidate bounding box(BBox) will be generated, which we define this process as depth based ROI initial location and BBox generation. In the last two steps, the BBox will be classified by fully connected neural networks and the defects types will be output in images level, and the prediction map in pixel-wise will be output in fully convolutional neural networks for delineation.

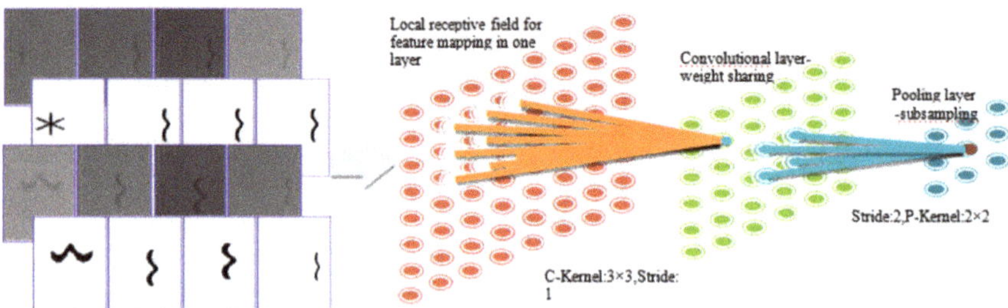

Figure 7. The convolution and pooling process in CNNs for locality perception and parameters sharing mechanism.

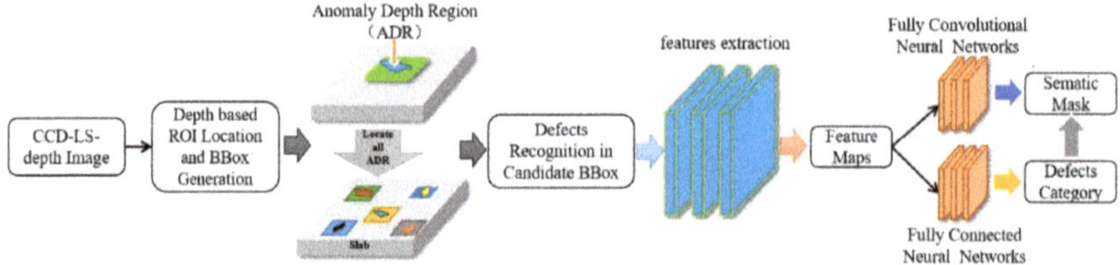

Figure 8. Defects recognition and delineation process based on deep CNN modeling mechanism.

A significant characteristic of DL strategies is the automatic feature learning for data representations through an end-to-end training process. To realize the two-step defects recognition and delineation in 3D-LDS, we constructed a novel network architecture by integrating the blocks of Resnet [32] and Unet [33]. The aim is to take advantages of the deep CNN merits in classifier design and fuzzy ROI delineation. Thereinto, ResNet were designed to enable training of very deep networks due to the residual block is introduced. Ronneberger's full convolution idea is a breakthrough towards automatic image segmentation. In fact, the ROI segmentation can be expressed as an auto encoder and decoder process. It consists of a contracting and an expanding branch and enables multi-resolution analysis. Figure 9 indicates the schematic network architectures for defects classification(recognition) and ROI segmentation (delineation). A novel idea here is the devised multi-model-based recognition and delineation that in the defects inspection process the system will according to the input images size automatically select different training models. Usually, the detected candidate ROI will have different sizes to reduce the computational complexity in 3D-LDS only the BBox will be input into system as shown in Figure 8. In the experimental testing process, we trained five different sizes of BBoxes for classifier and delineation DL models (input sizes: 32*32,48*48,64*64,80*80,128*128), the candidate depth ROI based BBoxes will be resized to one of the 5 sizes according to its size proximity. Note that the images will be reconstructed after the recognition and delineation are finished because the real location in CC products surface will be predicted through the system measurement calibration parameters.

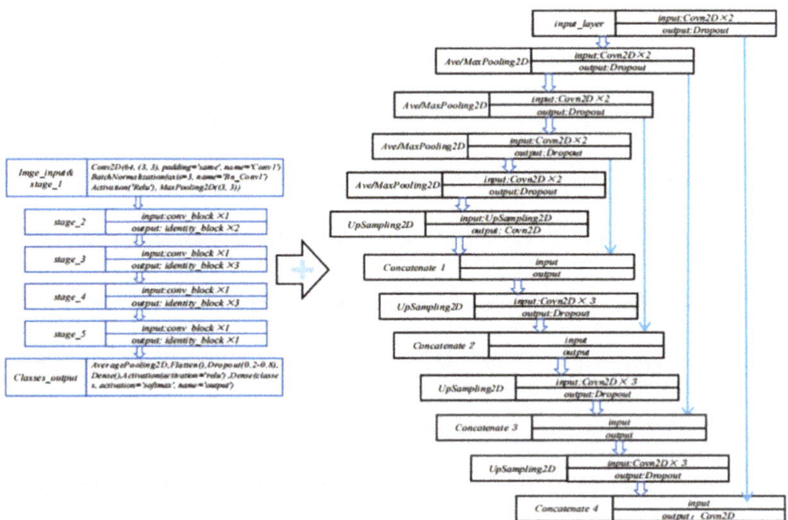

Figure 9. The schematic network architectures for defects recognition and ROI delineation.

3.2. Model Training Strategies in 3D-LDS

Generally, a CNN network consists of convolutional layers, pooling layers, full connection layers and loss layers, etc., among them, the algorithms in the full connection layer and yjr loss layer are basic parts of the network. CNN based recognition methods have been widely used in image analysis fields. CNN based modeling capability is gradually strengthened owing to the improvement of loss function and optimization algorithm in model training process. In this work, as shown in Figure 10 we utilized softmax function to train multi-classification model [34]:

$$P(y = j|z^{(i)}) = \phi_{softmax}(z^{(i)}) = \frac{e^{z^{(i)}}}{\sum_{k=1}^{t} e^{z_k^{(i)}}}, \tag{8}$$

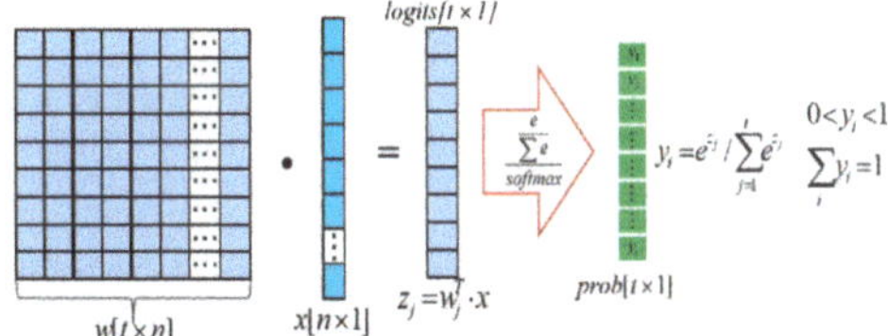

Figure 10. The system multi-classification method for different defects types in training and testing process.

We can see that the range of this function value is defined in [0,1], where, $z = w_0 x_0 + w_1 x_1 + \cdots + w_n x_n = \sum_{i=0}^{n} w_i x_i = \mathbf{w}^T \mathbf{x}$, t represents the total number of defects categories, w is the weight vector, x is the feature vector of a training sample, and w_0s the bias unit. z_k denotes the value of the output of class k, in the experimental process we basically tested five classifications of defects for transversal cracks, longitudinal cracks, star cracks, hole-shaped defect and others respectively. The softmax function computes the probability that the current training sample $x^{(i)}$ belongs to class j given the weight and net input $z^{(i)}$. Therefore, we compute the probability $(y = j|x^{(i)}; w_j)$ for each class label in $j = 1, \ldots .k$.

Note that the normalization term in the denominator causes the whole class probabilities sum up to one under the assumption that the training samples are independent of each other.

Based on the softmax function we can introduce the softmax *loss* as formulated as below:

$$L = -\sum_{j=1}^{T} y_j log s_j,$$

(9)

Here, s_j is the $j-th$ value of the output vector s from softmax function, which indicates the probability that the testing sample belongs to the $j-th$ category. y_j is a vector of *1*T* that only the value of the position corresponding to the real label is equal to 1. Therefore, this formula actually has a simpler form when j is the real label that points to the current sample:

$$L = -log s_j,$$

(10)

Next, we can give the concept of cross entropy which it is formulated as below:

$$E = -\sum_{j=1}^{T} y_j log p_j,$$

(11)

Here, *cross entry* is equal to softmax loss while the input p_j of cross entry is the output of *softmax*. In our work, we set the activation function as softmax in dense layer. Based on the above discussion, we can define the function of the optimization to minimize (or maximize) the loss function E in training process. Basically, gradient descent is one of the most popular algorithms to perform optimization and up to now the most common way to optimize neural networks. Moreover, there are three basic variants of gradient descent which differ in how much data we use to compute the gradient of the objective function, which include [35] batch gradient descent (GGD), stochastic gradient descent (SGD) and mini-batch gradient descent (MBGD). In fact, there are some challenges need to be solved in allusion to the above three optimization methods. However, these methods are often used to test the effectiveness of the network training process. We will not pay too much attention to these issues because of the focus of this paper. In these experiments, we utilized the adaptive moment estimation (Adam) optimization to compute adaptive learning rates for network parameters. Adam keeps an exponentially decaying average of past gradients similar to momentum besides storing an exponentially decaying average of past squared gradients like Adadelta and RMSprop [36]. Adam prefers flat minima in the error surface and the decaying averages of past and past squared gradients m_t and v_t are computed separately as follows [37]:

$$\begin{aligned} m_t &= \beta_1 m_{t-1} + (1-\beta_1)g_t \\ v_t &= \beta_2 v_{t-1} + (1-\beta_2)g_t^2 \end{aligned}$$

(12)

where m_t and v_t are estimates of the first moment and the second moment of the gradients respectively, if the m_t and v_t are initialized as vectors of 0, they counteract these biases by computing bias-corrected first and second moment estimates:

$$\hat{m}_t = \frac{m_t}{1-\beta_1^t}, \; \hat{v}_t = \frac{v_t}{1-\beta_2^t},$$

(13)

Therefore, based on the bias-corrected estimates, the Adam gradient update rule is generated as below:

$$\theta_{t+1} = \theta_t - \eta\frac{\hat{m}_t}{\sqrt{\hat{v}_t} + \varepsilon},$$

(14)

The authors propose default values of 0.9 for β_1, 0.999 for β_2, and 10^{-8} for ε.

3.3. Experimental Results Analysis

Due to the all-pervading oxide scales on CC products surface have similar characteristics with real defects, especially in 2D images while it is processed by imaging processing algorithms. We call it pseudo defects interference in inspection process as presented in Figure 11b. The steel plate displays confusing ROI with a crack and also some other outliers. This will make ROI extraction very challenging even in room temperature. In Figure 11b we clustered the ROI and finally found 1400 candidate ROIs. Figure 11c shows the laser scanning image for the Figure 11b, by the same way the counterpart of Figure 11b given by Figure 11c contains 3 candidate ROIs. Therefore, the selective patches given by the location of the candidate ROI will be computed and returned by the recognition model in 3D-LDS. Basically, region proposal algorithms are often employed to identify prospective objects in an image such as the proposed methods of objectness, randomized prim or selective search and so on. In this paper we referred to the region proposal method but devised a more effective way by referring to the laser scanning images depth location as given in Figure 11c. The candidate bounding boxes for defects recognition will be proposed and resized to the closest image patch for recognition.

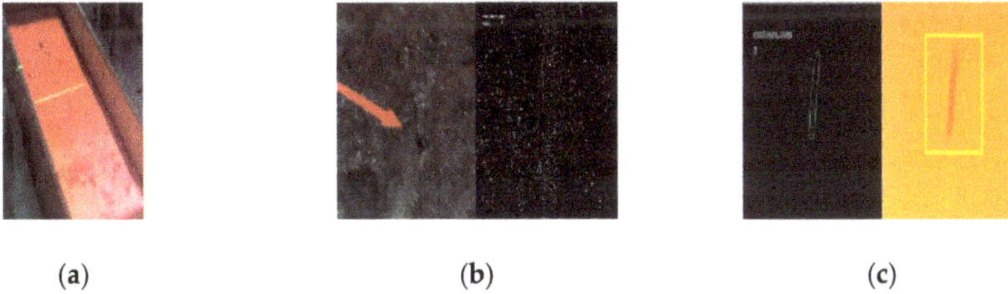

(a)   (b)   (c)

Figure 11. Searching for ROI in optical image and laser scanning (depth) image in 3D-LDS. (a) CCD image (Object depth:3mm); (b) Searching results for ROI; (c) Laser scanning image and search results for ROI.

Figure 12 denotes the ROI depth location method. For abnormal depth areas we only extract the centroid line as the position depth values and 3D image reconstruction in scanning process. Figure 12a is the artificial defect that for convenience of calculation we made some samples of different depths and sizes for four defect types and others (made randomly). Figure 12b,c are the laser location process that pixels offset reflected on the image. Figure 12d is the ROI depth based candidate bounding box generation method.

Figure 13a shows the training samples for L crack generated in 3D-LDS in different scanning angles, distances and optical integral times. The labels (ground truth) in second row are mainly delineated manually and generated by an interactive method to ensure accuracy. In this work, the data augmentation strategy was utilized, the parameters we used for generating a new image are as follow:rotation_range,translation_shift_range,zoom_range and blur operation. Roughly, the training and testing data sets were split in 7:3 separately from different original data. Figure 13b shows the testing results that actually is a reconstructed image from the mapping pixels' prediction values. We can set a different classification number for the softmax *function* to obtain different output. However, the final binary image will be segmented by a fixed threshold.

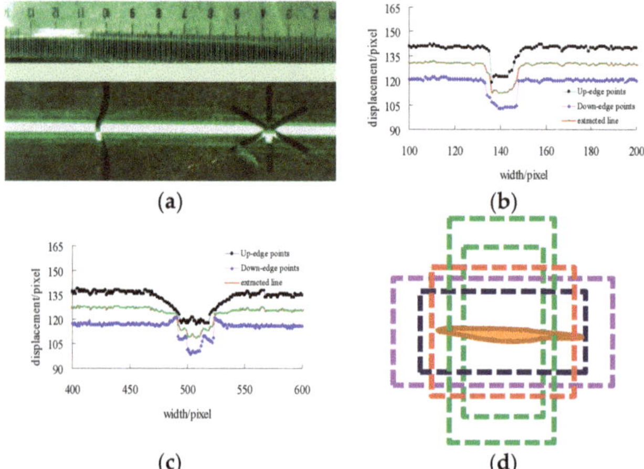

Figure 12. The candidate ROI location in different areas. (**a**) The testing sample defects; (**b**) ROI location in linear defect; (**c**) ROI extraction in star defect; (**d**) The candidate bounding box.

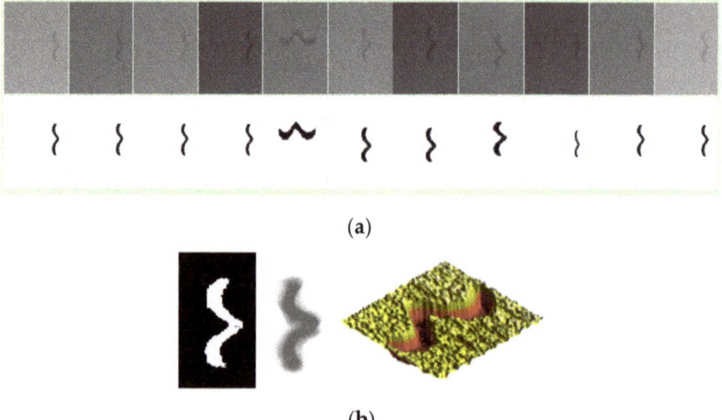

Figure 13. Experimental results for training data sets and predicted reconstruction images. (**a**) Training samples for crack based on the devised data augmentation in 3D-LDS; (**b**) Predication map (left) with its binary image (middle) and the 3D visualization (right) for the inspected L-crack defect.

In the 3D-LDS defect inspection process, there is a sensitive parameter: the radius of the candidate bounding box(BBox-R), which determines the size of the ROI relative to the size of BBox. Generally, in order to ensure the candidate BBox includes the ROI accurately. We can set a relatively large radius to locate the ROI. However, this will lead to regional imbalances (RI) and consequently, bring about two main issues, especially in full convolutional networks training and the testing process:

(i)   In the training process, the RI problem will make CNN-DL model training more challenging to converge and become time consuming because of the unbalance of positive and negative pixel samples.

(ii) In the testing process, RI defects always get undesirable segmentation results by automatic strategies due to the inaccurate positioning by traditional bounding box.

Table 1 is the testing results for five types of defects, thereinto, L-110(440) means the type is longitudinal cracks and training and testing samples are 440 and 110 respectively. T means transverse crack, S denotes star shape defects and H means hole defects. To facilitate the quantitative analysis, we employed image segmentation evaluation methods to test validation in delineation step that includes dice coefficient (DICE), false positive (FP), false negative (FN) and mean hausdorff distance(M-HD). Dice is twice the area of overlap between ground truth(A) and prediction(B) divided by the total number of pixels in both regions [38]:

$$Dice = \frac{2|A \cap B|}{|A| + |B|} \times 100\%,\tag{15}$$

Dice value ranges from 0 to 1 with 1 signifying the greatest similarity between the predicted and truth.

Table 1. Experimental quantitative results.

Type-Total(AUG)	DICE	FP	FN	M-HD
L-110(440)	0.93	0.05	0.08	0.32
T-121(484)	0.87	0.04	0.21	0.17
S_82(328)	0.85	0.03	0.24	0.07
H_102(408)	0.82	0.05	0.21	0.29
Others-12	0.83	0.07	0.25	0.33

We also used the FP and FN to give us an overall understanding for the predicted results. Because both of the FP and FN are errors in data reporting in which a test result improperly indicates presence of a condition. In general, we will get under segmentation results if the FP is greater than FN and and vice versa. Meanwhile, we utilized M-HD to check the predicted boundary as it is sensitive to it. However, we use the *mean* computing way instead of the *max* method to prevent isolated point noise interference:

$$d_H(X,Y) = mean\{d_{XY}, d_{YX}\} = mean\{\underset{x \in X, y \in Y}{maxmind(x,y)}, \underset{y \in Y, x \in X}{maxmind(x,y)}\},\tag{16}$$

In the model training process, we utilized the basic quantitative quality indicators ACC to validate the system:

$$ACC = (TP + TN)/(TP + TN + FP + FN),\tag{17}$$

ACC reflects the classifier's overall prediction correctness that TP represents the number of observations correctly assigned to the positive class. TN is the number of observations correctly assigned to the negative class. FP denotes the number of observations assigned by the model to the positive class. FN is the number of observations assigned to the negative class, which in reality belong to the positive class. Figure 14 is the validation process for training error and testing error. Table 1 shows the quantitative experimental results that we used the extra FP and FN to get feedback for over-segmentation and under-segmentation so that we can adjust the model parameters.

In allusion to the running time, we tested on the computer with two GPU cards: GEforce GTX 1080 and GEforce RTX 2080Ti, the 2080Ti was used to do the delineation and tested on the maximum BBox(320 × 320). It can perform 15 image segmentation tasks per second that meets the CC production online detection process. With regard to the image scanning speed, we tested image size: 1200 × 600 (the selected CCD cameras is 14fps in full resolution: 2592 × 1944). The system can finish 45fps laser scanning because only laser ROI will be processed in the image. Therefore, the casting speed should be less than 0.8 m/min if the scanning spacing is 0.3 mm. Actually, in real application, the high-performance image workstation or multi-machine distributed processing is preferred. The quantitative experimental results are given in Table 1.

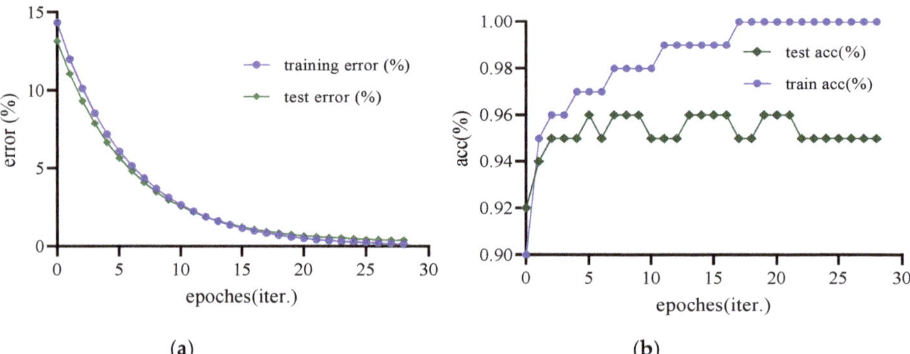

Figure 14. Validation on model training and testing; (**a**) Training and testing errors; (**b**) ACC results on training and testing data sets.

4. Conclusions and Future Work

In this paper, an improved binocular vision-based 3D laser image scanning deep-learning system (3D-LDS) was established for CC products surface evaluation. The main work is as below:

1) An optimal CCD laser image scanning method was designed in different high-temperature experiments.
2) In allusion to defects precise recognition, we proposed a novel region proposal method based on the 3D ROI initial location that can effectively suppress redundant candidate bounding boxes generated by pseudo-defects in a real-time recognition process.
3) To improve the inspection accuracy, a deep CNN architecture combined fully connected networks (for defects classification) and fully convolutional network (for defects delineation) was proposed to robustly make the whole inspection methodology defined as a two-step process.
4) The applicability of the presented methods is mainly tied to the surface quality inspection for slab, strip and billet products etcetera. Systematically, A 3D-LDS deep learning system is devised for CC products surface quality evaluation that allows an automatic way of AI algorithms to be applied to the MV inspection field in modern industries.

Future work: Based on the experimental analysis, it is found that the optimization of network architecture is a long-term job. There is no unified network model for different detection tasks and targets. Therefore, it is essential to conduct field experimental studies to improve and construct a more robust network architecture especially for the defects classification network. The aim is to solve the common over-fitting problem of current networks and to reduce the dependence on data source quality in model training process. Furthermore, the improvement method of optimization algorithm for deep CNN model training should be further studied through the deep neural network mechanism research in the specific application context. In the following work, we will carry out field experiments and application research in the continuous casting production line.

Author Contributions: Conceptualization and methodology, L.Z.; data analysis and writing, F.L. and L.Z.; CNN networks architectural design and algorithm Improvement, Y.Z.; design and improvement of motion control system, X.X.; mechanical structure design of high precision experimental platform, H.X.; literature search and system validation, Y.F. All authors have read and agreed to the published version of the manuscript.

Funding: The work was sponsored by National Natural Science Foundation of China (51604056 and 51605064) and Chongqing Science & Technology commission Foundation (cstc2016jcyjA0537). Chongqing Science and Technology Commission Industrial Application demonstration Project (cstc2017zdcy-zdyfX0025).

Conflicts of Interest: The authors declare no conflict of interest.

References

1. Thomas, B.G. Review on modeling and simulation of continuous casting. *Steel Res. Int.* **2018**, *89*, 1700312. [CrossRef]
2. Santos, C.A.; Spim, J.A.; Garcia, A. Mathematical modeling and optimization strategies (genetic algorithm and knowledge base) applied to the continuous casting of steel. *Eng. Appl. of Artif. Intell.* **2003**, *16*, 511–527. [CrossRef]
3. Popa, E.M.; Kiss, I. Assessment of Surface Defects in the Continuously Cast Steel. *Acta Tech. Corviniensis-Bull. Eng.* **2011**, *4*, 109–115.
4. Song, K.; Yan, Y. A noise robust method based on completed local binary patterns for hot-rolled steel strip surface defects. *Appl. Surf. Sci.* **2013**, *285*, 858–864. [CrossRef]
5. AI, Y.; Xu, K. Surface Detection of Continuous Casting Slabs Based on Curvelet Transform and Kernel Locality Preserving Projections. *Int. J. Iron Steel Res.* **2013**, *20*, 80–86. [CrossRef]
6. Zhao, L.; Ouyang, Q.; Chen, D.; Udupa, J.K.; Wang, H.; Zeng, Y. Defect detection in slab surface: A novel dual charge-coupled device imaging-based fuzzy connectedness strategy. *Rev. Sci. Instrum.* **2014**, *85*, 1–8. [CrossRef]
7. Ouyang, Q.; Zhao, L.M.; Ma, F.J.; Zhang, L.Z. Experiment study of surface defects in continuous casting using developed laser scanning system. *Ironmak. Steelmak.* **2011**, *38*, 12–16. [CrossRef]
8. Zhao, L.M.; Ouyang, Q.; Chen, D.F.; Wen, L.Y. Surface Defects Inspection Method in Hot Slab Continuous Casting Process. *Ironmak. Steelmak.* **2011**, *38*, 464–470. [CrossRef]
9. Hsu, C.Y.; Huang, J.W.; Kang, L.W.; Weng, M.F. Fast image stitching for continuous casting steel billet images. In Proceedings of the IEEE International Conference on Consumer Electronics-Asia (ICCE-Asia), Seoul, Korea, 26–28 October 2016.
10. Hsu, C.Y.; Kang, L.W.; Lin, C.Y.; Yeh, C.H.; Lin, C.T. *Vision-Based Detection of Steel Billet Surface Defects via Fusion of Multiple Image Features*; IOS Press: Amsterdam, The Netherlands, 2014; pp. 1239–1247.
11. Zhao, Q.-J.; Cao, P.; Tu, D.-W. Toward intelligent manufacturing: Label characters marking and recognition method for steel products with machine vision. *Adv. Manuf.* **2014**, *2*, 3–12. [CrossRef]
12. Ouyang, Q.; Zhang, L.Z.; Zhao, L.M.; Zhang, X.L.; Chen, D.F. Experimental study on quantitative surface defect depth detection based on laser scanning technology in continuous casting. *Ironmak. Steelmak.* **2011**, *38*, 363–368. [CrossRef]
13. Ai, Y.; Xu, K. Feature extraction based on contourlet transform and its application to surface inspection of metals. *Opt. Eng.* **2012**, *11*, 113605. [CrossRef]
14. Xu, K.; Yang, C.; Zhou, P.; Liang, J. 3D Detection Technique of Surface Defects for Steel Rails Based on Linear Lasers. *J. Mech. Eng.* **2010**, *46*, 1–5. [CrossRef]
15. He, Y.; Song, K.; Meng, Q.; Yan, Y. An End-to-end Steel Surface Defect Detection Approach via Fusing Multiple Hierarchical Features. *IEEE Trans. Instrum. Meas.* **2019**. [CrossRef]
16. Jiang, Z.; Zhang, W.; Cui, L. Research of three dimensional laser scanning coordinate measuring machine. In *MATEC Web of Conferences*; EDP Sciences: Les Ulis, France, 2018.
17. Marc, G.; Li, G. Inspection of Aircraft Engine Components Using Induction Thermography. In Proceedings of the IEEE Canadian Conference on Electrical & Computer Engineering (CCECE), Quebec City, QC, Canada, 13–16 May 2018.
18. Peng, Z.; Ke, X.; Chaolin, Y. Surface defect recognition for moderately thick plates based on a SIFT operator. *J. Tsinghua Univ. (Sci. Technol.)* **2018**, *58*, 881–887.
19. Zhao, L.; Zhang, Y.; Xu, X.; Xiao, H.; Huang, C. Defect inspection in hot slab surface: Multi-source CCD imaging based fuzzy-rough sets method. In *Proc. SPIE 9971, Applications of Digital Image Processing*; SPIE: Bellingham, WA, USA, 2016; Volume 9971.
20. Ferguson, M.; Ak, R.; Lee, Y.T.T.; Law, K.H. Automatic localization of casting defects with convolutional neural networks. In Proceedings of the IEEE International Conference on Big Data (Big Data), Boston, MA, USA, 11–14 December 2017.
21. Lee, J.H.; Oh, H.M.; Kim, M.Y. Deep learning based 3D defect detection system using photometric stereo illumination. In Proceedings of the International Conference on Artificial Intelligence in Information and Communication (ICAIIC), Okinawa, Japan, 11–13 February 2019.

22. Di, H.; Ke, X.; Peng, Z.; Dongdong, Z. Surface defect classification of steels with a new semi-supervised learning method. *Opt. Lasers Eng.* **2019**, *117*, 40–48. [CrossRef]
23. Du, W.; Shen, H.; Fu, J.; Zhang, G.; He, Q. Approaches for improvement of the X-ray image defect detection of automobile casting aluminum parts based on deep learning. *NDT E Int.* **2019**, *107*, 102144. [CrossRef]
24. Veitch-Michaelis, J.; Tao, Y.; Walton, D.; Muller, J.P.; Crutchley, B.; Storey, J.; Paterson, C.; Chown, A. Crack Detection in "As-Cast" Steel Using Laser Triangulation and Machine Learning. In Proceedings of the 13th Conference on Computer and Robot Vision (CRV), Victoria, BC, Canada, 1 June 2016.
25. Dong, H.; Song, K.; He, Y.; Xu, J.; Yan, Y.; Meng, Q. PGA-Net: Pyramid Feature Fusion and Global Context Attention Network for Automated Surface Defect Detection. *IEEE Trans. Ind. Inform.* **2020**. [CrossRef]
26. Saiz, F.A.; Serrano, I.; Barandiarán, I.; Sánchez, J.R. A Robust and Fast Deep Learning-Based Method for Defect Classification in Steel Surfaces. In Proceedings of the International Conference on Intelligent Systems (IS), Funchal-Madeira, Portugal, 25 September 2018.
27. Ouyang, Q.; Zhao, L.M.; Wen, L.Y.; Bai, C.G. Simulation study on radiative imaging of pulverised coal combustion in blast furnace raceway. *Ironmak. Steelmak.* **2011**, *38*, 181–184. [CrossRef]
28. Fabijanska, A.; Sankowski, D. Computer vision system for high temperature measurements of surface properties. *Mach. Vis. Appl.* **2009**, *20*, 411–421. [CrossRef]
29. Roy, M.; Seo, D.; Oh, S.; Yang, J.W.; Seo, S. A review of recent progress in lens-free imaging and sensing. *Biosens. Bioelectron.* **2017**, *88*, 130–143. [CrossRef]
30. Otsu, N. A threshold selection method from gray-level histograms. *IEEE Trans. Syst. Man Cybern.* **1979**, *9*, 62–66. [CrossRef]
31. Shin, H.C.; Roth, H.R.; Gao, M.; Lu, L.; Xu, Z.; Nogues, I.; Yao, J.; Mollura, D.; Summers, R.M. Deep convolutional neural networks for computer-aided detection: CNN architectures, dataset characteristics and transfer learning. *IEEE Trans. Med. Imaging* **2016**, *35*, 1285–1298. [CrossRef] [PubMed]
32. He, K.; Zhang, X.; Ren, S.; Sun, J. Deep residual learning for image recognition. In Proceedings of the IEEE Conference on Computer Vision and Pattern Recognition, Las Vegas, NV, USA, June 2016.
33. Olaf, R.; Fischer, P.; Brox, T. U-net: Convolutional networks for biomedical image segmentation. In *International Conference on Medical Image Computing and Computer-Assisted Intervention*; Springer: Cham, Switzerland, 2015.
34. Jiang, M.; Liang, Y.; Feng, X.; Fan, X.; Pei, Z.; Xue, Y.; Guan, R. Text classification based on deep belief network and softmax regression. *Neural Comput. Appl.* **2018**, *1*, 61–70. [CrossRef]
35. Jiwoong, I.D.; Tao, M.; Branson, K. An empirical analysis of the optimization of deep network loss surfaces. *arXiv* **2016**, arXiv:1612.04010.
36. Ruder, S. An overview of gradient descent optimization algorithms. *arXiv* **2016**, arXiv:1609.04747.
37. Kingma, D.P.; Ba, J. Adam: A method for stochastic optimization. *arXiv* **2014**, arXiv:1412.6980.
38. Shamir, R.R.; Duchin, Y.; Kim, J.; Sapiro, G.; Harel, N. Continuous dice coefficient: A method for evaluating probabilistic segmentations. *arXiv* **2019**, arXiv:1906.11031.

Article

Automatic Fabric Defect Detection Using Cascaded Mixed Feature Pyramid with Guided Localization

You Wu, Xiaodong Zhang and Fengzhou Fang *

State Key Laboratory of Precision Measuring Technology & Instruments, Centre of Micro/Nano Manufacturing Technology—MNMT, Tianjin University, Tianjin 300072, China; 13943173135@163.com (Y.W.); zhangxd@tju.edu.cn (X.Z.)
* Correspondence: fzfang@tju.edu.cn

Received: 17 January 2020; Accepted: 4 February 2020; Published: 6 February 2020

Abstract: Generic object detection algorithms for natural images have been proven to have excellent performance. In this paper, fabric defect detection on optical image datasets is systematically studied. In contrast to generic datasets, defect images are multi-scale, noise-filled, and blurred. Back-light intensity would also be sensitive for visual perception. Large-scale fabric defect datasets are collected, selected, and employed to fulfill the requirements of detection in industrial practice in order to address these imbalanced issues. An improved two-stage defect detector is constructed for achieving better generalization. Stacked feature pyramid networks are set up to aggregate cross-scale defect patterns on interpolating mixed depth-wise block in stage one. By sharing feature maps, center-ness and shape branches merges cascaded modules with deformable convolution to filter and refine the proposed guided anchors. After balanced sampling, the proposals are down-sampled by position-sensitive pooling for region of interest, in order to characterize interactions among fabric defect images in stage two. The experiments show that the end-to-end architecture improves the occluded defect performance of region-based object detectors as compared with the current detectors.

Keywords: fabric defect; object detection; mixed kernels; cross-scale; cascaded center-ness; deformable localization

---

1. Introduction

Industrial defect detection is important in manufacturing. Specifically, fabric defect control is the main content of quality control in the textile industry, which would significantly increase the additional processing costs of the fabric. The cost is derived from manual positioning and the detection of defects and suspending to remove them. On the one hand, manual quality inspections are inefficient and they must often be seen under good backlighting. On the other hand, there is no quantitative defect classification indicator or boundary. This can result in false or mis-detection, and it is not conducive to the late repair of defects or the removal of defects before they occur.

With the popularization of artificial intelligence, the automatic detection algorithm is gradually replaced by the data-based intelligent learning algorithm from the traditional extraction method based on feature values and low-dimensional pixel features. When compared with the traditional algorithm, the heuristic learning algorithm has the advantages of high recognition precision, strong generalization ability, no need to construct complex analytical relations, and small sensitivity range for hyper-parameters. The intelligent detecting methods are divided into unsupervised learning and supervised learning, both of which have gained good performance in defect detection. For the former, Ahmed et al. [1] proposed to conduct the low rank and sparse decomposition jointly and extract weaker defects feature based on wavelet integrated alternating dictionary matrix transformation; Gao et al. [2] utilized an unsupervised sparse component extraction algorithm to detect micro defects in a thermography imaging system by building an internal sub-sparse grouping mechanism and adaptive fine-tuning strategy; Wang et al. [3] established a successive optical flow for projecting the thermal diffusion and constructed principal component analysis to further mine the spatial-transient patterns for strengthening the detectability and sensitivity; Hamdi et al. [4] utilized non-extensive standard deviation filtering and K-means to cluster fabric defect block; Mei et al. [5] reconstruct fabric defect image patches with a convolutional denoising auto-encoder network at multiple Gaussian pyramid levels and to synthesize the detection results from the corresponding resolution channels. For the latter, detecting methods that are mainly based on convolutional deep learning [6] and researchers take an effort in optimizing architecture. Li et al. introduced DetNet [7], which was specifically designed to keep high resolution feature maps for prediction with dilated convolutions to increase receptive fields. Ren et al. proposed the Region Proposal Network (RPN) [8] to generate proposals in a supervised way based on sliding convolution filters. For each position, anchors (or initial estimates of bounding boxes) of varying size and aspect ratios were proposed. Liu et al. [9] learned a lightweight scale aware network to resize images, such that all objects were in a similar scale. Singh et al. [10] conducted comprehensive experiments on small and blurred object detection. Girshick et al. [11] proposed the Region of Interest Pooling (RoI-Pooling) layer to encode region features, which is similar to max-pooling, but across (potentially) different sized regions. He et al. [12] proposed RoI-Align layer, which addressed the quantization issue by bilinear interpolation at fractionally sampled positions within each grid due to the misalignment of the object position from down-sampling operation. Based on RoI-Align, Jiang et al. [13] presented PrRoI-Pooling, which avoided any quantization of coordinates and it had a continuous gradient on the bounding box coordinates.

Although generic models are simple and easy to deploy, they are either over-fitting or feature extracting insufficiency. More targeted detectors for defect images need to be re-established, owing to the dependence on the quality and manifold distribution of the dataset. A variety of imbalances in the image data structure lead to the difficulty of defect identification. The main contributions of this work lie in several aspects due to the misalignment of the object position from down-sampling operation. Firstly, a large imbalanced fabric defect dataset is collected and especially selected for training and validating a robust detector. Secondly, an efficient architecture for defect detection is well-designed, along with improved sub-modules from generic architectures for object detection. Third, the experiments reveal that overfitting and feature extracting sufficiency are the main causes for the accuracy loss of defect detection. By simplifying models, not only is the accuracy improved, but parameter space is also compressed, which reduces the inference delay and lays the foundation of industrial real-time defect detection.

The remainder of this paper is structured, as follows: Section 2 introduces a dataset well-designed on principles of imbalanced variance of real-scene fabric defect images aiming at exploring a robust detector framework and validating its generalization when compared with others. Section 3 reveals the general network configuration for fabric defect detection. A simplified backbone with mixed convolution is proposed for avoiding over-fitting, a composite interpolating pyramid is used for deep feature fusion and a cascaded center-ness refining block is provided for localization regression. Section 4 states the experimental settings and related work in the comparison between proposed network and other generic ones. The proposed modules for fabric defect are estimated effective on location by an ablation study. Finally, Section 5 concludes the paper.

2. Data Space

The feature unwrapping of high-dimensional abstract data space is the core task of deep learning. The data set connects the real scene and semantic algorithm. Therefore, the sample space approximation considers the real space, according to the maximum likelihood principle. This study trains the generalization ability in the corresponding unbalance of real data by constructing the imbalance of sample data. To date, there is still no fabric defect dataset that is adequate and classic in the size, class, and background variations. This paper proposes large-scale optical fabric images, which is named as Fabric Defects (FBDF) and consists of 2575 images covered by 20 defect categories and 4193 instances, in order to address the aforementioned problems. All the raw defect instances and fabric images are collected from textile mills in Guangdong Province, China. These images are selected from hundreds of fabric products with classical defects and labeled in 20 classes according to product demand and expert experience. Details and access are available in https://github.com/WuYou950731/Fabric-Defect-Detection.

2.1. Defect Class Selection

Selecting appropriate texture defect classes is the first step of constructing the dataset. Ambiguous category and bounding is one of the major issues for industrial datasets, in other words, they are too blurred to label accurately, despite experts making sure that it is a defect. Therefore, defect classes selected need to be high-resolution and relatively salient when compared with background and other categories. Some categories that are not common in real-world applications are not included in FBDF and some fine-grained categories are considered as a child category. For example, some stains that are common, clear, and play an important role in textile manufacturing environment analysis, such as oil stains, rust strains, and dye stains, are labeled as the same. In addition, most of the defect categories in existing datasets are selected from gray cloth, which are from the substrate with primary colors rather than decorative patterns. However, defects appear not only in the production stage, but also in transportation, sorting, and even cutting. Therefore, different patterns and backgrounds would be taken into account in FBDF beside texture of fabric, which could be the interference of detectors. By overall selecting classes and image properties, Table 1 shows the samples of FBDF.

Table 1. Samples of different classes of Fabric Defects (FBDF).

Feet	Particles	Knots	Spandex	Rg-Warp	Stains

2.2. Characteristics of FBDF Dataset

A good detector should be of high generalization ability and a good dataset should be a benchmark and guidance in testing and training, respectively. When dataset is relatively interior balanced [14], especially for natural images, such as MS-COCO (Microsoft Common Objects in Context) [15] and Pascal-VOC (Pattern Analysis, Statistical Modelling and Computational Learning Visual Object Classes) [16], different architectures of detectors published have similar performance. However, for industrial dataset, which is seriously imbalanced in image quality, size, and background, some techniques do not work anymore. It is necessary to introduce the imbalance into FBDF in order to train the recognition capability of defect detector. Four key characteristics are highlighted.

- **Large scale**. FBDF consists of 2k optical fabric images and 4k defect instances that are manually labeled with axis-aligned bounding boxes. The size of images is all 2446 × 1000 pixels and the spatial resolution could be down to 0.5 mm. FBDF is collected from the Ali Cloud by the experts in the domain of textile engineering.
- **Instance size and number variations**. Spatial size variation represents actual feature of fabric defects in industrial scene. This is not only because of the spatial resolutions of sensors, but also due to between-class size variation (e.g., "Knots" vs. "Indentation Marks") and within-class size variation (e.g., "Rough Warp" vs. "Loose Warp"). There is a large range of size variations of defect instances in the proposed FBDF dataset, as shown in Figure 1. For each class of fabric, area, height-width ratio, and the number of instances are various and widely ranged. Few-shot recognition ability of detectors could be validated by number variation; multi-scale recognition ability is from area and height-width ratio variation.

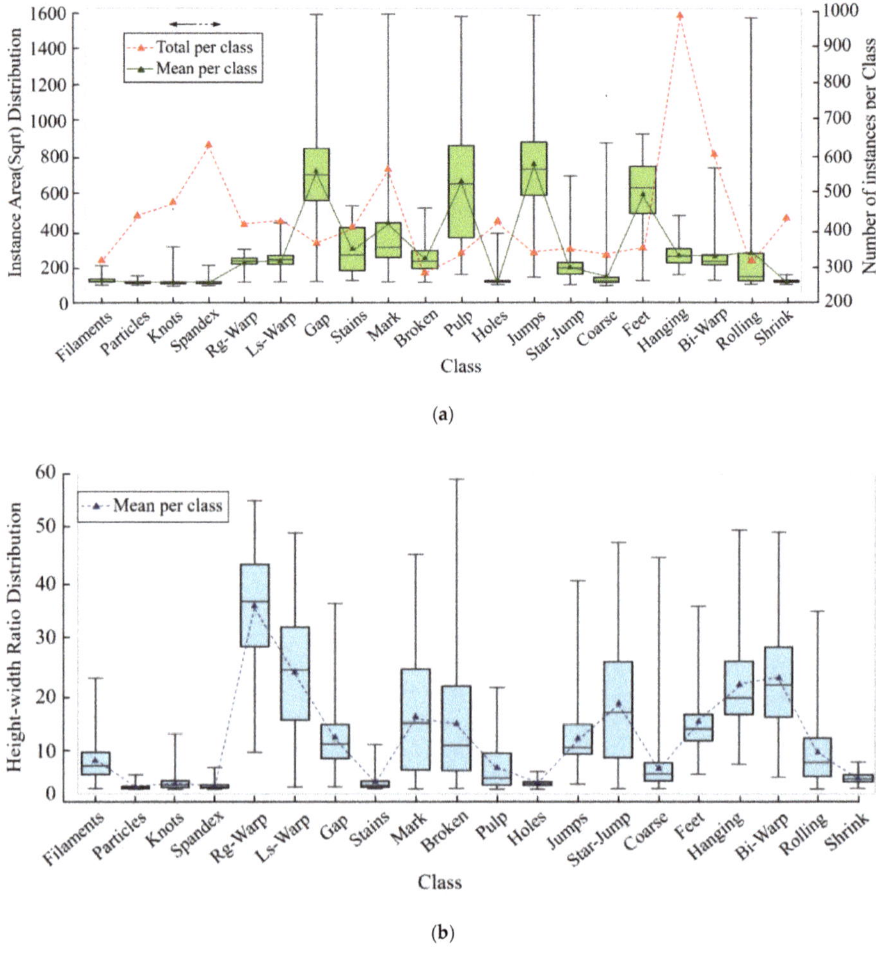

(a)

(b)

Figure 1. Imbalance distribution of instances for FBDF. (**a**) Area and number distribution of instances per class. (**b**) Size ratio distribution of bounding boxes per class.

- **Image variations.** A highly desired characteristic for any defect detection system is its robustness to image variations, concerning different textile, cloth pattern, back-light intensity, imaging conditions, etc. Textile is mainly from denim, muslin, satin and so on. Back-light is controlled to guarantee the sharpness of images. Because of the variations in viewpoint, translation is not that important as compared to illumination, background, and defect appearance for each defect class, so they are simplified in FBDF.

- **Inter-class similarity and intra-class diversity.** Inter-class similarity leads to False Positive (FP) and intra-class diversity leads to False Negative (FN) in classifying module of detectors. Comparable defect images in different class are collected without salient modification to obtain the former. To increase the latter, different defect colors, shapes, and scales are taken into account when selecting images. "Spandex" instances present distinguished shapes, and "Jumps" and "Star-jumps" instances are the opposite.

To sum up, FBDF are designed with the purpose of containing imbalanced image data, which are common and practical in textile scenes. FBDF provides a criterion data space for them to learn, fit, and represent to make detectors adapt to various environments, sizes, and classes.

3. Methodology

The state-of-the-art object detectors with deep learning can be mainly divided into two major categories: two-stage detectors and one-stage detectors. For a two-stage detector, in the first stage, a sparse set of proposals is generated; and, in the second stage, deep convolutional neural networks encode the feature vectors of generated proposals, followed by making object class predictions. A one-stage detector does not have a separate stage for proposal generation (or learning a proposal generation). They typically consider all positions on the image as potential objects, and they try to classify each region of interest as either background or a target object. Although two-stage detectors generally fall short in terms of lower inference speeds, they often reported state-of-the-art results on dark and low-saliency defect detection.

As shown in Figure 2, an end-to-end defect detector is composed by data input, backbone for feature extraction, neck for feature fusion and enhancement, RPN for anchor generation, and head for training or inference. In this section, more details of the framework and learning strategies of fabric defect detection application will be introduced.

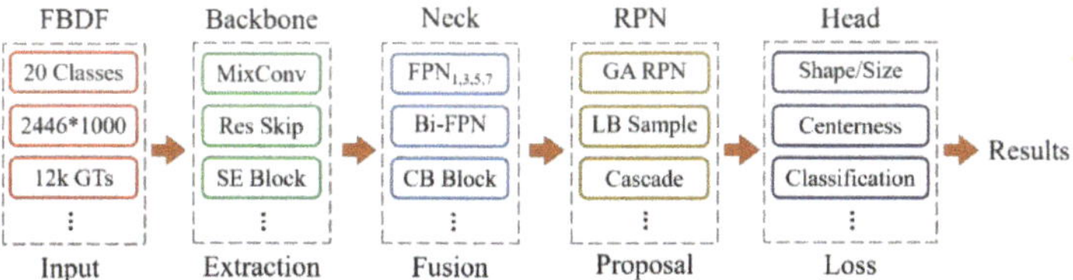

Figure 2. End-to-end fabric defect detection architecture.

3.1. Backbone for Feature Extraction

Detectors are usually trained based on high-dimension semantic information by adopting convolutional weights to reserve image spatial transformation. R-CNN (Region Convolutional Neural Network) [11] showed the classification ability of backbone is consistent with the location ability in detecting. Moreover, the amount of backbone parameters, which are positive correlation with detection performance, is majority of that in detectors. In this section, the trade-off between latency and accuracy of final forward inference is the main design principal. A lightweight mixed depth-wise convolutional network is introduced for enhancing feature extraction within one single layer, since the imbalance of multi-scale features derived from fabric defect.

Mixed convolution [17] utilized different receptive fields to fuse multi-scale local information by diverse kernel and group sizes. Squeezed and excited [18] branch distinguished the salience of feature layer with visual attention mechanism [19] and residual skip connection [20] deepened semantic extraction and decoding. Inspired by these works, the configuration of network would contain these sub-models and be adjusted for textile dataset in order to lower the burden of RoI extractor to recognize occluded and blurred objects.

Mixed convolution (MC) partitions channels into groups and applies different kernel sizes to each group, as shown in Figure 3. Group size g determines how many different types of kernels to use for a single input tensor. In the extreme case of g = 1, a mixed convolution becomes equivalent to a vanilla depth-wise convolution [21–23]. The experiments reveal that g = 5 is generally a safe choice for defect detection, which is size-imbalanced and the maximum area ratio is near 25, as illustrated in Figure 1a. For kernel size per group, if two groups share the same kernel size, it is equivalent to merging these two groups into a single one. Hence, each one should be restricted in different kernel size. Furthermore, kernel size is design to starts from 3×3, and monotonically increases by two per group, since small-size kernels generally possess less parameters and floating-point operations per second (FLOPS). Under this circumstance, the kernel size for each group is predefined for any group size g, thus simplifying the designing process. On the other hand, for channel size per group, exponential partition is more generalized than equal partition, since a smaller kernel size fuses less global information, but acquire more channels to compensate local details.

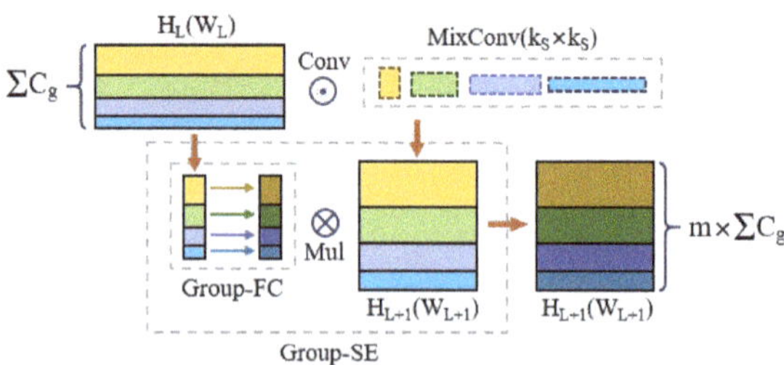

Figure 3. Mixed convolutional module for feature extraction.

Table 2 states the main specification for feature extraction backbone. SE denotes whether there is a squeezed excited module in that block. AF means the type of nonlinear activation function. Here, HS is for h-swish [24] and RE for ReLU. Batch normalization is used after convolution operations. The stride could be deduced from other information of layer and, here, would be passed over. EXP Size denotes the expansion of the convolution inherit from MobileNetv2 [23], which avoids the loss of pixel feature appeared in ResNet, and the number of elements in the list reveals the times while using Bneck [25]. The FPN column illustrates whether there is a head to introduce the feature map to FPN layers [26]. Operators are mainly mixed convolution parameters and $\{3 \times 3, 5 \times 5, 7 \times 7\}$ means that the group number is 3 and they are filtered by these kernels, respectively.

Table 2. Specification for mixed convolution (MC) backbone.

Input	Operator	EXP Size	AF	SE	FPN
$2446 \times 1000 \times 3$	3×3	-	RE	-	-
$896 \times 448 \times 16$	3×3	{40, 72}	RE	-	-
$448 \times 224 \times 24$	$3 \times 3, 5 \times 5$	{72, 72}	HS	√	√
$224 \times 112 \times 40$	3×3	{72, 120, 240}	HS	√	-
$112 \times 56 \times 80$	$3 \times 3, 5 \times 5, 7 \times 7$	{200, 240}	HS	√	√
$56 \times 28 \times 112$	$3 \times 3, 5 \times 5, 7 \times 7, 9 \times 9, 11 \times 11$	{240, 480}	RE	√	√
$28 \times 14 \times 160$	$3 \times 3, 5 \times 5$	{480, 672}	HS	√	√

3.2. Neck for Feature Integrating and Refining

Multi-scale feature fusion aims to aggregate features at different resolution necks. Formally, the multi-scale feature map of conventional FPN is defined as an iteration term:

$$P_i^{out} = Conv\left(P_i^{in} + Sample\left(P_{i+1}^{out}\right)\right) \tag{1}$$

in which,

$$\overrightarrow{P}^{in} = \left(P_{l_1}^{in}, P_{l_2}^{in}, \ldots\right) \tag{2}$$

Sample is an up-sampling or down-sampling operation for resolution matching, and *Conv* represents convolutional feature processing. In Equation (1), p_i^{in} is the i-th input feature layer with different pixel resolution. p_i^{out} is the i-th output feature layer in the other back propagation path with the same resolution as p_i^{in}. In Equation (2), $\overrightarrow{P}^{in}$ represents the parallel feature input flow with interpolating scales. As the foundation of scale fusion, FPN offers two crucial conclusions for defect detection use. Firstly, the defect instance of any scale could be unified to the same resolution as long as stride and kernel are well designed. Secondly, the sampling feature could reserve the most useful information of defect image and defect position is different from background mainly lies in its pixel brightness. This is slightly inconsistent with natural image dataset and more approaching to the principle of maximum pooling. However, the simple top-down FPN is inherently limited by the one-way feature flow.

Fusing layers need to be cross-scale connected with each other, which derived from compression or interpolation, to continue the strengthening features. Additionally, fusion operation focuses on two aspects, aggregation path, and expansion path. For the former, PANet [27] adds an extraction bottom-up path and CBNet [28] overlays parallel feature maps in different size. The latter outperforms the former for defect classes in detection, since CBNet possesses more parameters and more aggregating feature, as illustrated in Table 3. For the latter, as shown in Figure 4, NAS-FPN [29] treats up-sampling equally to convolution by employing neural architecture search and utilize large-ratio connection to deepen the above two operations. However, simplified configuration, especially unexplained topology of NAS [30] detectors and EfficientDet [31] series, takes the efficiency as the loss of semantic precision.

Table 3. Performance of the state-of-the-art FPN baselines on FBDF.

Baseline	Backbone	AP	AP_{50}	AP_{75}
FPN	ResNet-50	34.29	52.01	36.68
NAS-FPN	AmoebaNet	36.16	55.74	40.77
PANet	VGG-16	39.51	60.17	42.14
EfficientDet	EfficientNet-B4	46.39	65.68	50.27
CBNet	ResNet-50	44.22	60.38	46.96

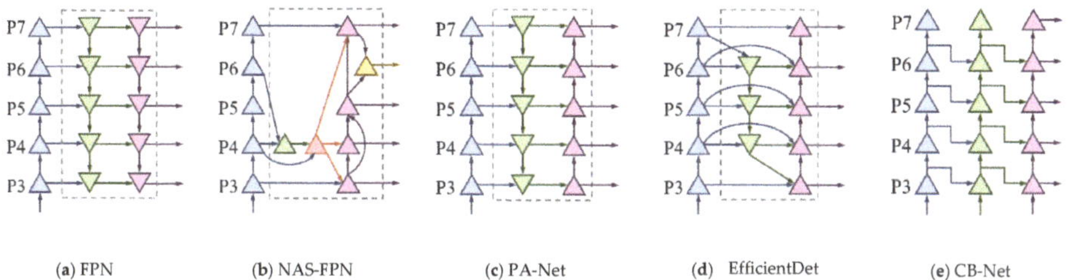

Figure 4. Feature pyramid network frameworks.

For low-salience defects, this paper proposes several designing principles for neck feature fusion:

(1) **Compress feature extraction maps.** Defect images share a common dark background and pixel values of object instance have no big difference, so there is no need to enlarge the number of kernels for a layer.

(2) **Add cross-scale fusion without extra computations.** Nodes derived from one input edge are supposed to remove for its low-level semantic representation and aggregation between input and output from the same level made defect region more visually clear.

(3) **Repeat bidirectional (top-down & bottom-up) block.** Unlike PANet, which only has one bidirectional block, the network that is proposed in this paper advices cascaded modules to enable more high-level feature fusion.

(4) **Keep no same lateral size in a repeated block.** Unlike EfficientDet (stacking Bi-FPN) that keep the lateral sizes of up-sampling and down-sampling the same, this paper applies the interpolating layer for approximately continuous extension of scale, as shown in Figure 5. Feature loss could be reduced with least amount of extra latency by this way.

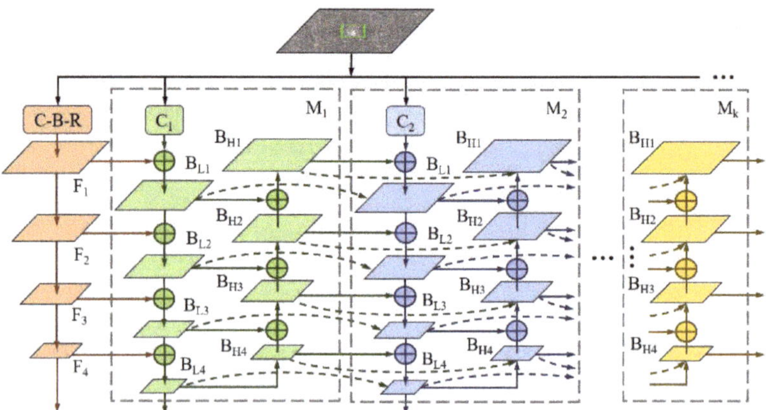

Figure 5. Composite interpolating feature pyramid (CI-FPN).

After setting the general fusion operation, weighted pixel analysis is necessary for feature refinement. Since different input layers are at distinguished resolutions, they usually unequally contribute to the output. Previous fusion methods treat all inputs equally without distinction and cause bounding regression drift. Based on conventional resolution resizing and summing up, the paper proposes to weight the salience of feature layer, as follows:

$$B_L^j(i) = Conv\left(\frac{w_1 \cdot B_L^{j-1}(i) + w_2 \cdot S_L^j\big(B_L^j(i)\big) + w_3 \cdot S_H^{j-1}\big(B_H^{j-1}(i-1)\big)}{w_1 + w_2 + w_3 + \epsilon}\right) \tag{3}$$

$$B_H^j(i) = Conv\left(\frac{w_1 \cdot S_L^j\big(B_L^j(i+1)\big) + w_2 \cdot S_H^j\big(B_H^j(i+1)\big)}{w_1 + w_2 + \epsilon}\right) \tag{4}$$

In Equations (3) and (4), the feature flow between M_i and M_{i+1} mainly build up on two kind of blocks B_L and B_H, which reveal the j-th module, i-th layer, and extraction path as well as interpolating path. Moreover, learnable weight w_i is scalar in the feature level, which is comparably accurate to tensor in pixel level, yet with minimal time costs. Normalization is resorted to bound the data fluctuation and h-swish replaces Softmax to assign probability to each weight here ranges from 0 to 1 and it alleviates the truncation error at the origin point. $\epsilon = 0.001$ is a disturbance constant to avoid numerical instability.

3.3. Anchor Sampling and Refining

Region anchors, which are the cornerstone of learning-based detection, play a role in predicting proposals from predefined fixed-size candidates. Selecting positive instances from a large set of densely distributed anchors manually is time-consuming and limited to finite size variance. Some defect instances contained extreme sizes and regression distance between ground truth and anchor may be great. Therefore, in the first stage, the detection pipelines of this paper focus on guided anchor (GA-Net) [32] mechanism to predict centers and sizes of proposals from FPN outputs and, in the second stage, regression and classification are conducted after feature fusion and alignment by position-sensitive (PS) RoI-Align [33], as shown in Figure 6.

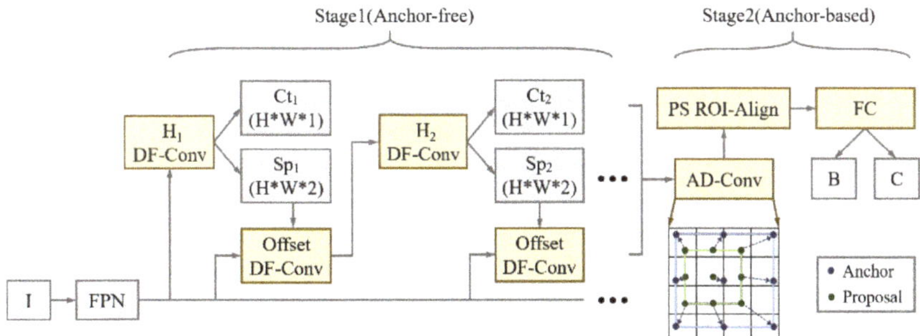

Figure 6. Cascaded Guided-Region Proposal Network (CG-RPN) with semantics-guided refinement.

3.3.1. Stage for Proposal Generation

All of the proposals would be regressed to bounding boxes of final prediction, thus the quality of generator is crucial. Following the paradigm of GA-Net, RPN comprised of two branches for location and shape prediction, respectively. Given a FPN input F, the location prediction branch Ct (Center-ness) yields a probability map that indicates the Sigmoid scores for centers of the objects by Conv1*1, while the shape prediction branch Sp (Shape) predicts the location-dependent sizes. This branch will lead shapes to the highest coverage with the nearest ground-truth bounding box. Two channels represent two variables height and width but it is necessary to be transformed by Equation (5) for the large range and instability of them.

$$
B_H^j(i) = Conv\left(\frac{w_1 \cdot S_L^j\big(B_L^j(i+1)\big) + w_2 \cdot S_H^j\big(B_H^j(i+1)\big)}{w_1 + w_2 + \epsilon} \right)
\tag{5}
$$

in which,

$$
\beta_w = \frac{w_{\max}}{w_{\min}} \cdot \prod_{i=1}^{L} s_i, \ \beta_h = \frac{h_{\max}}{h_{\min}} \cdot \prod_{i=1}^{L} s_i
\tag{6}
$$

where (w, h) are the output of shape prediction, s_i is the stride for different layer L, and β is a scale factor, depending on size of image data. The nonlinear mapping normalizes the shape space from approximately (0, 2000) to (0, 1), leading to an easier and stable learning target. Since it allows for arbitrary aspect ratios, our scheme can better capture those extremely tall or wide defect instances and encode them to a consistent representation.

A further size-adaptation offset map is introduced, as the anchor shapes are supposed to be changeable to capture defects within different ranges. With these branches, a set of anchors are generated by selecting the center-ness whose predicted probabilities are above a slightly lower threshold and several shapes with the top probability at each of the chosen feature position. Subsequently, the center-ness threshold is increased for the refinement of the next anchor-selection module and the policy for shapes of anchors is unchanged. Increasing thresholds are set in different sub-modules to include more probable central points and deal with the misalignment of extreme shaped defects. Center-ness is shifted and updated after DF convolution. By this way, with the aid of dense sampler, a large number of anchors are selected, suppressed, and regressed to 256 proposals for stage two. As shown in Figure 7, yellow boxes are the maximum-IOU (Intersection Over Union) candidates chosen after coarsely locating irregularly-shaped defect instances, which are named cascaded guided RPN (CG-RPN). The red points denote strongly semantic feature positions and blue triangles represent centroids of them.

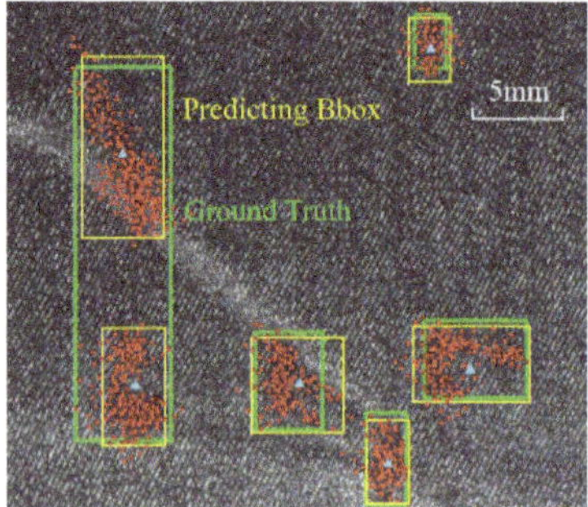

Figure 7. Effective receptive field from deformable extraction in guided localization.

3.3.2. Stage for Bounding Box Generation

The bounding boxes need to be regressed and filtered from a large amount of low-quality anchors and Non-maximum Suppression (NMS) [34] are operated to filter the overlaying ones by local maximum search, whose result is shown in Figure 8. On the other hand, multi-classifying branch fixes the output size of full-connection layer so RoI-Align along with adapting pooling aggregates different fields into shape-identical feature map.

Figure 8. From positive anchors to proposals and finally to bounding boxes.

Moreover, the loss of the proposed detector is divided into four parts: location branch and final classification branch are similar and Focal Loss (FL) [35] and Cross Entropy (CE) Loss optimize them. Shape branch, which compares IOUs by manually assigning central area ranges, use Bounded IOU Loss (BI) conducted by height and width and regression branch is common with Smooth L1 Loss (SM), as following:

$$L = FC_{loc} + BI_{shape} + CE_{cls} + SM_{reg} \tag{7}$$

where *CE* and *SM* are applied in stage two and smoothing factor is set 0.04 to avoid sensitivity to outliers and suppress gradient explosion. *BI* derived from *SM* along with parameters of bounding boxes. γ in *FC* loss is 2 for balancing positive and negative samples.

3.4. Evaluation Metrics for Imbalanced Detection for Defects

Imbalanced detection needs to be evaluated by average recognition precision and variance fluctuation among different categories. Similarity, between ground truth and predicting bounding boxes is proportional to the recognition ability of detectors. Similarity is denoted by IOU based on the Jaccard Index, which evaluates the overlap between two bounding boxes, as shown in Figure 9 and Equation (8), By comparing with confidence threshold, IOU of every instance in every category would divide the prediction results into three aspects: True Positive (TP) denotes a correct detection with IOU $\geq$ threshold, FP denotes a wrong detection with IOU $<$ threshold and FN reveals a ground truth not detected. After counting the number of instances in distinguished quality, a balanced metrics of AP (Average Precision) could be calculated and used for representing the average performance of detection. In Figure 1, the dashed curve is the Recall-Precision Curve, which is denoted by blue bins and whose area is no more than 1 for facilitating consistency with probability. AP could be calculated in Equation (9), in which b is the number of bins and here is 11. P and Δr are the height and width of each bin, respectively.

$$IOU_{(c,i)} = \frac{\text{area}\left(S_{Pred} \cap S_{GT}\right)}{\text{area}\left(S_{Pred} \cup S_{GT}\right)} \tag{8}$$

$$AP = \sum_{b=1}^{n} P(b)\Delta r(b) \tag{9}$$

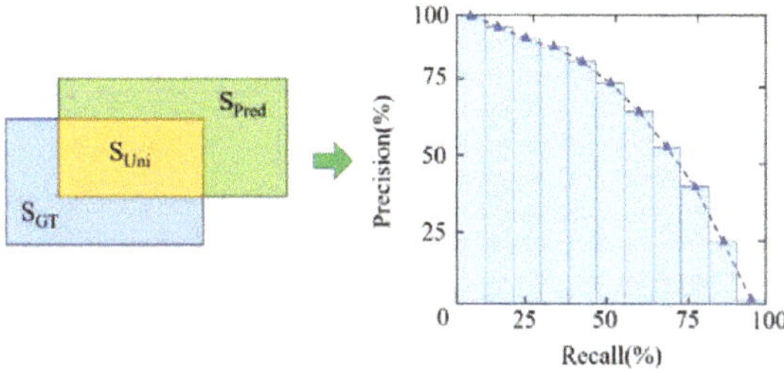

Figure 9. Metrics for performance of imbalanced detection.

For each single class, precision is the ability of a model to only identify the relevant objects. It is the percentage of correct positive predictions and it is given by TP/(TP + FP). Recall is the ability of a model to find all of the relevant cases (all ground truth bounding boxes). It is the percentage of true positive detected among all of the relevant ground truths and it is given by TP/(TP + FN). AP is different among every category due to the imbalance of fabric defects in inter-class and intra-class. Firstly, the mean AP of all categories could be used as the overall performance of detectors and it is named mAP (usually use AP as default). Secondly, for inter-class imbalance, which means that the data distribution of every class significantly differs from each other, the PR curve is better than ROC (Receiver Operating Characteristic), since ROC considers both positive and negative examples. AP focuses on positive ones and Variance Precision (VP) illustrates the inter-class accuracy stability, as in Equation (10). Thirdly, intra-class imbalance mainly lies in size-variance and the AP for small, medium, and large objects is divided by cross scale 96^2 and 256^2.

$$VP = \frac{1}{C} \sqrt{\sum_{k=1}^{C} (AP_k - \text{mAP})^2} \tag{10}$$

4. Experiments and Discussion

4.1. Experimental Settings

The experiments are performed on FBDF to validate whether the modules above could solve the imbalance of the textile industrial scenes. The validation dataset is evenly split from the whole at splitting ratio of 0.2. Additionally image size does not need to be resized and without changing the aspect ratio. Mini-batch stochastic gradient descent and batch normalization [36] are implemented over two TITAN RTX GPUs with 18 images per worker on one GPU. The training epochs are uniformed to 20 and learning rate is decreased every four epochs with a decreasing rate of 0.1. The evaluation metrics is AP at different IOU thresholds (from 0.5 to 0.95). 200 instances of every class from these images are randomly split as the pre-trained classification dataset, with which all backbones of the architectures are initialized, in order to further strengthen feature extraction.

4.2. Main Result

The proposed scheme can be evaluated with other state-of-the-art well-designed methods, as the comparison in Table 4. MC-Net and CI-FPN along with Mixed-16, which is short for Defect-Net and composite interpolating FPN along with mixed convolution network of 16 layers, achieves a remarkable improvement, especially for small defects. It reports a testing AP of 72.6%, an improvement of 12.1% over 60.5%, being achieved by cascaded FPN under the same setting. When using the light-weight backbone (i.e., Mixed-16), the AP improvement over Mixed-16 is 6.7% and AP_S improves 29.1%, which prove the availability of mixed convolution. The phenomenon of the increasing from 60.5% (Cascaded R-CNN) to 65.9% (DF-Net + CI-FPN + ResNet50) and 65.9% to 72.6% (MC-Net + CI-FPN + Mixed-16) prove the guessing of over-fitting from large-scale backbone, especially for small defects. The VP of generic baselines is larger than that of the proposed architecture on average. VP of MC-Net along with CI-FPN and Mixed-16 is the lowest and FCOS is the second one, which reveals that the range of AP for different categories is narrow and distribution is relatively even. However, it does not mean that the more VP is, the higher accuracy detector possesses. Take YOLOv3 as an example, the experiments show that the APs of 11 classes are less than 30% in spite of 10.8 in VP. Additionally, Cascaded-FPN gains 0.3 VP larger than Libra-FPN-RetinaNet, but 4.5% AP larger than that. Therefore, the ability for addressing the imbalance of detectors should be evaluated by a combination of AP and VP.

Table 4. Accuracy of different detectors on FBDF testing set.

Method	Backbone	AP	AP_{50}	AP_{75}	AP_S	AP_L	VP
Faster R-CNN	VGG-16	42.6	57.7	45.8	22.4	53.6	14.2
Faster + FPN	ResNet-50	53.3	69.0	57.7	39.3	64.7	13.8
RetinaNet + FPN	ResNet-50	55.7	73.3	60.9	42.7	66.5	12.6
YOLOv3 [37]	DarkNet-53	35.5	52.5	36.2	19.4	50.5	10.8
SSD-513 [38]	ResNet-101	32.9	54.9	34.1	12.2	48.4	12.9
Cascaded [39] + FPN	ResNet-50	60.5	75.2	66.3	47.1	71.0	11.5
CornerNet [40]	Hourglass-52	36.4	53.0	39.8	19.9	51.2	13.0
CenterNet [41]	Hourglass-52	39.5	57.5	40.6	22.7	54.3	12.6
Libra-FPN-RetinaNet [42]	ResNet-50	56.9	71.4	60.2	38.5	68.9	11.2
FCOS [43]	ResNet-50	33.0	49.8	34.4	19.2	44.3	10.3
MC-Net + CI-FPN	**ResNet-50**	**65.9**	**79.5**	**68.0**	**48.3**	**77.3**	**10.7**
MC-Net + CI-FPN	**Mixed-16**	**72.6**	**86.3**	**73.6**	**50.9**	**80.4**	**9.7**

4.3. Ablation Experiments

Backbone extraction. As MC-Net uses a light-weight powerful backbone, Figure 10 reveals how much each of them contributes to the accuracy and efficiency improvements. Faster R-CNN along with FPN is our baseline for comparison of different backbone. First, the RestNet series are heavy and low-efficiency, which achieve a relatively low accuracy and ResNet-101, along with ResNet-152 are even worse, and are thus are not shown in the figure. When replacing with MobileNet, AP increases from 53% to 61% over MobileNetv3-Large [44] without cropping the images. Mixed series achieve a similar performance with EfficientNet series and Mixed-16 gains the top AP of 67.2% based on FBDF and slightly decreasing from Mixed-20 is due to the redundancy of the weights.

Along with the variant improvement for Faster R-CNN, the MC module is still efficient for other detectors in defect detection. In Table 5, Cascaded FPN gains 6.0% promotion from 66.5% on the ResNet-50 backbone and the average improvement of small instances is 5.5%, which proves MC-Net could extract more and deep feature from the low-salience defects.

Neck fusion. In Figure 11, the AP of composite interpolating FPN is rising with the model complexity expanding. B_n is short for n blocks of two-way information flow. Notably, when three blocks are employed along with inter-layer and intra-block cross-scale connections, the scheme is the most accurate one, with 72.3% (AP), 50.9% (AP_S), and 36.4 MB training parameters.

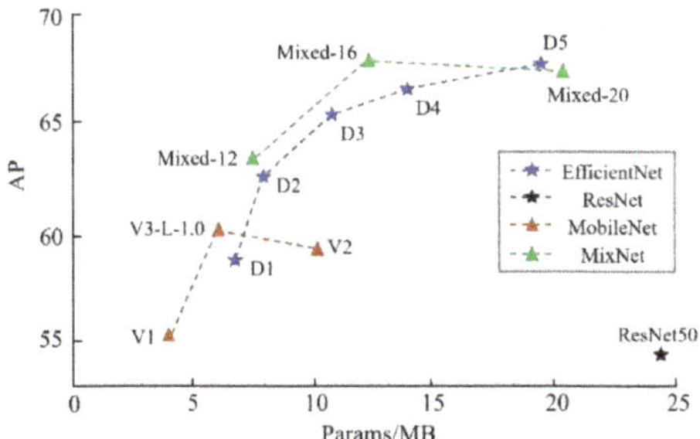

Figure 10. Model sizes with Average Precision (AP) for FBDF of different high-efficiency backbones upon Faster R-CNN and FPN.

Table 5. Performance of Mixed-16 applied in some generic detectors.

Method	Backbone	AP	AP$_{50}$	AP$_{75}$	AP$_S$	AP$_L$
Faster + FPN	Mixed-16	59.2(5.9)	74.2	63.6	43.0	70.1
RetinaNet + FPN	Mixed-16	61.9(6.2)	79.8	66.7	46.5	73.0
Cascaded R-CNN	Mixed-16	63.6(4.5)	78.4	80.1	47.5	72.8
Cascaded + FPN	Mixed-16	66.5(6.0)	83.5	72.5	59.8	77.9

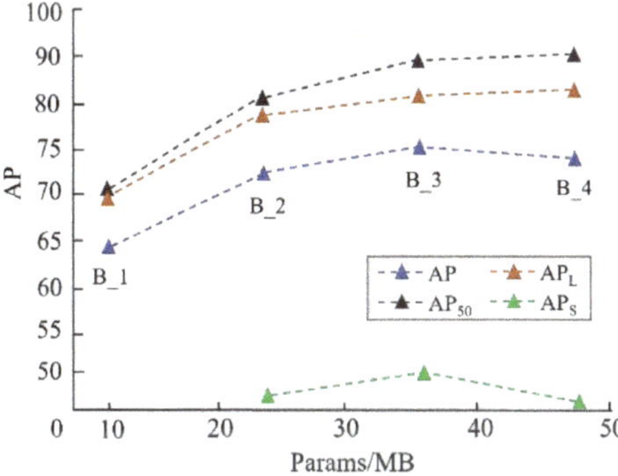

Figure 11. The performance of different configurations of CI-FPN with CG-RPN modules.

Proposals generation. With the deployment of CG-RPN, three feature adaption modules would refine the anchor centers and shapes in stage one. The center-ness thresholds are 0.3, 0.5, and 0.7 in different cycles and, in regression branch, every position in feature map choose three anchors with aspect ratios of 0.5, 1.0, and 2.0 to enlarge the search space. In Figure 11, the left one is from common Faster R-CNN and the right one is from CG-RPN and less low-quality proposals is reserved here. In Table 6, different center-ness configurations are displayed and the tuple (0.3, 0.5, 0.7) is better than the others in AP, since it introduces more computing parameters and relaxes the hard border of whether belonging to positive instances.

Table 6. Performance of CG-RPN with different center-ness threshold tuples.

Tuple	Mean IOU (Post)	AP	AP$_S$
(0.5)	46.5	71.4	39.1
(0.5, 0.7)	**59.7**	**75.8**	**44.8**
(0.3, 0.5)	53.6	72.9	41.3
(0.3, 0.5, 0.7)	**56.8**	**76.1**	**46.5**

Additionally, from the pre-trained model of MC-Net with CI-FPN, several bounding boxes and confidence values are shown in Figures 12 and 13.

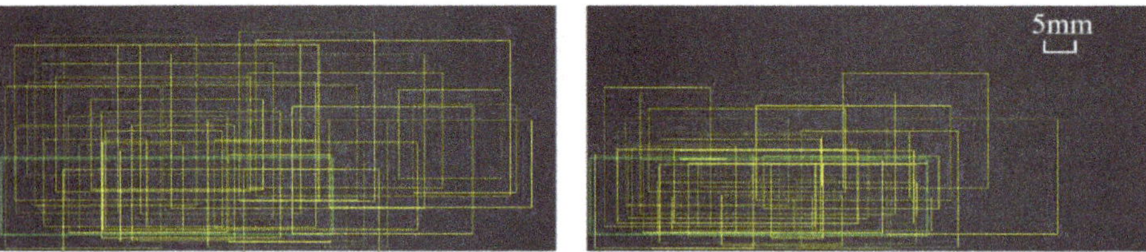

Figure 12. Sample RPN proposals from Faster R-CNN and C-G RPN.

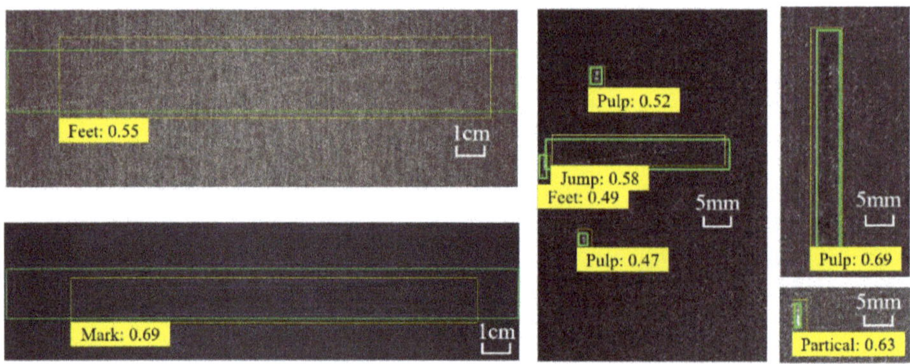

Figure 13. Visualization of the best bounding boxes of defects from MC-Net along with CI-FPN.

For Area Under Curve (AUC), the MC-Net along with Mixed-16 gains better performance than AP: 76.9% for mean AUC, 55.4% for small defects and 82.6% for large defects. For ConerNet and CenterNet, the mean AUC increases 3.5% and 4.3% and some small promotions in accuracy appear in other detecting systems. However, in textile industry, positive examples draw more attention than negative examples and a detector that is robust to sensitive metrics. When negative examples increase a lot, the curve does not change a lot, which is equivalent to generating a large number of FP. In the context of imbalanced categories, the large number of negative cases makes the growth of FPR (FPR = FP/(FP + TN); TPR = TP/(TP + FN)) not obvious, resulting in an ROC curve that shows an overly optimistic effect estimate. Finally, misdetection would lead to constant interruptions of machine tools, which results in low efficiency in manufacturing. Therefore, in this work, the ROC curve is replaced for the PR curve.

5. Conclusions

This study solves the problem of the imbalanced detection for fabric defect. Firstly, a large-scale, publicly available dataset for defect detection in optical fabric defect images is released, which enables the community to validate and develop data-driven defect detection methods. Secondly, several modules to refine traditional inefficient network are designed, including mix convolutional backbone, interpolating FPN, and cascaded guided anchor, etc., in order to improve recognition performance of occluded and size-variant defects Finally, the study shows the importance of these frameworks in defect detecting and provides a scheme for precisely meeting the needs of the textile industry.

Author Contributions: Conceptualization, Y.W. and X.Z.; methodology, Y.W.; software, Y.W.; validation, Y.W.; formal analysis, Y.W., X.Z. and F.F.; investigation, Y.W. and F.F.; resources, X.Z.; data curation, Y.W.; writing—original draft preparation, Y.W.; writing—review and editing, Y.W., X.Z. and F.F.; visualization, Y.W.; supervision, F.F.; project administration, X.Z. and F.F.; funding acquisition, F.F. All authors have read and agreed to the published version of the manuscript.

Funding: This research was funded by the "111" Project by the State Administration of Foreign Experts Affairs and the Ministry of Education of China (No. B07014).

Acknowledgments: The authors would like to thank the financial support received from National Natural Science Foundation of China (NSFC) (No. 51320105009 & 61635008) and the "111" Project by the State Administration of Foreign Experts Affairs and the Ministry of Education of China (No. B07014). Thanks go to the Alibaba Group for the dataset of fabric defect images.

Conflicts of Interest: The authors declare no conflict of interest.

References

1. Ahmed, J.; Gao, B.; Woo, W.L. Wavelet Integrated Alternating Sparse Dictionary Matrix Decomposition in Thermal Imaging CFRP Defect Detection. *IEEE Trans. Ind. Inform.* **2019**, *5*, 4033–4043. [CrossRef]
2. Gao, B.; Lu, P.; Woo, W.L.; Tian, G.Y.; Zhu, Y.; Johnston, M. Variational Bayesian Sub-group Adaptive Sparse Component Extraction for Diagnostic Imaging System. *IEEE Trans. Ind. Electron.* **2019**, *5*, 4033–4043.
3. Wang, Y.; Gao, B.; lok Woo, W.; Tian, G.; Maldague, X.; Zheng, L.; Guo, Z.; Zhu, Y. Thermal Pattern Contrast Diagnostic of Microcracks With Induction Thermography for Aircraft Braking Components. *IEEE Trans. Ind. Inform.* **2018**, *14*, 5563–5574. [CrossRef]
4. Hamdi, A.A.; Sayed, M.S.; Fouad, M.M.; Hadhoud, M.M. Unsupervised patterned fabric defect detection using texture filtering and K-means clustering. In Proceedings of the International Conference on Innovative Trends in Computer Engineering, Aswan, Egypt, 19–21 February 2018; pp. 130–144.
5. Mei, S.; Wang, Y.; Wen, G.J. Automatic Fabric Defect Detection with a Multi-Scale Convolutional Denoising Autoencoder Network Model. *Sensors* **2018**, *18*, 1064. [CrossRef] [PubMed]
6. Seker, A.; Peker, K.A.; Yuksek, A.G.; Delibaş, E. Fabric defect detection using deep learning. In Proceedings of the 24th Signal Processing and Communication Application Conference (SIU), Zonguldak, Turkey, 16–19 May 2016.
7. Li, Z.; Peng, C.; Yu, G.; Zhang, X.; Deng, Y.; Sun, J. Detnet: A backbone network for object detection. *arXiv* **2018**, arXiv:1804.06215.
8. Ren, S.; He, K.; Girshick, R.; Sun, J. Faster R-CNN: Towards Real-Time Object Detection with Region Proposal Networks. *IEEE Trans. Pattern Anal. Mach. Intell.* **2017**, *39*, 1137–1149. [CrossRef]
9. Liu, Y.; Li, H.; Yan, J.; Wei, F.; Wang, X.; Tang, X. Recurrent Scale Approximation for Object Detection in CNN. In Proceedings of the IEEE International Conference on Computer Vision (ICCV), Venice, Italy, 22–29 October 2017; pp. 571–579.
10. Singh, B.; Davis, L.S. An analysis of scale invariance in object detection snip. In Proceedings of the IEEE Conference on Computer Vision and Pattern Recognition, Salt Lake City, UT, USA, 18–23 June 2018; pp. 3578–3587.
11. Girshick, R. Fast R-CNN. In Proceedings of the 2015 IEEE International Conference on Computer Vision (ICCV), Santiago, Chile, 7–13 December 2015; pp. 1440–1448.
12. He, K.; Gkioxari, G.; Dollár, P.; Girshick, R. Mask R-CNN. In Proceedings of the IEEE International Conference on Computer Vision, Venice, Italy, 22–29 October 2017; pp. 2961–2969.
13. Jiang, B.; Luo, R.; Mao, J.; Xiao, T.; Jiang, Y. Acquisition of localization confidence for accurate object detection. In Proceedings of the European Conference on Computer Vision (ECCV), Munich, Germany, 8–14 September 2018; pp. 784–799.
14. Li, K.; Wan, G.; Cheng, G.; Meng, L.; Han, J. Object detection in optical remote sensing images: A survey and a new benchmark. *ISPRS J. Photogramm. Remote Sens.* **2018**, *159*, 296–307. [CrossRef]
15. Lin, T.Y.; Maire, M.; Belongie, S.; Hays, J.; Perona, P.; Ramanan, D.; Zitnick, C.L. Microsoft coco: Common objects in context. In Proceedings of the European Conference on Computer Vision, Zurich, Switzerland, 6–12 September 2014; pp. 740–755.
16. Everingham, M.; Van Gool, L.; Williams, C.K.; Winn, J.; Zisserman, A. The pascal visual object classes (voc) challenge. *Int. J. Comput. Vis.* **2010**, *88*, 303–338. [CrossRef]
17. Tan, M.; Le, Q.V. Mixconv: Mixed depthwise convolutional kernels. *arXiv* **2019**, arXiv:1907.09595.
18. Hu, J.; Shen, L.; Sun, G. Squeeze-and-excitation networks. In Proceedings of the IEEE Conference on Computer Vision and Pattern Recognition, Salt Lake City, UT, USA, 18–23 June 2018; pp. 7132–7141.

19. Chen, Y.; Dai, X.; Liu, M.; Chen, D.; Yuan, L.; Liu, Z. Dynamic Convolution: Attention over Convolution Kernels. *arXiv* **2019**, arXiv:1912.03458.

20. He, K.; Zhang, X.; Ren, S.; Sun, J. Deep residual learning for image recognition. In Proceedings of the IEEE Conference on Computer Vision and Pattern Recognition, Las Vegas, NV, USA, 27–30 June 2016; pp. 770–778.

21. Howard, A.G.; Zhu, M.; Chen, B.; Kalenichenko, D.; Wang, W.; Weyand, T.; Adam, H. Mobilenets: Efficient convolutional neural networks for mobile vision applications. *arXiv* **2017**, arXiv:1704.04861.

22. Zhang, X.; Zhou, X.; Lin, M.; Sun, J. Shufflenet: An extremely efficient convolutional neural network for mobile devices. In Proceedings of the IEEE Conference on Computer Vision and Pattern Recognition, Salt Lake City, UT, USA, 18–23 June 2018; pp. 6848–6856.

23. Sandler, M.; Howard, A.; Zhu, M.; Zhmoginov, A.; Chen, L.C. Mobilenetv2: Inverted residuals and linear bottlenecks. In Proceedings of the IEEE Conference on Computer Vision and Pattern Recognition, Salt Lake City, UT, USA, 18–23 June 2018; pp. 4510–4520.

24. Borji, A.; Cheng, M.M.; Jiang, H.; Li, J. Salient object detection: A benchmark. *IEEE Trans. Image Process.* **2015**, *24*, 5706–5722. [CrossRef] [PubMed]

25. Borji, A.; Itti, L. State-of-the-art in visual attention modeling. *IEEE Trans. Pattern Anal. Mach. Intell.* **2012**, *35*, 185–207. [CrossRef] [PubMed]

26. Lin, T.Y.; Dollár, P.; Girshick, R.; He, K.; Hariharan, B.; Belongie, S. Feature pyramid networks for object detection. In Proceedings of the IEEE Conference on Computer Vision and Pattern Recognition, Honolulu, HI, USA, 21–26 July 2017; pp. 2117–2125.

27. Liu, S.; Qi, L.; Qin, H.; Shi, J.; Jia, J. Path aggregation network for instance segmentation. In Proceedings of the IEEE Conference on Computer Vision and Pattern Recognition, Salt Lake City, UT, USA, 18–23 June 2018; pp. 8759–8768.

28. Liu, Y.; Wang, Y.; Wang, S.; Liang, T.; Zhao, Q.; Tang, Z.; Ling, H. CBNet: A Novel Composite Backbone Network Architecture for Object Detection. *arXiv* **2019**, arXiv:1909.03625.

29. Ghiasi, G.; Lin, T.Y.; Le, Q.V. NAS-FPN: Learning scalable feature pyramid architecture for object detection. In Proceedings of the IEEE Conference on Computer Vision and Pattern Recognition, Long Beach, CA, USA, 15–20 June 2019; pp. 7036–7045.

30. Zoph, B.; Vasudevan, V.; Shlens, J.; Le, Q.V. Learning transferable architectures for scalable image recognition. In Proceedings of the IEEE Conference on Computer Vision and Pattern Recognition, Salt Lake City, UT, USA, 18–23 June 2018; pp. 8697–8710.

31. Tan, M.; Pang, R.; Le, Q.V. Efficientdet: Scalable and efficient object detection. *arXiv* **2019**, arXiv:1911.09070.

32. Wang, J.; Chen, K.; Yang, S.; Loy, C.C.; Lin, D. Region proposal by guided anchoring. In Proceedings of the IEEE Conference on Computer Vision and Pattern Recognition, Long Beach, CA, USA, 15–20 June 2019; pp. 2965–2974.

33. Zhu, Y.; Zhao, C.; Wang, J.; Zhao, X.; Wu, Y.; Lu, H. CoupleNet: Coupling global structure with local parts for object detection. In Proceedings of the IEEE International Conference on Computer Vision, Venice, Italy, 22–29 October 2017; pp. 4126–4134.

34. Neubeck, A.; Van Gool, L. Efficient non-maximum suppression. In Proceedings of the 18th International Conference on Pattern Recognition (ICPR'06), Hong Kong, China, 20–24 August 2006; Volume 3, pp. 850–855.

35. Lin, T.Y.; Goyal, P.; Girshick, R.; He, K.; Dollár, P. Focal loss for dense object detection. In Proceedings of the IEEE International Conference on Computer Vision, Venice, Italy, 22–29 October 2017; pp. 2980–2988.

36. Ioffe, S.; Szegedy, C. Batch normalization: Accelerating deep network training by reducing internal covariate shift. *arXiv* **2015**, arXiv:1502.03167.

37. Redmon, J.; Farhadi, A. YOLOv3: An incremental improvement. *arXiv* **2018**, arXiv:1804.02767.

38. Liu, W.; Anguelov, D.; Erhan, D.; Szegedy, C.; Reed, S.; Fu, C.Y.; Berg, A.C. SSD: Single shot multibox detector. In Proceedings of the European Conference on Computer Vision, Amsterdam, The Netherlands, 11–14 October 2016; pp. 21–37.

39. Cai, Z.; Vasconcelos, N. Cascade R-CNN: Delving into high quality object detection. In Proceedings of the IEEE Conference on Computer Vision and Pattern Recognition, Salt Lake City, UT, USA, 18–23 June 2018; pp. 6154–6162.

40. Law, H.; Deng, J. Cornernet: Detecting objects as paired keypoints. In Proceedings of the European Conference on Computer Vision (ECCV), Munich, Germany, 8–14 September 2018; pp. 734–750.

41. Duan, K.; Bai, S.; Xie, L.; Qi, H.; Huang, Q.; Tian, Q. Centernet: Keypoint triplets for object detection. In Proceedings of the IEEE International Conference on Computer Vision, Long Beach, CA, USA, 15–20 June 2019; pp. 6569–6578.

42. Pang, J.; Chen, K.; Shi, J.; Feng, H.; Ouyang, W.; Lin, D. Libra R-CNN: Towards balanced learning for object detection. In Proceedings of the IEEE Conference on Computer Vision and Pattern Recognition, Long Beach, CA, USA, 15–20 June 2019; pp. 821–830.

43. Tian, Z.; Shen, C.; Chen, H.; He, T. FCOS: Fully Convolutional One-Stage Object Detection. *arXiv* **2019**, arXiv:1904.01355.

44. Howard, A.; Sandler, M.; Chu, G.; Chen, L.C.; Chen, B.; Tan, M.; Le, Q.V. Searching for mobilenetv3. *arXiv* **2019**, arXiv:1905.02244.

Article

Automatic Identification of Tool Wear Based on Convolutional Neural Network in Face Milling Process

Xuefeng Wu *, Yahui Liu, Xianliang Zhou and Aolei Mou

Key Laboratory of Advanced Manufacturing and Intelligent Technology, Ministry of Education, Harbin University of Science and Technology, Harbin 150080, China
* Correspondence: wuxuefeng@hrbust.edu.cn; Tel.: +86-451-8639-0538

Received: 13 August 2019; Accepted: 2 September 2019; Published: 4 September 2019

Abstract: Monitoring of tool wear in machining process has found its importance to predict tool life, reduce equipment downtime, and tool costs. Traditional visual methods require expert experience and human resources to obtain accurate tool wear information. With the development of charge-coupled device (CCD) image sensor and the deep learning algorithms, it has become possible to use the convolutional neural network (CNN) model to automatically identify the wear types of high-temperature alloy tools in the face milling process. In this paper, the CNN model is developed based on our image dataset. The convolutional automatic encoder (CAE) is used to pre-train the network model, and the model parameters are fine-tuned by back propagation (BP) algorithm combined with stochastic gradient descent (SGD) algorithm. The established ToolWearnet network model has the function of identifying the tool wear types. The experimental results show that the average recognition precision rate of the model can reach 96.20%. At the same time, the automatic detection algorithm of tool wear value is improved by combining the identified tool wear types. In order to verify the feasibility of the method, an experimental system is built on the machine tool. By matching the frame rate of the industrial camera and the machine tool spindle speed, the wear image information of all the inserts can be obtained in the machining gap. The automatic detection method of tool wear value is compared with the result of manual detection by high precision digital optical microscope, the mean absolute percentage error is 4.76%, which effectively verifies the effectiveness and practicality of the method.

Keywords: tool wear monitoring; superalloy tool; convolutional neural network; image recognition

---

1. Introduction of Tool Condition Monitoring (TCM)

In milling operations, the quality of machined workpiece is highly dependent on the state of the cutting insert. Factors such as wear, corrosion, or fatigue can affect tool wear. Therefore, monitoring of tool wear in machining process has found its importance to predict tool life, reduce equipment downtime, and optimize machining parameters [1]. The detection method of tool wear can be categorized into two groups: direct and indirect method [2]. Indirect measurement is the use of sensors to measure a signal related to tool wear and obtained by analyzing the signal. Applicable signals that are widely used for tool wear measurement include acoustic emission, force, vibration, current, and power signals, etc. For example, Yao et al. [3] used the acoustic emission signal, the spindle motor current signal, and the feed motor current signal to monitor the tool wear state. Li et al. [4] used acoustic emission (AE) signals to realize the TCM and tool life prediction. Jauregui et al. [5] used cutting force and vibration signals to monitor tool wear state in the high-speed micro-milling process. Prasad et al. [6] analyzed the sound and light-emitting signals during milling and obtained the relationship between tool wear and surface roughness. However, all of these signals are heavily contaminated by the inherent noise in the industrial environment, reducing their performance [7].

Recent advances in digital image processing have suggested the machine vision should be used for TCM. In this case, the direct method of measuring tool wear has higher accuracy and reliability than the indirect method [7]. An unsupervised classification was used to segment the tool wear area through an artificial neural network (ANN) and then used to predict tool wear life [8]. Garcia-Ordas et al. [9] used computer vision technology to extract the shape descriptor of the tool wear area, combined with the machine learning model to achieve TCM. D'Addona et al. [10] used ANN and DNA-based computing, the tool wear degree was predicted based on image information extracted from the pre-processing. Alegre-Gutierrez et al. [7] proposed a method based on image texture analysis for TCM in the edge contour milling. The extracted descriptor of the tool wear area and the cutting parameters as the input parameters of wavelet neural network (WNN) model, which can predict the tool wear degree [11]. Mikolajczyk et al. [12] developed a system that can automatically measure the wear degree of the cutting edge by analyzing its image. Therefore, applying machine vision method in TCM has become more and more mature.

Since deep learning is a network structure with multiple layers of nonlinear series processing units, it is possible to omit data preprocessing and use the original data for model training and testing directly [13]. With deep learning method, enormous breakthroughs have been made in image recognition and classification [14–16], fault diagnosis [17,18], and medical health [19,20], etc., and CNN has become one of the research focuses in artificial intelligence algorithm directions. In the mechanical manufacturing industry, CNN can be used for monitoring the operating conditions of gearboxes, bearings, etc., to identify the faults intelligently and classify the diagnosis [21–23]. In the aspect of TCM, Fatemeh et al. [24] used extracted features to train and test the Bayesian network, support vector machine, K nearest neighbor regression model, and established CNN model. The experimental results show that the CNN model has a higher recognition rate precision for TCM. Zhang et al. [25] converted the original vibration signal into an energy spectrum map by wavelet packet transform, which trained and tested the energy spectrum map based on the CNN to classify the tool wear.

The above methods are used to realize real-time monitoring of the tool wear, but the research is mainly concerned with whether it is worn is qualitative monitoring, and there are relatively few studies of quantitative determination, especially for each cutting edge of the milling tool. The precision of conventional measurement methods, such as using laser measuring instruments, optical projectors, etc. are high, and so is the evaluation criterion for tool wear, but the measurement process needs to be separated from the production process with the characteristics of low efficiency, significant labor intensity, and shut-down measurements. It is challenging to meet the modern production requirements of high efficiency and automated processing [26].

The purpose of this paper is to develop a detection system that automatically recognizes tool wear types and obtains the wear value. It can obtain the wear image information of all the inserts of face milling cutter in the machining gap, and no downtime measurement, just reducing the spindle speed. In the future, the entire system can be integrated into the machine tool computer numerical control (CNC) system to achieve high automation of the machining process and measurement process. We present an on-machine measurement method to measure the tool wear. A tool wear image acquiring device installed on the working platform of the CNC machine tool is used, and the images of each cutting edge are obtained through the frame rate matching of the rotating speed. This paper draws on the VGGNet-16 network model [27] with a strong classification ability and high recognition precision, combined with the characteristics of tool wear images and the actual processing environment, to construct a CNN model to identify wear types in the machining of superalloy tools. Based on the model, the image processing method is developed according to the type of wear to improve the wear value extraction algorithm. The Roberts operator is used to locate the wear boundary and refine the edge image coarsely, thus improving the precision of automatic tool wear value detection (named ATWVD). A sample set of the tool wear image is obtained by the milling process experiments, and the tool wear value can be measured by the method. Precision and efficiency are estimated to meet the requirement of automation and precision in the milling process.

2. Materials and Methods

The tool wear collection device consists of the Daheng image industrial digital camera (model: MER-125-30UC, Daheng company, Beijing, China) which has a resolution of 1292 (H) × 964 (V) and a frame rate of 30 fps. Telecentric lens with a distortion rate of less than 0.08% (model: BT-2336, BTOS company, Xian, China), ring light source (model: TY63HW, Bitejia Optoelectronics company, Suzhou, China) and regulator, multi-function adjustment bracket, and a laptop computer. Three-axis vertical machining center CNC milling machine (model: VDL-1000E, Dalian machine tool company, Dalian, China), 490 R series of CVD coated carbide tools, and PVD coated carbide inserts produced by Sandvik are used in the milling of Inconel 718. The dimensions of the workpiece are 160 mm × 100 mm × 80 mm. The chemical composition of Inconel 718 material is shown in Table 1. There is no coolant during processing. The parameters of the cutting tool are shown in Table 2.

Table 1. Chemical composition of Inconel 718 material.

Elements	C	Mn	Si	S	P	Ni	Cr	Ta
Chemical composition (%)	0.042	0.16	0.28	0.002	0.009	51.4	18.12	0.005
Mo	Al	Ti	B	Nb	Fe	Co	Cu	Pb + Bi
3.0	0.55	0.96	0.004	5.2	balance	0.2	0.11	0.001

Table 2. Parameters of the cutting tool.

Geometric Parameter					Material Properties		
Rake Angle, γ_0 deg	Clearance Angle, α_0 deg	Inclination Angle, λs deg	Corner Radius, r mm	Major Cutting Edge Angle, kr deg	Coating Materials	Basis Material	Hardness, HRC
30	6	5	0.8	75	TiAlN	C-W-Co	60

Since the CNC milling machine spindle usually does not have the function of angle control, the tool wear area cannot be accurately stopped at the captured position of the CCD, so we have to collect all the wear images of all the inserts while the spindle is rotating. Image acquisition is performed every two minutes in the milling process. It is necessary to reduce the spindle speed of the machine tool and move the face milling cutter to the coordinate measurement position. Wear images are automatically acquired by the tool wear collection device installed on the CNC machine, and then the generated image files are saved on the computer hard disk. After the automatic acquisition is completed, the inserts are removed in the process of downtime. Then the actual tool flank wear value is measured by using a precision digital optical microscope (model: Keyence VHX-1000).

The cutting parameters used for the face milling operation are shown in Table 3. There are a total of 18 possible combinations of experimental conditions (ADG, AEG, AFG, ADH, AEH, AFH, BDG, BEG, BFG, BDH, BEH, BFH, CDG, CEG, CFG, CDH, CEH, CFH) based on the given cutting parameters. Each image is labeled with a combination of cutting conditions and processing time. For example, the image "3ADG08" represents the third insert of face milling cutter, at a cutting speed of 60 m/min, a feed rate of 0.03 mm/z, a depth of cut of 0.3 mm, and a tool wear image is obtained after 8 min of machining. The above-acquired images are used as datasets.

Table 3. Cutting parameters for face milling operation.

Cutting Speed v_c (m/min)		Feed Per Tooth f_z (mm/z)		Depth of Cutting a_p (mm)	
A	60	D	0.03	G	0.3
B	109	E	0.05	H	0.5
C	148	F	0.07		

Each set of tests uses the same cutting parameters until the maximum flank wear of the cutting tool reaches 0.4 mm. Face milling cutters have six inserts. At the end of each set of milling tests, wear images of six inserts are acquired, and six sets of complete tool wear life cycle data are obtained. Both coated tools are used in this way. Experimental environment for milling and tool wear detection is shown in Figure 1. In the process of collecting the tool wear images, the trigger mode is a soft trigger, the exposure time is 10 ms, and the exposure mode is timed. If the spindle speed is 2 r/min, the time interval for saving the pictures is 250 ms, and the number of saved pictures is 120; if the spindle speed is 4 r/min, the time interval for saving the pictures is 150 ms, the number of saved pictures is 100, which guarantees a complete acquisition of the wear image of each cutting edge. The process of matching the camera frame rate with the machine spindle speed is shown in Figure 2. The matching between the rotational speed and the acquisition frame rate is studied, and the maximum speed of 4 r/min is obtained according to 30 fps of the industrial camera. If the spindle speed is increased, the acquired tool wear image will have different degrees of distortion. Improving the frame rate of the camera can increase the spindle speed to improve detection efficiency.

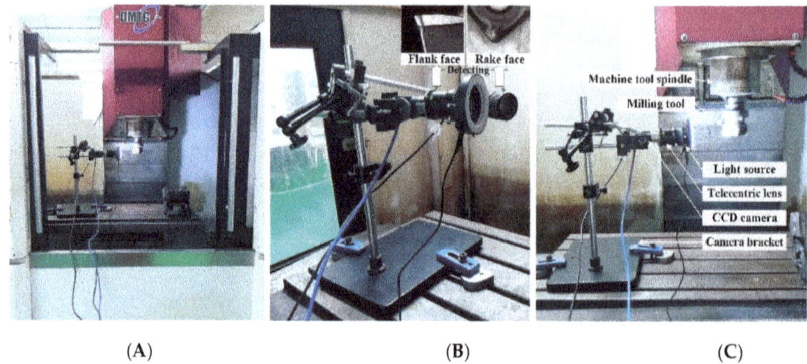

Figure 1. Process of matching the camera frame rate with the machine spindle speed. (**A**,**C**) Images of different numbered inserts at different speeds; (**B**) numbered picture of inserts for face milling cutter.

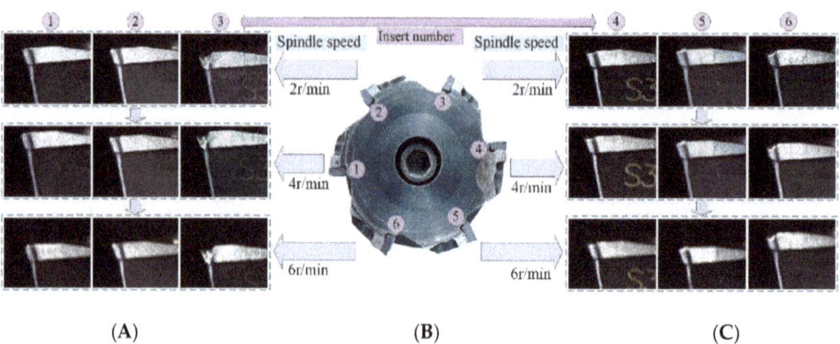

Figure 2. Experimental environment for milling and tool wear detection. (**A**,**C**) Actual detection images; (**B**) The overall image of camera bracket.

3. Tool Wear Types Identification Based on CNN

3.1. The Overall Process for Tool Wear Types Identification

The images of tool wear are the datasets after the image size normalization. Wear characteristics of tool wear images are extracted adaptively, some are used as the training data, and the other part is used as the test data. After the network model is pre-trained using a convolutional automated encoder (CAE), the output is set as the initial value of the CNN parameters. CNN is used to continue the training to obtain the optimal solution of the entire network parameters. Finally, the softmax classifier is used to identify and classify the wear types of milling tools. The whole network structure continuously iterates and feeds back the actual error results of the calculation. At the same time, the weight of the network is updated to make the whole network develop in the optimal direction, and the optimal wear type identification model is obtained finally. The overall process for tool wear identification is shown in Figure 3.

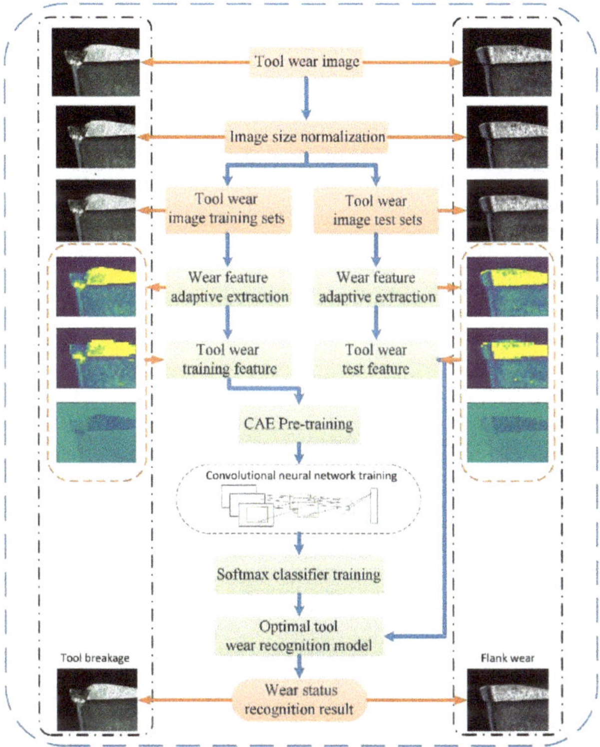

Figure 3. The overall process for tool wear identification.

3.2. Network Structure for Tool Wear Types Identification

The CNN structure is mainly composed of a convolutional layer, a pooled layer (also referred to as a down sampling layer), and a fully connected layer. The paper constructs a recognition model of tool wear types (named ToolWearnet) based on CNN concerning a VGGNet-16 model applied to image classification and developed by the visual geometry group of Oxford University. The parameters of the network are set according to the characteristics of the tool wear identification model. The ToolWearnet model consists of 12 layers, including 11 convolutional layers and one fully connected layer, and each convolution kernel is 3 × 3 in size.

Relative to the convolutional layer, the full connection layer has more connections, and more parameters will be generated. In order to reduce the network parameters, only one fully connected layer is set in the model, and the last pooling layer in front of the fully connected layer is set to the mean pooling layer. The kernel size of the mean pooling layer is 4 × 4. This configuration can effectively reduce network parameters while maintaining network feature extraction capabilities as much as possible. The acquired image size of the tool wear image is 256 × 256 × 3 and randomly cut into a uniform 224×224 resolution as the input data of the network model. The tool wear images are divided into four categories according to the wear types, and the fully connected layer is set as four types of output. The framework of tool wear identification method and network structure based on CNN are shown in Figure 4.

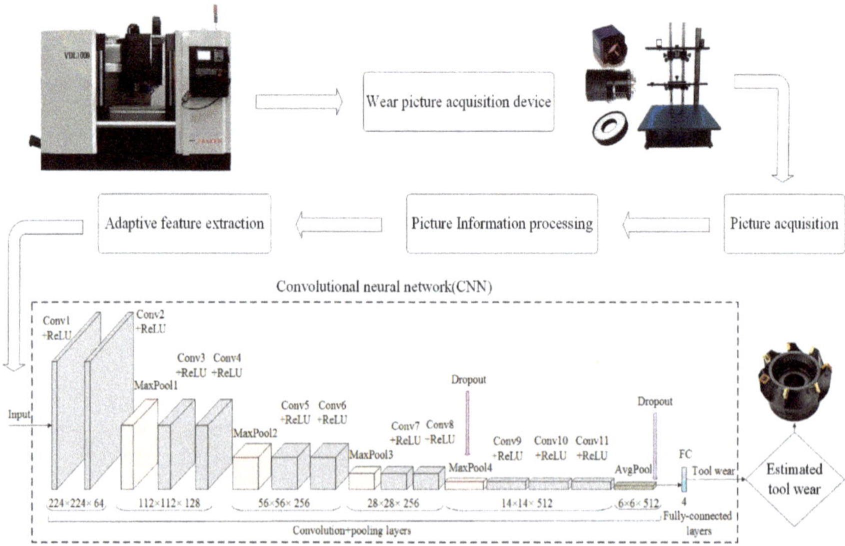

Figure 4. The framework of wear type identification method and network structure based on CNN.

3.2.1. Convolution Layer

The convolution operation is to use the convolution kernel to "slide" on the image at a set step size, thereby extracting features in the image to obtain feature maps. Batch normalization (BN) layer is add in the layers of Conv1, Conv3, Conv5, Conv7, Conv9, Conv10, and Conv11 to prevent over-fitting and speed up model convergence. Large-sized pictures x_l ($r_l \times c_l$) are operated through the convolution with the size of x_s ($r_s \times c_s$), the number of learned features is k_c, multiple features maps Cl_i ($i = 1, 2, 3... k_c$) can be obtained [24] as in the following Equation (1).

$$Cl_i = ReLu\left(W^{(1)}x_s + b^{(1)}\right) \tag{1}$$

where $W^{(1)}$ and $b^{(1)}$ are the weight and offset of the display unit to the hidden layer unit, respectively. $ReLu$ (x) is the Nonlinear mapping activation function. Then the actual size of the feature map Cl_i is:

$$S(Cl_i) = [((r_l + 2 \times pading - r_s)/stride) + 1] \times [((c_l + 2 \times pading - c_s)/stride) + 1] \times k_c \tag{2}$$

where k_c represents the number of convolution kernels; *padding* is the edge extension parameter; *stride* represents the step size. For example, in the CNN model of the tool wear identification, as shown in Figure 4, a tool wear image of $224 \times 224 \times 3$ is input, a convolution kernel of $3 \times 3 \times 64$ is used in the convolution layer Conv1 to perform convolution operation on the tool wear image. Set the step size as 1 and the edge expansion parameter as 1, and the actual size of the output feature map after convolution is $224 \times 224 \times 64$.

3.2.2. Pooling Layer

The pooling methods include maximum pooling, average pooling, and random value pooling. The dropout technology is added after the AvgPool layer and the MaxPool4 layer to reduce the complexity of interaction between neurons, so that neuron learning can obtain more robust features, improve the generalization, and avoid over-fitting of the network. Maximum pooling method and the average pooling method are used for pooling operation. Pooling kernel size is set as p_p ($r_p \times c_p$), and Cl ($r \times c$) is the feature map obtained after the convolution operation. Multiple feature maps p_i ($i = 1,2,3 \ldots k_p$) obtained [25] after the pooling operation are as shown in Equations (3) and (4).

$$p_i = \max_{r_p \times c_p}(Cl) \tag{3}$$

$$\text{or } p_i = \underset{r_p \times c_p}{average}(Cl) \tag{4}$$

Then the actual size of the feature map pi is: Type equation here.

$$S(p_i) = \left[\left(\left(r + 2 \times pading - r_p\right)/stride\right) + 1\right] \times \left[\left(\left(c + 2 \times pading - c_p\right)/stride\right) + 1\right] \times k_p \tag{5}$$

where k_p represents the number of pooling kernels. For example, input a feature map of $224 \times 224 \times 64$ in the pooling layer MaxPool1 as shown in Figure 4, pooling kernel of $2 \times 2 \times 128$ is used for pooling operation of the tool wear images. Set the step size as 2 and the edge expansion parameter as 0, and the actual size of the output feature map after pooling operation is $112 \times 112 \times 128$.

3.2.3. Fully Connected Layer

Each neuron in the fully connected layer is fully connected to all neurons in the upper layer. After the training of the tool wear image and the corresponding label, the classification results of tool wear types are output with the recognition precision rate and loss function value of the CNN. The ToolWearnet model is constructed using the network structure described above. The size of the input tool wear image is 224×224. The specific parameters of each layer are shown in Table 4. The ToolWearnet model, which includes a 12-layer network structure and does not include an input layer, has 11 convolutional layers, five pooling layers, and one fully connected layer. Conv1 to Conv11 are convolutional layers, MaxPool1 to MaxPool4 are the largest pooling layer, AvgPool is the average pooling layer, and FC layer is a fully connected layer.

Table 4. Structural parameters of the ToolWearnet model.

Number	Layer Name	Size of Input Feature Maps	Size of Filters	Padding	Stride	Size of Output Feature Maps
0	Input	$224 \times 224 \times 3$	-	-	-	$224 \times 224 \times 3$
1	Conv1	$224 \times 224 \times 3$	$3 \times 3 \times 64$	1	1	$224 \times 224 \times 64$
2	Conv2	$224 \times 224 \times 64$	$3 \times 3 \times 64$	1	1	$224 \times 224 \times 64$
3	MaxPool1	$224 \times 224 \times 64$	$2 \times 2 \times 128$	2	0	$112 \times 112 \times 128$
4	Conv3	$112 \times 112 \times 128$	$3 \times 3 \times 128$	1	1	$112 \times 112 \times 128$
5	Conv4	$112 \times 112 \times 128$	$3 \times 3 \times 128$	1	1	$112 \times 112 \times 128$
6	MaxPool2	$112 \times 112 \times 128$	$2 \times 2 \times 256$	2	0	$56 \times 56 \times 256$
7	Conv5	$56 \times 56 \times 256$	$3 \times 3 \times 256$	1	1	$56 \times 56 \times 256$
8	Conv6	$56 \times 56 \times 256$	$3 \times 3 \times 256$	1	1	$56 \times 56 \times 256$
9	MaxPool3	$56 \times 56 \times 256$	$2 \times 2 \times 256$	2	0	$28 \times 28 \times 256$
10	Conv7	$28 \times 28 \times 256$	$3 \times 3 \times 256$	1	1	$28 \times 28 \times 256$
11	Conv8	$28 \times 28 \times 256$	$3 \times 3 \times 256$	1	1	$28 \times 28 \times 256$
12	MaxPool4	$28 \times 28 \times 256$	$2 \times 2 \times 512$	2	0	$14 \times 14 \times 512$
13	Conv9	$14 \times 14 \times 512$	$3 \times 3 \times 512$	1	1	$14 \times 14 \times 512$
14	Conv10	$14 \times 14 \times 512$	$3 \times 3 \times 512$	1	1	$14 \times 14 \times 512$
15	Conv11	$14 \times 14 \times 512$	$3 \times 3 \times 512$	1	1	$14 \times 14 \times 512$
16	AvgPool	$14 \times 14 \times 512$	$4 \times 4 \times 512$	2	0	$6 \times 6 \times 512$
17	FC	$6 \times 6 \times 512$	-	-	-	4

3.2.4. Parameter Training and Fine Tuning

The training process of CNN is mainly composed of two stages: forward propagation and back propagation [28]. The network model is pre-trained by the convolutional automated encoder (CAE) and then is optimized and fine-tuned by the back propagation (BP) algorithm combined with the stochastic gradient descent (SGD) method. In the actual calculation, the batch optimization method decreases the value and gradient of the loss function in the entire training datasets. However, it is difficult to run on a PC for large data sets. The network may converge more quickly, and the above problem may be solved after applying the SGD optimization method in CNN.

The standard gradient descent algorithm updates the parameter θ of the objective function $J(\theta)$ using the following formula [24]:

$$\theta = \theta - \eta \nabla_\theta E[J(\theta)] \tag{6}$$

where expectation $E[J(\theta)]$ is the estimated value of loss function value and the gradient for all training datasets, and η is the learning rate.

The SGD method cancels the expected term when updating and calculating the gradient of the parameters. The formula is as follows:

$$\theta = \theta - \eta \nabla_\theta J\left(\theta; m^{(i)}, n^{(i)}\right) \tag{7}$$

where $(m^{(i)}, n^{(i)})$ is the training set data pair.

When updating the model parameters, the objective function of the deep network structure has a locally optimal form, and local convergence of the network may be led by the SGD method. Increasing the momentum parameter is an improved method that can speed up the global convergence of the network. The momentum parameter updated formula is as follows [18]:

$$v_t = \alpha v_{t-1} + \eta \nabla_\theta J\left(\theta; m^{(i)}, n^{(i)}\right) \tag{8}$$

$$\theta = \theta - v_t \tag{9}$$

where v_t represents the current velocity vector, and v_{t-1} represents the velocity vector of the previous moment, both having the same dimensions as the parameter θ; α is momentum factor, and $\alpha \in (0,1)$, where α is equal to 0.5.

3.3. Method of ATWVD Based on Tool Wear Types Identification

The minimum bounding rectangles (MBR) method [29] is used to study the tool wear and tool breakage in milling of superalloy mentioned above. For the tool wear in the case of abrasive wear or chipping, the value of tool wear can be detected accurately, but in the case of adhesive wear and sheet peeling, the value obtained by using the MBR method directly has a significant error. This is because when the tool wear is in adhesive wear or sheet peeling, the image obtained from the camera is affected by the chips adhered on the cutting edge, as shown in Figure 5a,d, resulting in a larger rectangle surrounding the wear zone than the actual one.

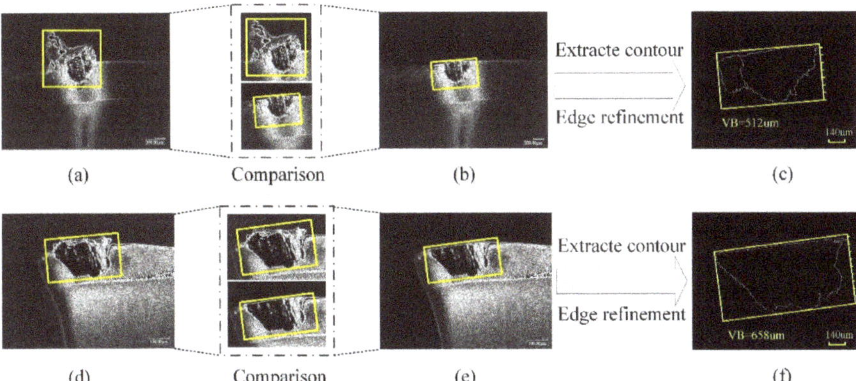

Figure 5. Comparison of detection methods for the value of tool wear. (**a,d**) Roughly processed images of adhesive wear and sheet peeling; (**b,e**) cropped images; (**c,f**) edge refined images.

In order to improve the deficiencies of the above detection methods, an ATWVD method based on tool wear types identification is proposed, the specific algorithm is as follows:

Step 1: The collected tool wear images are randomly cut into a uniform size of 224 × 224 and image size normalization was performed to achieve data normalization.

Step 2: ToolWearnet model based on CNN is used to test the tool wear images for intelligently identifying and classifying.

Step 3: After the tool wear images are normalized, the process of graying, median filter denoising, contrast stretch, segmentation boundary image using a watershed algorithm, coarse positioning using Roberts operator boundary and refining the edge image, extracting the contour of the tool wear area are conducted according to the tool wear types. If the tool wear type is adhesive wear (or the chip is sheet peeling, and the chip is bonded to the tool), the wear image is matched with the original image using image registration technology. Cropping algorithm is used to cut off the excess image to remove the contour of the chip bonded to the tool, as shown in Figure 5b,e. Finally, the area contour of effective wear is obtained, and the minimum circumscribed rectangle obtained after the edge refinement is shown in Figure 5c,f. Besides, other tool wear types are measured by directly extracting the contour of the wear area.

Step 4: According to the transformation between the image pixel value and the actual value, combined with the width of the minimum circumscribed rectangle, the actual value of tool wear value is obtained.

4. Results and Discussion

4.1. Dataset Description

The tool wear and tool failure in the milling of high superalloys are analyzed as an example. The tool wear is severe due to high hardness, excellent fatigue performance, and fracture toughness of superalloys. At a lower cutting speed, flank wear and rake face wear are the main forms of wear, as shown in Figure 6a,d; at a high-speed cutting, the heat generated causes significant chemical and adhesive wear on the tool, as shown in Figure 6c. In actual cutting, the wear form is often complicated, and multiple wears coexist. For example, flank wear and crater wear (rake face wear) can be observed simultaneously, and two wears that interact with each other will accelerate the wear of the tool and finally lead to breakage, as shown in Figure 6b. The adhesive wear at the edge will cause chips to adhere to the rake face, and finally, the sheet peeling will occur at the edge.

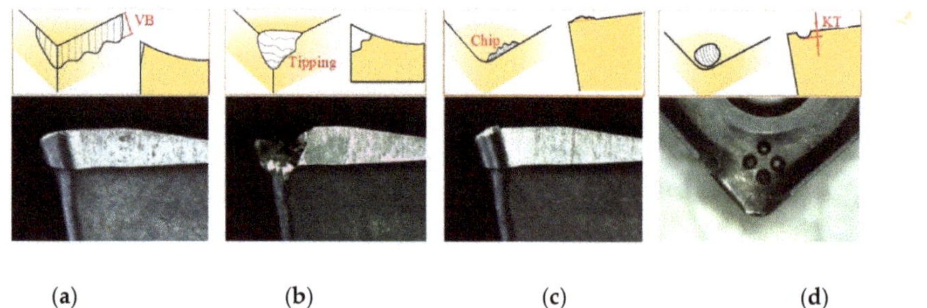

(a)   (b)   (c)   (d)

Figure 6. Form of tool wear and tool breakage (**a**) Flank wear; (**b**) tool breakage; (**c**) adhesive wear; (**d**) rake face wear.

After the tool wear images were acquired in the milling process using the above methods, the datasets were divided into three parts: 70% as the training set; 15% as the validation set; 15% as the test set. We randomly selected two different cutting parameters and obtained twelve sets of wear images with complete tool wear life as the test set. Then, twelve sets of wear images were selected in the same way as the verification set. The remaining tool wear images were all used as the training set. The test set and training set are mutually exclusive (the test samples do not appear in the training set and are not used in the training process). The training set was used for model training, the validation set was used to select model hyper parameters, and the test set was used to evaluate model performance.

The CNN model is prone to overfitting due to its powerful fitting ability, especially when the number of datasets is small [30]. Since the number of tool wear images collected in this paper is limited, and the ToolWearnet model has more layers, more tool wear images are needed for training. Methods of rotation, flipping transformation, scaling transformation, translation transformation, color transformation, and noise disturbance are used to extend the datasets by five times. There are 8400 images in the datasets. The tool wear images mentioned above are classified according to the adhesive wear, tool breakage, rake face wear, and flank wear. After randomly cutting into a uniform, the datasets of the images were obtained. The specific number of images for each wear type is shown in Table 5.

Table 5. Statistics of the tool wear image data.

Tool Wear Types	Training Set	Validation Set	Test Set	Total
Adhesive wear	1425	285	280	1990
Tool breakage	1455	315	305	2075
Rake face wear	1490	330	325	2145
Flank wear	1510	340	340	2190

The tool wear type identification is a multi-classification task. Since the tool wear types are identified in the milling machine during machining, it is likely to be affected by pollutants and fragment chips in the processing environment. Irregular chips adhere to the tool when it is adhesive wear. If the chip is small and not easily observed, it is likely that the classifier will recognize it as a rake face wear, as shown in Figure 7a. In order to reduce the probability of such errors, we collected images that are prone to recognition errors and used them as a sample set to re-train the CNN model. Similarly, if contaminants are attached to the main cutting edge of the rake face, it is likely to be recognized by the classifier as adhesive wear, as shown in Figure 7b. In order to avoid such errors, we will configure a high-pressure blowing device on one side of the machine spindle, using high pressure gas to blow away contaminants and fragment chips before performing tool wear type detection. In actual cutting, the wear form is often complicated, and multiple wears coexist. As shown in Figure 7c, it has the characteristics of adhesive wear and tool breakage, which will also cause the classifier to have recognition error. Due to the current small sample set, this mixed form of wear type has a lower occurrence probability. With the expansion of the sample set, the number of its occurrences will increase. In the future, we will further study a specialized algorithm to reduce the above error cases.

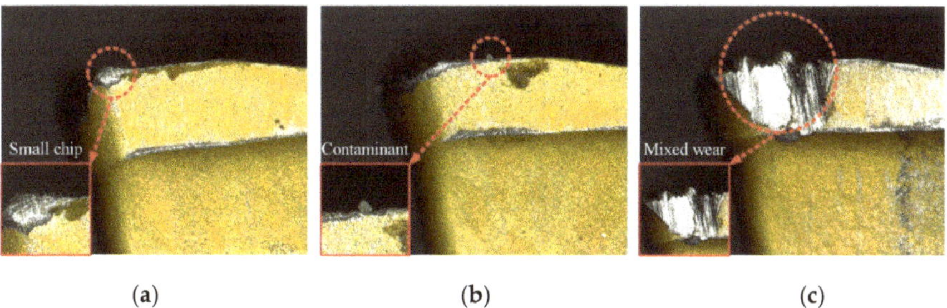

(a) (b) (c)

Figure 7. Images of the detection error cases. (**a**) adhesive wear, (**b**) flank wear, (**c**) multiple wear coexist.

The CNN used in the deep learning experiment is built using the deep learning development kit provided by MATLAB2017b combined with the Caffe deep learning framework. Keyence VHX-1000 (KEYENCE company, Osaka, Japan) high precision digital optical microscope was used to test the actual tool wear value and verify the feasibility of the ATWVD method. Images manually collected by high precision digital optical microscopes are also part of the datasets. The partial images of the datasets and the process of depth search are shown in Figure 8.

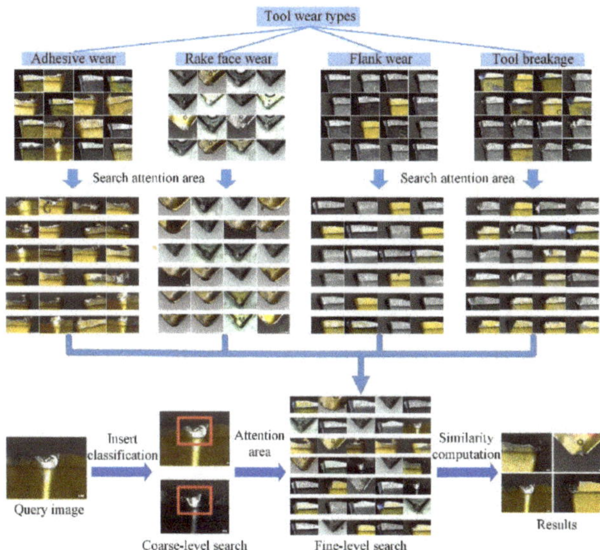

Figure 8. The partial images of the datasets and the process of depth search.

4.2. Evaluation Indications

After training the ToolWearnet network model, the test set was used to evaluate the precision of the ToolWearnet network model in identifying the tool wear image. We used the "precision and average recognition precision" metrics to perform the evaluation process [31], as shown in Equations (10) and

(11). The ToolWearnet network model constructed in this paper has an average recognition precision (*AP*) of about 96% for tool wear types.

$$precision = \frac{TP}{TP + FP} \times 100\% \tag{10}$$

$$AP = \left[\left(\frac{1}{n}\frac{TP}{TP + FP}\right)\right] \times 100\% \tag{11}$$

where *TP* is the number of true positive samples in the test samples, *FP* is the number of false positive samples in the test samples, *n* is the number of categories in the test sample, $n = 4$.

In order to test the performance of the automatic tool wear value detection (named ATWVD) method based on the tool wear types in detecting the tool wear degree, the parameter error rate (e_k) and mean absolute percentage error (*MAPE*) were used to evaluate, as shown in Equations (12) and (13). In this paper, the mean absolute percentage error (*MAPE*) of the ATWVD method is about 5%.

$$e_k = \frac{|F_k - A_k|}{A_k} \times 100\% \tag{12}$$

$$MAPE = \left[\left(\frac{1}{k}\sum_{k=1}^{k} e_k\right)\right] \times 100\% \tag{13}$$

where F_k represents the tool wear degree detected by the ATWVD method, A_k represents the wear value measured by the precision optical microscope, and *k* represents the number of measured tool wear images.

4.3. Analysis and Results

The initial learning rate of the training sets in the experimental phase is set to 10^{-3}, the maximum number of training iterations is 10^3, the dropout rate is 0.2, the weight attenuation is 5×10^{-4}, and the number of batch images is 128. The algorithm of the neural network parameter optimization is stochastic gradient descent (SGD). When training this network model, the batch normalization (BN) method is used to accelerate the training of the network. Because the BN method uses the normalized method to concentrate the data at the center, it can effectively avoid the problem of the gradient disappearing. When testing this network model, the training sets are tested once every six training sessions.

4.3.1. Comparison of Two CNN Models for Tool Wear Recognition

In order to reflect the reliability of the ToolWearnet network model built in this paper in tool wear types identification, and to make a reasonable evaluation of its recognition precision rate and efficiency, the VGGNet-16 network model was applied for the comparative experiment. VGGNet-16 is a CNN model developed by the VGG (visual geometry group) group of Oxford University for image classification. The VGGNet-16 network model has a 16-layer network structure with 5 maximum pooling layers, 13 convolutional layers, and 3 fully connected layers. The convolution kernels are all 3×3, and the pooling kernels are all 2×2. The first and second fully connected layers use dropout technology, and all hidden layers are equipped with ReLU layers. The model has strong image feature extraction ability. The detailed training parameters for the two models are shown in Table 6. After training and testing, different recognition precision rate and loss function values are obtained, as shown in Figure 9. The comparison of the precision rate of the recognition, training time, and recognition time of the different tool wear by the two models are shown in Table 7.

Table 6. The detailed training parameters for the two models.

Network Model	Training Parameter						
	Learning Rate	Maximum Iteration Times	Dropout	Weight Attenuation	Batch Size	Optimization Algorithm	Test Interval
ToolWearnet	10^{-3}	10^3	0.2	5×10^{-4}	128	SGD	6
VGGNet-16	10^{-3}	10^3	0.2	5×10^{-4}	128	SGD	6

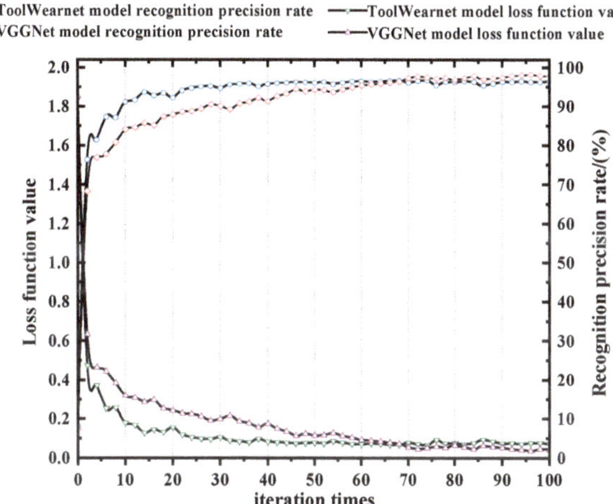

Figure 9. Comparison of two models testing precision rate and loss function values.

Table 7. The comparison of the recognition precision rate, training time, and recognition time of the two models.

Network Model	Recognition Precision Rate of Different Wear Types (%)				Training Time (h)	Recognition Time (s)
	Adhesive Wear	Tool Breakage	Rake Face Wear	Flank Wear		
VGGNet-16	97.85	97.10	96.56	95.81	32.6	174
ToolWearnet	97.27	96.35	95.78	95.41	1.8	11

The results show that the ToolWearnet network model achieves a high precision rate at the beginning of the test quickly. In the first 24 iteration periods, the recognition precision rate of the two models are continuously increasing, and the loss function is continuously decreasing; but when the iteration period exceeds 24 times, the recognition precision rate of ToolWearnet model tends to be stable, while the recognition precision rate of the VGGNet model is continuously increasing, and its loss function is continuously decreasing. It will be stable until the iteration period exceeds 76 times. Therefore, the convergence speed of this model is faster than the VGGNet-16 network model. The VGGNet-16 model has more network layers, complex network structure, and network parameters, resulting in an astronomical calculation amount and slow convergence speed. The trend of recognition precision rate of the two models increases with the number of iterations, and the trend of the loss function decreases with the number of iterations. When the recognition precision rate and loss function of both models tend to be stable, the recognition precision rate curve of the VGGNet model is above the ToolWearnet model. Therefore, the recognition precision rate of this model is slightly lower than the VGGNet model. Compared with the ToolWearnet model, it has the deeper network depth and more feature surfaces, the feature space that the network can represent is more extensive, which will make the learning ability of the network stronger and the deeper feature information can be mined, but it will also make the network structure more complicated. The storage memory and the training time will increase.

The ToolWearnet network model has an average recognition precision rate of 96.20% for tool wear types, the highest recognition precision rate for adhesion wear, with a value of 97.27%, and the lowest recognition precision rate for flank wear, which is 95.41%. Because irregular chips adhere to the cutting edge during adhesion wear, the contour features of the wear area are most apparent, while the contour features of the flank wear are not evident due to the difference of the shape of the wear area being small. Although the recognition precision rate of the ToolWearnet network model is slightly lower than the VGGNet-16 network model, the training time and recognition time are shorter, and the hardware has low computational power requirements, which can meet the needs of slightly fast and accurate cutting in the industry.

4.3.2. Tool Wear Value Detection Comparison Experiment

In the experiment, the spindle speed was matched with the frame rate of the camera, and the wear image of each insert on the face milling cutter was collected. Each acquisition took the corresponding insert directly below the label on the cutter as the starting point and sequentially numbers the six inserts in the direction while the tool rotates. The images with good quality and complete tool wear areas were acquired by filtering by the fuzzy and the insert position detection. Then, the flank wear value of each insert was obtained by the ATWVD method combined with the CNN model. After the automatic acquisition of the tool wear image, the inserts are detected by a high precision digital optical microscope to obtain the exact value of tool wear after stopping the CNC machine. The comparison of the pictures collected in two ways is shown in Figure 10. We randomly selected one set of tool wear images with complete tool wear life cycle in the test set for comparison, taking "2CEG" as an example. The maximum wear value VB of the ATWVD method detection and high precision digital microscope manual detection and the error rate of the ATWVD method were obtained, as shown in Figure 11.

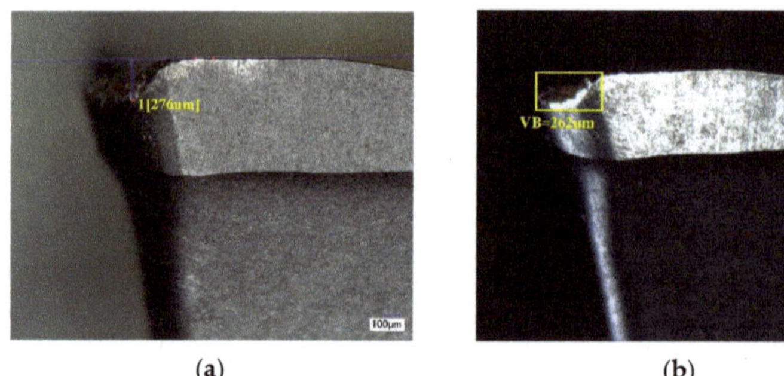

(a)   (b)

Figure 10. Comparison of two acquisition methods. (**a**) image is manually collected by high precision digital microscope (model: Keyence VHX-1000); (**b**) image is collected by the automatic acquisition device.

The results show that the wear value detected by the ATWVD method is close to the wear value of high precision digital optical microscope manual detection, and the error rate of the wear value is in the range of 1.48~7.39% and the mean absolute percentage error (*MAPE*) is 4.76%. Therefore, the method of tool wear types automatic identification based on CNN is verified. Practicality, the process is suitable for intermittent detection of the tool, such as collecting tool information while the tool is in the tool magazine or the tool change gap. It can automatically and effectively obtain the tool wear types and wear value, which can be used to provide data for the future tool life prediction, tool selection, and tool design.

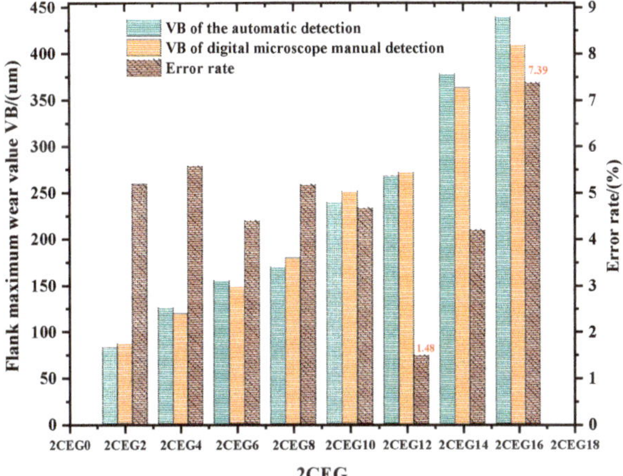

Figure 11. Comparison of the maximum wear value measured by the two methods and the error rate of the ATWVD method.

5. Conclusions

In this paper, an automatic recognition model of tool wear types based on CNN is proposed by using Caffe deep learning framework. The model considers the characteristics of the actual processing environment and tool wear images in milling of superalloy. The tool wear images obtained through the milling experiments of Nickel-based superalloy Inconel 718 are used as the datasets to train and test the ToolWearnet network model. The results show that the model has a robust feature extraction ability. The recognition precision rate of different wear types of high-temperature alloy tools is in the range of 95.41~97.27%, and the average recognition precision rate is 96.20%. It has the advantages of high recognition precision rate and robustness.

Furthermore, an ATWVD method is improved based on ToolWearnet network model, the value of tool wear obtained by this method is compared with the wear value detected by a high precision digital optical microscope. The error rate of this method is in the range of 1.48~7.39%, and the mean absolute percentage error is 4.76%, which proves the reliability of the method. Although the recognition precision rate of the network model is slightly lower than that of the VGGNet-16 network model, the training time and recognition time are shorter, and the network parameters are less. It can be applied to the application scenario that identifies tool wear types accurately and slightly faster with lower hardware consumption.

Practicality, the process is suitable for intermittent detection of the tool, such as collecting tool information while the tool is in the tool magazine or the tool change gap. It can automatically and effectively obtain the tool wear types and wear value, which can be used to provide data for the future tool life prediction, tool selection, and tool design. The proposed tool wear identification method can also be extended to other machining processes, such as drilling and turning. Meanwhile, future work will consider more extensive experimental data with different cutting tools, as well as extend the applications to predict tool life, optimizing the machining parameters with the method.

Author Contributions: Conceptualization, X.W. and Y.L.; Data curation, Y.L., A.M., and X.Z.; Formal analysis, X.W. and Y.L.; Investigation, X.W. and Y.L.; Methodology, X.W. and Y.L.; Project administration, X.W.; Resources, A.M. and X.Z.; Software, X.W., Y.L., and X.Z.; Supervision, X.W.; Writing—original draft, Y.L.; Writing—review and editing, X.W., Y.L., A.M., and X.Z.

Funding: This work was financially supported by the National Science Foundation of China (Grant No.51575144).

Conflicts of Interest: The authors declare no conflict of interest.

Nomenclature

a_p	Depth of cutting (mm)
f_z	Feed per tooth (mm/z)
v_c	Cutting speed (m/min)
padding	Edge extension parameter
stride	Step size
k_c	Number of convolution kernels
k_p	Number of pooling kernels
θ	Parameter of the objective function
η	Learning rate
$(m^{(i)}, n^{(i)})$	Training set data pair
v_t	Current velocity vector
v_{t-1}	Velocity vector of the previous moment
α	Momentum factor
n	Number of wear types
TP	Number of true positive samples
FP	Number of false positive samples
F_k	Wear degree detected by the ATWVD method
A_k	Wear value measured by the optical microscope
k	Number of measured tool wear images

References

1. Castro, J.R.; Castillo, O.; Melin, P.; Rodríguez-Díaz, A. A hybrid learning algorithm for a class of interval type-2 fuzzy neural networks. *Inf. Sci.* **2009**, *179*, 2175–2193. [CrossRef]
2. Rehorn, A.G.; Jiang, J.; Orban, P.E. State-of-the-art methods and results in tool condition monitoring: A review. *Int. J. Adv. Manuf. Technol.* **2005**, *26*, 693–710. [CrossRef]
3. Yao, Y.; Li, X.; Yuan, Z. Tool wear detection with fuzzy classification and wavelet fuzzy neural network. *Int. J. Mach. Tools Manuf.* **1999**, *39*, 1525–1538. [CrossRef]
4. Li, H.K.; Wang, Y.H.; Zhao, P.S.; Zhang, X.W.; Zhou, P.L. Cutting tool operational reliability prediction based on acoustic emission and logistic regression model. *J. Intell. Manuf.* **2015**, *26*, 923–931. [CrossRef]
5. Jauregui, J.C.; Resendiz, J.R.; Thenozhi, S.; Szalay, T.; Jacso, A.; Takacs, M. Frequency and Time-Frequency Analysis of Cutting Force and Vibration Signals for Tool Condition Monitoring. *IEEE Access* **2018**, *6*, 6400–6410. [CrossRef]
6. Prasad, B.S.; Sarcar, M.M.; Ben, B.S. Real-time tool condition monitoring of face milling using acousto-optic emission—An experimental approach. *Int. J. Comput. Appl. Technol.* **2011**, *41*, 317–325. [CrossRef]
7. Garcia-Ordas, M.T.; Alegre-Gutierrez, E.; Alaiz-Rodriguez, R.; Gonzalez-Castro, V. Tool wear monitoring using an online, automatic and low cost system based on local texture. *Mech. Syst. Signal Process.* **2018**, *112*, 98–112. [CrossRef]
8. Mikolajczyk, T.; Nowicki, K.; Bustillo, A.; Pimenov, D.Y. Predicting tool life in turning operations using neural networks and image processing. *Mech. Syst. Signal Process.* **2018**, *104*, 503–513. [CrossRef]
9. Garcia-Ordas, M.T.; Alegre, E.; Gonzalez-Castro, V.; Alaiz-Rodriguez, R. A computer vision approach to analyze and classify tool wear level in milling processes using shape descriptors and machine learning techniques. *Int. J. Adv. Manuf. Technol.* **2017**, *90*, 1947–1961. [CrossRef]
10. D'Addona, D.M.; Ullah, A.M.M.S.; Matarazzo, D. Tool-wear prediction and pattern-recognition using artificial neural network and DNA-based computing. *J. Intellig. Manuf.* **2017**, *28*, 1285–1301. [CrossRef]
11. Ong, P.; Lee, W.K.; Lau, R.J.H. Tool condition monitoring in CNC end milling using wavelet neural network based on machine vision. *Int. J. Adv. Manuf. Technol.* **2019**, *29*, 112–120. [CrossRef]
12. Mikolajczyk, T.; Nowicki, K.; Klodowski, A.; Pimenov, D.Y. Neural network approach for automatic image analysis of cutting edge wear. *Mech. Syst. Signal Process.* **2017**, *88*, 100–110. [CrossRef]
13. Bertrand, C.; Jayanti, M.; Manuela, S.; Taryn, J.; Earley, S.R.; Matt, J.; Petrus, A.E.; Dubin, A.P. Piezo1 and Piezo2 are essential components of distinct mechanically-activated cation channels. *Science* **2010**, *330*, 55–60.

14. Sa, I.; Ge, Z.; Dayoub, F.; Upcroft, B.; Perez, T.; McCool, C. DeepFruits: A Fruit Detection System Using Deep Neural Networks. *Sensors* **2016**, *16*, 1222. [CrossRef] [PubMed]
15. Fuentes, A.; Yoon, S.; Kim, S.C.; Park, D.S. A Robust Deep-Learning-Based Detector for Real-Time Tomato Plant Diseases and Pests Recognition. *Sensors* **2017**, *17*, 2022. [CrossRef]
16. Rahnemoonfar, M.; Sheppard, C. Deep Count: Fruit Counting Based on Deep Simulated Learning. *Sensors* **2017**, *17*, 905. [CrossRef]
17. Li, S.; Liu, G.; Tang, X.; Lu, J.; Hu, J. An Ensemble Deep Convolutional Neural Network Model with Improved D-S Evidence Fusion for Bearing Fault Diagnosis. *Sensors* **2017**, *17*, 1729. [CrossRef]
18. Guo, S.; Yang, T.; Gao, W.; Zhang, C. A Novel Fault Diagnosis Method for Rotating Machinery Based on a Convolutional Neural Network. *Sensors* **2018**, *18*, 1429. [CrossRef]
19. Dai, M.; Zheng, D.; Na, R.; Wang, S.; Zhang, S. EEG Classification of Motor Imagery Using a Novel Deep Learning Framework. *Sensors* **2019**, *19*, 551. [CrossRef]
20. Al Machot, F.; Elmachot, A.; Ali, M.; Al Machot, E.; Kyamakya, K. A Deep-Learning Model for Subject-Independent Human Emotion Recognition Using Electrodermal Activity Sensors. *Sensors* **2019**, *19*, 1659. [CrossRef]
21. Chen, Z.; Li, C.; Sanchez, R.V. Gearbox fault identification and classification with convolutional neural networks. *Shock Vib.* **2015**, *2015*, 1–10. [CrossRef]
22. Zhang, W.; Peng, G.; Li, C.; Chen, Y.H.; Zhang, Z.J. A New Deep Learning Model for Fault Diagnosis with Good Anti-Noise and Domain Adaptation Ability on Raw Vibration Signals. *Sensors* **2017**, *17*, 425. [CrossRef] [PubMed]
23. Jing, L.; Zhao, M.; Li, P.; Xu, X. A convolutional neural network based feature learning and fault diagnosis method for the condition monitoring of gearbox. *Measurement* **2017**, *111*, 1–10. [CrossRef]
24. Fatemeh, A.; Antoine, T.; Marc, T. Tool condition monitoring using spectral subtraction and convolutional neural networks in milling process. *Int. J. Adv. Manuf. Technol.* **2018**, *98*, 3217–3227.
25. Zhang, C.; Yao, X.; Zhang, J.; Liu, E. Tool wear monitoring based on deep learning. *Comput. Integr. Manuf. Syst.* **2017**, *23*, 2146–2155.
26. Zhao, R.; Yan, R.; Wang, J.; Mao, K. Learning to Monitor Machine Health with Convolutional Bi-Directional LSTM Networks. *Sensors* **2017**, *17*, 273. [CrossRef]
27. Ren, S.; He, K.; Girshick, R.; Sun, J. Faster R-CNN: Towards Real-Time Object Detection with Region Proposal Networks. *IEEE Trans. Pattern Anal. Mach. Intell.* **2017**, *39*, 1137–1149. [CrossRef]
28. Sainath, T.N.; Kingsbury, B.; Saon, G.; Soltau, H.; Mohamed, A.; Dahl, G.; Ramabhadran, B. Deep Convolutional Neural Networks for Large-scale Speech Tasks. *Neural Netw.* **2015**, *64*, 39–48. [CrossRef]
29. Amit, Y.; Geman, D. A Computational Model for Visual Selection. *Neural Comput.* **1999**, *11*, 1691–1715. [CrossRef]
30. Wen, S.; Chen, Z.; Li, C. Vision-Based Surface Inspection System for Bearing Rollers Using Convolutional Neural Networks. *Appl. Sci.* **2018**, *8*, 2565. [CrossRef]
31. Zhou, P.; Zhou, G.; Zhu, Z.; Tang, C.; He, Z.; Li, W. Health Monitoring for Balancing Tail Ropes of a Hoisting System Using a Convolutional Neural Network. *Appl. Sci.* **2018**, *8*, 1346. [CrossRef]

 *sensors*

Article

Citrus Pests and Diseases Recognition Model Using Weakly Dense Connected Convolution Network

Shuli Xing [1], Marely Lee [1,*] and Keun-kwang Lee [2,*]

[1] Center for Advanced Image and Information Technology, School of Electronics & Information Engineering, Chon Buk National University, Jeonju, Chon Buk 54896, Korea
[2] Department of Beauty Arts, Koguryeo College, Naju 520-930, Korea
* Correspondence: mrlee@chonbuk.ac.kr (M.L.); kklee7410@hanmail.net (K.-k.L.)

Received: 18 June 2019; Accepted: 16 July 2019; Published: 19 July 2019

Abstract: Pests and diseases can cause severe damage to citrus fruits. Farmers used to rely on experienced experts to recognize them, which is a time consuming and costly process. With the popularity of image sensors and the development of computer vision technology, using convolutional neural network (CNN) models to identify pests and diseases has become a recent trend in the field of agriculture. However, many researchers refer to pre-trained models of ImageNet to execute different recognition tasks without considering their own dataset scale, resulting in a waste of computational resources. In this paper, a simple but effective CNN model was developed based on our image dataset. The proposed network was designed from the aspect of parameter efficiency. To achieve this goal, the complexity of cross-channel operation was increased and the frequency of feature reuse was adapted to network depth. Experiment results showed that Weakly DenseNet-16 got the highest classification accuracy with fewer parameters. Because this network is lightweight, it can be used in mobile devices.

Keywords: citrus; pests and diseases identification; convolutional neural network; parameter efficiency

---

1. Introduction

Pests and diseases are the two most important factors affecting citrus yields. Types of citrus pests and diseases are numerous in nature. Some of them are similar in appearance, making it difficult for farmers to precisely recognize them in time. In recent years, developments of convolutional neural networks (CNNs) have dramatically improved the state-of-the-art in computer vision. These new structures of network have enabled researchers to obtain high accuracy for image classification, object detection, and semantic segmentation [1]. Therefore, some studies have adopted the CNN model to identify the category of pests or diseases based on image. Liang et al. [2] have proposed a novel network consisted of residual structure and shuffle units for plant diseases diagnosis and severity estimation. Cheng et al. [3] have compared the classification performance of different depths of CNN models for 10 classes of crop pests with complex shooting background. The highest classification accuracy in both studies was obtained with the deepest network. For detection tasks, people are also more willing to select a very deep network architecture instead of a shallow one. Shen et al. [4] have applied a faster R-CNN [5] framework with improved Inception-V3 [6] to detect stored-grain insects under field condition with impurities. The same feature extractor network and SSD [7] model have been utilized by Zhuang et al. [8] to evaluate the health status of farm broilers.

In theory, the complexity of the CNN model depends on the scale of dataset. However, deep convolutional networks mentioned above were all over-fitted because they were proposed based on ImageNet [9] initially. Although a fine-tuned method [10] can be used to reduce the divergence between training and testing, the space required for model storage is so large that they cannot be deployed on mobile devices with little memory.

In this paper, a simple but effective network architecture was developed to classify pictures of citrus pests and diseases. Our network design principles focused on improving the utilization of model parameters. There has been evidence suggesting that some feature maps generated by convolutions are not useful [11,12]. To decrease the impact of redundant features on classification, Hu et al. [13] and Woo et al. [14] have introduced an attention mechanism to suppress unnecessary channels. Their approaches are more adaptable than the Dropout [15] and stochastic depth [16]. However, the extra branch in each building block increases the overhead of a network. Unlike these approaches, the channel selection of this paper was implemented through the method of cross-channel feature fusion. In Network in Network [17], two consecutive 1×1 convolutional layers were regarded as a way to enhance model discriminability for local patches. From another perspective, this structure is also a good choice to refine feature maps. Highway network [18] first provided the idea of feature reuse to ease the optimization difficulty suffered by deep networks. ResNet [19] generalized it with identity mappings. DenseNet [20] further boosted the frequency of skip-connection. DenseNet has a better representation ability than ResNet because it can produce a higher accuracy with fewer parameters. The concatenation operation of DenseNet was followed but some connections between long-range layers were removed by us. Because of this weakly dense pattern, our network is called Weakly DenseNet.

Experiment results showed that Weakly DenseNet achieved the highest accuracy in classifying citrus pests and diseases. With regard to computational complexity, our proposed model is also lightweight. These phenomena indicate that the optimization of network structure is more important than blindly increasing the depth or width. The main contributions of this work are summarized as follows:

A specific image dataset of citrus pests and diseases is created. It is a relatively complete image dataset for the diagnosis of citrus pests and diseases.

A novel and lightweight convolutional neural network model is proposed to recognize the types of citrus pests and diseases. The network design is based on improving parameter efficiency.

A new data augmentation method is developed to reinforce model generalization ability, which can significantly reduce the similarity between generated images.

2. Related Work

Pests and diseases can cause great damage to crops if they are not controlled. To recognize them, farmers used to rely on experienced experts. With the popularity of image sensors, using computer vision methods to identify pests and diseases has become a trend. Boniecki et al. [21] have proposed to use image analysis techniques and artificial neural network model to classify images of apple pests in simple background. Their dataset included 12,000 images from six species of apple pests which are most commonly found in Polish orchards. For training and testing proposed artificial neural network model, seven selected coefficients of shape and 16 color characteristics were extracted from each pest image as inputs. Sun et al. [22] have combined SLIC (simple linear iterative cluster) with SVM (support vector machine) classifier to detect diseases on tea plant. Their algorithm improved the prediction accuracy of disease images taken with complex backgrounds but needed more pre-treatments to reduce interference. A total of 1308 pictures from five common tea plant diseases were included in their dataset. These images were divided into two parts with a ratio of 4:1 for training and testing. Ferentinos [23] has employed deep CNN models to perform plant disease detection and diagnosis. They used an open database which contains 87,848 photographs of leaves to train each model. Images without pre-processing were regarded as inputs in his study. Compared with other selected models, VGG achieved the highest success rate with 99.48%. These advantages of deep CNNs have encouraged more researchers to apply them in the agricultural field.

A wide range of CNN architectures has been proposed to improve performance. VGGNets [24] first use small size convolution filters to reduce parameters and increase depth. ResNet exploits a simple identity skip-connection to ease optimization issues of deep networks. WideResNet [25] replaces the bottleneck structure in ResNet with two broad 3×3 convolutional layers to reduce depth. DenseNet enhances deep supervision [26] by iteratively concatenating input features with output features. Xception [27] introduces a depthwise separable convolution to decrease the number of parameters in a regular convolution. In it, depthwise convolution is responsible for feature extraction and pointwise convolution (a regular 1×1 convolution) is used for cross-channel feature fusion. This new convolution operation has become a core component of many lightweight networks, such as MobileNets [28,29] and ShuffleNets [30,31]. The structure of MobileNet-v1 is similar to that of VGG. MobileNet-v2 develops an inverted residual block to increase memory efficiency. To maintain the representational power of narrow layer in each inverted residual block, ReLU activation [32] behind it is removed. ShuffleNet-v1 employs group convolution [33] to further reduce the computational cost of depthwise separable convolution, a channel shuffle operation is adopted to enhance the information exchange of subgroups. ShuffleNet-v2 is constructed based on ShuffleNet-v1. However, it suggests splitting channels into two equal parts and using concatenation instead of addition to execute feature reuse. People tend to use their architectures designed for the ImageNet without considering their own dataset scale. This behavior may lead to overfitting problems and waste of computing resources. Different from previous approaches, a novel, and lightweight network was constructed to classify images in our dataset.

3. Dataset

The dataset used in our experiment included 17 species of citrus pests and seven types of citrus diseases. Pests' images were mainly collected from the Internet. Images of diseases were taken in a tangerine orchard of Jeju Island using a high-resolution camera. Our image dataset is available at the website of Appendix B. Table 1 shows the name and number of images of each kind of pest and disease.

Table 1. The description of citrus pests and diseases image dataset.

Class ID	Common Name	Scientific Name	Number of Samples
	Citrus Pests		
8	Mediterranean fruit fly	Ceratitis capitata	558
0	Asian citrus psyllid	Diaphorina citri Kuwayama	359
5	Citrus longicorn beetle	Anoplophora chinensis	597
7	Brown marmorated stink bug	Halyomorpha halys	606
3	Southern green stink bug	Nezara viridula	488
4	Fruit sucking moth	Othreis fullonica	600
1	Citrus swallowtail	Papilio demodocus	600
15	Citrus flatid planthopper	Metcalfa pruinosa (Say)	555
9	Citrus mealybug	Planococcus citri	495
13	Aphids	Toxoptera citricida	514
11	Citrus soft scale	Hemiptera: Coccidae	497
12	False codling moth	Thaumatotibia leucotreta	511
14	Root weevil	Diaprepes abbreviatus, Pachnaeus opalus	378
2	Forktailed bush katydid	Scudderia furcata	600
10	Cicada	Cicadoidea	508
6	Garden snail	Cornu aspersum	618
16	Glassy-winged sharpshooter	Homalodisca vitripennis	567
	Total		9051
	Citrus Diseases		
17	Anthracnose	Colletotrichum gloeosporioides	467
18	Canker	Xanthomonas axonopodis	598
20	Melanose	Diaporthe citri	532
21	Scab	Elsinoë fawcettii	503
19	Leaf miner	Liriomyza brassicae	427
22	Sooty mold	Capnodium spp	568
23	Pest hole		415
	Total		3510

3.1. Image Collection of Citrus Pests

Insect pests have metamorphosis properties. We focused on images of adults. This is because other stages in their life cycles are short and rare to observe. The appearance of the same pest can vary significantly from one viewing angle to another (refer to Figure 1). To reduce the effect of shooting angle on classification accuracy, photos of pests taken from different angles were gathered. Some citrus pests have small sizes, such as aphid, mealybug, and scale. It is difficult to capture images of their individuals and most of them live by groups to resist predators. For these species, pictures of their group living on a tree were collected (refer to Figure 2).

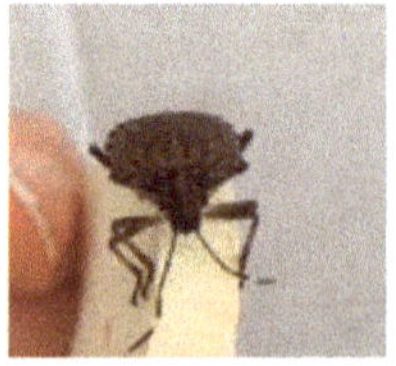

Figure 1. Pictures of brown marmorated stink bug taken from different angles.

Figure 2. Examples of small size citrus pests: (**a**), (**c**), and (**e**) are the images of their individuals, (**b**), (**d**), and (**f**) are those of their groups.

3.2. Image Collection of Citrus Diseases

Compared with pests, features of citrus diseases are more regular. Pictures of citrus diseases were mainly taken in the summer after a heavy rain because the incidence was higher than usual. To keep more details, the distance between camera and diseases was close. Some diseases will cause leaf holes at a later phase. To enhance comparison, images of the leaf holes created by pests (PH) were included as a disease label. Figure 3 displays sample images of each disease.

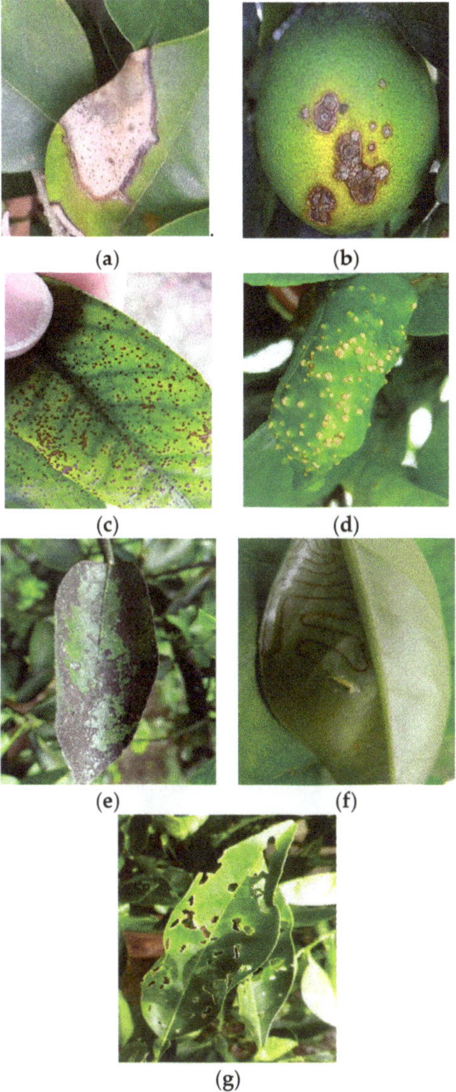

Figure 3. Representative images of citrus diseases. (**a**) citrus anthracnose, (**b**) citrus canker, (**c**) citrus melanose, (**d**) citrus scab, (**e**) sooty mold, (**f**) leaf miner, and (**g**) pest holes.

3.3. Data Augmentation

The problem of imbalanced data classification has been discussed by Das et al. [34]. It prompted us to increase the number of images in the class whose dataset scale was smaller than that of others. For augmenting image data, the generic practice is to perform geometric transformations, such as rotation, reflection, shift, and flip. However, images generated by a single type of operation are similar to each other. They increase the probability of overfitting. To avoid this situation, a new data augmentation method was proposed, which could randomly select three kinds of operations and combine them together to produce new images. Available operations and values of them are shown in Table 2. Figure 4 presents pictures obtained from this approach.

Table 2. Parameter set for data augmentation.

Operation	Value
Rotation	$[0^\circ, 15^\circ]$
Width shift	$[0, 0.2]$
Height shift	$[0, 0.2]$
Shear	$[0, 0.2]$
Zoom	$[0.8, 1.2]$
Horizontal flip	-

Figure 4. Comparison between different data augmentation methods. (**a**) Original image, (**b**), (**c**), and (**d**) images generated from proposed algorithm, (**e**), (**f**), and (**g**) pictures produced by a single rotation operation.

4. Weakly DenseNet Architecture

Convolutional layers in the CNN model are responsible for feature extraction and generation. Therefore, many researchers have focused on increasing depth and width to improve classification accuracy. In contrast, the proposed Weakly DenseNet was created to improve parameters' utilization. To reach this goal, a complex cross-channel operation was adopted to refine feature maps and concatenation method was used for feature reuse.

4.1. The 1 × 1 Convolution for Feature Refinement

A regular convolution contains two aspects: Local receptive field and weight share. From the local receptive field point of view, a 1 × 1 convolution regards each pixel of a feature map as input. However, when weight share is considered, it was equivalent to the whole feature map multiplied by a learnable weight. Therefore, this kind of convolution has the function of refining feature maps.

One layer of 1 × 1 convolution only implements a linear transformation. Many network architectures just use it to alter channel dimension [19,20]. To extend the functionality of 1 × 1 convolution, two layers of it were stacked after each 3 × 3 convolutional layer. The proposed structure takes each whole feature map as an input and thus does not need an extra branch to execute feature recalibration. This reduces the optimization difficulty in contrast with SENet [13]. Figure 5 illustrates the difference between them.

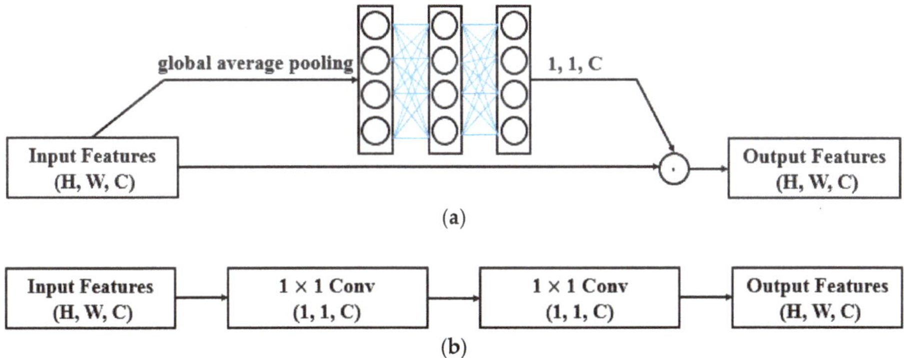

Figure 5. Feature refinement. (**a**) Squeeze-and-excitation block, (**b**) the proposed method.

4.2. Feature Reuse

As network depth increases, gradient propagation becomes more difficult. ResNet addresses this issue by adding input features to output features across a few layers (1). DenseNet simplifies the addition operation by concatenation, which allows feature maps from different depths to combine along channel dimension (2). Considering operation efficiency, the concatenation method of DenseNet was selected for feature reuse

$$\widetilde{X} = X + H(X) \tag{1}$$

$$\widetilde{X} = \{X, H(X)\} \tag{2}$$

where X represents inputs, $H(X)$ is defined as the underlying mapping, $\widetilde{X}$ denotes outputs. In the addition operation, X and $\widetilde{X}$ should have the same dimension.

However, overuse of features of previous layers can increase network overhead. To solve this problem, DenseNet employs a transition layer to reduce the number of input features for a dense block. The dense connectivity pattern of DenseNet also made low-level features to be repeatedly used many times. Yosinski et al. [35] have proved that features generated by convolutions far from the classification layer are general, thus contributing little to classification accuracy. According to their conclusion, some connections between long-range layers were removed.

4.3. Network Architecture

The network architecture was divided into three parts during design. Figure 6 demonstrates the building block of each part. As for the frequency (v) of feature reuse, it was adapted to the depth of the network:

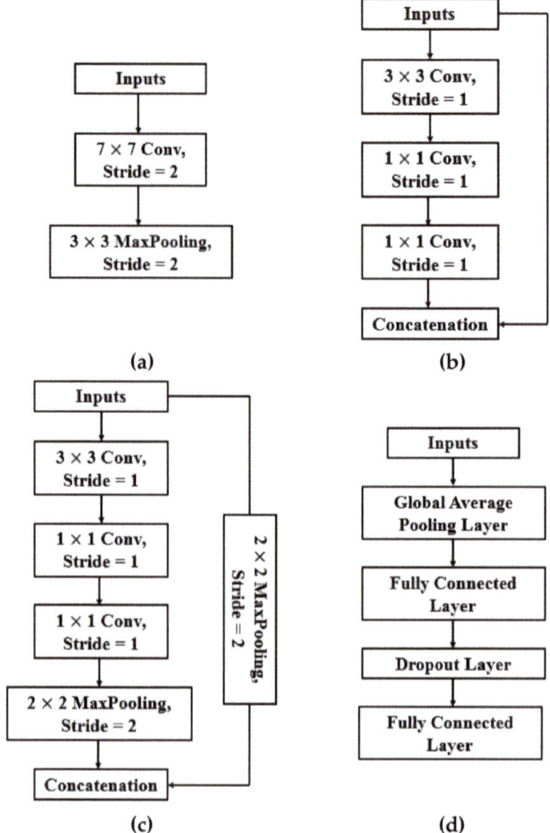

Figure 6. Building blocks of the Weakly DenseNet. (**a**) initial building block, (**b**) and (**c**) intermediate building blocks, (**d**) final classification building block.

Features generated by adjacent convolutions are highly correlated [6]. A final classification layer concentrates on using high-level features [20]. Therefore, keeping connections between short-distance layers and reducing the combination of low-level features to the classification layer are necessary.

Middle layers produce features are between general and specific [35]. It is assumed that if the network depth is increased, the value of v in the middle layers should be enlarged. Figure 7 illustrates the building block of DenseNet for fitting ImageNet dataset. In it, $v > 1$.

We set v to 1, because our network was constructed not very deeply based on image data scale of citrus pests and diseases. Table 3 summarizes the architecture of the network.

Each convolution in the building block is followed by a batch normalization layer [36] and a ReLU layer [32].

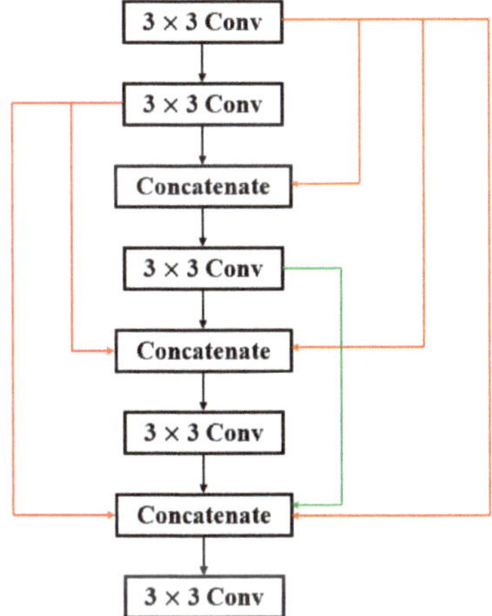

Figure 7. Building block of DenseNet.

Table 3. Network architecture for citrus pests and diseases.

Block	Output Size
Initial Block (a)	$56 \times 56 \times 32$
Intermediate Block (b)	$56 \times 56 \times 96$
Intermediate Block (c)	$28 \times 28 \times 192$
Intermediate Block (b)	$28 \times 28 \times 384$
Intermediate Block (c)	$14 \times 14 \times 768$
1×1 conv, stride 1	$14 \times 14 \times 512$
1×1 conv, stride 1	$14 \times 14 \times 512$
2×2 max pool, stride 2	$7 \times 7 \times 512$
Classification Block (d)	$1 \times 1 \times 24$

5. Experiments and Results

The original dataset was split into three parts: Training set, validation set, and test set. The ratio between them was 4:1:1. The images in each set were resized to 224×224 by a bilinear interpolation approach. To evaluate the effectiveness of Weakly DenseNet, it was compared with several baseline networks from different aspects. The software implementation was based on Keras with TensorFlow backend. The hardware foundation was GPU, 1080Ti. Code and models are provided in Appendix C.

5.1. Training

All the networks were trained with SGD and a Nesterov momentum [37] of 0.9 was introduced to accelerate convergence. To improve models' generalization performance, a small batch size of 16 was selected during training [38]. The initial learning rate was set to 0.001 and it can be adjusted based on Algorithm 1.

Algorithm 1. Learning Rate Schedule

Input: Patience P, decay θ, validation loss L
Output: Learning rate γ
1: Initialize $L = L_0$, $\gamma = \gamma^0$
2: $i \leftarrow 0$
3: **while** $i < P$ **do**
4:    **if** $L \leq L_i$ **then**
5:       $i = i + 1$
6:    **else**
7:       $L = L_i$
8:       $i = i + 1$
9:    **end if**
10: **end while**
11: **if** $L = L_0$ **then**
12:    $\gamma = \gamma * \theta$
13: **end if**

In the experiment, $P = 5$ and $\theta = 0.8$. Weight initialization proposed by He et al. [39] was followed, and a weight decay of 10^{-4} was used to alleviate the overfitting problem. The maximum training epoch of each model was 300. The rate of Dropout was set to be 0.5.

The VGG-16 of this paper used a global average pooling layer [17] to reduce the number of parameters in fully connected layers. In the original SE block, the size of the hidden layer was reduced by a ratio ($r > 1$) to limit model complexity. To keep the same computation cost as NIN-16, the value of r was set to 1 by us.

Table 4 shows the training results of each model. With regard to classification accuracy, WeaklyDenseNet-16 was the highest, followed by VGG-16, and NIN-16. The higher accuracy of NIN-16 than SENet-16 indicates that two layers of 1×1 convolution have better performance of refining feature maps than SE block. By concatenating previous layers' features, the recognition performance of NIN-16 was significantly strengthened: The accuracy of WeaklyDenseNet-16 was 1.58 percent higher than that of NIN-16. As for computation cost, VGG-16 was the largest while that of SENet-16 was the least. It should be noticed that ShuffleNets and MobileNets overfitted citrus pests and diseases image dataset. Even though they had similar model sizes to WeaklyDenseNet-16, their larger values of depth brought bigger error rate on validation dataset. As for training speed per batch size, bigger size model took more time except for ShuffleNet-v2 [31]. The accuracy training plots of benchmark models are displayed in Figure 8. It can be seen that each model completely converges in 300 epochs.

Table 4. Training performance of selected models.

Model Name	Training Accuracy	Validation Accuracy	Model Size (MB)	Training Time (ms)/Batch Size
MobileNet-v1	99.23	85.45	25	152
MobileNet-v2	99.28	87.97	33.9	198
ShuffleNet-v1	99.13	83.58	28.8	145
ShuffleNet-v2	98.72	83.58	42	144
VGG-16	99.82	93	120.2	303
SENet-16	99.10	88.71	19.5	138
NIN-16	99.63	91.84	19.6	137
WeaklyDenseNet-16	99.83	93.42	30.5	138

'NIN' represents Network in Network.

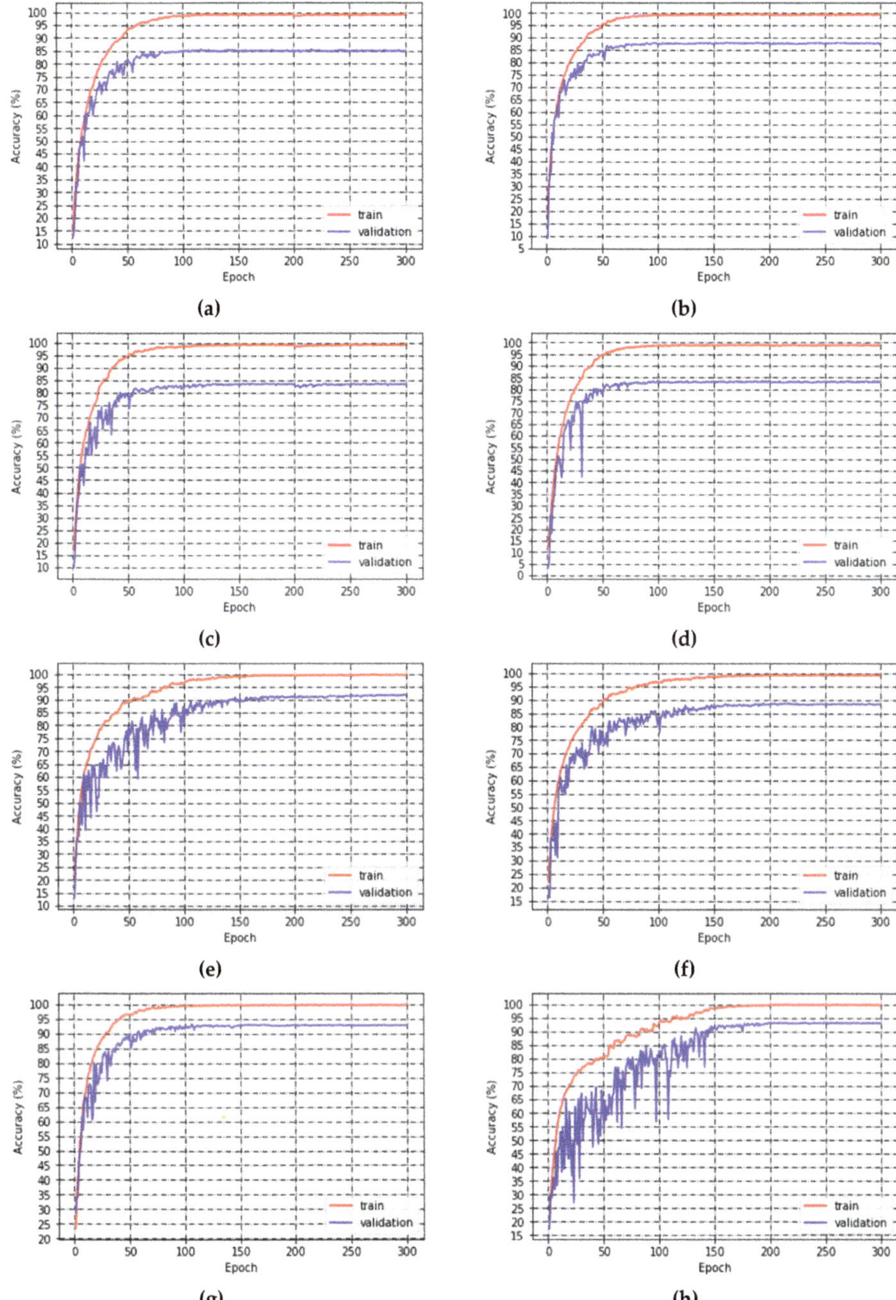

Figure 8. Training plot of each model. (**a**) MobileNet-v1, (**b**) MobileNet-v2, (**c**) ShuffleNet-v1, (**d**) ShuffleNet-v2, (**e**) NIN-16, (**f**) SENet-16, (**g**) VGG-16, (**h**) WeaklyDenseNet-16.

5.2. Test

Test accuracy results of selected models are shown in Figure 9 which shows the same accuracy trend as Table 4. The confusion matrix of Weakly DenseNet-16 on the test dataset is presented in Figure A1 (refer to Appendix A). Figure A1a shows that the recall rate (3) of citrus root weevil is the lowest and that of the citrus swallowtail is the highest. Among misclassified images, citrus anthracnose and citrus canker are the most easily confused by the proposed model: Nine images of citrus anthracnose were considered as the class of citrus canker and four images of citrus canker were regarded as citrus anthracnose by the network. The two diseases at later phase show a similar appearance on leaves. Thus, Weakly DenseNet-16 gave some incorrect predictions. The precision rate (4) of the PH is the largest while that for citrus flatid planthopper is the lowest. Figure A1b displays the wrong predictions between citrus pest and disease. Ten images of citrus disease were misclassified into pest labels and seventeen pictures of citrus pest were identified as diseases by mistake. In them, the probability that citrus soft scale is falsely regarded as citrus sooty mold is the highest. Adult citrus soft scales can secrete honeydew sooty mold around them for growth, which shows a similar pattern to the symptom of sooty mold.

$$Recall = \frac{Number\ of\ true\ positive\ samples}{Number\ of\ true\ positive\ samples\ +\ Number\ of\ false\ negative\ samples} \times 100\% \qquad (3)$$

$$Precision = \frac{Number\ of\ true\ positive\ samples}{Number\ of\ true\ positive\ samples\ +\ Number\ of\ false\ positive\ samples} \times 100\% \qquad (4)$$

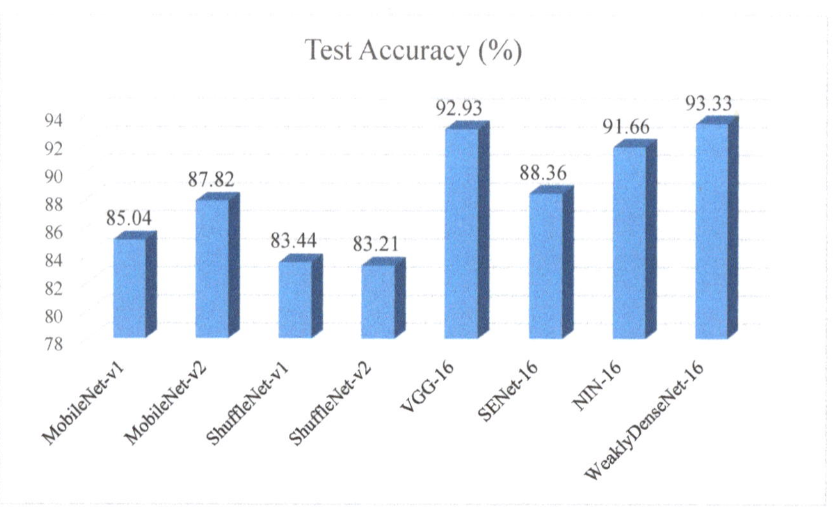

Figure 9. Comparison of the test accuracy.

The hierarchical structure of the CNN model allows features generated from layers of different depths to show significant differences [40]. To better understand the learning capacity of intermediate building blocks of Weakly DenseNet-16, several important feature maps of them were visualized and compared. From Figure 10, it can be noticed that:

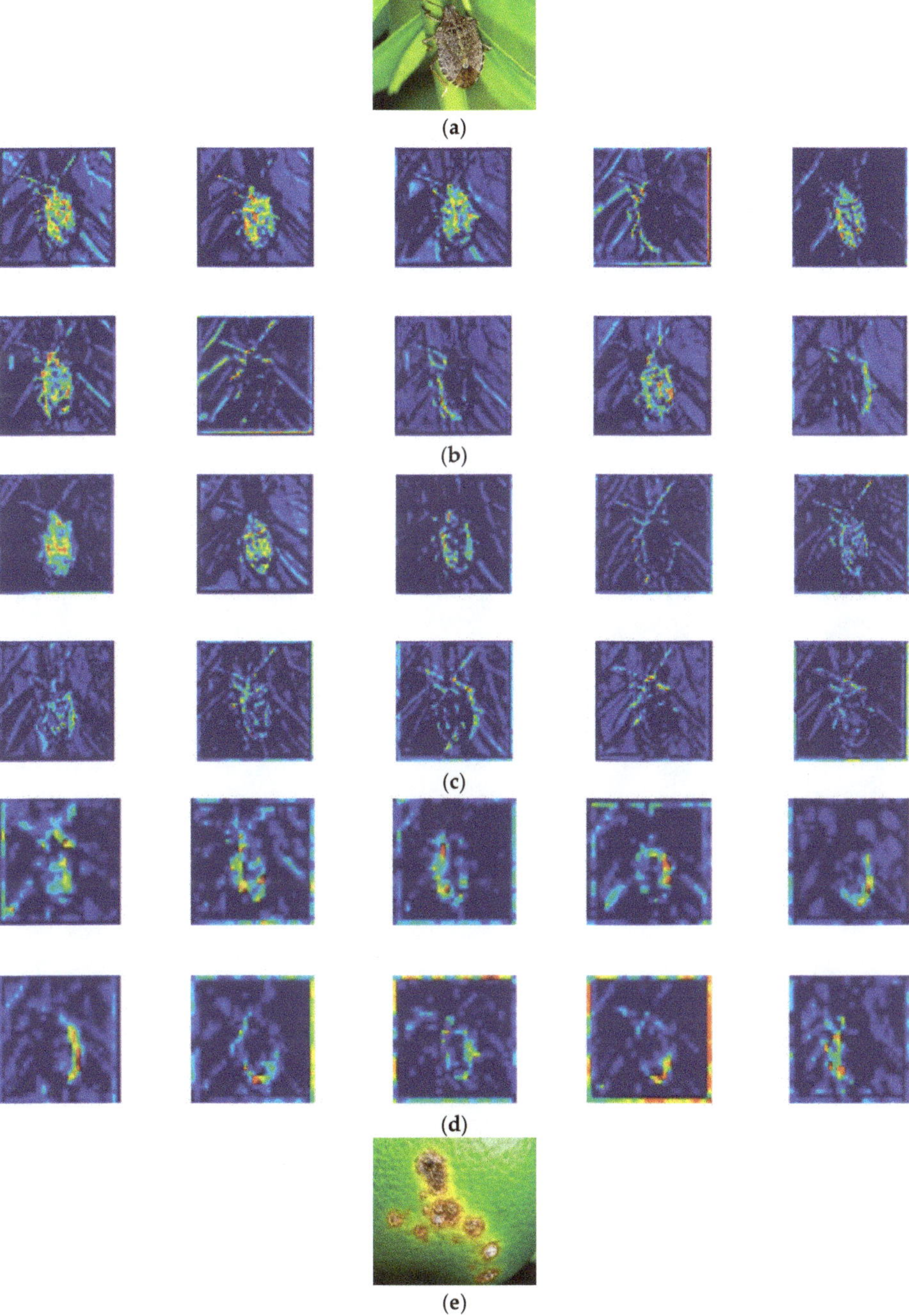

Figure 10. *Cont.*

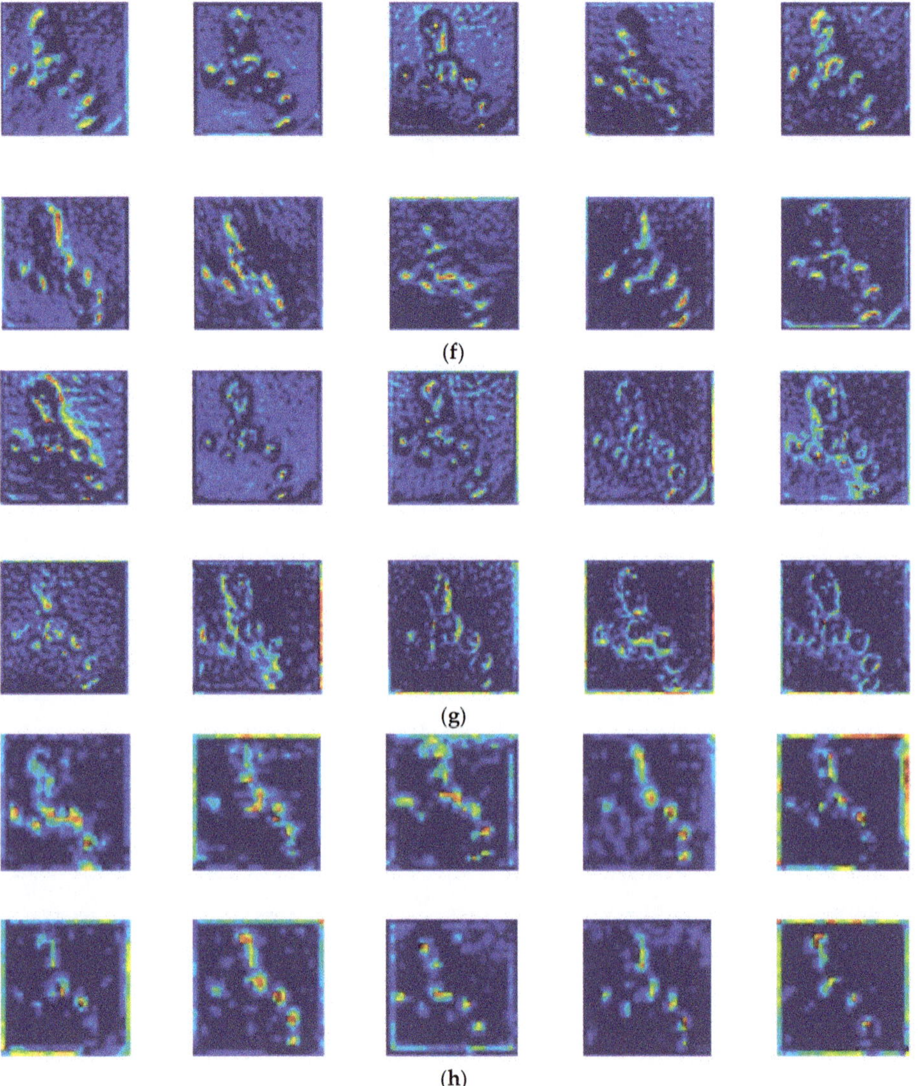

Figure 10. Visualization of features. (**a**) and (**e**) input images, (**b**) and (**f**) output features of the intermediate building block 1, (**c**) and (**g**) sampled features of the intermediate building block 2, (**d**) and (**h**) examples of the feature maps in the intermediate building block 3. Brighter color in images corresponds to higher value.

A bank of convolutional filters in the same layer can extract features of different parts of the target object. This feature extraction method allows the CNN model to acquire sufficient visual information for subsequent analysis.

With increasing depth, the background features become less visible. Therefore, CNN models do not require additional pre-processing techniques to reduce background noise. They are more convenient to use than conventional machine learning algorithms.

Features of the deeper layer are more abstract than those of the shallow layer. More convolution and max pooling operations are performed on shallow layer features in the deeper layer, resulting in higher-level features that are more suitable for classification.

6. Conclusions and Future Work

Pests and diseases can reduce citrus output. To control their impact, a new image dataset about citrus pests and diseases was created and a novel CNN architecture was proposed to recognize them. The network was constructed from the aspect of improving parameters' utilization instead of depth and width. The structure of two 1×1 convolutional layers was revisited and applied to refine feature maps. To relieve the optimization difficulty in the deep network, the idea of feature reuse was followed. Considering operation efficiency, the concatenation method of DenseNet was employed. However, the high frequency of feature reuse increased the overhead of network. To save computation cost, the value of feature reuse frequency was set based on network depth. To further improve the robustness of the CNN model, a new data augmentation algorithm was provided, which can significantly lessen the similarity between generated images. In experimental studies, NIN-16 got a test accuracy of 91.66% which was much higher than that of SENet-16 (88.36%). This phenomenon indicates that two-layer 1×1 convolution has better performance of refining feature maps than SE block. The higher accuracy of WeaklyDenseNet-16 (93.33%) than NIN-16 indicates that feature reuse method can further enhance network performance. VGG-16 achieved the second-highest classification accuracy (93%) but consumed the most computing resources: Model size is 120.2 MB. This fact implies the importance of network structure optimization on fitting different datasets.

The object scale in the image is an essential factor that influences classification accuracy of the CNN model. Using extremely deep networks to identify big scale objects will cause a waste of computational resources. For the identification of small size objects, shallow networks cannot give accurate results. Future work is to build a CNN model that can adapt to the size of the object in the image.

Author Contributions: Conceptualization, S.X. and M.L.; methodology, S.X.; software, S.X.; validation, M.L., S.X., K.-k.L.; formal analysis, M.L., K.-k.L.; data curation, S.X.; writing—original draft preparation, S.X.; writing—review and editing, all authors.

Funding: This work was supported by a grant (No: 2017R1A2B4006667) of National Research Foundation of Korea (NRF) funded by the Korea government (MSP).

Conflicts of Interest: The authors declare no conflict of interest.

Appendix A

	0	1	2	3	4	5	6	7	8	9	10	11	12	13	14	15	16	17	18	19	20	21	22	23	Recall
0	58	0	0	0	0	0	0	0	0	0	0	0	0	1	0	0	0	0	0	0	0	0	0	0	98.31
1	0	100	0	0	0	0	0	0	0	0	0	0	0	0	0	0	0	0	0	0	0	0	0	0	100
2	0	0	96	2	0	0	0	0	0	0	1	0	0	1	0	0	0	0	0	0	0	0	0	0	96
3	0	0	3	84	1	0	0	0	0	0	0	0	0	0	0	0	0	0	0	0	0	0	0	0	94.32
4	0	1	2	0	95	0	0	0	0	0	0	1	1	0	0	0	0	0	0	0	0	0	0	0	95
5	0	0	0	0	0	96	0	0	0	0	1	0	0	0	0	0	0	0	0	0	0	0	0	0	98.97
6	0	0	1	0	0	0	90	1	0	0	3	0	0	1	0	3	0	0	0	0	0	0	1	0	90
7	0	0	0	0	0	0	0	97	0	0	1	0	0	0	1	0	1	0	0	0	0	0	0	0	97
8	0	0	0	0	0	0	0	0	87	0	0	0	0	0	0	0	0	0	0	0	0	1	0	0	98.86
9	1	0	0	0	0	0	0	0	1	86	0	0	0	2	0	2	1	0	0	0	0	2	0	0	90.53
10	0	0	0	0	0	0	0	0	0	0	94	0	2	1	2	0	0	1	0	0	0	0	0	0	94
11	0	0	0	0	0	0	0	0	0	4	0	82	0	2	0	3	0	0	0	0	0	1	5	0	84.54
12	0	0	0	0	0	0	1	2	0	0	0	0	97	0	0	0	0	0	0	0	0	0	0	0	97
13	0	0	4	3	0	1	1	0	0	1	0	2	0	84	0	0	1	0	1	1	1	0	0	0	84
14	1	0	0	3	0	1	2	4	1	2	0	0	1	3	56	0	1	0	1	1	0	1	0	0	71.79
15	0	0	1	0	0	0	1	0	0	1	0	1	0	0	2	79	0	0	0	0	0	0	0	0	92.94
16	2	0	0	1	0	0	0	0	1	0	3	0	1	0	0	1	78	0	0	0	0	0	0	0	89.66
17	0	0	0	0	0	0	0	0	0	0	0	0	0	0	0	0	0	58	9	0	0	0	0	0	86.57
18	1	0	0	0	0	0	0	0	1	0	0	0	0	0	0	0	0	4	92	0	0	0	0	0	93.88
19	0	0	0	0	0	0	2	0	0	0	0	0	0	0	0	0	0	0	0	65	0	0	0	0	97.01
20	0	0	0	0	0	0	0	0	1	0	0	0	0	0	0	1	0	0	0	0	96	0	2	0	96
21	0	0	0	0	0	0	0	0	0	0	0	0	0	1	0	1	0	1	0	1	1	95	0	0	95
22	0	0	0	0	0	0	0	0	0	0	0	0	0	0	0	1	0	0	0	1	0	0	86	0	97.73
23	0	0	0	0	0	0	0	0	0	0	0	0	0	0	0	1	0	1	0	1	0	0	0	62	95.38
Precision	92.06	99.01	89.72	90.32	98.96	97.96	92.78	93.27	94.57	91.49	91.26	95.35	95.1	87.5	91.8	85.87	95.12	89.23	89.32	92.86	97.96	95	91.49	100	

The topmost header spans columns 0–23 as **Predicted Class**; the row-label column is **True Class**.

(**a**)

		Predicted Class	
		Pest	**Disease**
True Class	**Pest**	1557	17
	Disease	10	575

(**b**)

Figure A1. The confusion matrix for the test set. (**a**) test result of each class, (**b**) prediction for pest and disease label.

Appendix B

Image dataset is available at: https://files.mycloud.com/home.php?brand=webfiles#23a3c71/device_30757105/ARlab/xingshuli600/.

Appendix C

Models and code are available at: https://github.com/xingshulicc/xingshulicc/tree/master/citrus_pest_diseases_recognition.

References

1. LeCun, Y.; Bengio, Y.; Hinton, G. Deep learning. *Nature* **2015**, *521*, 436–444. [CrossRef] [PubMed]
2. Liang, Q.; Xiang, S.; Hu, Y.; Coppola, G.; Zhang, D.; Sun, W. PD2SE-Net: Computer-assisted plant disease diagnosis and severity estimation network. *Comput. Electron. Agric.* **2019**, *157*, 518–529. [CrossRef]
3. Cheng, X.; Zhang, Y.; Chen, Y.; Wu, Y.; Yue, Y. Pest identification via deep residual learning in complex background. *Comput. Electron. Agric.* **2017**, *141*, 351–356. [CrossRef]

4. Shen, Y.; Zhou, H.; Li, J.; Jian, F.; Jayas, D.S. Detection of stored-grain insects using deep learning. *Comput. Electron. Agric.* **2018**, *145*, 319–325. [CrossRef]

5. Ren, S.Q.; He, K.M.; Girshick, R.; Sun, J. Faster R-CNN: Towards Real-Time Object Detection with Region Proposal Networks. *IEEE Trans. Pattern Anal. Mach. Intell.* **2017**, *39*, 1137–1149. [CrossRef] [PubMed]

6. Szegedy, C.; Vanhoucke, V.; Ioffe, S.; Shlens, J.; Wojna, Z. Rethinking the inception architecture for computer vision. *arXiv* **2015**, arXiv:1512.00567.

7. Jeong, J.; Park, H.; Kwak, N. Enhancement of SSD by concatenating feature maps for object detection. *arXiv* **2017**, arXiv:1705.09587.

8. Zhuang, X.; Zhang, T. Detection of sick broilers by digital image processing and deep learning. *Biosyst. Eng.* **2019**, *179*, 106–116. [CrossRef]

9. Deng, J.; Dong, W.; Socher, R.; Li, L.J.; Li, K.; Li, F.F. ImageNet: A large-scale hierarchical image database. In Proceedings of the IEEE Conference on Computer Vision and Pattern Recognition (*CVPR*), Miami, FL, USA, 20–25 June 2009.

10. Reyes, A.K.; Caicedo, J.C.; Camargo, J.E. Fine—Tuning Deep Convolutional Networks for Plant Recognition. In Proceedings of the CLEF (Working Notes), Toulouse, France, 8–11 September 2015.

11. Li, H.; Kadav, A.; Durdanovic, I.; Samet, H.; Graf, H.P. Pruning Filters for Efficient ConvNets. *arXiv* **2016**, arXiv:1608.08710v3.

12. Molchanov, P.; Tyree, S.; Karras, T.; Aila, T.; Kautz, J. Pruning Convolutional Neural Networks for Resource Efficient Inference. *arXiv* **2016**, arXiv:1611.06440v2.

13. Hu, J.; Shen, L.; Albanie, S.; Sun, G.; Wu, E. Squeeze-and-Excitation Networks. *arXiv* **2017**, arXiv:1709.01507.

14. Woo, S.; Park, J.; Lee, J.Y.; Kweon, I.S. CBAM: Convolutional Block Attention Module. *arXiv* **2018**, arXiv:1807.06521.

15. Srivastava, N.; Hinton, G.; Krizhevsky, A.; Sutskever, I.; Salakhutdinov, R. Dropout: A simple way to prevent neural networks from overfitting. *J. Mach. Learn. Res.* **2014**, *15*, 1929–1958.

16. Huang, G.; Sun, Y.; Liu, Z.; Sedra, D.; Weinberger, K. Deep Networks with Stochastic Depth. *arXiv* **2016**, arXiv:1603.09382v3.

17. Lin, M.; Chen, Q.; Yan, S. Network in network. *arXiv* **2013**, arXiv:1312.4400v3.

18. Srivastava, R.K.; Greff, K.; Schmidhuber, J. Highway Networks. *arXiv* **2015**, arXiv:1505.00387.

19. He, K.; Zhang, X.; Ren, S.; Sun, J. Deep residual learning for image recognition. *arXiv* **2015**, arXiv:1512.03385.

20. Huang, G.; Liu, Z.; van der Maaten, L.; Weinberger, K.Q. Densely Connected Convolutional Networks. *arXiv* **2016**, arXiv:1608.06993v5.

21. Boniecki, P.; Koszela, K.; Piekarska Boniecki, H.; Weres, M.; Zaborowcz, M.; Kujawa, A.; Majewski, A.; Raba, B. Neural identification of selected apple pests. *Comput. Electron. Agric.* **2015**, *110*, 9–16. [CrossRef]

22. Sun, Y.; Jiang, Z.; Zhang, L.; Dong, W.; Rao, Y. SLIC_SVM based leaf diseases saliency map extraction of tea plant. *Comput. Electron. Agric.* **2019**, *157*, 102–109. [CrossRef]

23. Ferentinos, K.P. Deep learning models for plant disease detection and diagnosis. *Comput. Electron. Agric.* **2018**, *145*, 311–318. [CrossRef]

24. Simonyan, K.; Zisserman, A. Very deep convolutional networks for large-scale image recognition. *arXiv* **2014**, arXiv:1409.1556.

25. Zagoruyko, S.; Komodakis, N. Wide residual networks. In Proceedings of the British Machine Vision Conference (*BMVC*), York, UK, 19–22 September 2016.

26. Lee, C.Y.; Xie, S.; Gallagher, P.; Zhang, Z.; Tu, Z. Deeply-supervised nets. In Proceedings of the 18th International Conference on Artificial Intelligence and Statistics, San Diego, CA, USA, 9–12 May 2015.

27. Chollet, F. Xception: Deep Learning with Depthwise Separable Convolutions. In Proceedings of the 30th IEEE Conference on Computer Vision and Pattern Recognition (*CVPR*), Honolulu, HI, USA, 22–25 July 2017.

28. Howard, A.G.; Zhu, M.; Chen, B.; Kalenichenko, D.; Wang, W.; Weyand, T.; Andreetto, M.; Adam, H. MobileNets: Efficient Convolutional Neural Networks for Mobile Vision Applications. *arXiv* **2017**, arXiv:704.04861.

29. Sandler, M.; Howard, A.; Zhu, M.; Zhmoginov, A.; Chen, L. MobileNetV2: Inverted Residuals and Linear Bottlenecks. *arXiv* **2018**, arXiv:1801.04381v4.

30. Zhang, X.; Zhou, X.; Lin, M.; Sun, J. ShuffleNet: An Extremely Efficient Convolutional Neural Network for Mobile Devices. *arXiv* **2017**, arXiv:1707.01083.

31. Ma, N.; Zhang, X.; Zheng, H.; Sun, J. ShuffleNet V2: Practical Guidelines for Efficient CNN Architecture Design. *arXiv* **2018**, arXiv:1807.11164.

32. Nair, V.; Hinton, G.E. Rectified linear units improve restricted boltzmann machines. In Proceedings of the 27th International Conference on Machine Learning (*ICML*), Haifa, Israel, 21–24 June 2010.

33. Xie, S.; Girshick, R.; Dollar, P.; Tu, Z.; He, K. Aggregated Residual Transformations for Deep Neural Networks. *arXiv* **2016**, arXiv:1611.05431.

34. Das, S.; Datta, S.; Chaudhuri, B.B. Handling data irregularities in classification: Foundations, trends, and future challenges. *Pattern Recognit.* **2018**, *81*, 674–693. [CrossRef]

35. Yosinski, J.; Clune, J.; Bengio, Y.; Lipson, H. How transferable are features in deep neural networks? In Proceedings of the 28th International Conference on Neural Information Processing Systems (*NeurIPS*), Montreal, PQ, Canada, 8–13 December 2014.

36. Ioffe, S.; Szegedy, C. Batch normalization: Accelerating deep network training by reducing internal covariate shift. *arXiv* **2015**, arXiv:1502.03167.

37. Sutskever, I.; Martens, J.; Dahl, G.E.; Hinton, G.E. On the importance of initialization and momentum in deep learning. In Proceedings of the 30th International Conference on Machine Learning (*ICML*), Atlanta, GA, USA, 16–21 June 2013.

38. Masters, D.; Luschi, C. Revisiting Small Batch Training for Deep Neural Networks. *arXiv* **2018**, arXiv:1804.07612.

39. He, K.; Zhang, X.; Ren, S.; Sun, J. Delving Deep into Rectifiers: Surpassing Human-Level Performance on ImageNet Classification. In Proceedings of the 2015 International Conference on Computer Vision (*ICCV*), Santiago, Chile, 11–18 December 2015.

40. Zeiler, M.; Fergus, R. Visualizing and Understanding Convolutional Networks. *arXiv* **2013**, arXiv:1311.2901.

Article

Vision-Based Novelty Detection Using Deep Features and Evolved Novelty Filters for Specific Robotic Exploration and Inspection Tasks

Marco Antonio Contreras-Cruz [1], Juan Pablo Ramirez-Paredes [1],
Uriel Haile Hernandez-Belmonte [2] and Victor Ayala-Ramirez [1],*

[1] Department of Electronics Engineering, University of Guanajuato, Campus Irapuato-Salamanca, Carr.
Salamanca-Valle de Santiago Km 3.5 + 1.8, Comunidad de Palo Blanco, Salamanca 36885, Mexico
[2] Department of Art and Enterprise, University of Guanajuato, Campus Irapuato-Salamanca, Carr.
Salamanca-Valle de Santiago Km 3.5 + 1.8, Comunidad de Palo Blanco, Salamanca 36885, Mexico
* Correspondence: ayalav@ugto.mx; Tel.: +52-464-647-9940 (ext. 2444); Fax: +52-464-647-9940 (ext. 2413)

Received: 8 May 2019; Accepted: 28 June 2019; Published: 5 July 2019

Abstract: One of the essential abilities in animals is to detect novelties within their environment. From the computational point of view, novelty detection consists of finding data that are different in some aspect to the known data. In robotics, researchers have incorporated novelty modules in robots to develop automatic exploration and inspection tasks. The visual sensor is one of the preferred sensors to perform this task. However, there exist problems as illumination changes, occlusion, and scale, among others. Besides, novelty detectors vary their performance depending on the specific application scenario. In this work, we propose a visual novelty detection framework for specific exploration and inspection tasks based on evolved novelty detectors. The system uses deep features to represent the visual information captured by the robots and applies a global optimization technique to design novelty detectors for specific robotics applications. We verified the performance of the proposed system against well-established state-of-the-art methods in a challenging scenario. This scenario was an outdoor environment covering typical problems in computer vision such as illumination changes, occlusion, and geometric transformations. The proposed framework presented high-novelty detection accuracy with competitive or even better results than the baseline methods.

Keywords: visual inspection; one-class classifier; grow-when-required neural network; evolving connectionist systems; automatic design; bio-inspired techniques; artificial bee colony

---

1. Introduction

Novelty detection is the task of recognizing data that are different in some aspects from the already known data [1]. This is a challenging problem because the datasets may have a large number of examples of the normal class and an insufficient number of examples of the novel class (in almost all cases, no novelty examples are available). Having robust methods for this type of problem is of great importance in practical applications such as fraud detection [2,3], fault detection [4], medical diagnosis [5–7], video surveillance [8,9], and robotic tasks [10–12], among others. For these applications, it is not common to have access to data labeled as novel. Another complication is that even when using the same type of information across different applications (e.g., visual information), the concept of novelty varies among them. For these reasons, multi-class classifiers are infeasible for novelty detection. As an alternative, there are dedicated methods for novelty detection that provide all the elements to solve the problem.

In general, the novelty detection methods construct a model with the examples of the normal class and use this model with unknown data to compute novelties. The methods can be classified into five categories [1]: probabilistic, distance-based, reconstruction-based, domain-based, and information-theoretic techniques. One-class classification techniques have been broadly applied for novelty detection with successful results in environments where no dynamic adaptation of the models is required. Recently, advances in deep learning algorithms have shown a new open area into novelty detection [9,13]. The deep-learning-based methods for novelty detection combine the ability of deep neural networks to extract features with the ability of one-class classifiers to model the normal data. The main drawback of these techniques is the need for large-scale datasets and high computational load to train the models.

Inspired by the ability of animals to detect novelties and to respond to changes in their environment [14], researchers have tried to incorporate novelty detection methods into robots to improve their adaptation capability to the dynamic environments that are often present in real-world robotic tasks. Presently, it is possible to capture useful information to perform this process with the use of sensors incorporated into the robots (e.g., sonar, laser, camera, GPS, etc.). Among them, visual sensors are one of the most popular devices to extract information for novelty detection [10,11,15], perhaps because humans use visual information unconsciously as a central component to detect novelties.

In robotics, a novelty detection module is beneficial for several applications (e.g., exploration, inspection, vigilance, etc.). Specifically, in exploration and inspection tasks [11], the robot should explore its environment, building a model of normality using the sensed information. After the model construction, the robot patrols (inspection phase) the same route of the exploration phase in order to detect novelties. It is worth noting that the number of path executions is limited. Although the routes are the same in both phases, due to the operating conditions it is not possible to ensure the same robot positions between different path executions.

For the above problem, the robot needs online novelty detectors to cope with dynamic environments and approaches with fast learning capabilities to detect novelties in scenarios with a reduced amount of information. Most of the traditional one-class classifiers operate offline, which means that it is difficult to adapt these methods to dynamic environments. Meanwhile, deep-learning approaches need large-scale datasets and a huge computation load to train the models. Alternatively, online approaches are based on evolving connectionist systems [11] and grow when required neural networks [16] meet the above conditions. These methods not only build a model of normality incrementally, but they also adapt the model to dynamic changes of the input data—that is, they can insert new information and forget old information. However, we still see challenges in the application of the online novelty detectors into exploration and inspection tasks based on visual information. First, current robotic applications use low-level visual features that are sensitive to illumination changes, occlusion, or geometric transformations. Some visual features used in robotic applications are RGB histograms [11], color angular indexing [17], GIST descriptor [15], and others. Second, in different exploration and inspection tasks, the robots use the same parameters in the novelty detection module, without considering that the performance of the detector depends on the specific task to be solved. These reasons have restricted the applications of the above online novelty detectors to indoor environments where many conditions have been controlled.

Motivated by the previous issues, in this work we propose the application novelty detectors based on evolutionary connectionist systems and grow when required neural networks with visual descriptions drawn from deep convolutional networks for exploration and visual inspection tasks. In contrast with existing deep learning approaches for novelty detection, we propose the use of already-trained networks to extract visual features, instead of learning new visual features, in order to reduce the computational load in the feature extraction phase. We prefer deep descriptions over traditional visual description due to its reliability in generating robust features for classification tasks. Additionally, we propose a framework to design novelty detectors automatically via the selection of the best parameters, depending on the specific robotic exploration and inspection task. This framework uses a global optimization technique as the main component to find the most appropriate parameters for the task. We verified the utility of the proposed visual novelty detection system in outdoor applications, where an unmanned aerial vehicle (UAV) captured images in challenging environments (i.e., environments with illumination changes, geometric transformations in the objects of the environment, and occlusions). In summary, this proposed work presents the following contributions:

1. We extend the application of the above online novelty detectors to outdoor environments where illumination changes, occlusions, and geometric transformations are presented.
2. Most of the existing visual novelty detectors involve humans to select the appropriate parameters for a specific visual exploration and inspection task. In contrast to these previous works, we propose a framework for the automatic design of novelty detectors.
3. In contrast with previous deep-learning and one-class classifiers, our proposal uses a pre-trained convolutional neural network to extract features from images to reduce the computational load. That enables the system to operate online (sample rate of 4 Hz).
4. As far as we know, this is the first time that online novelty detectors based on evolving connectionist systems or grow-when-required neural networks have been applied in unmanned aerial vehicles for detecting novelties in visual exploration and inspection tasks.

The rest of this document is structured as follows. Section 2 reviews some works related to visual novelty detection in robotics. Section 3 presents our visual-based novelty detection approach. Section 4 describes the experimental setup and compares our experimental results against traditional visual novelty detectors. In Section 5, we discuss the results and limitations of this work. Finally, in Section 6 we share our main conclusions and perspectives for future work.

2. Related Work

Marsland et al. [14] proposed a self-organizing map (SOM) with a habituation model embedded into the nodes to detect novelty. The system uses sonar readings as inputs, and the nodes habituate to similar inputs. The habituation level of the nodes represents the novelty value of the input. Crook and Hayes [18] developed a novelty detection system based on the Hopfield network—a type of fully-connected recurrent neural network. They implemented the novelty detector in a robot to detect cards in a gallery. The robot captures a color image and through simple processing finds the orange cards. The binary image (detection of the orange color) enters the network to perform the novelty detection process. The operation of the detector consists of updating the weights of the network every time a new input is fed into the network. The system uses a threshold value and the energy level of the network to decide if the input is novel.

Both detectors have restrictions in their operation because they keep a fixed network structure. Therefore, they cannot adapt their behaviors to dynamic changes in the inputs. For this reason, Marsland et al. [16] proposed a novelty detection system for mobile robots based on a grow-when-required (GWR) neural network. The GWR network topologically connects nodes subject to habituation and incorporates new nodes based on their habituation level and the activation level of the nearest node to the given input. Besides, the GWR network can forget patterns, deleting nodes without topological connections. Crook et al. [19] compared the Hopfield-based novelty detector against the GWR network for novelty detection. In this study, they performed two experiments: the first experiment used sonar readings as input, and the second one used images (the problem of card detection in galleries). The results showed that both approaches could construct an appropriate model of the environment. However, the GWR-based approach produced more precise models because of its lower sensitivity to noise, more flexible representation of the inputs, and ability to adapt to dynamic changes in the inputs.

Afterwards, Neto et al. [20] applied a GWR network with visual information as input. They proposed a framework that combines a visual attention model and a visual description of the more salient points in the image based on color angular indexing and the standard deviation of the intensity.

This type of description is invariant to illumination changes; however, it is infeasible to detect new objects outside the attention regions. Neto and Nehmzow [17] used the novelty detectors based on GWR and incremental principal component analysis (IPCA) with two interest point detectors: the detection based on saliency and the Harris detector. They compared two ways of representing the patches in the visual input (raw pixels of the image). The first method was to keep a fixed size of the patch, while the second was to find the size of the patch automatically. The results often showed that the fixed-size approach presented the best results. Inspired by the evolving connectionist systems [21] and the habituation model proposed in the GWR networks, Özbielge [11] proposed a recurrent neural network for novelty detection for exploration and inspection tasks. This method predicts the next input and computes a novelty threshold value during its operation. This information is used and compared to the observed input to decide if it is novel. The system uses laser readings, motor outputs, and RGB color histograms as input information. Also, Özbielge [22] proposed a dynamic neural network for static and dynamic environments. The method computes the novelty in a similar way to the previous approach—it computes the error between the input observation and the prediction of the network, and if the error is higher than the evolved threshold, then the object is considered a novelty.

Apart from the above detectors, Kato et al. [15] implemented a system based on reconstruction that takes advantage of the position where the robot captured the images. The novelty detector used the GIST descriptor and a reconstruction-based approach to generate a system invariant to illumination changes. A principal limitation of their system is the absence of a threshold value to detect novelties (no optimization is provided for tuning the threshold). Gonzalez-Pacheco et al. [23] developed a novelty filter to detect new human poses. The system uses visual information of the Kinect sensor and four one-class classifiers: Gaussian mixture model, K-means, one-class support vector machines, and lLeast suares anomaly detection. For this task, the Gaussian mixture model performed better than the other novelty detectors. However, the performance of the method depended on the number of specified Gaussians (the user defined this value in the experiment). Recently, Gatsoulis and McGinnity [24] proposed an online expandable neural network similar to the GWR network. The method uses speeded-up robust features (SURF) and an ownership vector. The main difference between the GRW approach and this method is that the habituation is defined by the object and not by the feature vectors.

All the above novelty detectors have been applied for indoor environments, and few works have been proposed for outdoor environments. For instance, Wang et al. [25] implemented an approximation to the nearest neighbor via search trees to detect novelties in indoor and outdoor environments (they used a static camera for the outdoor environment). The inputs were visual features extracted from patches—for example, color histograms in the HSV space (hue, saturation, value) and texture information (Gabor filters). They compared the performance of their system against the GWR network. The results showed that their proposed approach was better than the GWR network in their particular experiments. Ross et al. [12] presented a vision system for obstacle detection based on novelty for field robotics. The motivation in the use of novelty is that in agricultural applications, it is infeasible to train a system with all types of obstacles. The inputs of the detector were color, texture, and position of the patches in stereo images. The system detects novelty by using the probability density estimated by a weighted version of Parzen windows.

Previous works have explored low-level visual features for image description such as color angular indexing, GIST descriptor, RGB raw values, RGB color histograms, HSV histograms, and Gabor filters, among others. Few efforts have been made to take advantage of emerging deep convolutional neural networks for feature description in visual novelty detection. One such effort is the robotic system proposed by Ritcher and Roy [26]. The objective of this work was to develop a robot with a safe navigation module. An autoencoder network composes the novelty detection module with three hidden layers that automatically find a compressed representation of the image captured by the robot. The goal of the network is to reconstruct the input image, and if the input image cannot be reconstructed (i.e., the error between the input and the output is higher than an error tolerance) then the system will detect the novelty and use it to maintain the safety of the robot.

In summary, most of the existing visual novelty detectors have been configured manually by humans, or no specific procedure for the configuration of the detector has been provided. Also, most of the visual novelty detectors use traditional feature extraction techniques. There are few explorations applying the recent advances in convolutional neural networks as visual feature descriptors. Both the lack of automatic configuration of novelty detectors and the use of low-level traditional visual features have restricted the exploration and inspection task for indoor environments, with controlled conditions (e.g., illumination), and with simple visual novelty detection problems (i.e., conspicuous objects). The proposed work presents an approach to addresses these issues.

3. Materials and Methods

In this section, we describe the proposed system for visual exploration and inspection tasks. In this work, we used images captured by a UAV operating in outdoor environments. Figure 1 illustrates the proposed system. In the exploration phase, the UAV follows a fixed trajectory and captures images of the environment. The system represents the captured images via deep features by using a pre-trained convolutional neural network called MobileNetV2 [27]. The novelty detector processes the feature vector and constructs a model of the environment. The user can select between two detectors: simple evolving connectionist systems (SECoS) or GWR network. Finally, in the inspection phase, the UAV again executes its path and searches for novel objects. The UAV uses the above model to identify novelties. Then, we describe in more detail the components of the proposed visual novelty detection system.

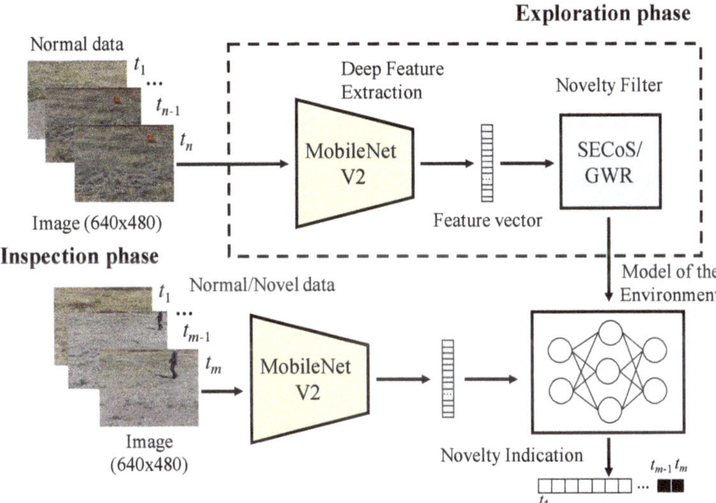

Figure 1. Graphical description of the proposed system for visual exploration and inspection tasks. SECoS: simple evolving connectionist systems.

3.1. Visual Feature Extraction

One way to represent the images is via visual feature vectors. Among the visual features, traditional features such as RGB color histograms [11], color angular indexing [10], and the GIST descriptor [15] have been applied for visual novelty detection in robotics. However, traditional visual features are highly sensitive to illumination changes, noise, occlusion, or geometric transformations.

Recently, convolutional neural networks have been applied successfully as powerful tools to extract features from images [28], having robust performances in a wide variety of classification tasks.

Motivated by the success of convolutional neural networks as feature extraction methods, we propose the application of a convolutional neural network to extract features from images for the task of visual novelty detection in robotics. In this work, we selected MobileNetV2 [27] because it is the network with the lowest number of parameters in the Keras API and the TensorFlow engine. In our implementation, we used a pre-trained network with the weights trained on the ImageNet dataset. In order to extract the visual features, we resized the input image to the default size in the Keras API of 224×224 pixels. We also deactivated the classification layer and activated the average pooling mode for feature extraction. We obtained visual feature vectors of 1280 elements.

3.2. Novelty Detectors

We selected two online novelty detection methods that are used as the base to develop exploration and inspection tasks with real robots [10,11,16]. Both techniques are constructive and can evolve the structures of the models and their parameters during their operation. We selected the SECoS and the GWR network.

3.2.1. Simple Evolving Connectionist Systems

The evolving connectionist systems (ECoS) proposed by Kasabov [21] are a type of neural network that can evolve their parameters and their structure over time. Below, we show the characteristics of the ECoS that make them attractive to address the problem of visual novelty detection in robotics [29]:

- Fast learning capabilities (one-pass learning).
- Online learning and incremental adaptation to new data.
- The model is evolved to adapt to the input information, and the examples are added to the model when they are different in some aspects from the current model of the data.

The SECoS conserve these characteristics [30], but they present two advantages concerning the other ECoS implementations. The SECoS are easy to implement because they have a low number of layers to learn the input data, and they work directly on the input space. Figure 2 shows a graphical description of the SECoS network. Three layers compose the network: the input layer, which transfers the inputs to the nodes of the next layer; the hidden layer (evolving layer), which incorporates new nodes to represent novel data; and the output layer, which uses saturation linear activation functions to compute the output. In a SECoS network, there are two connection layers: the connections between the nodes of the input layer and the nodes of the evolving layer (incoming connections), and the connections between the nodes of the evolving layer and the nodes of the output layer (outgoing connections).

In this work, we used the SECoS learning algorithm proposed by Watts and Kasabov [30]. The algorithm receives as input the weights of the connections in the network, the input features, and the desired output. The proposed approach uses a SECoS implementation with the same number of nodes in the input layer and the output layer. The objective of the approach is to generate a system able to reconstruct the input vector. When the model generated by the SECoS implementation is not able to represent an input, it should add a new node in the evolving layer with the incoming weight values equal to the input vector and the outgoing weight values equal to the desired output. Also, it should add a new node to the model when the reconstructed output is significantly different from the desired output, that is, when the Euclidean distance between the desired output and the current output of the network is greater than the threshold E_{thr}. When the model can represent a given input successfully, the SECoS implementation only updates the model (updating of the connection weights) to better represent the input data. The parameters of this learning model include the learning coefficients (η_1, η_2), the sensitivity threshold (S_{thr}), and the error threshold (E_{thr}). For more details about this learning algorithm, the readers can refer to the work by Watts and Kasabov [30].

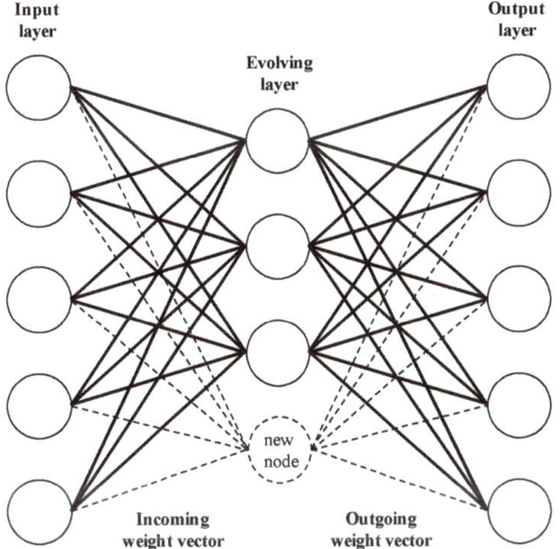

Figure 2. Graphical description of the SECoS network. Adaptation of the general ECoS representation from Watts [29].

3.2.2. Grow-When-Required Neural Network

GWR is an online self-organized neural network proposed to solve the novelty detection problem [31]. Figure 3 shows a graphical representation of the GWR neural network. A clustering layer of nodes and a single output node compose the network. The nodes in the clustering layer use weight vectors to represent the centers of the clusters. The GWR network can add and remove nodes to its structure, specifically in the clustering layer, to adapt to the changes of the inputs. The connection synapses to the clustering layer in the network are subject to a habituation model, which is a reduction in response to similar inputs.

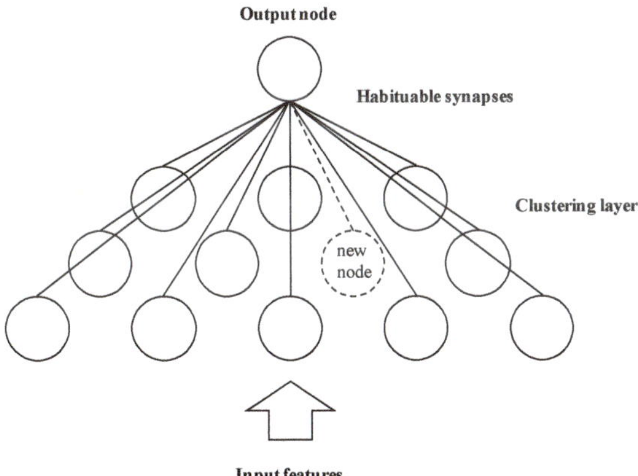

Figure 3. Graphical representation of the grow-when-required (GWR) neural network. Adaptation of the network architecture presented by Neto et al. [20].

In the proposed framework, we use the algorithm of the GWR network for novelty detection as described by Neto [10]. The network starts with two dishabituated nodes with weight vectors initialized to the positions of the first two input vectors. At the beginning, there are no topological connections between both nodes. From the third input vector, the best matching node s and the second best matching node t of the clustering layer are found (i.e., the nearest nodes to the input vector). If there is a topological connection between both nodes, its age is set to zero; otherwise, the connection between both nodes is created with age zero. The GWR network uses the activation and habituation levels of node s to decide if the input is novel or not. If the input vector is novel, a new node in the clustering layer is created with its weight vector initialized to the average position between the input vector and the best matching node. Also, the topological connections of the nodes in the clustering layer are updated by removing the connection between the best matching nodes and inserting new connections between the best matching nodes and the created node. Then, the best matching and its topological neighbors update their positions in the direction of the input vector and also update their habituation levels. Finally, all the connections increase their ages and all connections with ages higher than the maximum age are removed. A node is also removed when it has no topological connections (i.e., ability to forget). The parameters that impact the behavior of the network are the parameters of the habituation model, the activation threshold (a_T), the habituation threshold (h_T), the proportionality factor (η), and the learning rate (ϵ). A detailed description of the learning algorithm of the GWR neural network can be found in [10].

3.3. Global Optimization of Novelty Detectors

One of the main problems in the application of novelty detectors is the proper selection of their parameters in order to obtain the best results regarding the detection accuracy. With this in mind, we propose a framework to tune the novelty detectors automatically for a specific task (see Figure 4). Our optimization approach not only searches for parameters of the novelty detectors, but also finds the best size of the visual feature vector.

In this work, we propose the use of the artificial bee colony algorithm (ABC) [32] as the optimization tool. Note that although in this work we show the use of the ABC algorithm, in the proposed framework we can incorporate different algorithms to find the more appropriate parameters of the filters to solve specific tasks. The ABC algorithm offers a population-based approach for numerical optimization. In the ABC algorithm, artificial bees update their position over time to find the best food sources. This algorithm has shown to be better than or competitive to other bio-inspired optimization techniques. Besides, we can find applications of the ABC algorithm for a wide variety of engineering problems, such as image processing, data mining, control, and mobile robotics [32]. The implementation details of the algorithm can be found in Mernik et al. [33]. In the proposed methodology, we use an implementation with a termination condition based on the number of iterations, also known as ABC_{imp1}.

In our implementation of the ABC algorithm, each food position represents a set of parameter values of the novelty detector. Table 1 shows the parameters that should be adjusted by using the ABC algorithm. The search range of all the decision variables is within $[0, 1]$. In the case of the GWR novelty filter, we set the parameters of the habituation model to the default values, and we also keep the maximum age value constant. For the ABC algorithm, we used a population of 20 food positions and a total number of 100 iterations.

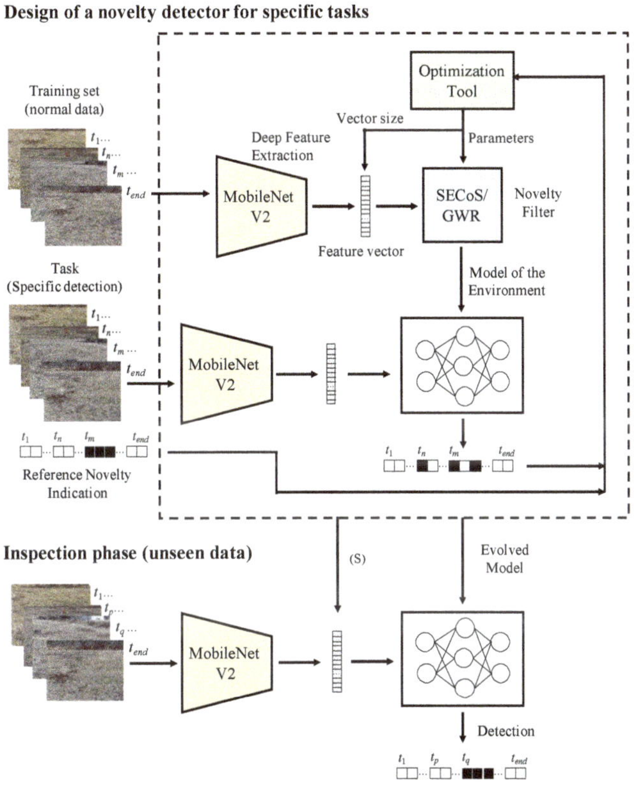

Design of a novelty detector for specific tasks

Figure 4. Flowchart of the visual novelty detection for specific tasks. In the training phase, the novelty filter learns to detect a specific object. In the inspection phase, the evolved model is used to detect the object(s) in the environment.

Table 1. Parameters to be tuned for each novelty detector.

Novelty Detector	Parameter	Description
SECoS	η_1	Learning rate 1
	η_2	Learning rate 2
	S_{thr}	Sensitivity threshold
	E_{thr}	Error threshold
GWR	a_T	Activation threshold
	h_T	Habituation threshold
	η	Proportionality factor
	ϵ	Learning rate

4. Experimental Preparation

We validated the performance of the proposed method using images captured by a real robot in outdoor environments. We constructed the datasets using these images to train and test the novelty-detection system. We designed an experiment to compare the deep visual feature extraction technique against commonly used visual features for the problem of visual exploration and inspection. In this section, we describe the datasets, the methods for comparison, the experimental setup, and the evaluation metrics.

4.1. Datasets

In this work, we constructed a dataset with images captured by the visual sensor of a UAV. For this purpose, we used a Parrot Bebop 2 Drone with a 14-Mpx flight camera. The captured images had a dimension of 1920 × 1080 pixels, but we constrained the search in the center region of the images with a reduced field-of-view of 640 × 480 pixels. Figure 5 shows the UAV used for data acquisition. Note that the novelty detector system received images of the environment every 250 ms.

Figure 5. Parrot Bebop 2 Drone with a 14-Mpx flight camera. In the bottom-left corner, we show its visual sensor system.

Figure 6 illustrates the outdoor environment used in this experiment. The UAV executed its default execution control module to fly over the environment in a rectangular shape. In order to generate the datasets, the UAV executed the same path several times with different environment setups.

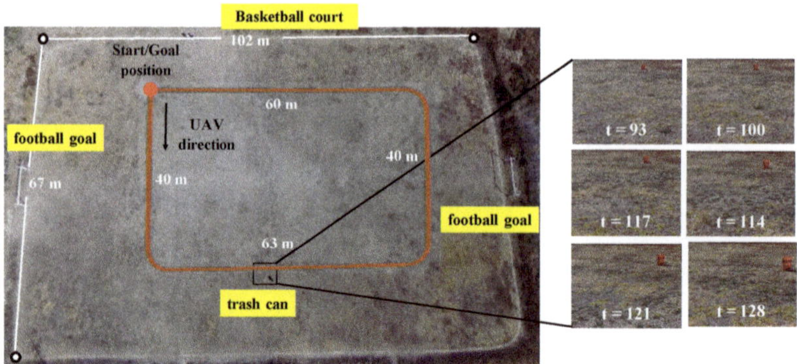

Figure 6. Experimental setup: the outdoor environment, and some sample captured images. UAV: unmanned aerial vehicle.

In the first set of experiments, the UAV flew at 2 m above the ground with morning light conditions (around 11:00 and 12:00). The original environment contained an orange trash can (we called this environment "O-1"). First, the UAV explored the O-1 environment, executing its path two times. The UAV captured a total of 896 images—448 for each execution. Then, it executed the inspection phase and captured another 896 images. In this inspection phase, a person appeared in the environment (we denoted this new environment as O-2). The sequence contains 60 frames with the person. In the second experiment, we added a tire to the O-1 environment (we denoted this environment as O-3). The UAV captured a total of 896 images. The tire is present in 58 frames. Finally, the UAV executed its path in the environment with the person and the tire at the same time. The UAV captured another

896 images in its two path executions. In total, the person is present in 37 frames, and the tire is present in 64 frames. We identified this environment as O-4.

We developed a second set of experiments to test the robustness of the proposed method, considering different scales, types of occlusions, novel objects, and light conditions. In this new set, the UAV flew 4 m above the ground with afternoon light conditions (around 16:00 and 17:00). The methodology to capture the image sequences was similar to the first set of experiments, but with some differences in the settings of the environments. We introduced environment O-5, where the orange trash can was removed. We designed another environment with a person in a different position, and named it O-6. To test the robustness of the proposed method, we added inconspicuous novel objects to environment O-5 (brown boxes). We denoted this environment as O-7. Finally, we set a new environment O-8, where the UAV could visualize how the person occluded the boxes in the environment.

Figure 7 shows some sample images of the above environments. Table 2 summarizes the environments used for novelty detection, and Table 3 reports the data partition of the environments to perform the training and test phases.

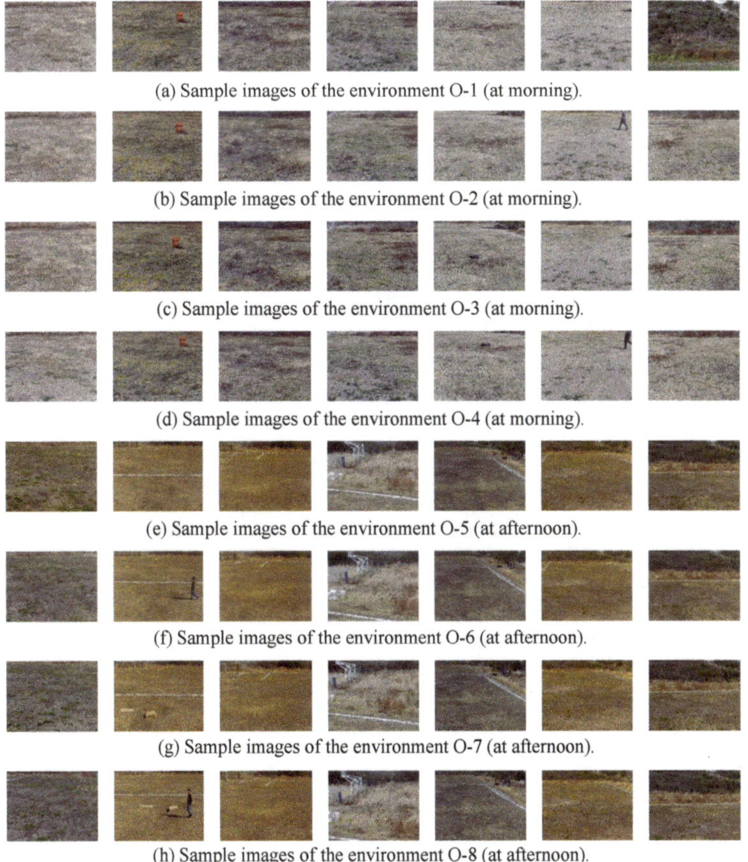

(a) Sample images of the environment O-1 (at morning).

(b) Sample images of the environment O-2 (at morning).

(c) Sample images of the environment O-3 (at morning).

(d) Sample images of the environment O-4 (at morning).

(e) Sample images of the environment O-5 (at afternoon).

(f) Sample images of the environment O-6 (at afternoon).

(g) Sample images of the environment O-7 (at afternoon).

(h) Sample images of the environment O-8 (at afternoon).

Figure 7. Sample images captured by the UAV in the environments: (**a**) original in the morning (O-1), (**b**) the person in the morning (O-2), (**c**) the tire in the morning (O-3), (**d**) the person and the tire in the morning (O-4), (**e**) empty environment in the afternoon (O-5), (**f**) the person in the afternoon (O-6), (**g**) the boxes in the afternoon (O-7), and (**h**) the person and the boxes in the afternoon (O-8).

Table 2. Summary of the environments used in the experiments for novelty detection.

Environment	Description	Loops	#Normal	#Novel
O-1	Original setup of the environment (morning).	2	896	0
O-2	A person in the O-1 environment (morning).	2	836	60
O-3	Inclusion of a tire to the O-1 environment (morning).	2	838	58
O-4	A person and tire in the O-1 environment (morning).	2	795	101
O-5	Empty environment (afternoon).	2	896	0
O-6	A person in the O-5 environment (afternoon).	2	822	74
O-7	Inclusion of brown boxes to the O-5 environment (afternoon).	2	835	61
O-8	A person and boxes in the O-5 environment (afternoon).	2	825	71

Table 3. Data partition for novelty detection.

Dataset	Exploration	Inspection	Test Case (Novelty)
D-1	O-1	O-2	A dynamic object (person).
D-2	O-1	O-3	A small conspicuous object (black tire).
D-3	O-2	O-4	A conspicuous object in a dynamic environment (black tire).
D-4	O-3	O-4	A dynamic object in an environment with a conspicuous object (person).
D-5	O-1	O-4	Multiple novel objects (person and tire).
D-6	O-5	O-7	Inconspicuous objects (brown boxes).
D-7	O-6	O-8	Occlusion of inconspicuous objects (brown boxes).

In all the experiments, the novelty detectors used the images of the training environment of both loops for exploration while only using one loop of the test environment for inspection. The other loop of the test environment was used to evolve the novelty detectors.

4.2. Evaluation Metrics

To measure the performance of the novelty detectors, we used the confusion matrix shown in Table 4. *TP* represents the number of true positives (normal data labeled as normal), *TN* represents the number of true negatives (novel data labeled as novel), *FP* represents the number of false positives (novel data labeled as normal), and *FN* represents the number of false negatives (normal data labeled as novel).

Table 4. Confusion matrix to evaluate the performance of the novelty detectors. *FN*: false negative; *FP*: false positive; *TN*: true negative; *TP*: true positive.

Class/Prediction	Normal	Novel
Normal	*TP*	*FN*
Novel	*FP*	*TN*

Different metrics have been proposed to reflect the performance reached by the novelty detectors in a single quantity. Three of the most commonly adopted are the F_1 score, accuracy (*ACC*), and Matthews correlation coefficient (*MCC*). Similar to Özbielge [11], we used these three metrics to evaluate the performance of the novelty detectors. These metrics are respectively defined as:

$$F_1 = \frac{2 \cdot TP}{2 \cdot TP + FP + FN}, \tag{1}$$

$$ACC = \frac{TP + TN}{TP + TN + FP + FN}, \tag{2}$$

$$MCC = \frac{TP \times TN - FP \times FN}{\sqrt{(TP + FP)(TP + FN)(TN + FP)(TN + FN)}}. \tag{3}$$

In the problem of novelty detection, it is important to correctly label all the novel data as novel. Also, it is tolerable to label normal data as novel data, but it is inadmissible to label a novel data as normal. For example, suppose a thief, representing novel data, enters a warehouse. In our novelty detection system, we prefer a system that can detect the thief all the time in order to prevent theft. If the system detects a thief and there is no thief in the scene, there is no problem concerning theft. In order to reflect the desired behavior of novelty detectors, we also incorporated two additional metrics: the true negative rate (*TNR*) and the true positive rate (*TPR*).

To establish the quality of a detector with a single number, we used the average ranking of the measures in all the metrics, inspired by Bianco et al. [34]. Let us consider a set of detectors to be compared, denoted as $\mathcal{M} = \{M_1, M_2, \ldots, M_m\}$, where m is the number of detectors; a set of test images denoted as $\mathcal{T}$; and a set of P performance metrics, in this study $P = 5$. We can compute the average ranking of a detector M_i as:

$$R_i = \frac{1}{P} \sum_{j=1}^{P} rank\Big(M_i; measure_j(M_k(\mathcal{T})), \forall k \neq i \Big), \tag{4}$$

where $rank(M_i; \cdot)$ computes the rank of the detector M_i considering the results of the rest of the detectors in the measure $measure_j$.

4.3. Experimental Setup

All the algorithms for novelty detection under study can operate online. However, to compare the detectors, they used the same data partition shown in Table 3. We implemented the SECoS, GWR, and ABC algorithms in the C++ programming language. The developed ABC library used the Mersenne Twister pseudo-random generator of 32-bit numbers. In the case of the deep feature extraction technique, we used the pre-trained MobileNetV2 available in the Keras API and the TensorFlow engine. The experiments were developed in a computer with an Intel Core i5 processor, running at 2.9 GHz and with 16 GB of RAM.

To verify the performance of the detectors, we used three traditional visual feature extraction techniques: the RGB color histograms used by Özbilge [11], the color angular indexing used by Neto [10], and the GIST descriptor used by Kato et al. [15]. We compared the performance of the detectors with these feature extraction techniques against the features extracted by the MobileNetV2 network. In this experiment, the system for automatic design used the two image sequences in the exploration phase as training and one sequence of the inspection phase as a validation. The goal of the optimization process was to maximize the performance of the detector concerning the F_1 score, the *ACC*, and the *MCC*. Therefore, we used the following fitness function:

$$f = 1 - \frac{1}{3}\Big(F_1 + ACC + \frac{1 + MCC}{2}\Big), \tag{5}$$

where $f \in [0,1]$, $f = 1$ represents the worst case with no data classified correctly and $f = 0$ indicates that the novelty detector under study classifies all the data from the validation correctly. In this experiment, we executed 30 simulations for each novelty detector, and we report the average results to perform the comparison.

5. Results and Discussion

This section shows and discusses the results of the experiments. We designed the specific novelty detectors for each visual feature independently. We found the most suitable size of the feature vector and the parameters of the novelty detection methods for the particular visual exploration and inspection tasks. In the first part of this section, we compare the results of the proposed feature extraction technique against the well-established feature extraction techniques in the problem of visual novelty detection. Then, we present an analysis of the optimization process of the novelty detectors that use the MobileNetV2 feature extractor. We also show some sample novelty detectors (evolved detectors) generated by the proposed framework and their visual results. Finally, we discuss some limitations of the proposed methodology.

5.1. Deep Features and Traditional Visual Features in Novelty Detection

We used well-known visual feature extraction techniques in the problem of novelty detection to compare the performance of the MobileNetV2. We used as reference the RGB color histograms used by Özbilge [11], the color angular indexing applied by Neto [10], and the GIST descriptor implemented by Kato et al. [15]. Table 5 reports the average performance of the novelty detectors in the inspection phase for each dataset, where CAI represents the color angular indexing technique, hRGB represents the RGB color histograms, and MNF represents the feature extraction method based on MobileNetV2. In the table, we also report the average vector size of the features (*VSize*) and the average size of the learned models of the environment (*MSize*)—that is, the average number of nodes in the models. Note that the CAI descriptor produces feature vectors of four elements. In the rest of the descriptors, the optimization process can produce feature vectors of different sizes. In the table, we mark the best-performing method for each metric, according to the specific detector and the particular dataset. The ranking metric uses the *TPR*, *TNR*, F_1, *ACC*, and *MCC* values to compare the different descriptors for each dataset and detector.

For the D-1 dataset, the objective was to learn a model of the original environment O-1, and to detect a dynamic object represented by a person. In this dataset, the feature extraction technique MNF showed the best performance compared to all other visual extraction techniques. The detectors that used the MNF descriptor could generate compact models of the environment and keep higher performance. They showed accuracies greater than 98%, and *MCC* near 0.9. On the second dataset (D-2), the novelty detectors had to learn a model of the environment O-1 and identify the black tire as the new object. The proposed method achieved the best performance over all others in this dataset—see the ranking of the D-2 dataset in Table 5. The average *ACC* by using both detectors with the MNF technique was around 98%, and the *MCC* was 0.87. Dataset D-3 presents a more challenging situation because the detector was required to learn a model of the environment with a person and detect a black tire. The environment in the inspection phase included both the person and the black tire. Under this situation, the novelty detectors that used the MNF also achieved the best performance, with *ACC* values around 96% for both detectors, and *MCC* values of 0.79 and 0.76 for the SECoS and GWR detectors, respectively. On dataset D-4, the objective was to learn a model of the environment with a tire. In the inspection phase, the person represented the novel object and the black tire represented a normal object. The results indicate that the MNF technique was the second best (the first was the GIST descriptor) with 96% *ACC* and 0.6 *MCC* for both detectors. On dataset D-5, the novelty detectors were required to learn a model of environment O-1 and detect multiple novel objects (both the tire and the person). The MNF description achieved the best performance, with *ACC* values around 97% for both novelty detectors and *MCC* values of 0.89 and 0.88 for the SECoS and GWR detectors, respectively.

On the above datasets, the novelty detectors were tested with novel objects that were highly different from the environment. This could facilitate their detection. In the following, we tested the detectors in more challenging situations. To this end, we used datasets D-6 and D-7, generated by the UAV at a different height (4 m) and with a different light condition (images captured in the afternoon). In the inspection phase of dataset D-6, we used inconspicuous brown boxes to represent the novel objects. In this dataset, the detectors with MNF feature extraction were the best methods to detect novelties, with a ranking of 1.2. Finally, we show the results of the detectors on dataset D-7. The objective in this dataset was to learn a model of an environment with a person and tire and to detect the brown boxes that were occluded by the person in some frames. The results show the superiority of the MNF descriptor for novelty detection, with *MCC* values above of 0.9 and *ACC* values around of 98%, for both detectors.

Table 5. Average results in the inspection phase over the 30 runs. Bold values indicate the best result for each metric according to the specific dataset and the specific novelty detector. CAI: color angular indexing; hRGB: RGB color histogram; MNF: feature extraction based on MobileNetV2; $VSize$: average vector size; $MSize$: average model size; TPR: true positive rate; TNR: true negative rate; F_1: F_1 score; ACC: accuracy; MCC: Matthews correlation coefficient; R: ranking of the detector.

Dataset	Detector	Descriptor	$VSize$	$MSize$	TPR	TNR	F_1	ACC	MCC	R
D-1	SECoS	CAI	4.0	17.5	0.9692	0.2750	0.9607	0.9258	0.2865	4.0
		hRGB	305.0	12.4	0.9738	0.4571	0.9689	0.9415	0.4673	3.0
		GIST	350.5	47.5	0.9867	0.8571	0.9886	0.9786	0.8312	2.0
		MNF	169.1	7.1	**0.9922**	**0.9000**	**0.9928**	**0.9865**	**0.8859**	**1.0**
	GWR	CAI	4.0	29.1	0.9520	0.3238	0.9532	0.9127	0.2530	3.6
		hRGB	357.3	20.5	0.9757	0.2393	0.9628	0.9297	0.2317	3.4
		GIST	398.1	46.7	**0.9900**	0.8452	0.9898	0.9810	0.8418	1.8
		MNF	153.3	13.8	0.9899	**0.8869**	**0.9912**	**0.9835**	**0.8646**	**1.2**
D-2	SECoS	CAI	4.0	13.6	**0.9879**	0.0155	0.9620	0.9271	0.0076	2.6
		hRGB	337.0	25.3	0.9734	0.0857	0.9567	0.9179	0.0655	3.0
		GIST	384.9	37.7	0.8444	0.8333	0.9084	0.8438	0.4295	3.2
		MNF	143.3	16.6	0.9806	**0.9976**	**0.9901**	**0.9817**	**0.8729**	**1.2**
	GWR	CAI	4.0	2.4	0.9943	0.0000	0.9649	0.9321	−0.0104	3.0
		hRGB	365.4	2.0	**1.0000**	0.0000	0.9677	0.9375	0.0000	2.2
		GIST	334.3	79.8	0.8300	0.7821	0.8976	0.8270	0.3758	3.2
		MNF	180.8	23.7	0.9852	**0.9548**	**0.9910**	**0.9833**	**0.8729**	**1.4**
D-3	SECoS	CAI	4.0	11.8	0.9426	0.6086	0.9561	0.9195	0.4765	2.2
		hRGB	427.2	29.3	0.9642	0.1452	0.9507	0.9075	0.0914	3.2
		GIST	269.0	50.2	0.9019	0.6022	0.9323	0.8812	0.3742	3.6
		MNF	184.1	27.8	**0.9788**	**0.8484**	**0.9836**	**0.9698**	**0.7881**	**1.0**
	GWR	CAI	4.0	6.5	0.9905	0.1118	0.9632	0.9297	0.1111	2.4
		hRGB	445.0	6.1	**0.9922**	0.0118	0.9602	0.9244	0.0024	3.0
		GIST	353.4	152.2	0.9117	0.4645	0.9317	0.8807	0.2852	3.2
		MNF	216.0	34.2	0.9723	**0.8710**	**0.9812**	**0.9653**	**0.7653**	**1.4**
D-4	SECoS	CAI	4.0	16.9	0.9790	0.0157	0.9703	0.9424	−0.0072	3.6
		hRGB	303.2	29.5	0.9745	0.3000	0.9733	0.9489	0.3008	3.0
		GIST	315.0	2.2	**0.9912**	**0.8706**	**0.9930**	**0.9866**	**0.8259**	**1.0**
		MNF	147.2	15.8	0.9729	0.8098	0.9825	0.9667	0.6585	2.4
	GWR	CAI	4.0	6.1	**0.9947**	0.0000	0.9780	0.9570	−0.0105	3.0
		hRGB	289.0	51.9	0.9552	0.3176	0.9633	0.9310	0.2046	3.6
		GIST	334.0	15.3	0.9690	**0.9039**	0.9821	0.9665	**0.7279**	1.8
		MNF	173.8	15.4	0.9770	0.7784	**0.9840**	**0.9695**	0.6578	1.6
D-5	SECoS	CAI	4.0	7.5	0.9765	0.0778	0.9356	0.8802	0.0945	3.8
		hRGB	276.5	40.4	**0.9823**	0.1299	0.9414	0.8910	0.1976	2.6
		GIST	306.7	24.6	0.9536	0.5764	0.9512	0.9132	0.5516	2.4
		MNF	180.0	26.9	0.9813	**0.9472**	**0.9874**	**0.9776**	**0.8916**	**1.2**
	GWR	CAI	4.0	7.8	0.9749	0.1049	0.9361	0.8817	0.1565	3.2
		hRGB	331.4	36.6	0.9833	0.0660	0.8978	0.8850	0.0795	3.4
		GIST	385.9	46.7	0.9305	0.6403	0.9420	0.8994	0.5475	2.4
		MNF	221.8	26.9	**0.9917**	**0.8681**	**0.9880**	**0.9784**	**0.8847**	**1.0**
D-6	SECoS	CAI	4.0	7.9	0.9560	0.0344	0.9439	0.8943	−0.0206	3.2
		hRGB	213.2	6.0	0.9270	0.8900	0.9580	0.9246	0.6233	2.4
		GIST	245.4	9.5	0.8352	**0.9167**	0.9059	0.8407	0.4707	3.2
		MNF	150.2	20.4	**0.9750**	0.8911	**0.9834**	**0.9693**	**0.7950**	**1.2**
	GWR	CAI	4.0	11.6	0.9761	0.0200	0.9535	0.9121	−0.0045	2.8
		hRGB	304.3	61.3	0.8946	0.8622	0.9388	0.8924	0.5277	2.8
		GIST	289.0	11.5	0.7977	**0.9111**	0.8825	0.8053	0.4194	3.2
		MNF	210.5	16.0	**0.9796**	0.8878	**0.9857**	**0.9734**	**0.8107**	**1.2**

Table 5. *Cont.*

Dataset	Detector	Descriptor	VSize	MSize	TPR	TNR	F_1	ACC	MCC	R
D-7	SECoS	CAI	4.0	8.5	0.9730	0.0192	0.9487	0.9028	−0.0085	3.8
		hRGB	276.6	6.3	0.9482	**0.9939**	0.9731	0.9516	0.7605	2.8
		GIST	444.9	9.9	**0.9867**	0.9364	0.9907	0.9829	0.8831	2.0
		MNF	165.2	17.6	0.9855	0.9848	**0.9921**	**0.9855**	**0.9065**	**1.4**
	GWR	CAI	4.0	2.8	**0.9982**	0.0000	0.9609	0.9247	−0.0021	3.2
		hRGB	255.7	12.2	0.9369	0.8424	0.9607	0.9300	0.6307	3.4
		GIST	370.7	14.4	0.9654	0.9030	0.9783	0.9608	0.7757	2.2
		MNF	175.7	8.1	0.9862	**0.9960**	**0.9929**	**0.9869**	**0.9162**	**1.2**

We then compared the average CPU time to generate the visual features per image on all the
datasets. The average time excludes the reading of the image and the post-processing of the visual
features. The post-processing only consisted of reducing the vector size to the size found by the
optimization process. The reduction was through the average of sectors of equal elements. Figure 8
shows the average time to generate visual features in all the datasets. hRGB was the fastest method,
mainly because it only needs to count the number of pixels that belong to a given intensity value.
The CAI method was the second fastest method because its computation consists of simple image
operations such as average, standard deviation, inverse cosine, and dot product. Meanwhile, the GIST
descriptor involves more advanced operations. It includes convolution between the image and Gabor
filters at different scales and orientations. The MNF was the slowest feature extraction technique
because it includes more complex operations in the image (i.e., it is a deep structure with different
convolutional layers). However, all the feature extraction techniques in this work could generate
visual features in less than 200 ms—a time that is acceptable for the proposed visual exploration and
inspection tasks.

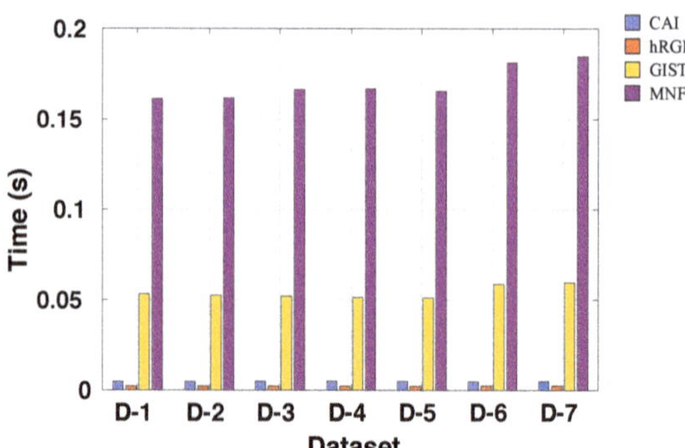

Figure 8. Average time (seconds) to generate the visual features using different descriptors on
all datasets.

Overall, MNF had balanced results in contrast with the baseline methods. The models found by
the MNF descriptor and the novelty detectors were compact, with no more than 35 nodes. In most
cases, MNF worked better in detecting novelties than the traditional visual descriptors. Besides, we
found that traditional visual features required a low number of nodes to represent the environment.
However, their low performance concerning the *ACC* and the *MCC* indicates that the extracted features
were insufficient to differentiate the image in the sequences.

5.2. Analysis of the Optimization Process

Figure 9 presents the average fitness value of the best-evolved novelty detectors per iteration in the 30 runs on dataset D-2. We will show the optimization processes of both novelty detectors that use the MNF feature extraction technique. In this figure, we also present the standard deviation of the fitness values through bars. At the beginning, the best detectors in the different runs had more variations among them, and this variation was reduced according to the increment in the number of iterations. Analyzing the curve, we can observe that detectors evolved easily on the dataset because they reached fitness values near to the perfect score (zero values), that is, the optimization process found the appropriate parameter values of the detector for the specific novelty detection task. For the GWR, from the initial to the final iteration, it had a decrement of 0.2591 in the average fitness. The more notable change occurred in the first 20 iterations with a change of 0.2532. For the SECoS detector, the optimization process showed a decrease of 0.3386 in the average fitness from the initial to the final iteration. The more significant change occurred in the first 14 iterations, with a change in the average fitness of 0.3341. For the rest of the datasets, the results showed similar behaviors in the optimization process.

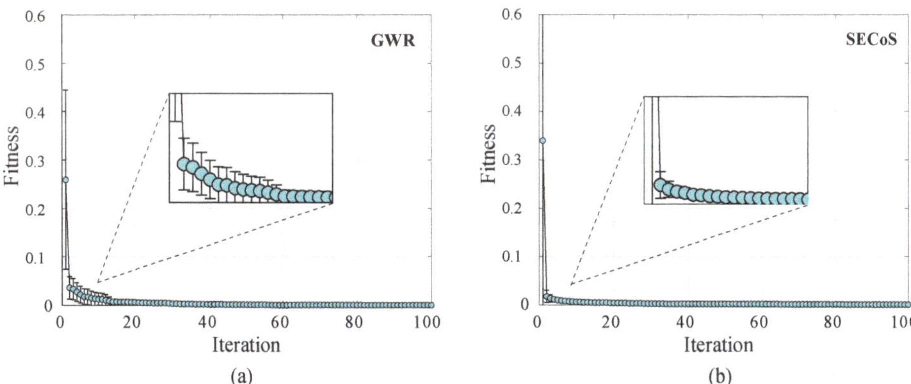

Figure 9. Average fitness value of the best-evolved detectors by using the artificial bee colony (ABC) algorithm in the 30 independent runs on dataset D-2. The detectors used the MNF feature extraction technique: (**a**) GWR detector; (**b**) SECoS detector.

Now, we compare the CPU time used in evolving the novelty detectors for specific exploration and inspection tasks of the different feature extraction techniques. Figure 10 shows the average CPU time to evolve the novelty detectors in all the datasets. The search cost excludes the feature extraction phase and includes the post-processing time of the feature vectors. In the figure, we can observe that the GWR detector evolved faster than the SECoS detector. One reason is that the SECoS detectors need to reconstruct the input data and compute the distance to the nearest neighbor node in the novelty detection process, while the GWR method only requires the computation of the distance between the input data and the closest node and the habituation level of this node (without reconstruction).

It is not surprising that the CAI descriptor was the fastest method to evolve the detectors because it keeps the number of inputs in the detector fixed (4 data points) during the entire optimization process. For the rest of the approaches, the vector size varied during the optimization. The maximum number of elements was 778 (3 channels with 256 intensity values), 512, and 256 features for the hRGB, GIST, and MNF descriptors, respectively.

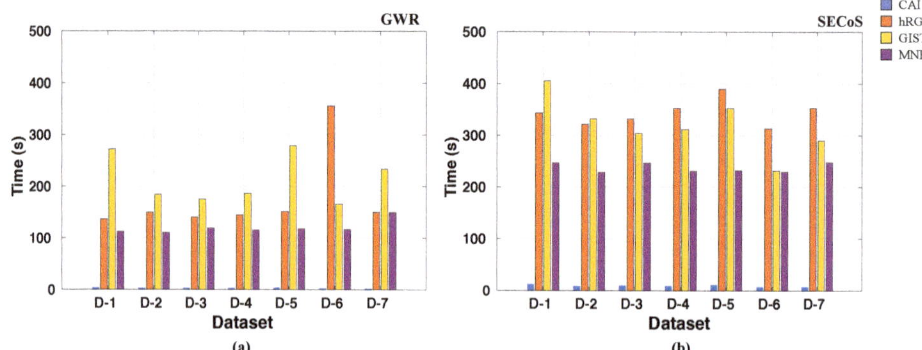

Figure 10. Average CPU time (seconds) to generate a specific novelty detector for each dataset by using different feature extraction techniques: (**a**) GWR detectors; (**b**) SECoS detectors.

5.3. Evolved Novelty Detectors

We used an evolved SECoS detector with deep features on dataset D-3 to illustrate the effects of task-specific novelty detectors. The evolved detector had the following characteristics: $\eta_1 = 0.0183574$, $\eta_2 = 0.4830270$, $A_{thr} = 0.4651190$, $E_{thr} = 0.7776980$, and $VSize = 256$. The proposed global optimization process obtained these parameters. In dataset D-3, the training of the detector consisted of generating a model of the O-2 environment (an environment with a person) and the objective was to detect a black tire in an environment with a tire and a person (this new environment was called "O-4").

Figure 11 presents the exploration and inspection phases by using the evolved SECoS novelty detector. In the exploration phase, the detector constructs the model of the environment finding the most relevant information as the football goal, the orange trash can, the basketball court, and the person. It is commonly adopted for novelty detectors that the first input will be part of the learned model. The image to the left of the football goal in Loop 1 represents the first input image. We used two loops of the same normal environment (O-2) to train the detector. The evolved detector found a model of 18 nodes to represent the O-2 environment. In the inspection phase, the detector uses this model on environment O-4 to detect novelties. In this new environment, the detector found the tire as the novel object in almost all cases, with a single false novelty detection. The performance of this particular detector was $TPR = 0.9976$, $TNR = 0.9677$, $F_1 = 0.9976$, $ACC = 0.9955$, and $MCC = 0.9653$.

Figure 12 shows some image frames captured by the UAV at different time steps, where the evolved SECoS detector classified these images as normal images in the inspection phase. The first row represents some sample images in the exploration phase, and the second row represents the corresponding image frames in the inspection phase. Although there was considerable variation with the dynamic object and slightly different perspective changes in the images, the evolved detector could classify both situations as part of the normal class. Figure 13 shows some image frames where the evolved SECoS detected novelty: image frames used in the exploration phase at different time steps (see Figure 13a), and some sample images captured in the inspection phase where the detector found the novelty (see Figure 13b). We can observe the black tire at different scale in the images captured in the inspection phases.

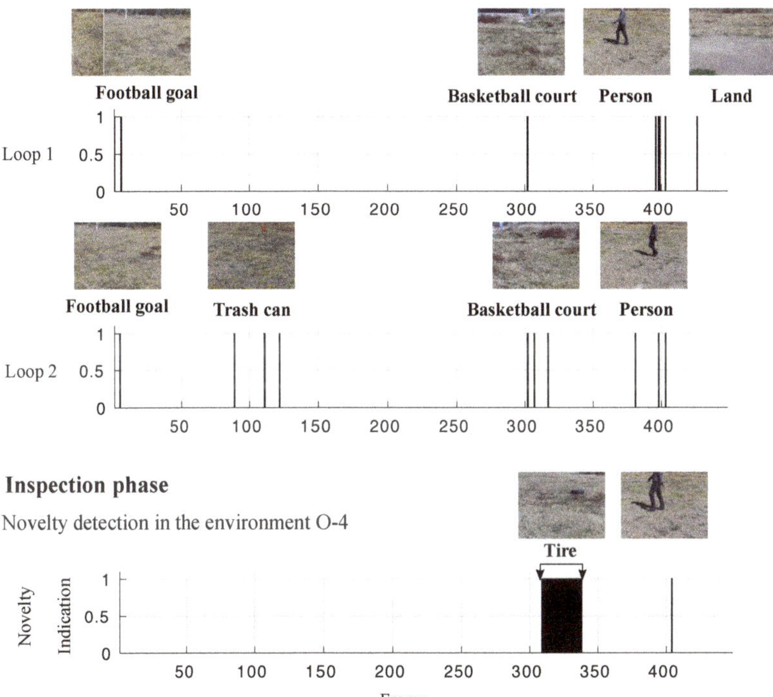

Figure 11. Illustration of the visual exploration and inspection task on dataset D-3 to detect the black tire as the novel object. In the exploration phase, the SECoS detector constructs a model of the environment with the person. In the inspection phase, the detector uses this model to detect the black tire.

Figure 12. Sample image frames labeled as normal images by the evolved SECoS detector in the inspection phase: (**a**) sample image frames used to learn the model of the environment, and (**b**) sample images detected as normal images in the inspection phase.

Figure 13. Sample image frames labeled as novelty images by the evolved SECoS detector in the inspection phase: (**a**) sample images frames used to learn the model of the environment, and (**b**) sample images detected as novelty in the inspection phase.

Table 6 presents a set of sample novelty detectors generated by the proposed framework for each dataset. We show the parameter values of η_1, η_2, S_{thr}, and E_{thr} for the SECoS detectors, and the parameter values of a_T, h_T, η, and ϵ for the GWR detectors. The table also reports the found vector size of the deep features for each detector.

Table 6. Set of sample evolved detectors generated by the proposed global optimization framework on all the datasets.

Detector	Dataset	η_1	η_2	A_{thr}	E_{thr}	*VSize*
	D-1	0.2002440	0.2428720	0.0545579	0.4578700	170
	D-2	0.2045570	0.2697980	0.5078360	0.2089900	74
SECoS	D-3	0.0183574	0.4830270	0.4651190	0.7776980	256
	D-4	0.1456960	0.3827950	0.1083890	0.2627370	75
	D-5	0.0000000	0.0109682	0.2810780	0.5950580	242
	D-6	0.0000000	0.0000000	0.6285940	0.5863690	144
	D-7	0.6577200	0.2922090	0.1750940	0.4164290	96
		a_T	h_T	η	ϵ	*VSize*
	D-1	0.6827340	0.6826510	0.0706664	0.0490785	152
	D-2	0.7888710	0.2963600	0.3931710	0.0000000	101
GWR	D-3	0.5653500	0.3496060	0.4179080	0.0631437	249
	D-4	0.5521850	0.4037900	0.2024040	0.0000000	216
	D-5	0.5756130	0.8404430	0.0000000	0.0000000	256
	D-6	0.7806360	0.7388830	0.2143130	0.0000000	67
	D-7	0.5295850	0.6676220	0.0790152	0.7237070	135

In Table 7, we report the performance of the above-evolved detectors. We can observe that the SECoS detectors had similar behavior to the GWR detectors concerning the novelty detection (see the *TNR* values), except on dataset D-5, where the SECOS detector outperformed the GWR. Besides, on datasets D-1, D-3, D-4, D-6, and D-7, the SECoS detectors exceeded the GWR concerning the *TPR* values.

Table 7. Results in the inspection phase (unseen data) of the sample evolved detectors. Bold values indicate the best result for each metric.

Dataset	Detector	*MSize*	*TPR*	*TNR*	F_1	*ACC*	*MCC*
D-1	SECoS	6	**0.9976**	0.9643	**0.9976**	**0.9955**	**0.9619**
	GWR	9	0.9952	0.9643	0.9964	0.9933	0.9440
D-2	SECoS	12	0.9929	1.0000	0.9964	0.9933	0.9470
	GWR	27	0.9929	1.0000	0.9964	0.9933	0.9470
D-3	SECoS	18	**0.9976**	0.9677	**0.9976**	**0.9955**	**0.9653**
	GWR	21	0.9856	0.9677	0.9915	0.9843	0.8900
D-4	SECoS	11	**0.9930**	0.7647	**0.9919**	**0.9844**	**0.7802**
	GWR	19	0.9861	0.7647	0.9884	0.9777	0.7118
D-5	SECoS	23	0.9975	**0.9792**	**0.9975**	**0.9955**	**0.9767**
	GWR	32	0.9975	0.9375	0.9950	0.9911	0.9527
D-6	SECoS	17	**0.9952**	0.9000	**0.9940**	**0.9888**	**0.9094**
	GWR	14	0.9904	0.9000	0.9916	0.9844	0.8770
D-7	SECoS	9	**0.9952**	1.0000	**0.9976**	**0.9955**	**0.9687**
	GWR	4	0.9928	1.0000	0.9964	0.9933	0.9540

Now, we introduce some visual results of the evolved detectors in the environments in the morning. In Figure 14, the novelty detectors learned a model of the original environment O-1, and detected the person as the novel object. The figure shows the novelty indication of both methods, an image frame in the exploration phase (picture in the upper left corner), and a picture at the same time step in the inspection phase. We mark the novel object with a yellow ellipse. This figure also presents some successful novelty detections on the right side. From these samples, we can observe the advantage of the evolved detectors, which is that they could detect the person at different scales, perspectives, and occlusion levels.

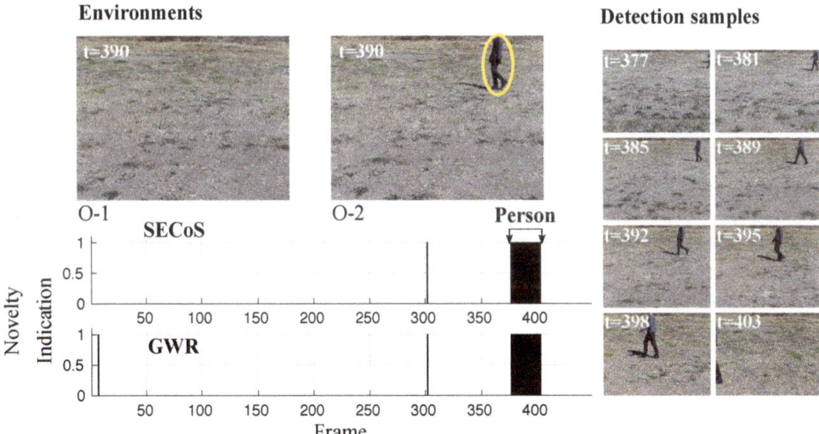

Figure 14. Visual results in novelty detection on dataset D-1, with the person as the novel object.

Figure 15 shows another example of the visual exploration and inspection task. The task consists of learning a model of the original environment O-1 and to detect the black tire in the inspection phase in environment O-3. The detectors found the tire as the novel object in all cases, the methods could even detect novelties with occlusion; see the last detection sample ($t = 334$), where the tire is almost incomplete.

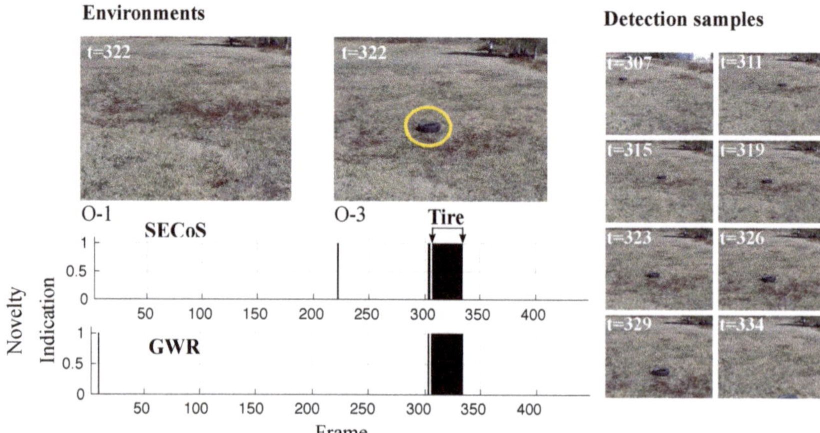

Figure 15. Visual results in novelty detection on dataset D-2, with the tire as the novel object.

A more challenging example is presented in Figure 16. In this figure, the detectors should have found that the black tire was the novel object and the person was the normal object. In almost all cases, the methods could detect the novel object. However, some false novelty detections appeared with the person. The SECoS was less sensitive to this phenomenon than the GWR. Another challenging problem is to detect the person as the novel object and the tire as the normal object. Figure 17 illustrates the performance of both detectors in this situation. Like the above example, the methods could detect the person in almost all cases and discover false novelties in the tire.

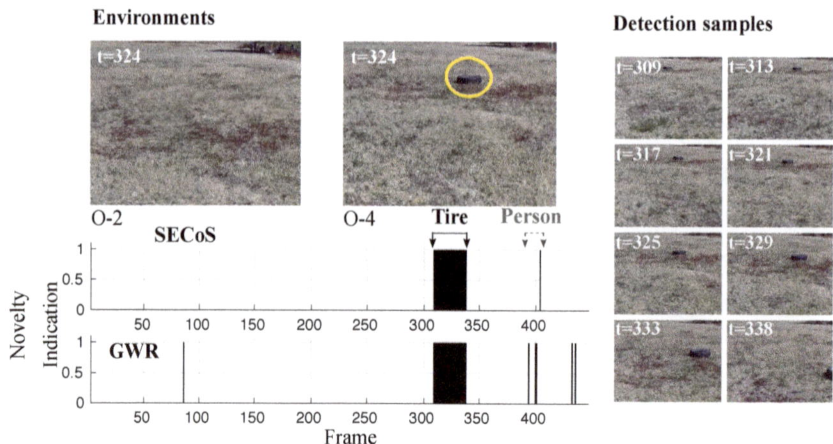

Figure 16. Visual results in novelty detection on dataset D-3 (the tire as the novel object, and the person as the normal object).

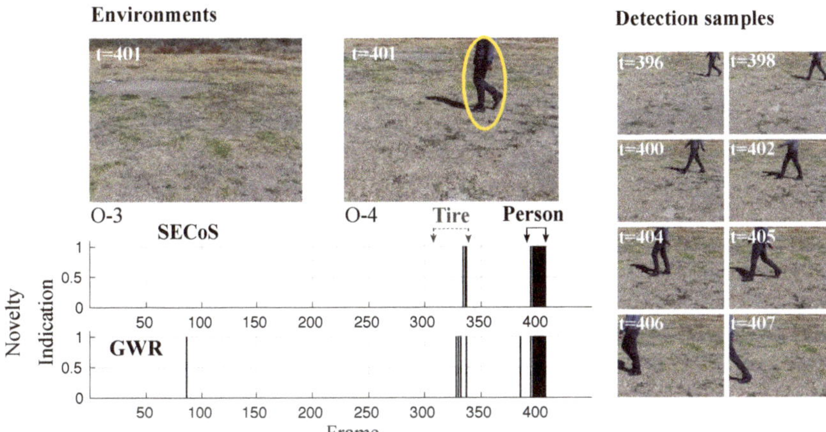

Figure 17. Visual results in novelty detection on dataset D-4 (the person as the novel object and the tire as the normal object).

We then present the visual results in detecting both the tire and the person as the novel objects (multiple novel object detection). In this case, both methods could identify the tire and the person with only one false novelty detection; see Figure 18.

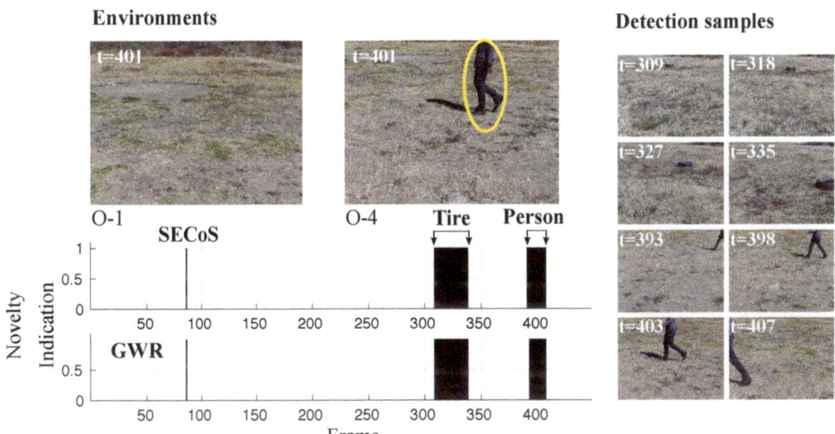

Figure 18. Visual results in novelty detection on dataset D-5, with the person and the tire as the novel objects.

While the previous cases showed results on novel objects that were different from the environment, the next cases show visual exploration and inspection tasks with inconspicuous novel objects (i.e., brown boxes in this experiment). To capture the image frames, the UAV flew at a 4 m height with afternoon light conditions. In Figure 19, the problem was to detect the images with the brown boxes through a learned model of the empty environment in the afternoon (called environment "O-5"). We can observe that the evolved detector detected the brown boxes in almost all cases, with only two false novelty indications.

Finally, we show the results of the evolved detectors when a person occluded the brown boxes. Figure 20 presents this situation. The results show that the evolved detectors learned a model of the environment with the person and detected the images with the brown boxes, even if the person occluded them.

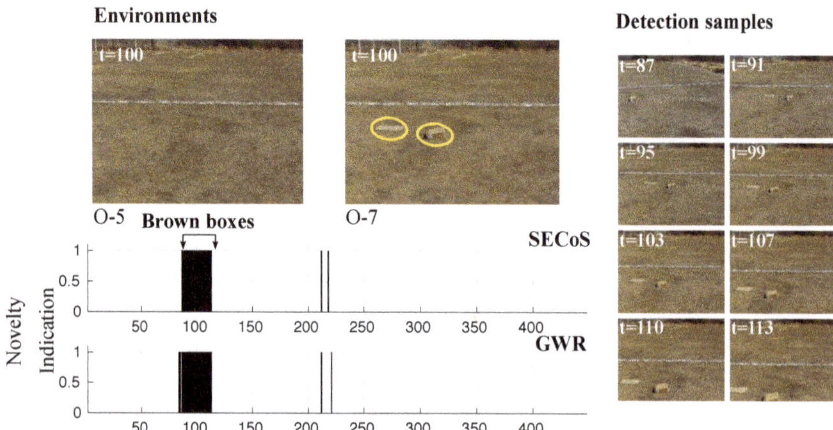

Figure 19. Visual results in novelty detection on dataset D-6 (the brown boxes as the novel objects).

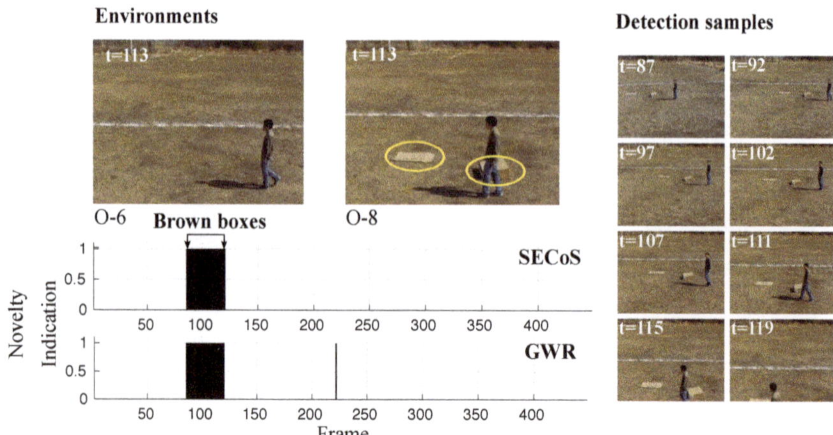

Figure 20. Visual results in novelty detection on dataset D-7 (the brown boxes as the novel objects).

In summary, the visual results show that the evolved detectors could identify the novelty in almost all cases. The detectors presented some false novelty detections. However, it is more critical in this type of problem to detect the novelties than to miss the novelties and detect all the normal data. Furthermore, the proposed detectors had excellent capabilities in challenging scenarios with illumination changes, scales, and occlusions.

5.4. Limitations

The proposed framework addresses the visual novelty detection in exploration and inspection tasks. Although our proposed method was robust to illumination changes, scale, and occlusion, the evolved detectors presented some issues with abrupt perspective changes in images induced by the flight control of the UAV.

Figure 21 shows some failure samples of novelty detections. In the first row, we present some sample images for the training of the evolved novelty detector (GWR in this case). In the second row, we show some sample images in the inspection phase, with a change in the perspective induced by the flight control of the UAV. In the exploration phase, the GWR system builds a model of normality of the environment with the tire (environment O-3). In the inspection phase, the system should detect the person as the novelty in the environment with the tire and the person (environment O-4). Due to the change in perspectives in the image frames in the inspection phase induced by the flight control module of the UAV, these frames were encoded by information that was not currently represented in the learned model of normality. Therefore, the system detected them as novelty. A possible solution to the problem is to evolve the novelty detectors online to adapt to dynamic changes in the environment. Another possible solution is to learn ad-hoc visual features for the problem. We could also explore the incorporation of information from several UAV sensors in order to complement the visual information. With this new information, we could detect new types of novelty, such as novelty based on the object position. All these issues will be the subject of future studies.

Figure 21. Failure cases in the evolved GWR detector on dataset D-4: (**a**) sample image frames in the exploration phase, and (**b**) false novelty indications in the inspection phase. In the exploration phase, the UAV explores environment O-3. Then, it should detect the person as the novelty in environment O-4. In the inspection phase, due to changes in perspective in the frames induced by the UAV's flight, some false novelty detections were presented because the information of the frame encoding was too different from the learned model.

6. Conclusions

The proposed methodology addresses the problem of automatic design of novelty detectors in visual exploration and inspection tasks, facing the challenge of unbalanced data. We proposed a new framework that uses deep features extracted by a pre-trained neural convolutional network. The methodology exploited the robust capabilities of the deep features to represent the images. A significant contribution of the work is the design of novelty detectors for specific tasks based on a global optimization technique. The proposed methodology simultaneously finds the size of the feature vector and the parameters of the novelty detectors. The methodology was tested on an outdoor environment with images captured by an unmanned aerial vehicle. We considered different types of novelties to verify the performance of the proposed methodology, including conspicuous or inconspicuous novel objects, static or dynamic novel objects, and multiple novel objects. We also considered two different light conditions in the outdoor environment (morning and afternoon), and two different flight heights of 2 m and 4 m, respectively. We performed a comparison with well-established feature extraction techniques in the problem of visual exploration and inspection tasks in the above conditions. The results showed that the proposed methodology is competitive or even better than these traditional techniques. Based on the results, we observed that the evolved detectors are robust to illumination changes, scale changes, and some levels of occlusion. Although they presented some problems with perspective changes produced by the flight control module of the unmanned aerial vehicle, the proposed evolved methods could detect the novelties in almost all cases, which is a desirable characteristic of novelty detection methods.

As future work, we will develop an online technique to design novelty detectors to address dynamic changes in the environment. More studies must be done to test the performance of the methodology with abrupt perspective changes of the objects. Another exciting research direction would be to use sensor fusion to detect novelties when it is difficult to do so with visual information alone.

Author Contributions: Conceptualization, M.A.C.-C. and V.A.-R.; Methodology, M.A.C.-C.; Software, M.A.C.-C.; Validation, V.A.-R., U.H.H.-B., and J.P.R.-P.; Investigation, M.A.C.-C.; Resources, J.P.R.-P. and U.H.H.-B.; Data Curation, U.H.H.-B. and J.P.R.-P.; Writing—Original Draft Preparation, M.A.C.-C.; Writing—Review and Editing, M.A.C.-C., V.A.-R., U.H.H.-B., and J.P.R.-P.; Supervision, V.A.-R.

Funding: This research received no external funding.

Acknowledgments: Marco A. Contreras-Cruz thanks the National Council of Science and Technology (CONACYT) for the scholarship with identification number 568675. The authors thank the Program for the Strengthening of Educational Quality (PFCE) 2019 of the University of Guanajuato for providing the publication costs.

Conflicts of Interest: The authors declare no conflicts of interest.

References

1. Pimentel, M.A.; Clifton, D.A.; Clifton, L.; Tarassenko, L. A review of novelty detection. *Signal Process.* **2014**, *99*, 215–249. [CrossRef]
2. Verma, A.; Taneja, A.; Arora, A. Fraud detection and frequent pattern matching in insurance claims using data mining techniques. In Proceedings of the 2017 Tenth International Conference on Contemporary Computing (IC3), Noida, India, 10–12 August 2017; pp. 1–7.
3. Stripling, E.; Baesens, B.; Chizi, B.; vanden Broucke, S. Isolation-based conditional anomaly detection on mixed-attribute data to uncover workers' compensation fraud. *Decis. Support Syst.* **2018**, *111*, 13–26. [CrossRef]
4. Ziaja, A.; Antoniadou, I.; Barszcz, T.; Staszewski, W.J.; Worden, K. Fault detection in rolling element bearings using wavelet-based variance analysis and novelty detection. *J. Vib. Control* **2016**, *22*, 396–411. [CrossRef]
5. Mohammadian Rad, N.; van Laarhoven, T.; Furlanello, C.; Marchiori, E. Novelty Detection using Deep Normative Modeling for IMU-Based Abnormal Movement Monitoring in Parkinson's Disease and Autism Spectrum Disorders. *Sensors* **2018**, *18*, 3533. [CrossRef] [PubMed]
6. Burlina, P.; Joshi, N.; Billings, S.; Wang, I.J.; Albayda, J. Deep embeddings for novelty detection in myopathy. *Comput. Biol. Med.* **2018**, *105*, 46–53. [CrossRef] [PubMed]
7. Bogaarts, J.; Hilkman, D.; Gommer, E.D.; van Kranen-Mastenbroek, V.; Reulen, J.P. Improved epileptic seizure detection combining dynamic feature normalization with EEG novelty detection. *Med. Biol. Eng. Comput.* **2016**, *54*, 1883–1892. [CrossRef]
8. Emami, A.; Harandi, M.T.; Dadgostar, F.; Lovell, B.C. Novelty detection in human tracking based on spatiotemporal oriented energies. *Pattern Recognit.* **2015**, *48*, 812–826. [CrossRef]
9. Ribeiro, M.; Lazzaretti, A.E.; Lopes, H.S. A study of deep convolutional auto-encoders for anomaly detection in videos. *Pattern Recognit. Lett.* **2018**, *105*, 13–22. [CrossRef]
10. Neto, H.V. On-line visual novelty detection in autonomous mobile robots. *Introd. Mordern Robot.* **2011**, *2*, 241–265.
11. Özbilge, E. On-line expectation-based novelty detection for mobile robots. *Robot. Auton. Syst.* **2016**, *81*, 33–47. [CrossRef]
12. Ross, P.; English, A.; Ball, D.; Upcroft, B.; Corke, P. Online novelty-based visual obstacle detection for field robotics. In Proceedings of the 2015 IEEE International Conference on Robotics and Automation (ICRA), Washington, DC, USA, 26–30 May 2015; pp. 3935–3940.
13. Chalapathy, R.; Menon, A.K.; Chawla, S. Anomaly Detection using One-Class Neural Networks. *arXiv* **2018**, arXiv:1802.06360.
14. Marsland, S.; Nehmzow, U.; Shapiro, J. Detecting novel features of an environment using habituation. In Proceedings of the Simulation of Adaptive Behavior, Paris, France, 11–15 September 2000.

15. Kato, H.; Harada, T.; Kuniyoshi, Y. Visual anomaly detection from small samples for mobile robots. In Proceedings of the 2012 IEEE/RSJ International Conference on Intelligent Robots and Systems (IROS), Algarve, Portugal, 7–12 October 2012; pp. 3171–3178.

16. Marsland, S.; Nehmzow, U.; Shapiro, J. On-line novelty detection for autonomous mobile robots. *Robot. Auton. Syst.* **2005**, *51*, 191–206. [CrossRef]

17. Neto, H.V.; Nehmzow, U. Visual novelty detection with automatic scale selection. *Robot. Auton. Syst.* **2007**, *55*, 693–701. [CrossRef]

18. Crook, P.; Hayes, G. A robot implementation of a biologically inspired method for novelty detection. In Proceedings of the Towards Intelligent Mobile Robots Conference, Maui, HI, USA, 29 October–3 November 2001.

19. Crook, P.A.; Marsland, S.; Hayes, G.; Nehmzow, U. A tale of two filters-on-line novelty detection. In Proceedings of the IEEE International Conference on Robotics and Automation, Washington, DC, USA, 10–17 May 2002; Volume 4, pp. 3894–3899.

20. Neto, H.V.; Nehmzow, U. Real-time automated visual inspection using mobile robots. *J. Intell. Robot. Syst.* **2007**, *49*, 293–307. [CrossRef]

21. Kasabov, N. ECOS: Evolving Connectionist Systems and the ECO Learning Paradigm. In Proceedings of the International Conference on Neural Information Processing, Kitakyushu, Japan, 21–23 October 1998; Volume 98, pp. 1232–1235.

22. Özbilge, E. Detecting static and dynamic novelties using dynamic neural network. *Procedia Comput. Sci.* **2017**, *120*, 877–886. [CrossRef]

23. Gonzalez-Pacheco, V.; Sanz, A.; Malfaz, M.; Salichs, M.A. Using novelty detection in HRI: Enabling robots to detect new poses and actively ask for their labels. In Proceedings of the 2014 14th IEEE-RAS International Conference on Humanoid Robots (Humanoids), Madrid, Spain, 18–20 November 2014; pp. 1110–1115.

24. Gatsoulis, Y.; McGinnity, T.M. Intrinsically motivated learning systems based on biologically-inspired novelty detection. *Robot. Auton. Syst.* **2015**, *68*, 12–20. [CrossRef]

25. Wang, X.; Wang, X.L.; Wilkes, D.M. An automated vision based on-line novel percept detection method for a mobile robot. *Robot. Auton. Syst.* **2012**, *60*, 1279–1294. [CrossRef]

26. Richter, C.; Roy, N. *Safe Visual Navigation via Deep Learning and Novelty Detection*; Science and Systems Foundation: Boston, MA, USA, 2017.

27. Sandler, M.; Howard, A.; Zhu, M.; Zhmoginov, A.; Chen, L.C. MobileNetV2: Inverted Residuals and Linear Bottlenecks. *arXiv* **2018**, arXiv:1801.04381.

28. Nguyen, T.V.; Kankanhalli, M. As-similar-as-possible saliency fusion. *Multimed. Tools Appl.* **2017**, *76*, 10501–10519. [CrossRef]

29. Watts, M.J. A decade of Kasabov's evolving connectionist systems: A review. *IEEE Trans. Syst. Man Cybern. Part C* **2009**, *39*, 253–269. [CrossRef]

30. Watts, M.; Kasabov, N. Simple evolving connectionist systems and experiments on isolated phoneme recognition. In Proceedings of the 2000 IEEE Symposium on Combinations of Evolutionary Computation and Neural Networks, San Antonio, TX, USA, 11–13 May 2000; pp. 232–239.

31. Marsland, S.; Shapiro, J.; Nehmzow, U. A self-organising network that grows when required. *Neural Netw.* **2002**, *15*, 1041–1058. [CrossRef]

32. Karaboga, D.; Gorkemli, B.; Ozturk, C.; Karaboga, N. A comprehensive survey: artificial bee colony (ABC) algorithm and applications. *Artif. Intell. Rev.* **2014**, *42*, 21–57. [CrossRef]

33. Mernik, M.; Liu, S.H.; Karaboga, D.; Črepinšek, M. On clarifying misconceptions when comparing variants of the Artificial Bee Colony Algorithm by offering a new implementation. *Inf. Sci.* **2015**, *291*, 115–127. [CrossRef]

34. Bianco, S.; Ciocca, G.; Schettini, R. Combination of video change detection algorithms by genetic programming. *IEEE Trans. Evolut. Comput.* **2017**, *21*, 914–928. [CrossRef]

Article

Training Data Extraction and Object Detection in Surveillance Scenario †

Artur Wilkowski *, Maciej Stefańczyk and Włodzimierz Kasprzak

Institute of Control and Computation Engineering, Warsaw University of Technology, Nowowiejska 15/19, 00-665 Warszawa, Poland; maciej.stefanczyk@pw.edu.pl (M.S.); wlodzimierz.kasprzak@pw.edu.pl (W.K.)
*   Correspondence: artur.wilkowski@pw.edu.pl; Tel.: +48-22-234-7276
†   This paper is an extended version of our paper published in: Wilkowski, A.; Kasprzak, W.; Stefańczyk, M. Object detection in the police surveillance scenario. In Proceedings of the 14th Federated Conference on Computer Science and Information Systems, Leipzig, Germany, 1–4 September 2019.

Received: 31 March 2020; Accepted: 5 May 2020; Published: 8 May 2020

Abstract: Police and various security services use video analysis for securing public space, mass events, and when investigating criminal activity. Due to a huge amount of data supplied to surveillance systems, some automatic data processing is a necessity. In one typical scenario, an operator marks an object in an image frame and searches for all occurrences of the object in other frames or even image sequences. This problem is hard in general. Algorithms supporting this scenario must reconcile several seemingly contradicting factors: training and detection speed, detection reliability, and learning from small data sets. In the system proposed here, we use a two-stage detector. The first region proposal stage is based on a Cascade Classifier while the second classification stage is based either on a Support Vector Machines (SVMs) or Convolutional Neural Networks (CNNs). The proposed configuration ensures both speed and detection reliability. In addition to this, an object tracking and background-foreground separation algorithm is used, supported by the GrabCut algorithm and a sample synthesis procedure, in order to collect rich training data for the detector. Experiments show that the system is effective, useful, and applicable to practical surveillance tasks.

Keywords: object detection; few shot learning; SVM; CNN; cascade classifier; video surveillance

---

1. Introduction

Police and various security services use video analysis when investigating criminal activity. Long surveillance videos are increasingly searched by dedicated image analysis software to detect criminal events, to store them, and to initiate proper security actions. One of the prominent examples is the P-REACT project [1] (Petty cRiminality diminution through sEarch and Analysis in multi-source video Capturing and archiving plaTform). Solutions to automatic analysis of surveillance videos seem already to be mature enough, as the research community is recently also involved in significant benchmark initiatives [2,3]. The computer vision research focus is now shifted to the analysis of video data coming from handheld, body-worn, and dashboard cameras and on the integration of such analysis results with police- and public-databases.

In typical object detection scenarios, there are much data to learn from and a major objective is to use them effectively. In a security-oriented environment, the user interaction should be kept as simple as possible. The optimal solution would be marking only a single object in a selected image frame and initiating a search to find occurrences of similar objects in other frames of the processed sequence or different sequences. This imposes several constraints on the Machine Vision solution that need to be addressed.

First of all, the system should learn on-line or nearly on-line. The system must also perform per-frame detection quickly and provide approximate results in a short time, but also should be tuned in such a way, that no occurrence of the interesting object is skipped. Last but not least, the system must be able to learn from small data sets.

This paper is an extension of our conference paper [4] describing an effective and time-efficient algorithm for instance search and detection in images from handheld video cameras. The system described there uses a discriminant approach to differentiate the object from its foreground. To do so, a combined Haar–Cascade detector and Histogram of Oriented Gradients–Support Vector Machine (HOG-SVM) classifier are used. We argued that this provided a desirable trade-off between detection quality and training/detection times. Both the positive, as well as negative samples, are extracted only from training images.

The extended version presented here includes new system elements and experiments. In particular, the additions are as follows:

- introduction of a new foreground/background segmentation procedure,
- incorporation and evaluation of a CNN classifier in the detector framework,
- additional experiments covering new and previous features,
- extended state-of-the-art analysis.

The main components of the system that are affected by the additions were marked bold in Figure 1.

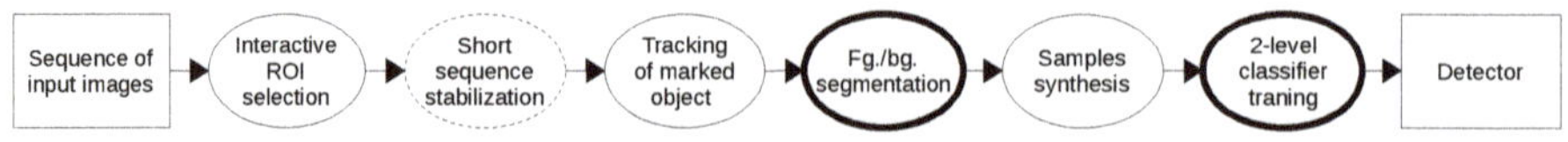

Figure 1. Structure of the training procedure.

Comparable detector solutions based on CNNs provide excellent detection performance [5]. Such solutions, however, rely on off-line training and the training/detection speed is still a bottleneck for such systems. This effect is, to some extent, ameliorated by GPU utilization. Recent developments aim at the reduction of detection times by cascading CNNs [6] or by detecting salient regions first using fuzzy logic [7]. However, a significant reduction of training time is still an open area of research.

This work can be categorized as the few-shot learning system. The most popular approaches in this field focus on distance metric learning (which, on its own, has a long history [8]) with further data clustering using the L_2 metric. In recent years, new solutions based on deep learning emerged [9], with relational networks being the popular choice for few-shot learning [10]. A popular choice for the loss function, in that case, is triplet loss [11]. The application of distance metric learning for classification is straightforward. Metric is produced based on training samples, and the queries embed-dings are compared with the class representatives, with L_2 metric and/or k-Nearest Neighbor. This approach proved to be successful in multiple classification tasks [12–15].

Application of few-shot learning for detection is a harder problem, as usually the marked object in training set occupies only a small portion of the image. Hence, the training data set is heavily unbalanced. One of the possibilities is to transfer the knowledge from existing detectors [16]. Another alternative is to use semi-supervised learning, with few labelled samples and a larger set of unlabelled data, iteratively used in training [17]. A different approach is described in [18], where the imitations are used in the training process for a robot-grasping task. Finally, the direct application of distance metric learning was proposed in [19], where the InceptionV3 network is used as a backbone for metric learning.

Using the taxonomy proposed in [20] (data-, model-, or algorithm-based), our solution is data-based, with handcrafted rules for transforming the training dataset.

One contribution of the paper is the procedure of collecting as much realistic training data as possible, providing limited user interaction. Another contribution involves the proposition of a complete computer-aided video surveillance procedure and a thorough evaluation of the applicability of particular computer vision methods for specific stages of processing, having in mind the system requirements such as the ability to learn from short video sequences, performing quick learning and reaching fast detection time. The evaluation comprised both classic computer vision methods as well as contemporary CNN approaches.

Ideally, the modeling stage should be able to start from a single selection of a Region of Interest (ROI), and all additional examples should be obtained automatically. Such least-user-effort approaches were already discussed e.g., for semi-automatic video annotation and detection systems, such as [21,22]. In the cited method, however, the user may be asked to annotate video several times (to decide about samples lying on decision boundary), which is not necessarily acceptable for all end users. An example of another successful detector that works on a single selection is given in [23]. The detector operates on sparse image representation (collection of Scale Invariant Feature Transform (SIFT) descriptors), so it is very time efficient. Our initial experiments have shown that descriptor-based approaches work the best for highly textured and fairly complex objects that occupy a large part of the image, which is not always true in the surveillance scenarios.

The procedure of collecting training data, given in this paper, combines object tracking and background subtraction methods for semi-supervised collection of training windows and their foreground masks. This step is supported by the GrabCut algorithm [24], with the possibility of smoothly mixing the results of both. The samples collected during tracking are further synthetically generalized (augmented) to enrich the training set. Scenarios, where tracking results are utilized for the collection of detector's training data, were already covered in literature, especially regarding tracking, with prominent examples [25,26] or more recent CNN approaches [27,28]. In such approaches, the exact foreground–background separation (which is crucial for effective synthesis of samples) is often neglected, since the algorithms typically have enough frames to collect rich training data.

The proposed methods were evaluated on a corpus of surveillance videos and proved that its efficiency is good enough to be effective in supporting a user (police officer or security official) in their everyday working tasks. The proposed two-level detector architecture was evaluated using alternative methods for feature calculation, namely Histogram of Oriented Gradients (HOG) features and VGGNet-based deep features. In the former case the SVM classifier was used, in the latter, the neural network classification layers were consistently utilized. For final evaluation we compared the accuracy of our proposed system with the state-of-the-art Faster-RCNN [29] detector trained on the same data.

The paper is organized as follows: in Section 2 there is given a technical background and methodology used in our system, Section 3 provides experimental results, and Section 4 contains conclusions. For the reader's convenience, a short dictionary of abbreviations used in the paper is presented at the end.

2. Methods

2.1. Detector Overview

In the system described in this paper, we utilize a classic detection framework, where a sliding window with varying sizes is moved over each frame and, for each location, the selected image part is evaluated against information gathered from training samples. A crucial part of the detector is formed by a classifier (SVM or NN), which is responsible for the evaluation of each selected image part. A pure classifier, when applied to hundreds of thousands of candidate areas, would be too slow to learn and detect. In our scenario, a pre-classification step utilizing Haar-like feature-based cascade classifier is applied to limit the number of candidate windows to several hundred. We claim that this simple structure combines a good detection rate with acceptable detection speed (about ten full-HD frames per second on modest computer) as well as acceptable training speed in typical scenarios (less than a few minutes per pattern).

In essence, the two-stage detector architecture resembles some significant modern CNN approaches, where the detection is divided into the region proposal part and the region recognition part [29]. In our approach, region proposal is performed by the cascade classifier, and the (SVM/NN) classifier does the final classification. Both methods offer reasonable training and detection speeds required for this application.

In our scenario, sources of data are naturally sparse. Depending on user decision, the detector can be trained either on one or a short sequence of training images. Therefore a critical part of our system is a set of tools aiding the user in an effortless collection of training examples from short image sequences as well as methods for artificial synthesis and generalization of training samples to provide the detector with the training data as rich as possible. These tools and methods are discussed in subsequent sections. The overall structure of the training procedure is given in Figure 1.

The first stage of processing is and interactive collection of training samples (*Interactive ROI selection*). The user marks an object in the image and initializes the tracking procedure to collect more training data for the selected object (*Tracking of marked object*). In case of un-stabilized images (e.g., from hand-held camera) additional stabilization step can be applied (*Short sequence stabilization*). The tracking is open to user intervention e.g., correction of the ROI in a single frame or frames.

The next step of the procedure is foreground–background segmentation (*Fg./Bg. segmentation*), which helps to recover the precise outline of the object from the rectangular ROI basing on motion and color information. After this comes *Samples synthesis*, here, the collected training samples are jittered in different ways to enrich the training set. Both steps enable the operator to correct algorithm parameters with visual feedback. The last step of detector generation is the training of a 2-level classifier (*2-level classifier training*).

2.2. Collection of Positive Training Samples

Although for some patterns (which include e.g., flat patterns) good detection results can be obtained using only one selected sample that is further generalized and synthesized into a set with larger variability, in most cases detection results highly depend on size and diversity of input training set. In the scenario discussed in this paper, these properties of the training set can (at least partially) be achieved by collecting samples from a short sequence of input images. Our scenario is organized as follows: (1) a user selects an object of interest using a rectangular area, (2) the application tracks the object in subsequent frames of the sequence (with optional manual reinitialization), (3) object foreground masks are established using motion information and image region properties.

2.2.1. Object Tracking and Foreground–Background Separation Using Motion Information

For tracking of rectangular area an optimized version of Circulant Structure of Kernels (CSK) tracker [30] that utilizes color-names features [31] is used. As a result of the tracking procedure, we obtain a sequence of rectangular areas that encompass the object of interest in subsequent frames. In most cases both object foreground, as well as background, will be present in the tracked rectangle. However, if the object is moving against a moderately static background, we can exploit motion information to effectively separate object foreground from background by background subtraction.

Let the tracking results be described by a sequence of rectangular areas $\{R^1, \ldots, R^T\}$ and let us denote coordinates of pixel i as p_i, color attributes for pixel i at time t as c_i^t, and a mean of color attributes in the background as:

$$\bar{c}_i = \frac{1}{n_i} \sum_{t:p_i \notin R^t} c_i^t \tag{1}$$

where averaging factor n_i is the number of frames where tracking window does not contain pixel i and can be computed as $n_i = |\{t : p_i \notin R^t\}|$.

Now we can specify a background training sequence for each pixel $\{\hat{c}_i^t\}$

$$\hat{c}_i^t = \begin{cases} c_i^t & \text{if } p_i \notin R^t \\ \bar{c}_i & \text{if } p_i \in R^t \end{cases} \tag{2}$$

Following the rule above, only pixels that at given time-step do not belong to the tracked area contribute to the background model computed for the image. Each pixel that always belongs to the tracked area is conservatively treated as a foreground as we are unable to establish a background model for these areas.

The background model adopted here follows algorithms from [32]. In this method, scene color is represented independently for all pixels. The color for each pixel (both from the background and foreground $BG + FG$) given the training sequence C_T, is modeled as:

$$p(c_i|C_T, BG + FG) = \sum_{m=1}^{M} \hat{\pi}_m \mathcal{N}(c_i; \hat{\mu}_m, \hat{\sigma}_m) \tag{3}$$

where $\hat{\mu}_m, \hat{\sigma}_m$ are estimated means and standard deviation of color mixture components, $\hat{\pi}_m$ are mixing coefficients, M is the total number of mixtures, and $\mathcal{N}(c_i; \hat{\mu}_m, \hat{\sigma}_m)$ denotes Gaussian density function evaluated in c_i.

In the algorithm a (generally correct) assumption is made that background pixels, as appearing most often, will dominate the mixture. Therefore the background model (BG) is built from the selected number of largest clusters in the color mixture:

$$p(c_i|C_T, BG) = \sum_{m=1}^{B} \hat{\pi}_m \mathcal{N}(c_i; \hat{\mu}_m, \hat{\sigma}_m) \tag{4}$$

where B is the selected number of background components. The pixel is decided to belong to the background when

$$p(c_i|C_T, BG) > c_{thr} \tag{5}$$

Threshold c_{thr} can be interactively adjusted by the user. Exact algorithms for updating mixture parameters are given in [32]. Sample result of background subtraction procedure is given in Figure 2.

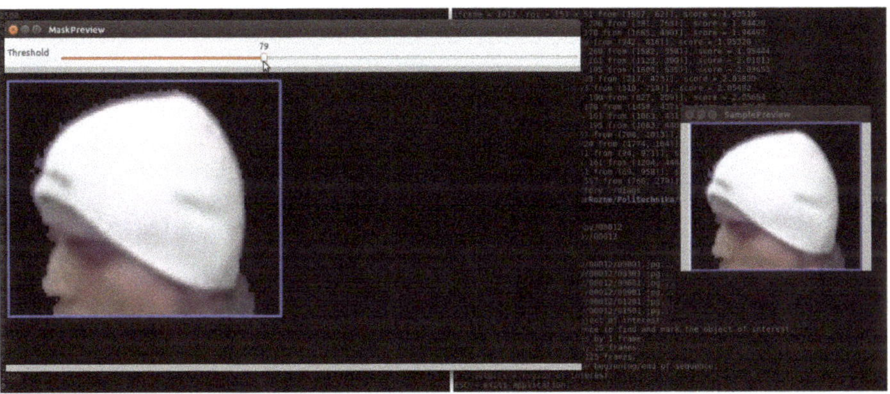

Figure 2. Results of automatic foreground–background separation.

Some modern developments in foreground–background separation using the Robust Principal Component Analysis (RPCA) approach were proposed e.g., in [33,34]. They are founded on extensive optimization in 3D spatio-temporal volumes and offer excellent accuracy at the expense of some processing speed. Since our system relies on the interaction between the human operator and computer, the processing speed is very important, so purely local methods seem to be currently the best choice. However, this topic will be investigated in the future versions of the system since some ideas from [33,34] are likely to be complementary to our developments presented in Section 2.2.3.

2.2.2. GrabCut Algorithm

GrabCut [24] is a widely acclaimed method for (semi)automatic foreground/background segmentation. The method takes into account several properties or image regions: color distribution, coherence of regions, contrast between regions. These factors are described in the form of image-wide energy function to optimize, that assumes the following form:

$$\mathbf{E}(\underline{\alpha}, \underline{\theta}, \mathbf{z}) = U(\underline{\alpha}, \underline{\theta}, \mathbf{z}) + V(\alpha, \mathbf{z})$$

where α_n describes segmentation information for pixel n (can be binary), $\underline{\theta}$ is the set of parameters of the Gaussian Mixture Model representing color distribution in background and foreground and $\mathbf{z}$ are observed image pixels. Estimated parameters are underlined. The energy term U describes, how well current estimation of foreground and background pixels matches the assumed color distribution of foreground and background, while the term V evaluates spatial consistency of regions by penalizing discontinuities (except for the areas of high contrast).

The best configuration of parameters is the one minimizing the term $\mathbf{E}(\underline{\alpha}, \underline{\theta}, \mathbf{z})$. The components are designed in a way that the energy term can be minimized using an effective graph-cut algorithm. The optimization algorithm is iterative and switches between (re)estimation of region color distribution and (re)estimation of segmentation.

The input of the algorithm is defined in [24] as a trimap $\{T_B, T_U, T_F\}$. T_B stands for sure background, T_F stands for sure foreground, and T_U is an unknown area (to estimate). T_B and T_F are fixed and cannot change during the algorithm. Typical initialization is to set T_B to the area outside of object ROI, set T_F to $\varnothing$, and T_U is the remaining part of the image. In the first iteration, all pixels from T_B are initialized as background and all pixels from the unknown area T_U as foreground (which is subject to change). Implementations like [35], however, allow to specify the additional areas within T_U: the likely foreground $\hat{T}_F$ and likely background $\hat{T}_B$ as a convenient starting point for optimization.

2.2.3. Object Tracking and Foreground–Background Separation Using Hybrid Motion Information and GrabCut

While pure object motion information is sufficient to perform foreground–background segmentation in most cases, it fails altogether for static objects. In addition, the precision of such an approach varies from case to case and strongly depends on the manual threshold selection for background subtraction c_{thr}. Therefore we propose a modified procedure of fine-tuning results of background subtraction using GrabCut.

1. The object is tracked and its foreground mask is obtained using methods from Section 2.2.1.
2. For each frame the current foreground–background segmentation results are used to initialize a GrabCut trimap, specifically:

   - the foreground region is used to initialize the G-C T_F, with an exception for the area for which no background model could be reliably established (areas that belong to each collected tracking ROI),
   - the G-C T_B is initialized outside the tracked ROI border, to provide enough pixels for background estimation the tracking area is scaled uniformly by 50%,
   - the remaining area of ROI becomes the T_U.

We found our solution somewhat similar to the one proposed quite recently in [36]. However, in the cited approach the trimap is initialized differently. The G-C T_U area is limited only to the area of the morphological gradient (difference between dilation and erosion) of the foreground area established by background subtraction. In our solution we safely assume that T_B is always outside the tracked ROI, so there is no risk to incorporate the foreground object in T_B. Additionally, in contrast to our approach, in [36] the background subtraction is not discussed within the tracking context.

The solution proposed here is parametrized by a single background subtraction threshold c_{thr} and provides a smooth user experience when transiting between different threshold values. By specifying very low thresholds, the user selects a foreground mask covering the whole tracked ROI. For larger thresholds, we obtain the results of G-C algorithm with foreground constrained to be at least the mask generated by background subtractor. For very large, extreme values of c_{thr}, the foreground seeds from the background subtractor become small, and the method converges to the output of the vanilla G-C algorithm for static images.

This is interesting to note that for static ROIs we use exactly the same procedure. For such ROIs, the area of uncertain background model (the area where no background model could be reliably established) is very large and covers the whole ROI. In such a situation, the entire ROI area is simply a subject to the classic G-C algorithm. Note also, that using only methods from Section 2.2.1 the whole ROI area would be inevitably labelled as foreground.

2.2.4. Image Stabilization in a Short Sequence

The foreground–background segmentation procedure works best when the stable camera position is available (or image sequence is stabilized before segmentation). The system proposed here uses a stabilization procedure basing on matching of SURF features [37] and computation of homography transformation between pairs of images. The stabilization works on short subsequences of the original sequence. The first frame to stabilize is the one used for marking the initial region of interest. The procedure then aligns all subsequent frames to the first frame by evaluating homography relating two images. In order to do so, matching methods from [38] and the Least Median of Squares principle [39] are utilized. To increase stabilization efficiency, GPU-accelerated procedures for keypoints/descriptors extraction and matching from OpenCV library are utilized [35].

2.3. Collection of Negative Training Samples

Negative samples that are used in detector training are extracted from the same sequence images that positive samples originated from. For each training image, one fragment is used to extract a positive sample, while the remaining part of the image is divided into at most four sources of negative samples, as given in Figure 3. Thus, an assumption is made that these remaining parts of the training sequence images do not contain positive samples. This assumption is not always valid, but may be strengthened by asking a user to mark all positive examples in the training sequence.

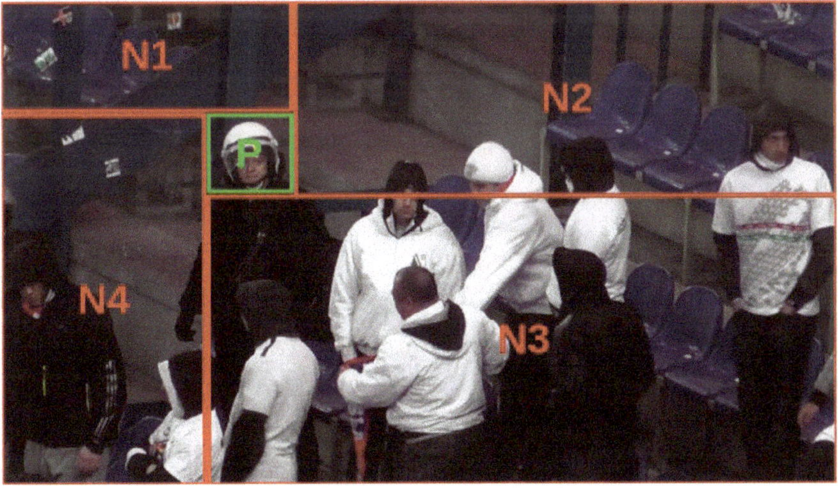

Figure 3. Division into positive (P) and negative (N1–N4) examples.

2.4. Positive Samples Generalization and Synthesis

2.4.1. Geometric Generalization

In this step, 3D rotations are applied to collected pattern images and their masks. It is assumed that patterns are planar, so this generalization method can be useful only to some extent for non-planar objects. The rotation effect is obtained by applying a homography transformation, imitating application of three rotation matrices $R_x(\alpha)$, $R_y(\beta)$, $R_z(\gamma)$ to a 3D object. The matrices correspond to rotations around x, y, and z axes correspondingly. 3D rotation matrices are defined classically:

$$
\begin{aligned}
R_x(\theta) &= \begin{pmatrix} 1 & 0 & 0 \\ 0 & \cos\theta & -\sin\theta \\ 0 & \sin\theta & \cos\theta \end{pmatrix} \\
R_y(\theta) &= \begin{pmatrix} \cos\theta & 0 & \sin\theta \\ 0 & 1 & 0 \\ -\sin\theta & 0 & \cos\theta \end{pmatrix} \\
R_z(\theta) &= \begin{pmatrix} \cos\theta & -\sin\theta & 0 \\ \sin\theta & \cos\theta & 0 \\ 0 & 0 & 1 \end{pmatrix}
\end{aligned}
\tag{6}
$$

To compute the transformation, first a homography matrix is computed using formula:

$$
H = R - \frac{\mathbf{t}\mathbf{n}^T}{d}
\tag{7}
$$

where $\mathbf{n}$ is a vector normal to the pattern plane (we set it to $\mathbf{n} = (0,0,1)^T$), d is the distance from the virtual camera to the pattern (we set it arbitrarily to $d = 1$, since it only scales 'real-world' units of measurement) and R is the 3D rotation matrix that is decomposed as:

$$
R = (R_x(\alpha) \cdot R_y(\beta) \cdot R_z(\gamma))^{-1}
\tag{8}
$$

In order for the image center (having world coordinates $C = (0,0,d)^T$) to remain intact during transformation we define 'correcting' translation vector as:

$$
\mathbf{t} = -RC + C
\tag{9}
$$

Then we can specify artificial camera matrices as K_1 and K_2

$$
K_1 = \begin{pmatrix} f & 0 & c_{in}^x \\ 0 & f & c_{in}^y \\ 0 & 0 & 1 \end{pmatrix}, \quad K_2 = \begin{pmatrix} f & 0 & c_{out}^x \\ 0 & f & c_{out}^y \\ 0 & 0 & 1 \end{pmatrix}
\tag{10}
$$

where $(c_{in}^x, c_{in}^y)^T$ and $(c_{out}^x, c_{out}^y)^T$ are pixel coordinates of input and output image correspondingly, while f is the artificial camera focal length given in pixels. In this application we set f to be f_{mul} times larger input image dimension. Multiplier f_{mul} decides about the virtual distance of our virtual camera to the object. Smaller values introduce larger perspective distortions of the transformation, larger values introduce smaller distortions. We arbitrarily set f_{mul} to 10 implying only slight perspective distortions.

The final homography transformation applied to the pixels of the input image is given by

$$
P = K_2 H K_1^{-1}
\tag{11}
$$

Rotation angles α, β, and γ are selected randomly from the uniform distribution (denoted here as $\mathcal{U}$). The amount of rotation around axes y is twice times the amount of rotation around remaining axes to better reflect dominant rotations in human movement.

$$\alpha \sim \mathcal{U}(-1,1) \cdot \delta_{max} \cdot 0.5,$$
$$\beta \sim \mathcal{U}(-1,1) \cdot \delta_{max},$$
$$\gamma \sim \mathcal{U}(-1,1) \cdot \delta_{max} \cdot 0.5$$

and δ_{max} is the parameter specifying the maximum extent of allowed rotation.

2.4.2. Intensity and Contrast Synthesis

In the proposed approach image intensity and contrast synthesis is applied in addition to geometric transformations. It is especially important for Haar-like features that lack intensity normalization.

First, intensity values of pixels I_{in} are retrieved from RGB image by extracting V component from HSV representation of the image and setting $I_{in} = $ V. The intensity and contrast adjustment affects only V channel. After adjustment, the RGB image is reconstructed from HSV' where V' $= I_{out}$.

For adjustment, a simple linear formula is used. For each pixel gray value I_{in} we have

$$I_{out} = a * I_{in} + b \tag{12}$$

where

$$a = 1 + c_{dev}, \, b = I_{dev} - \mu_I \cdot c_{dev} \tag{13}$$

where μ_I is the average intensity of the sample. Contrast deviation c_{dev} as well as intensity deviation I_{dev} are sampled from the uniform distribution $c_{dev} \sim \mathcal{U}(-1,1) \cdot c_{max}$ and $I_{dev} \sim \mathcal{U}(-1,1) \cdot I_{max}$. c_{max} is a parameter denoting the maximum allowed contrast change and I_{max} is a parameter denoting the maximum allowed intensity change. Changes in contrast preserve mean intensity of an image. After application of the formula its results are appropriately saturated.

2.4.3. Application of Blur

Training and test samples may differ in terms of quality of image details due to different factors such as deficiencies of optics used, motion blur or distance. In our case, we apply a simple Gaussian filter to simulate natural blur effects

$$\sigma = \mathcal{U}(0,1) \cdot \sigma_{max} \cdot \min(I_{width}, I_{height}) \tag{14}$$

where I_{width} and I_{height} are image sample sizes and σ_{max} controls the maximum size of the Gaussian kernel.

2.4.4. Merging with the Background

Generalized training images are superimposed on background samples extracted from negative examples of size ranging from about 0.25 to 4 times the positive sample size. Gray-level masks are used for the seamless incorporation of positive samples into background images.

2.5. Detector Training

The detector training procedure is divided into two steps. In the first step, the cascade classifier using HAAR-like features is trained. The classifier is trained on training samples resampled to a fixed size of 24 $\times$ 24 pixels. In our scenario, for each cascade stage, 300 positive samples and 100 negative samples are utilized. The minimum true positive rate for each cascade level is set to 0.995, and the maximum false positive rate is set to 0.5. The classifier is trained for a maximum of 15 stages or until reaching ≈ 0.00003 FPR. The expected TPR is at least $0.995^{15} \approx 0.93$. By using these settings, up to about 1000 detections are generated for each Full-HD test image.

2.6. Detector Training Using HOG+SVM

During the second stage of training an SVM classifier is trained to handle samples that passed the first cascade classification. For most experiments, the SVM classifier is trained on 300 positive and 300 negative samples or 600 and 600 samples accordingly. The SVM classifier uses the Gaussian RBF kernel.

$$K(\mathbf{x}, \mathbf{y}) = \exp(-\gamma ||\mathbf{x} - \mathbf{y}||^2) \tag{15}$$

The Gaussian kernel size γ and SVM regularization parameter C are adjusted using automatic cross-validation procedure performed on the training data. For SVM classification Histogram of Oriented Gradients features [40] are extracted. There were used 2 resolutions of training images: 24×24 and 32×32. For each sample a 9-element histogram in 4×4 cells is created with 16×16 histogram normalization window overlapping by 8 pixels, thus giving $4 * 16 * 9 = 576$ HOG features in total in 24×24 case and $9 * 16 * 9 = 1296$ in 32×32 case.

Negative samples are extracted from the Cascade Classifier decision boundary (containing samples that were positively verified by CC but still negative) if possible. If not, image fragments used as background images for positive samples or (as the last resort) other randomly selected samples are used. In all experiments OpenCV 3.1 [35] Cascade Classifier and SVM implementation are utilized.

Given our test data, the number of resulting support vectors in the SVM classifier varies between 200 and 400 for 24×24 case but can be twice as large for 32×32 case. Let us review one specific configuration: 'hat' pattern trained on 55 24×24 images with masks and pattern generalization settings $\sigma_{max} = c_{max} = 0$, $\delta_{max} = 0.7$, $I_{max} = 50$. After SVM metaparameter optimization we obtain SVM regularization parameter $C = 2.5$, RBF kernel size $\gamma = 0.5$, and the number of support vectors 233.

2.7. Detector Training Using Tuned VGG16 Network

The number of features processed by our detector is limited to HOG and Haar-like features. There exist quite a few other solutions that use a much richer set of features for object detection e.g., [41], however, at the expense of increased computational complexity, which is an important practical concern of our system. As an alternative to evaluating complex sets of hand-crafted features, we decided to resort to the state-of-the-art methods automatic feature computation based on Convolutional Neural Networks.

Therefore in addition to the classifier described in Section 2.6, we evaluate a VGG16 network [42] trained on the ImageNet dataset [43] tuned to our problem using transfer-learning principle [44]. Transfer learning is a common technique to overcome the problem of the availability of training data. In case of a limited access to the training data for the given problem, one can use an existing network trained on another dataset and subsequently fine-tune (retrain) this network to accommodate the specific problem samples. Since usually the original network was trained on millions of examples and hundreds of classes, it is capable of extracting robust and usually quite universal features. Thus, the new network can benefit from pre-trained feature layers and train only classification layers.

For our task, we utilize a convolutional layer of VGG16 network (which is obtained from the original network by removing fully-connected layers). The network is augmented with two fully connected layers (1024 neurons with ReLu activation and a single neuron with sigmoid activation correspondingly) and one dropout layer (with dropout rate 0.5) to avoid overfitting. The VGG16 network was originally trained on a sample image of 224×224, however, the convolutional part would accept any multiplicity of 32 for image width and height. In this paper, we verify classifier performance for input image sizes 32×32, 64×64, 128×128, and finally 224×224 pixels. The total number of parameters in fully connected layers ranges from 522 337 to 25 692 161 depending on input image size. The overall network structure is given in Figure 4.

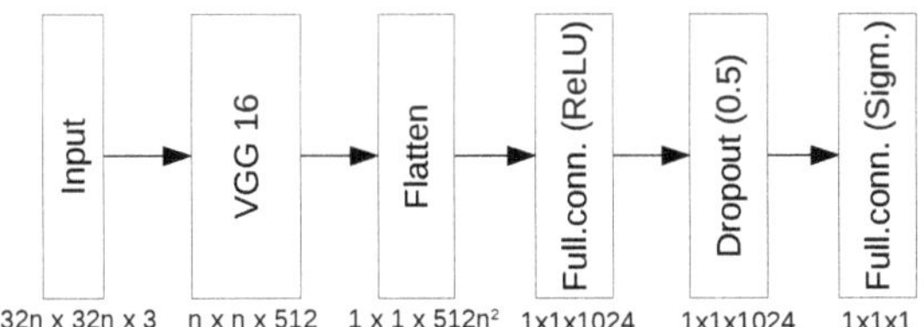

Figure 4. Neural Network Structure, $n \in \{1, 2, 4, 7\}$.

The selected loss function is a binary cross-entropy for binary classification. During training, all the convolutional layers are frozen and only the parameters of fully connected layers are adapted.

2.8. Detection and Post-Processing

During the detection phase, each test image is first processed by the cascade classifier, typically returning several hundreds of candidate areas. After this, each candidate area is examined by the classifier and a score is assigned to each detection. In case of the SVM, the score is computed as the signed distance from the separating plane in support vector space with the lowest negative scores treated as best matches and high positive scores as worst matches. For the VGG16-based neural network, the output of the sigmoid function is negated and used as the score.

For each image, only the best score area is considered for further processing. Frames from the test sequence are sampled and processed with increasing density (first, last, and middle frame for a start, and then intermittent frames), to quickly produce some results for the user to review (non-minima suppression is used to reduce clutter).

3. Experiments

3.1. Experimental Setup

In order to evaluate the detector performance the following tools are utilized.

- ROC curve: Receiver Operator Characteristic curve is the curve relating the True Positive Rate (TPR) and the False Positive Rate (FPR). The True Positive Rate (TPR) is defined as the number of correctly detected examples (TP) by the total number of positive occurrences (P), the False Positive Rate (FPR) is defined as the number of incorrectly detected examples (FP) by the total number of negative samples (N). The ROC curve gives comparable results even for the imbalanced datasets.
- PR curve: Precision-Recall curve is the curve relating Precision (the number of truly positive samples among all positive detections) and the Recall (another name for True Positive Rate). The PR curve is useful for establishing how many good hits can be expected among those best ranked by the detector.
- AUC: Area Under Curve is a useful single measure to summarize ROC curve, computed by the integration of TPR over FPR
- AVGPR: Average Precision-Recall is another a useful measure to summarize PR curve, computed by integration of the Precision over Recall
- EER: Equal Error Rate denotes a line on ROC plot where TPR + FPR = 1 or a point on ROC curve meeting this requirement

In the subsequent sections, each ROC and PR curve was plotted basing on a single run of the detector, whereas the majority of tables included in the subsequent sections contain aggregated data from multiple runs of the algorithm to compensate the stochastic nature of Machine Learning algorithms used. Computational performance experiments were performed on AMD Ryzen 5 2600X processor (32 GB of memory) and GeForce RTX 2070 GPU (8 GB of memory) unless stated otherwise.

3.2. Preliminary Experiments

During the first stage of experiments there was selected a single test sequence '00012' with 1776 Full-HD frames. Using this sequence, various parameter configurations were evaluated in order to assess the basic properties of the solution proposed. Basing on these experiments, some answers can be given regarding problems such as the impact of utilization of a two-layer detector on detection results and detection/training speed, the impact of the method of selection of training samples on detection accuracy or influence of values of image synthesis parameters on overall quality. Above questions will be discussed in the following paragraphs. During the first three experiments, one sample pattern 'hat' was utilized, and in the last experiment three other patterns: 'logo', 'helmet', and 'shirt' were introduced. Examples of training samples are given in Figure 5, and samples marked in the full-frame image are given in Figure 6. Filtered detection results for one test sequence presented in the form of a simple GUI are given in Figure 7.

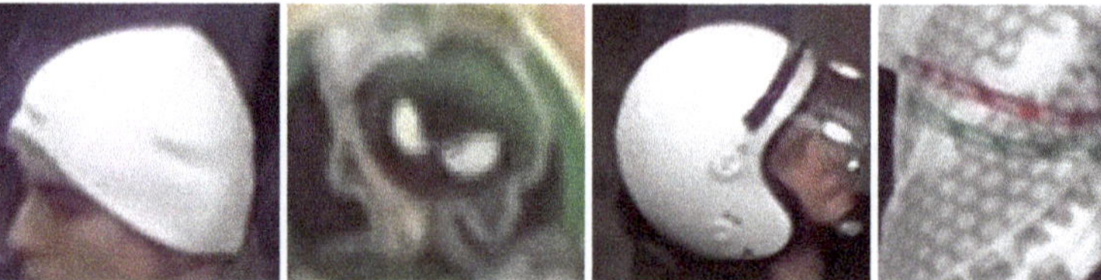

Figure 5. Example training samples of hat, logo, helmet, and shirt.

Figure 6. Frame with marked hat, logo, helmet, and shirt samples.

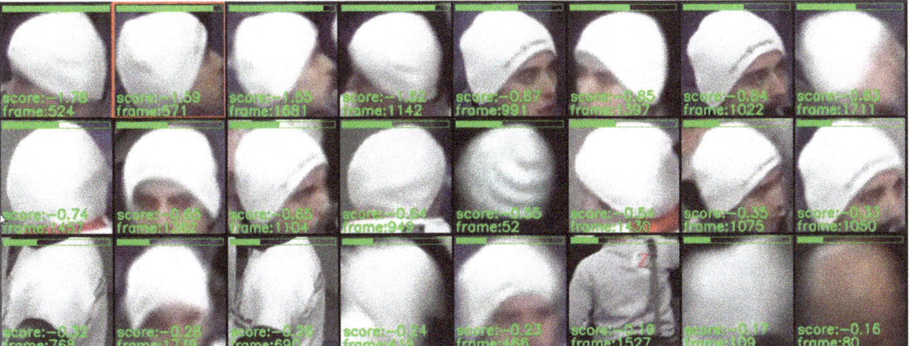

Figure 7. Detection results filtered by minimum distance (25 frames) between hits.

3.2.1. Two-Layer Detector

In the first experiment a trade-off between the detection and training speed for different number of expected cascade stages k was evaluated (Figure 8a). In this experiment HOG+SVM second stage classifier was used. Identical parameters were used for all k values, except for the number of the SVM training samples. For $k < 15$, a larger number of 900 positive and negative samples were used. For $k \geq 15$, the default of 300 positive and negative samples were utilized. This was used to balance the total number of training samples consumed by the detector (for smaller k the task of SVM is harder, since more negative samples pass the first stage of detection).

The experiment shows, that for low k training time is dominated by SVM training, and for large k cascade training dominates. A good compromise for our data can be obtained for value of $k = 15$. Larger k obviously means faster detection (Figure 8a), but also slightly worse detection results (Figure 8b), likely due to utilization of more robust HOG features in the second stage. This statement is also generally supported by the shape of corresponding ROC curves in Figure 9. These preliminary performance experiments were performed on Intel Core i5 computer.

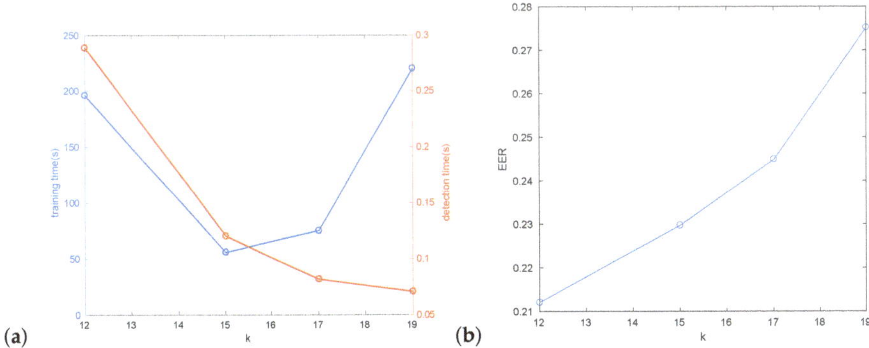

Figure 8. Impact of different number of cascades (k): (**a**) training/detection time; (**b**) hat detection EER.

3.2.2. Collection of Training Samples

In the next experiments, detector performance for different training data collection methods was evaluated. In the first place the data samples were collected using the automatic tracking and foreground–background separation methods given in this paper. In the process 55 data samples of hat pattern were collected, together with their automatically generated masks (using methods from Section 2.2.1). The data consisted of images of a white hat on top of a head, while the head was making full 180 degrees rotation around central axis. For comparison, a short sequence of training samples representing only 3 extreme head positions (*en-face* and two profiles) was utilized. For both sequences either appropriate foreground–background masks or no masks were used given 4 different combinations of settings. The detection results are given in Figure 10.

Not surprisingly, the richest possible data source (55 frames with generated masks) gives the best results. It is valuable to note that for our data, applying both object tracking and automatic mask generation is substantial to get optimal results.

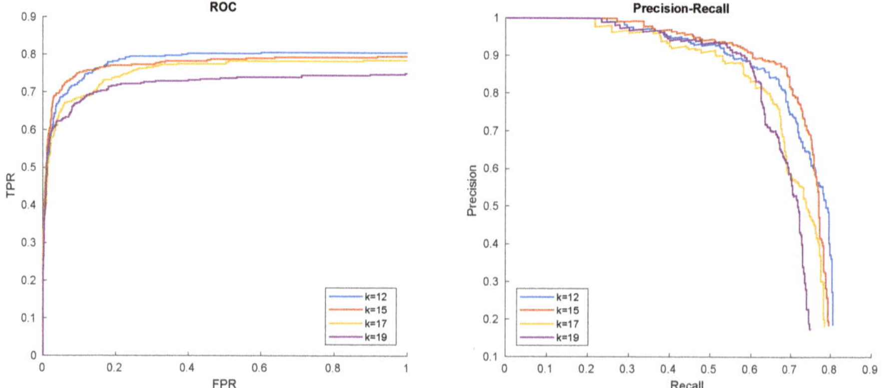

Figure 9. Hat in '00012' detection results with respect to number of the requested cascade stages.

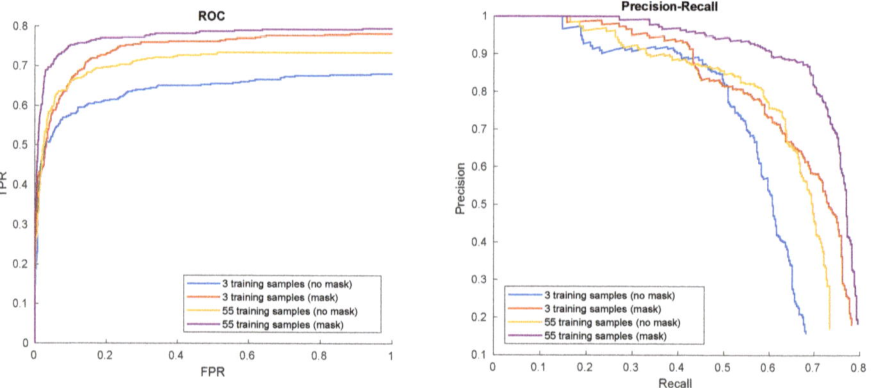

Figure 10. Hat in '00012' detection results for different training data collection methods.

3.2.3. Application of the GrabCut Algorithm

In this paragraph, we evaluate the hybrid method for foreground/background segmentation utilizing background subtractor and GrabCut that was described in Section 2.2.3. It is evident that for single training images, the proposed segmentation method is reduced to pure GrabCut, and as such GrabCut becomes the only segmentation option, which should be beneficial in most cases (see: discussion on masks in the previous paragraph). However, the more interesting question is if GrabCut can effectively cooperate with background subtractor and enhance its results. Intuitively, it should be the case since both methods operate on different principles.

Taking a quick glance at Figure 11 shows us that this is actually the case, so the GrabCut building on the initial foreground/background segmentation can effectively smooth out imperfections in the initial estimation. In order to evaluate it quantitatively, we propose an experiment where the detection results are evaluated by a GrabCut module turned on and off for different values of background cut-off threshold c_{thr}). The results are summarized in the Tables 1 and 2, and Figure 12.

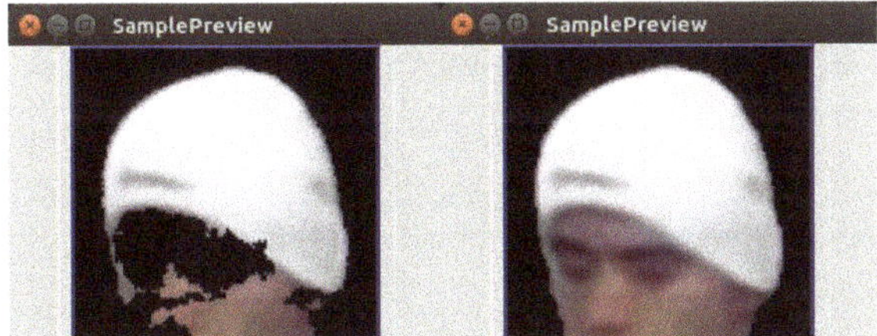

Figure 11. Foreground/background segmentation results using only background subtraction (on the left) and background subtraction + GrabCut (on the right) for background cut-off threshold $c_{thr} = 600$.

Table 1. Detection results with application of GrabCut.

	$c_{thr} = 30$	$c_{thr} = 130$	$c_{thr} = 300$	$c_{thr} = 600$
AUC avg.	0.76 ± 0.02	0.77 ± 0.01	0.78 ± 0.03	0.76 ± 0.07
AUC st.dev.	0.02	0.01	0.02	0.06
AVGPR avg.	0.69 ± 0.03	0.73 ± 0.01	0.73 ± 0.02	0.72 ± 0.07
AVGPR st.dev.	0.02	0.00	0.02	0.06

Table 2. Detection results without application of GrabCut.

	$c_{thr} = 30$	$c_{thr} = 130$	$c_{thr} = 300$	$c_{thr} = 600$
AUC avg.	0.73 ± 0.08	0.74 ± 0.05	0.71 ± 0.04	0.64 ± 0.07
AUC st.dev.	0.07	0.04	0.04	0.06
AVGPR avg.	0.65 ± 0.11	0.66 ± 0.05	0.62 ± 0.08	0.51 ± 0.08
AVGPR st.dev.	0.10	0.05	0.07	0.07

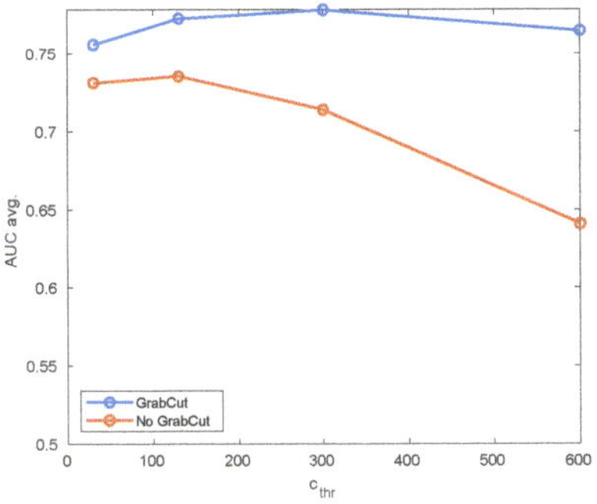

Figure 12. Comparison of detector accuracy with GrabCut turned on or off for different thresholds of background cut-off (pattern hat).

The results can be summarized two-fold. Firstly, the compound algorithm (mixing background subtraction and GrabCut) offers better accuracy for all thresholds analyzed, especially for larger background cut-off thresholds. However, since GrabCut results must always contain the area initialized by B-S, this does not need to hold for extremely small thresholds (where the initial area is already very large).

Secondly, the performance of the pure background subtractor strongly depends on the cut-off threshold, while GrabCut seems to loosen this correlation. This may enable (in the future version of the system) to remove the threshold c_{thr} altogether and to fix some reasonable default.

Finally we decided to evaluate the performance of 3 methods: pure background subtractor, the mixture of a background subtractor with GrabCut, and a successful contemporary tracker and mask generator [45]. The latter method relies on Siamese Network (a popular method for tracking applications), extended with the ability to generate binary foreground/background masks for the tracked object. In our experiments we used a network trained on a DAVIS dataset [46]. Comparison of three methods is given in Table 3.

Table 3. Comparison of detection results using Background Subtraction + GrabCut (BS + GC), pure Background Subtraction (BS), and SiamMask [45] method for collecting training samples.

	BS + GC: $c_{thr} = 300$	BS: $c_{thr} = 130$	**SiamMask**
AUC avg.	0.78 ± 0.03	0.74 ± 0.05	0.66 ± 0.03
AUC st.dev.	0.02	0.04	0.03
AVGPR avg.	0.73 ± 0.02	0.66 ± 0.05	0.54 ± 0.05
AVGPR st.dev.	0.02	0.05	0.04

Although the results are not fully comparable (e.g., we had to approximate more complex ROI returned by the SiamMask algorithm with a rectangular ROI for further utilization in training), we can observe that the performance of SiamMask is somewhat worse than the performance of the other two methods. Although the tracking process is fine, it can be observed that the generated mask does not correspond well to ground truth data. This effect could be easily attributed to a natural bias of the method towards specific classes of objects that it was trained on. However, if the method was trained on the data from a domain similar to our test data, the results could be much better. This is an open area for further research.

3.2.4. Synthetic Generalization of Training Data

In these experiments, different measures and intensity of synthetic samples augmentation were evaluated. The results are given in Figures 13 and 14. The results show that moderate geometric, as well as contrast and sharpness generalization, provides the best results. However, the selection of appropriate parameters is object and sequence-specific. It may be observed that near-flat surfaces (like logo) benefits from aggressive geometric distortions (i.e., larger rotation angles). In addition, the reduction of sharpness proved to work best for computer-graphics-generated patterns.

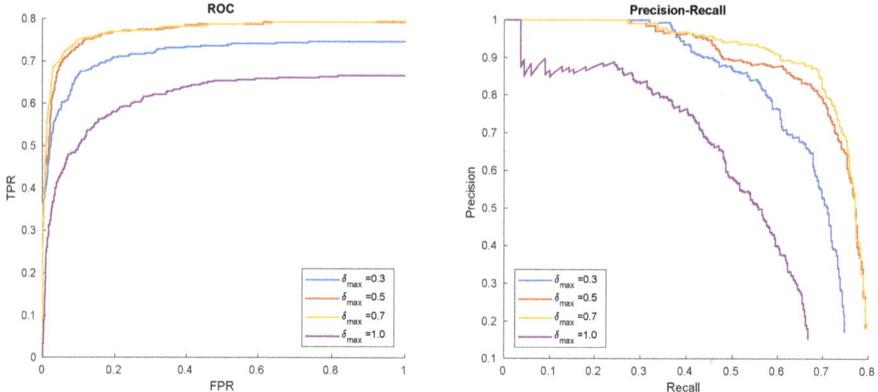

Figure 13. Hat in '00012' detection results for different levels of geometric synthesis.

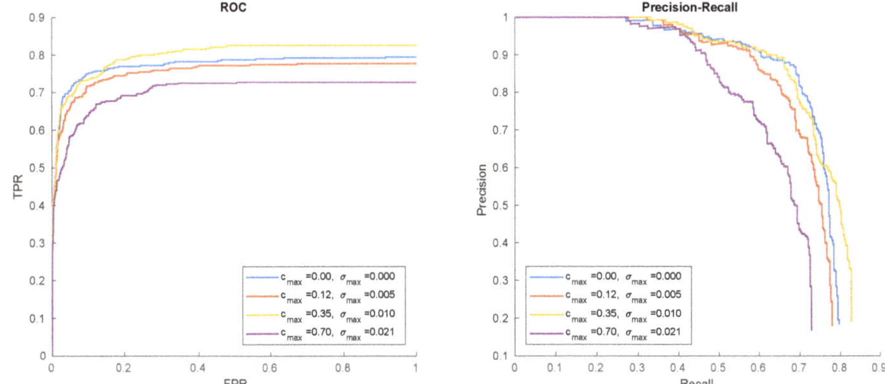

Figure 14. Hat in '00012' detection results for different contrast and sharpness synthesis levels.

3.2.5. Selection of Training Data Size

The selection of the appropriate size of the training dataset is crucial for detection accuracy as well as time performance. It applies especially to the 2-nd stage of detection since this is the layer that performs the final evaluation of samples. To tune our system with this respect, we performed an analysis using one pattern: hat. There was evaluated a HOG+SVM classifier. The input data was augmented basing on data resulting from foreground–background segmentation from Section 2.2.3 ($c_{thr} = 130$). The evaluated input image size were 24×24 and 32×32. The tested number of training images was 300, 600, 1200, and 2400. In all cases the SVM hyperparameters were trained using a 10-fold cross-validation procedure. The results obtained are given in the Tables 4 and 5. Visual comparison of accuracies is given in Figure 15.

Table 4. Detector performance for 24×24 HOG features and different number of training samples per class.

	HOG24(300)	HOG24(600)	HOG24(1200)	HOG24(2400)
AUC avg.	0.77 ± 0.05	0.77 ± 0.01	0.80 ± 0.01	0.81 ± 0.01
AUC st.dev.	0.04	0.01	0.01	0.01
AVGPR avg.	0.73 ± 0.06	0.73 ± 0.01	0.75 ± 0.01	0.77 ± 0.02
AVGPR st.dev.	0.05	0.00	0.01	0.01
training (sec.)	33	49	92	170
detection(msec/fr.)	58	63	65	63

Table 5. Detector performance for 32×32 HOG features and different number of training samples per class.

	HOG32(300)	HOG32(600)	HOG32(1200)	HOG32(2400)
AUC avg.	0.69 ± 0.06	0.73 ± 0.05	0.80 ± 0.02	0.80 ± 0.01
AUC st.dev.	0.05	0.04	0.02	0.01
AVGPR avg.	0.59 ± 0.12	0.66 ± 0.05	0.76 ± 0.02	0.77 ± 0.00
AVGPR st.dev.	0.10	0.04	0.02	0.00
training (sec.)	44	68	171	353
detection (msec/fr.)	70	77	107	100

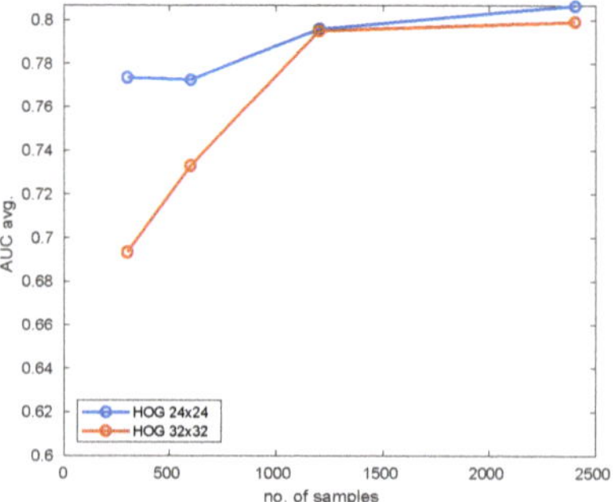

Figure 15. Comparison of detector performance using 24 × 24 and 32 × 32 HOG features for different sizes of the training set and hat pattern.

As can be deduced from the figure, the 24 × 24 HOG descriptor performs significantly better for smaller dataset sizes (like 300/300). This is quite obvious, taking into account the difference in the size of space parameters. For 1200/1200 samples and above, both resolutions offer similar performance. The accuracy (measured by AUC) saturates at about 1200/1200 samples, so for this pattern there is little justification to use larger training sets. An attractive choice of parameters seems to be 24 × 24 HOG descriptor with 600/600 training samples. In this experiment this solution offered almost top accuracy accompanied by very quick training (only 49 s.) and detection (63 ms. per frame). The 24 × 24 HOG descriptor with 300/300 samples may also look attractive, but it more depends on the actual data selection (high standard deviation of results for different runs), and the performance is expected to differ more between consecutive runs.

It should be noted that the number of required samples strongly depends on the complexity of the pattern and its uniqueness, so it is tough to establish the value in general. Therefore this discussion should be treated rather as a proposition of some sane default values for the dataset size parameter in the system.

3.2.6. VGG16 as 2-nd Stage Classifier

In this part of the experiments, the VGG16 network was utilized as the last stage classifier. After pre-detection by the cascade classifier, the results were fed to the VGG. The classifier, pre-trained on the ImageNet dataset, was immediately trained on-line on the set of gathered training samples. The number of training samples was set to 2400 for each class. There were evaluated four variants of the classifier: accepting images of resolution 32 × 32, 64 × 64, 128 × 128, and 224 × 224. The selection of a particular image size induced different numbers of layers. The training set of images was divided into training and validation using proportion 5:1. Only the fully-connected layers in the network were trained. The network was trained for a maximum of 6 epochs (for 224 × 224 resolution) or 10 epochs for smaller resolutions using Stochastic Gradient Descent or up to the moment when the validation error started to increase (using early-stopping principle). Practically in all cases, the network was able to train up to 100% of accuracy on the validation set.

The results obtained are given in the Table 6. The overall performance of the detector using VGG16 network can be regarded as acceptable, with the value of AUC greater than 0.7. It should be noted that the AUC value is quite similar for different network configurations, but the differences between them manifest in the *average precision-recall* measure. As could be expected, the full-width version, accepting images 224 × 224, where the output of the convolution part is 7 × 7 × 512 (as in the original paper), offers the best performance. The worst performance is offered by the network version accepting input 32 × 32, where the output of the convolution part is 1 × 1 × 512. The performance hit is probably due to getting rid of the information of feature location (the output is only 1 pixel wide). In such a configuration, the network is able to take into account the *presence* of features but not their position.

The full-width version of the network additionally has a prohibitive computational complexity of detection (even on a good GPU) consuming almost 1.5 s per each frame (compare it to classifier HOG+SVM classifier also given in the Table 6). The overall accuracy for VGG16 turns out to be worse than HOG+SVM. The VGG network suffers from the hidden overfitting problem (not manifesting itself in validation set error) due to incomplete training data. On the other hand, it seems that SVM+HOG classifier benefits more from the fully generic HOG feature extractor. The VGG16 network should perform better after some retraining of the specialized convolutional layer using enough quality data. Our unreported experiments show that the training data available in our problem is not sufficient for this purpose.

Table 6. Performance of VGG16 network as a second stage classifier. Last column contains HOG results for reference.

	VGG16(32)	**VGG16(64)**	**VGG16(128)**	**VGG16(224)**	**HOG(2400)**
AUC avg.	0.69 ± 0.05	0.72 ± 0.01	0.72 ± 0.02	0.72 ± 0.05	0.80 ± 0.01
AUC st.dev.	0.06	0.01	0.01	0.05	0.01
AVGPR avg.	0.38 ± 0.07	0.45 ± 0.03	0.54 ± 0.04	0.61 ± 0.06	0.77 ± 0.00
AVGPR st.dev.	0.08	0.03	0.03	0.06	0.00
training (sec.)	60	74	140	223	353
detection(msec./fr.)	118	225	466	1457	100

3.2.7. Application of Faster-RCNN to Preprocessed Data

As the final experiment involving neural networks, we evaluated the Faster-RCNN [29] detector on our test sequence '00012' using as the training input the data collected and preprocessed using methods described in this paper. In order to prepare data acceptable by RCNN, augmented training images for hat pattern and their masks were superimposed on the negative training images in a regular grid-like fashion using masks for effective blending. The ROIs of input images were defined accordingly.

The original Faster-RCNN network was then re-trained with respect to the new data for 3000 training loops (which took 15 minutes in the Google Colab [47] environment). Then, the detection was performed, which took about 2 s per frame.

The Faster-RCNN network was able to achieve AUC = 0.67 and AVGPR = 0.56. The results are comparable to the ones obtained using VGG16 classifier and worse than the pair HOG+SVM.

3.2.8. Detection of Various Patterns

In the last of our preliminary experiments we evaluated how the detector handles different types of patterns. Therefore, the pattern logo was trained on a single training example with no mask, the pattern shirt was trained on a sequence of 30 samples without a mask and the pattern helmet was trained on 41 samples also without a mask using HOG+SVM for the second stage classifier. The results are given in Figure 16.

The relatively worse performance for the shirt pattern is mainly due to numerous occlusions. Even in the case of the 'shirt' pattern we still have about 90% of successful hits for recall rates of 0.3. For best patterns, such as helmet, we have about 50% of positive examples with still 0 false positives!

In the course of the experiments, it was observed that motion blur (inherent or originating from de-interlacing) is the most destructive type of noise regarding both the training and detection phase. In addition, due to quite severe subsampling of the pattern (down to 24×24), the detector may suffer from problems in distinguishing between patterns differing only in small details. On the other hand, due to this property, the detector should well handle also small patterns—only slightly bigger than the nominal 24×24 pattern size.

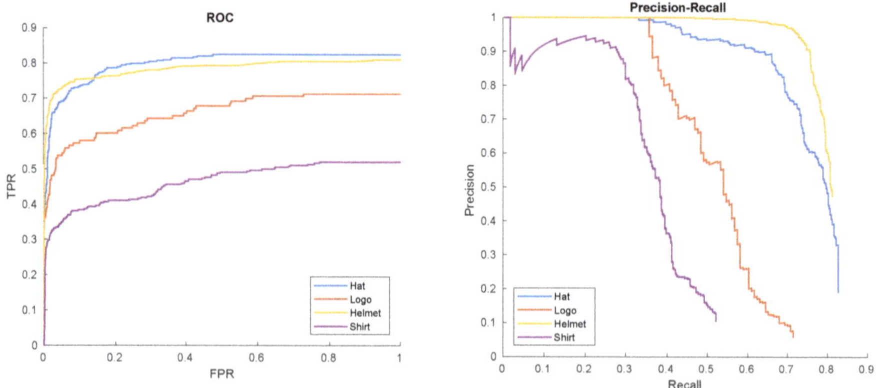

Figure 16. Receiver Operator Characteristic (ROC) and Precision-Recall (PR) curves of hat, logo, helmet, and shirt detections in '00012' sequence.

3.3. Large-Scale Experiments

Tests of the presented algorithm were conducted on a dataset containing 11 recordings, with nearly 30 thousand frames in total, with full HD resolution. Three patterns were created (Figure 17), and all sequences were carefully labeled by hand to create ground-truth data. All patterns were created based on a single frame (one positive sample). As training data, a high quality still picture was used, with resolution scaled down to full HD.

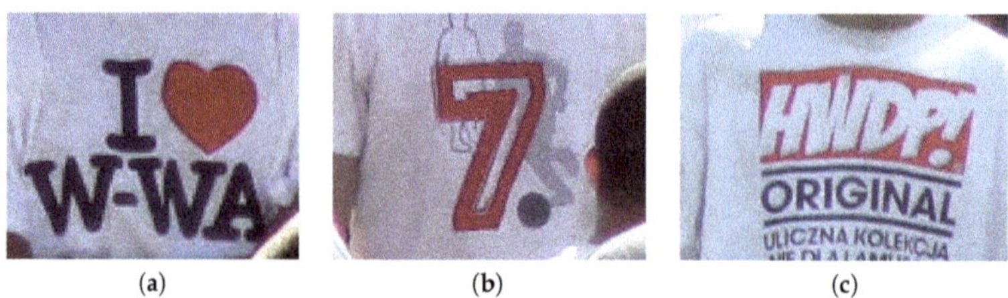

(a) (b) (c)

Figure 17. Three tested t-shirt logo patterns: (a) pattern P1, (b) pattern P2, (c) pattern P3.

Results of the experiments (ROC curve) for the selected pattern P1 are presented in Figure 18a. EER is similar for all patterns P1–P3, and is equal to 25.3%, 28.3%, and 28.0% for each pattern, respectively. Accumulated EER equals to 27.4%. Obtained results resemble those from smaller dataset. Even though the training sample and query images were obtained from different devices and had a different quality, the algorithm gave satisfactory results.

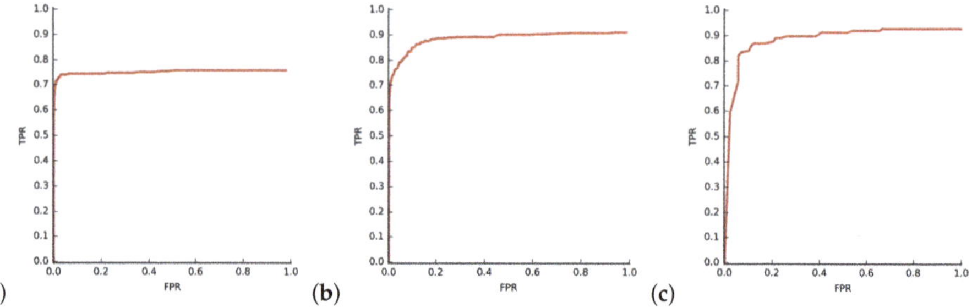

(a) (b) (c)

Figure 18. (a) ROC curve for the pattern P1. (b) Accumulated ROC curve for 5-elements sequence analysis. (c) ROC curve for cereal_1 object in desk_3 sequence.

The final addition to the testing scenario was the utilization of short sequences. For every short time window, from all the results, only the one with the best response was taken as a final detection and passed to further processing. Accumulated results for the windows of length 5 is presented in Figure 18b. EER for them are, respectively: 27.4%, 15.1%, and 14.3%. It was observed that the longer the window the smaller is the quality gain.

From the point of view of a surveillance system operator, the system does not need to correctly annotate each and every frame in the sequence. In reality, what the operator expects is at least a single detection per continuous segment of appearances of the object. If the system can issue an alert, that the object is present in a single frame from such sub-sequence, the operator can perform further investigation manually.

To evaluate this, the experiments were performed concerning the detection of two selected patterns in seven different sequences. The detection was evaluated on per-segment and not per-frame basis. The ground (GT) for the movie sequence was composed of positive and negative segments. A positive segment was the continuous segment of positive samples; negative segment was a continuous segment of negative samples. Similar positive and negative segments were extracted from the detection result by applying hysteresis thresholds to the detector response. If the positively evaluated segment overlapped the true segment, it was considered ((H)it). In other case, it was considered ((M)iss). If the negatively evaluated segment overlapped the true positive segment, it was a ((S)kip).

Results are presented in Table 7. The Fragmentation describes the ratio of good Hits (multiple hits in a single segment are possible) to the number of segments in the ground truth data (GT). For each sequence, we had the information about the total number of times the target appears (number of continuous time segments, GT).

Table 7. Continuous sequence detection results.

Sequence	GT	Hit	Skip	Miss	Fragmentation
146	1	3	0	0	3
154	3	3	0	2	1
162 (A)	1	1	0	0	1
162 (B)	8	14	0	3	1.75
163	2	2	0	0	1
164	3	5	0	0	1.67
168	1	1	0	3	1
total	19	29	0	8	1.53

The most important conclusion here is that every continuous segment with the query pattern was hit at least once. The total fragmentation of 1.53 means, that the operator, on average, will be notified only slightly more often than he or she expects. Number of Misses, although not very small (0.42 of expected detections), is balanced by no Skips, which is the most important factor in the surveillance systems.

Additional experiments were conducted using one of the publicly available datasets. As our system relies on the masks, and one of the key concepts is to model the 3D object based on the small number of 2D views, we decided to use the Washington RGB-D dataset [48]. This dataset contains training sequences with a single object with labels (masks) and testing sequences with complex scenes containing multiple objects. In the experiments, the operator used the training sequence to pick the object of interest, and our system continued with the creation of the masks from the provided short sequence. Masks available in the dataset were used as a measure of the accuracy impact of automatic mask generation. After creating the model from a few images object was searched in the complex scenes. Figure 18c presents sample results obtained for the cereal_1 object in desk_3 sequence. The model was created using only 7 views of the object in this case, without using the GrabCut for mask creation.

4. Conclusions

In this paper, we presented a solution that can support the work of surveillance system operators. The system proved to positively address difficult task requirements concerning small training data sets, quick learning, and fast and reliable detection. An attractive training/detection speed and recognition rate trade-off was obtained by the application of a 2-layer cascade/SVM classifier. Additionally, a combination of cascade classifier and CNN was evaluated in the paper. The system proposed can learn from a single training sample but also can collect samples from short image sequences with only small user supervision in order to obtain rich training data. The addition of background subtraction and GrabCut for pattern generation made the process even more reliable. The performance of the system varies depending on the type and quality of training/test data, but we argue that, on average, results are satisfactory, and even not-the-best results provide sufficient information to be useful in practical surveillance scenarios.

Author Contributions: Conceptualization, A.W. and W.K.; methodology, A.W. and M.S.; software, A.W., M.S., and W.K.; validation, A.W., W.K., and M.S.; formal analysis, A.W. and W.K.; investigation, A.W., M.S., and W.K.; resources, W.K.; data curation, M.S. and A.W.; writing—original draft preparation, A.W. and M.S.; writing—review and editing, A.W., M.S., and W.K.; visualization, A.W. and M.S.; supervision, W.K.; project administration, W.K.; funding acquisition, W.K. All authors have read and agreed to the published version of the manuscript.

Funding: Work done as part of the CYBERSECIDENT/369195/I/NCBR/2017 project supported by the National Centre of Research and Development in the frame of CyberSecIdent Programme.

Conflicts of Interest: The authors declare no conflict of interest. The funders had no role in the design of the study; in the collection, analyses, or interpretation of data; in the writing of the manuscript, or in the decision to publish the results.

Abbreviations

The following abbreviations are used in this manuscript:

AUC	Area Under Curve (ROC)
AVGPR	Average Precision-Recall
CC	Cascade Classifier
CNN	Convolutional Neural Network
CSK	Circulant Structure of Kernels
EER	Equal Error Rate
FPR	False Positive Rate
GPU	Graphic Processing Unit
HD	High Definition
HOG	Histogram of Oriented Gradients
PR	Precision-Recal (curve)
P-REACT	Petty cRiminality diminution through sEarch and Analysis in multi-source video Capturing and archiving plaTform
RBF	Radial Basis Function
RGB-D	Red Green Blue-Depth
ROC	Receiver Operator Characteristics
ROI	Region of Interest
RPCA	Robust Principal Component Analysis
SIFT	Scale Invariant Feature Transform
SURF	Speeded-Up Robust Features
SVM	Support Vector Machine
TPR	True Positive Rate

References

1. Arraiza, J.; Aginako, N.; Kioumourtzis, G.; Leventakis, G.; Stavropoulos, G.; Tzovaras, D.; Zotos, N.; Sideris, A.; Charalambous, E.; Koutras, N. Fighting Volume Crime: An Intelligent, Scalable, and Low Cost Approach. In Proceedings of the 9th Summer Safety & Reliability Seminars, SSARS 2015, Gdansk/Sopot, Poland, 21–27 June 2015.

2. Blunsden, S.; Fisher, R. The BEHAVE video dataset: Ground truthed video for multi-person behavior classification. *Ann. BMVA* **2010**, *2010*, 1–11.

3. Awad, G.; Snoek, C.G.M.; Smeaton, A.F.; Quénot, G. TRECVid Semantic Indexing of Video: A 6-Year Retrospective. *ITE Trans. Media Technol. Appl.* **2016**, *4*, 187–208. [CrossRef]

4. Wilkowski, A.; Kasprzak, W.; Stefańczyk, M. Object detection in the police surveillance scenario. In Proceedings of the 2019 Federated Conference on Computer Science and Information Systems, Leipzig, Germany, 1–4 September 2019; Volume 18, pp. 363–372.

5. Redmon, J.; Divvala, S.K.; Girshick, R.B.; Farhadi, A. You Only Look Once: Unified, Real-Time Object Detection. In Proceedings of the IEEE Conference on Computer Vision and Pattern Recognition, Boston, MA, USA, 7–12 June 2015.

6. Zeng, D.; Zhao, F.; Ge, S.; Shen, W. Fast cascade face detection with pyramid network. *Pattern Recognit. Lett.* **2019**, *119*, 180–186. [CrossRef]

7. Woźniak, M.; Połap, D. Object detection and recognition via clustered features. *Neurocomputing* **2018**, *320*, 76–84. [CrossRef]

8. Yang, L.; Jin, R. Distance metric learning: A comprehensive survey. *Mich. State Univ.* **2006**, *2*, 4.

9. Sohn, K. Improved deep metric learning with multi-class N-pair loss objective. In *Advances in Neural Information Processing Systems*; Curran Associates Inc.: Red Hook, NY, USA, 2016; pp. 1857–1865.

10. Sung, F.; Yang, Y.; Zhang, L.; Xiang, T.; Torr, P.H.; Hospedales, T.M. Learning to compare: Relation network for few-shot learning. In Proceedings of the IEEE Conference on Computer Vision and Pattern Recognition, Salt Lake City, UT, USA, 18–23 June 2018; pp. 1199–1208.

11. Wang, J.; Zhou, F.; Wen, S.; Liu, X.; Lin, Y. Deep metric learning with angular loss. In Proceedings of the IEEE International Conference on Computer Vision, Venice, Italy, 22–29 October 2017; pp. 2593–2601.

12. Zhou, F.; Wu, B.; Li, Z. Deep meta-learning: Learning to learn in the concept space. *arXiv* **2018**, arXiv:1802.03596.

13. Wang, Y.X.; Girshick, R.; Hebert, M.; Hariharan, B. Low-shot learning from imaginary data. In Proceedings of the IEEE Conference on Computer Vision and Pattern Recognition, Salt Lake City, UT, USA, 18–23 June 2018; pp. 7278–7286.

14. Hariharan, B.; Girshick, R. Low-shot visual recognition by shrinking and hallucinating features. In Proceedings of the IEEE International Conference on Computer Vision, Venice, Italy, 22–29 October 2017; pp. 3018–3027.

15. Chiatti, A.; Bardaro, G.; Bastianelli, E.; Tiddi, I.; Mitra, P.; Motta, E. Task-agnostic object recognition for mobile robots through few-shot image matching. *Electronics* **2020**, *9*, 380. [CrossRef]

16. Chen, H.; Wang, Y.; Wang, G.; Qiao, Y. Lstd: A low-shot transfer detector for object detection. In Proceedings of the Thirty-Second AAAI Conference on Artificial Intelligence, New Orleans, LA, USA, 2–7 February 2018.

17. Dong, X.; Zheng, L.; Ma, F.; Yang, Y.; Meng, D. Few-example object detection with model communication. *IEEE Trans. Pattern Anal. Mach. Intell.* **2018**, *41*, 1641–1654. [CrossRef] [PubMed]

18. Shao, Q.; Qi, J.; Ma, J.; Fang, Y.; Wang, W.; Hu, J. Object Detection-Based One-Shot Imitation Learning with an RGB-D Camera. *Appl. Sci.* **2020**, *10*, 803. [CrossRef]

19. Karlinsky, L.; Shtok, J.; Harary, S.; Schwartz, E.; Aides, A.; Feris, R.; Giryes, R.; Bronstein, A.M. RepMet: Representative-based metric learning for classification and few-shot object detection. In Proceedings of the IEEE Conference on Computer Vision and Pattern Recognition, Long Beach, CA, USA, 16–20 June 2019; pp. 5197–5206.

20. Wang, Y.; Yao, Q.; Kwok, J.; Ni, L.M. Generalizing from a Few Examples: A Survey on Few-Shot Learning. *arXiv* **2019**, arXiv:cs.LG/1904.05046.

21. Abramson, Y.; Freund, Y. *Active Learning for Visual Object Detection*; Technical Report; UCSD: San Diego, CA, USA, 2006.

22. Abramson, Y.; Freund, Y. SEmi-automatic VIsual LEarning (SEVILLE): Tutorial on active learning for visual object recognition. In Proceedings of the IEEE Conference on Computer Vision and Pattern Recognition, San Diego, CA, USA, 25 July 2005.

23. Sivic, J.; Zisserman, A. *Video Google: A Text Retrieval Approach to Object Matching in Videos*; IEEE Computer Society: Washington, DC, USA, 2003; Volume 2, p. 1470.

24. Rother, C.; Kolmogorov, V.; Blake, A. *"GrabCut": Interactive Foreground Extraction Using Iterated Graph Cuts*; ACM SIGGRAPH 2004 Papers; Association for Computing Machinery: New York, NY, USA, 2004; pp. 309–314.

25. Kalal, Z.; Mikolajczyk, K.; Matas, J. Tracking-Learning-Detection. *IEEE Trans. Pattern Anal. Mach. Intell.* **2012**, *34*, 1409–1422. [CrossRef]

26. Andriluka, M.; Roth, S.; Schiele, B. People-tracking-by-detection and people-detection-by-tracking. In Proceedings of the 2008 IEEE Conference on Computer Vision and Pattern Recognition, Anchorage, AK, USA, 24–26 June 2008; pp. 1–8.

27. Feichtenhofer, C.; Pinz, A.; Zisserman, A. Detect to Track and Track to Detect. In Proceedings of the IEEE International Conference on Computer Vision, Venice, Italy, 22–29 October 2017; pp. 3057–3065.

28. Kang, K.; Ouyang, W.; Li, H.; Wang, X. Object Detection from Video Tubelets with Convolutional Neural Networks. In Proceedings of the 2016 IEEE Conference on Computer Vision and Pattern Recognition, Las Vegas, NV, USA, 26–30 June 2016; pp. 817–825.

29. Ren, S.; He, K.; Girshick, R.B.; Sun, J. Faster R-CNN: Towards Real-Time Object Detection with Region Proposal Networks. *IEEE Trans. Pattern Anal. Mach. Intell.* **2017** *39*, 1137–1149. [CrossRef]

30. Henriques, J.F.; Caseiro, R.; Martins, P.; Batista, J. Exploiting the Circulant Structure of Tracking-by-Detection with Kernels. In Proceedings of the 12th European Conference on Computer Vision, Florence, Italy, 7–13 October 2012.

31. Danelljan, M.; Khan, F.S.; Felsberg, M.; Van de Weijer, J. Adaptive Color Attributes for Real-Time Visual Tracking. In Proceedings of the 2014 IEEE Conference on Computer Vision and Pattern Recognition, Columbus, OH, USA, 23–28 June 2014; pp. 1090–1097.

32. Zivkovic, Z. Improved adaptive Gaussian mixture model for background subtraction. In Proceedings of the 17th International Conference on Pattern Recognition, Cambridge, UK, 23–26 August 2004; Volume 2, pp. 28–31.

33. Chen, B.; Shi, L.; Ke, X. A Robust Moving Object Detection in Multi-Scenario Big Data for Video Surveillance. *IEEE Trans. Circuits Syst. Video Technol.* **2019**, *29*, 982–995. [CrossRef]

34. Cao, X.; Yang, L.; Guo, X. Total Variation Regularized RPCA for Irregularly Moving Object Detection Under Dynamic Background. *IEEE Trans. Cybern.* **2016**, *46*, 1014–1027. [CrossRef] [PubMed]

35. Itseez. Open Source Computer Vision Library. 2015. Available online: https://github.com/itseez/opencv (accessed on 7 May 2020).

36. Posłuszny, T.; Putz, B. An Improved Extraction Process of Moving Objects' Silhouettes in Video Sequences. In *Advanced Mechatronics Solutions*; Jabłoński, R., Brezina, T., Eds.; Springer International Publishing: Cham, Switzerland, 2016; pp. 57–65.

37. Bay, H.; Tuytelaars, T.; Van Gool, L., SURF: Speeded Up Robust Features. In Proceedings of the Computer Vision–ECCV 2006: 9th European Conference on Computer Vision, Graz, Austria, 7–13 May 2006.

38. Lowe, D.G. Distinctive Image Features from Scale-Invariant Keypoints. *Int. J. Comput. Vis.* **2004**, *60*, 91–110. [CrossRef]

39. Rousseeuw, P.J.; Leroy, A.M., Robust Regression and Outlier Detection. In *Robust Regression and Outlier Detection*; John Wiley & Sons, Inc.: Hoboken, NJ, USA, 2005; pp. 197–215.

40. Dalal, N.; Triggs, B. Histograms of oriented gradients for human detection. In Proceedings of the 2005 IEEE Computer Society Conference on Computer Vision and Pattern Recognition (CVPR'05), San Diego, CA, USA, 20–26 June 2005; Volume 1, pp. 886–893.

41. Hu, Q.; Paisitkriangkrai, S.; Shen, C.; van den Hengel, A.; Porikli, F. Fast Detection of Multiple Objects in Traffic Scenes With a Common Detection Framework. *IEEE Trans. Intell. Transp. Syst.* **2016**, *17*, 1002–1014. [CrossRef]

42. Simonyan, K.; Zisserman, A. Very Deep Convolutional Networks for Large-Scale Image Recognition. *arXiv* **2014**, arXiv:1409.1556.

43. Deng, J.; Dong, W.; Socher, R.; Li, L.J.; Li, K.; Fei-Fei, L. ImageNet: A Large-Scale Hierarchical Image Database. In Proceedings of the 2009 IEEE Conference on Computer Vision and Pattern Recognition, Miami Beach, FL, USA, 20–25 June 2009.

44. Tan, C.; Sun, F.; Kong, T.; Zhang, W.; Yang, C.; Liu, C. A Survey on Deep Transfer Learning. *arXiv* **2018**, arXiv:cs.LG/1808.01974.

45. Wang, Q.; Zhang, L.; Bertinetto, L.; Hu, W.; Torr, P.H. Fast online object tracking and segmentation: A unifying approach. In Proceedings of the IEEE Conference on Computer Vision and Pattern Recognition, Long Beach, CA, USA, 16–20 June 2019; pp. 1328–1338.
46. Perazzi, F.; Pont-Tuset, J.; McWilliams, B.; Van Gool, L.; Gross, M.; Sorkine-Hornung, A. A Benchmark Dataset and Evaluation Methodology for Video Object Segmentation. In Proceedings of the Computer Vision and Pattern Recognition, Las Vegas, NV, USA, 26–30 June 2016.
47. Bisong, E. Google Colaboratory. In *Building Machine Learning and Deep Learning Models on Google Cloud Platform*; Apress: Berkeley, CA, USA 2019; pp. 59–64.
48. Lai, K.; Bo, L.; Ren, X.; Fox, D. A large-scale hierarchical multi-view RGB-D object dataset. In Proceedings of the 2011 IEEE International Conference on Robotics and Automation (ICRA), Shanghai, China, 9–13 May 2011; pp. 1817–1824.

Article

Active Player Detection in Handball Scenes Based on Activity Measures

Miran Pobar and Marina Ivasic-Kos *

Department of Informatics University of Rijeka, Rijeka 51000, Croatia; mpobar@uniri.hr
* Correspondence: marinai@uniri.hr; Tel.: +385-51-584-700

Received: 31 January 2020; Accepted: 5 March 2020; Published: 8 March 2020

Abstract: In team sports training scenes, it is common to have many players on the court, each with his own ball performing different actions. Our goal is to detect all players in the handball court and determine the most active player who performs the given handball technique. This is a very challenging task, for which, apart from an accurate object detector, which is able to deal with complex cluttered scenes, additional information is needed to determine the active player. We propose an active player detection method that combines the Yolo object detector, activity measures, and tracking methods to detect and track active players in time. Different ways of computing player activity were considered and three activity measures are proposed based on optical flow, spatiotemporal interest points, and convolutional neural networks. For tracking, we consider the use of the Hungarian assignment algorithm and the more complex Deep SORT tracker that uses additional visual appearance features to assist the assignment process. We have proposed the evaluation measure to evaluate the performance of the proposed active player detection method. The method is successfully tested on a custom handball video dataset that was acquired in the wild and on basketball video sequences. The results are commented on and some of the typical cases and issues are shown.

Keywords: object detector; object tracking; activity measure; Yolo; deep sort; Hungarian algorithm; optical flows; spatiotemporal interest points; sports scene

---

1. Introduction

Many applications of computer vision, such as action recognition, security, surveillance, or image retrieval, depend on object detection in images or videos. The task of object detection is to find the instances of real-world objects, such as people, cars, or faces. Detecting an object includes determining the location of that object and predicting the class to which it belongs. Object location and object classification problems both pose individual challenges, and the choice of the right object detection method can depend on the problem that needs to be solved.

In our case, object detection, or specifically, leading player detection, can be very helpful for the recognition of actions in scenes from the handball domain. Handball is a team sport played in the hall on a handball court. Two teams consisting of a goalkeeper and six players participate in the game. The goal of the game is to score more points than the opposing team by throwing the ball into the goal of the opposing team, or by defending the goal, so that the opposing team does not score. The game is very fast and dynamic, with each team trying to arrange the action to defeat the defense and score a goal or to prevent the attack and defend the goal. All of the players can move all over the court except in the space 6 m in front of both goals where the goalkeepers are. The players move fast, change positions, and combine different techniques and handball actions, depending on their current role, which might be attack or defense. A player can shoot the ball toward the goal, dribble it, or pass it to a teammate during the attack, or block the ball or a player when playing defense.

Different techniques and actions that players perform, as well as the frequent changes of position and occlusions, cause the shapes and appearances of players to change significantly during the game, which makes the detection and tracking process more challenging. The indoor court environment with relatively reflective floors is often illuminated with harsh lighting so that the shadows and reflections of players are usually present on the recordings and, due to the speed of action performance, motion blur occurs, which further complicates the tracking and detection problems. Additionally, the positions of the player and the distance to the fixed-mounted camera are constantly changing, so that players can be close to the camera, covering most of the image, or at a distance from the camera and occupy just a few pixels. The ball is also an important object for recognizing the handball scene, but it is even more complex to detect and track than the player, because it is even smaller, moves quickly, in different directions, and it is often covered by the player's body or not visible because it is in the player's hands [1].

Handball rules are well-defined and prescribe permitted actions and techniques, but during training, when adopting and practicing different techniques or game elements, the coach changes the rules to maintain a high level of player activity and a lot of repetition. Players then preform several sets of exercises in a row to train and improve certain techniques or actions, and to learn to position themselves properly in situations that imitate real game situations. In training sessions, each player also usually has his own ball, to reduce the waiting time to practice techniques and to train handball skills as quickly as possible. For example, a shooting technique involves a player that shoots the ball towards the goal and the goalkeeper that moves to defend the goal. Other players are waiting their turn to perform this activity or they gather their balls on the playground and then run to their position in the queue, as shown in Figure 1.

Figure 1. A typical training situation. The goalkeeper and the player in the center are performing the current task, while the rest are waiting, returning to queue, collecting the ball, etc.

Although all of the players move and interact in a certain moment, not all players contribute to the action, such as passing the ball or shooting at the goal, which is the most relevant for some exercise or for the interpretation of a handball situation. Thus, the focus should be on the players that are responsible for the currently most relevant handball action, and whose label can then be used to mark the whole relevant part on the video. Here, these players are referred to as active players.

The goal of the proposed method is to determine which of the players that are present in the video sequences are active at a certain time of the game or at the training sessions, and who of them performs the requested action at a given moment.

Detecting the player who performs an action of interest is particularly demanding during training sessions, because there are more players on the field than allowed during a normal game, there is more than one ball, because each player has their own ball and certain actions take place in parallel to make the players get the most out of the time to adopt or perfect the technique. The given goal is too complex for simple methods for people segmentation, such as background subtraction or Chroma keying, so, for this purpose, we suggest an active player detection method that combines player detection and tracking with an activity measure to determine the leading player(s).

For the player detection, we suggest the use of deep convolution neural networks (CNNs) that have proven to be successful in classification and detection tasks in real-world images. In this paper, we will use Yolo V3 [2] which achieves great accuracy when detecting objects in the images.

We propose three activity measures to determine the level of activity of a particular player. The one, named DT+STIP, is based on density of spatiotemporal interest points (STIPs) that are detected within the area of a player, the second, DT+OF, is computationally simpler and based on optical flow (OF) motion field between consecutive video frames, and the third, DT+Y, uses convolutional neural network (CNN) for determining player activity and classifying active and inactive players.

We consider the use of Hungarian assignment algorithm based on the bounding box positions of detected players, and the more complex Deep SORT tracker [3] that uses additional visual appearance features to assist the assignment process to track the detected players along the time dimension.

We have defined the evaluation measure for evaluate the performance of the proposed active player detection method. The proposed method is tested on videos that were taken in the gym during several training sessions, in real, uncontrolled indoor environment with cluttered background, complex light conditions and shadows, and with multiple players, on average 12, who practice and repeat different handball actions and dynamically change positions and occlude each other.

The rest of the paper is organized, as follows: in Section 2 we will briefly review the object action localization, sports player tracking and salient object detection. In Section 3, the proposed method that combines the Yolo object detector, three proposed activity measures and tracking methods to determine the most active player in sports scenes is described. We have defined the evaluation measure and applied the proposed method on a custom dataset that consists of handball scenes recorded during handball training lessons. The performance evaluation of the proposed active player detection method and comparison of different method setups with respect to the three proposed activity measures and two player tracking methods are given and discussed in Section 4 The paper ends with a conclusion and the proposal for future research.

2. Related Work

The goal of detecting the active or leading player in a sports video is to find the sequence of bounding boxes that correspond to the player currently performing the most relevant action. The task can be separated into player detection and player tracking, which produces sequences of bounding boxes that correspond to individual players, combined with saliency detection to determine which of the sequences belong to the active player. Alternatively, the first part of the problem can be viewed as a task of action localization in general, where the goal is to find the sequences of bounding boxes in videos that likely contain actions, not necessarily by explicitly trying to detect players, but possibly using other cues, such as only the foreground motion.

Visual object tracking, including player tracking, is a very active research area that attracts dozens of papers to computer vision conferences annually [4], and numerous approaches have been developed for both the problem of multiple object tracking in general [5] and specifically for player tracking in sports videos. For the sports domain, player tracking is usually considered together with detection, e.g. in [6] for the case of hockey, [7] for handball, [8,9] for indoor, and [10–13] for outdoor soccer. These methods commonly attempt to exploit the specific knowledge regarding the domain of the given sport and about the video conditions, such as the distribution of the colors of the playing field or players to isolate potential areas that contain players [6–8,11–13], or the layout of the playing field to recover depth information [8–11]. Player detection ranges from template matching with handcrafted features [7] to machine learning approaches using e.g., a SVM classifier [13] or Adaboost [12], often with particle filter based tracking [6,9,13].

Recently, deep learning-based approaches to player detection, e.g. [14], are becoming increasingly attractive due to the increased detection performance as well as the reduced need for domain-specific knowledge. With the increased performance of convolutional neural network-based object detectors, relatively simple tracking-by-detection schemes that use the Hungarian algorithm to assign the detected bounding boxes to tracks only based on box dimensions have achieved good performance in the multiple object tracking task [15], such as tracking the leading player [16]. This work uses a similar approach.

Determining who is the leading player among the detected players is related to the problem of salient object detection, where the aim is to detect and segment the object in an image that visually stands out from its surroundings, which would be in the center of attention in the human system. In addition to finding naturally (visually) distinct regions (bottom-up-saliency), the criteria for saliency may be based on a specific task, expectations, prior knowledge, etc. (top-down saliency) [17]. In this sense, finding the active player among the detected players can be regarded as a top-down saliency task. Different cues have been used for saliency detection, such as those that are based on intensity or color contrast, informed by the sensitivity of human perception to color, or those that are based on location information, e.g., treating center region of an image as more likely to contain salient objects [18]. With the popularization of deep learning models for object detection, similar models are also increasingly used for salient object detection [19]. In the video context, motion-based saliency cues can be used, e.g., based on by directional masks applied to the image [18] or on optical flow estimation [20,21]. In this work, motion cues that are based on optical flow and STIPS are used to inform saliency of a player. A survey of salient object detection work can be found in [17,19].

In the area of action localization, the goal is to find the sequence of bounding boxes in a sequence of video frames that correspond to an action. Generating a set of candidate spatiotemporal regions, followed by the classification of the candidate region into an action class or the non-action class, is commonly utilized to perform this. Jain et al. [22] first segment a video sequence into supervoxels, and then use a notion of independent motion evidence to find the supervoxels where motion deviates from the background motion, which signals the potential area of interest where action is likely going to be detected. Similarly, in [23], a motion saliency measure that is based on optical flow is used along with image region proposal based on selective search to select candidate spatiotemporal regions in videos with actions and to disregard areas with little movement where no action is likely. Kläser et al. [24] explicitly split the action localization task into first detecting the persons in video while using a HOG [25] descriptor and a sliding window support vector machine classifier and then tracking the detected persons.

In [26], basketball players are detected and tracked in the video for the task of event detection, and the most relevant player is implicitly learned from the correspondence of video frames to the event label in a LSTM model trained to classify these events.

3. Proposed Method for Active Player Detection

The goal of the proposed method is to automatically determine the player or players in the scene that are responsible for the current action, among all of the players present in the scene. Figure 2 presents the overview of the proposed method for active player detection.

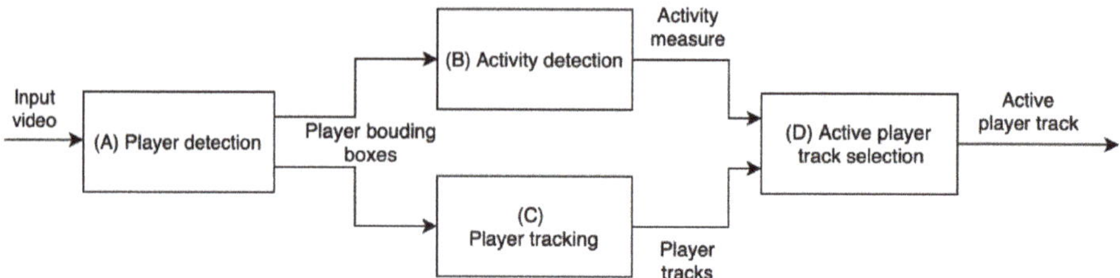

Figure 2. An overview of the active player track detection. Player detector provides bounding boxes for each input video frame (**A**). Player activity measure is computed for each detected player bounding box in each frame (**B**) and players are tracked across frames (**C**). Combining the track information from (**C**) and activity measure from (**B**), final decision about the active player track is made (**D**).

First, the players are detected in the input video while using a CNN-based object detector. Here, the YOLOv3 detector pre-trained on the MS-COCO dataset was used to mark the players on each frame of the video sequence with their bounding boxes (Figure 2A), using the detector's general person class. In this case, the object detector does not distinguish between active and inactive players, i.e. those who perform the action of interest from those who do something else. The activity measures for each detected player in a frame are computed based on motion features that are extracted within each player's bounding box in that video frame (Figure 2B). To determine a player's overall activity in a time period, it is necessary to know their activity measure in each frame, but to also uniquely identify that player throughout the period of interest, i.e., to track them through the sequence. While considering that the object detector only determines the class of a detected object (e.g., person), without distinguishing between objects of the same class, the tracking of player's identities between frames is completed outside the detection step (Figure 2C). Finally, the information about player activity in each frame and the track information is joined to determine the track of the most active or leading player in the whole video (Figure 2D).

Each of the steps is described in more detail below.

The Hungarian assignment algorithm was used to track the players, with a distance function based on properties of bounding boxes as the baseline method. As an alternative, the Deep SORT [3] method, which uses additional visual features to aid the tracking, is considered. Since Deep SORT gave better tracking results, only this method was used when testing the activity measures.

Different ways of computing player activity were considered and three activity measures are proposed. The first measure, named DT+OF, is based on the optical flow calculated within a player's bounding box and considers the velocity component of the motion, while the second measure, named DT+STIP, uses the density of the spatiotemporal interest points (STIPs) within the bounding boxes to determine the activity measure. The third method, called DT+Y, uses the YOLOv3 network to classify a player into an active or inactive class.

The input videos used as input data were recorded during the handball training involving 15 and more players. Each video sequence was selected and trimmed, so that it contains all the steps of the given action.

3.1. Player Detection

The goal of object detectors is to locate and classify objects on the scene. The detected objects are typically marked with a bounding box and labeled with corresponding class labels.

The current focus in object detection is on convolutional neural networks (CNNs) that were, at first, used for image classification, but have later been extended to be able to both detect and localize individual objects on the scene. This can be achieved by the independent processing of image parts that are isolated while using a sliding window that is moved over the image [27]. Multiple scales of windows are used in order to take into account different sizes of objects in the image. If the classifier recognizes the contents of the window as an object, the window is used as the bounding box of the object and the corresponding class label is used to label the window. After processing the entire image, the result is a set of bounding boxes and corresponding class labels. As a certain object can be partially or fully contained within multiple sliding windows, a large number of duplicate predictions can be generated, which can then be rejected while using non-maximum suppression and a confidence threshold. Since the sliding window approach essentially makes an image classification for each window position, a naive implementation can be much slower and computationally expensive when compared to simple image classification.

Here, the role of the object is to detect the players in each frame of the handball videos. Ideally, the detector should be as precise as possible, yet less computationally demanding, so that in can detect objects in real time. It has been previously shown [28,29] that Mask R-CNN [30] and Yolo have comparable performances of player detection in handball scenes, but, since Yolo was significantly faster and was successful for player detection in previous work [31], it is used for this experiment.

The YOLOv3 detector was used here using the pre-trained parameters on the COCO dataset with the standard Resnet-101-FPN network configuration, with no additional training.

YOLOv3 is the third iteration of the YOLO object detector [32], which performs a single pass through a neural network for both detecting the potential regions in the image where certain objects are present, and for classifying those regions into object classes. The object detection task is framed as a problem of regression from image pixels to objects' bounding box coordinates and associated class probabilities.

The architecture of the YOLOv3 detector is a convolutional network that consists of 53 convolutional layers of 3×3 and 1×1 filters with shortcut connections between layers (residual blocks) is used for feature extraction. The last convolutional layer of the network predicts the bounding boxes, the confidence scores, and the prediction class. YoloV3 predicts candidate bounding boxes at three different scales using a structure that is similar to feature pyramid networks, so three sets of boxes are predicted at each feature map cell for each scale, to improve the detection of objects that may appear at different sizes.

For the task of active player detection, only the bounding boxes and confidence values for the objects of the class "person" were used. An experimentally determined confidence threshold value of 0.55 was used to filter out the unwanted detections and obtain good balance of high detection and low false positives rates. Figure 3 shows an example of detection results for the "person" class.

Figure 3. Bounding boxes with confidence values as results of person detection with YOLOv3.

3.2. Player Tracking

The detections that were obtained with the Yolo detector are independent of each processed frame and, thus, contain no information about which bounding boxes correspond to the same objects in consecutive frames. Finding this correspondence between bounding boxes and player identities across frames is required to obtain the player trajectories. An additional post-processing step for player tracking is used for this purpose. Two methods are considered for this task. The first is a simpler approach utilizing the Hungarian algorithm, which assigns player bounding boxes detected in different frames into the player track while taking the dimensions and positions of the bounding box into account. In the second approach, the Deep SORT method [3] is used, which uses additional visual features to determine which detection in a frame matches a previously detected player.

3.2.1. Tracking Using the Hungarian Algorithm

In the first video frame, the player tracks are initialized, so that a track ID is assigned to each detected player bounding box. In the next frames, the individual detected bounding boxes are assigned to previously initialized tracks while using the Munkres' version of the Hungarian algorithm [33], whose objective is to minimize of the total cost of assigning the detections to tracks. Assigning a bounding box to a track has a cost value that depends on the scale difference and the relative position of the candidate bounding box and the box that was already assigned to the track in the previous frame to take simplified visual and spatial distances into account.

More precisely, a new bounding box B_b will be assigned to a player track $\mathcal{T}_{b-1}$ if it has the minimal cost computed as a sum of the linear combination of Euclidean distance between the centroids (C_{b-1}) of the last bounding box assigned to the player track $\mathcal{T}_{b-1}$ and the detected centroids (C_b) and the absolute area difference of the detected bounding boxes area (P_b) and the last bounding box area (P_{b-1}) assigned to the track $\mathcal{T}_{b-1}$ (1):

$$B_b(C_b, P_b) : \ b = \operatorname*{argmin}_i \sum_{i \in F}(wd_2(C_{b-1}, C_i) + (1-w)|P_i - P_{b-1}|); w \in [0,1]; \ d_2(C_{b-1}, C_i) < T. \quad (1)$$

where the bounding box $B_b(C_b, P_b)$ is represented with its centroids C_b and the area P_b, $T \in \mathbb{R}$ is a distance threshold between centroids in consecutive frames, w is the adjustable parameter that determines the relative influence of the displacement and the change of bounding boxes area in consecutive frames and F is the set of all the bounding boxes within the current frame.

The number of tracks can also change throughout the video, since players can enter or exit the camera field of view at any time. Additionally, the detection of players is not perfect, so some tracks should resume after a period where no detection was assigned. The distance threshold T is used to control the maximum allowed distance between a detected bounding box and a last bounding box assigned to the track. Any box whose distance to a track is greater than this threshold cannot be assigned to that track, even if it is the closest one to the track. If for $M \in \mathbb{N}$ consecutive frames no detections are assigned to a track, the track is considered to be completed and no further detections can be added to it. The values of M and T are experimentally determined and set to 20 and 100, respectively.

The Figure 4 shows an example of a sequence of frames from a video, with player bounding boxes being additionally associated with a track IDs across five different frames, such that, on different sequential frames, the same player has the same ID. It can be noted that the player with track ID 6 is not detected in several frames, but the tracking continues with the correct ID after he was detected again in the last shown frame.

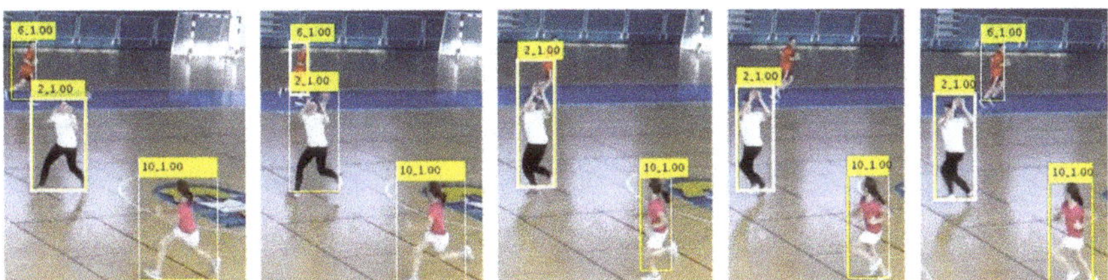

Figure 4. Player tracking across frames.

3.2.2. Tracking Algorithm—DeepSORT

DeepSORT [3] is a tracking algorithm that is based on the Hungarian algorithm that, in addition to the parameters of the detected bounding boxes, also considers the appearance information regarding the tracked objects to associate new detections with previously tracked objects. The appearance information should be useful, in particular, with re-identifying players that were occluded or have temporarily left the scene. As with the basic application of Hungarian algorithm, it is able to perform the tracking online, i.e. it does not need to process the whole video at once, but only needs to consider the information about the current and the previous frames to assign detections to tracks.

Like in the previous case, in the first frame, a unique track ID is assigned to each bounding box that represents a player and has a confidence value higher than a set threshold and the Hungarian algorithm is used to assign the new detections to existing tracks, so that the assignment cost function reaches the global minimum. The cost function entails the spatial (Mahalanobis) distance $d^{(1)}$ of a detected bounding box from the position that is predicted from its previously known position, and a visual distance $d^{(2)}$ that compares the appearance of a detected object with a history of appearances of the tracked objects. The cost of associating a detected bounding box B_b to player track $\mathcal{T}_{b-1}$ that ends with the bounding box B_{b-1} is given by the expression:

$$c_{b,b-1} = \lambda d^{(1)}(B_{b-1}, B_i) + (1-\lambda)d^{(2)}(B_{b-1}, B_b),\tag{2}$$

where λ is a settable parameter that determines the relative influence of the spatial and the visual distances $d^{(1)}$ and $d^{(2)}$.

The distance $d^{(1)}$ is given by the expression:

$$d^{(1)}(B_{b-1}, B_b) = \left(d_b - y_{b-1}\right)^T S_{b-1}^{-1}\left(d_b - y_{b-1}\right),\tag{3}$$

where y_{b-1} and S_{b-1} are the mean and the covariance matrix of the last bounding box observation assigned to player track $\mathcal{T}_{b-1}$, and d_b is the detected bounding box.

The visual distance $d^{(2)}$ is given by the expression:

$$d^{(2)}(B_{b-1}, B_b) = \min\left\{1 - r_b^T r_k^{(b-1)} \middle| r_k^{(b-1)} \in \mathcal{T}_{b-1}\right\},\tag{4}$$

where r_b is the appearance descriptor obtained from the part of the image within the detected bounding box B_b and $\mathcal{T}_{b-1}$ is the set of last 100 appearance descriptors $r_k^{(b-1)}$ that are associated with the $\mathcal{T}_{b-1}$ track.

The goal of the $d^{(2)}$ measure is to select the track where visually the most similar detection was previously found to the current detection. The similarity is computed via the cosine distance between the appearance descriptors of the B_b detected frame and $\mathcal{T}_{b-1}$ track. The 128-element descriptors are extracted while using a wide residual neural network with two convolutional layers and six residual blocks that was pre-trained on a person re-identification dataset of more than one-million images of 1261 pedestrians. The appearance descriptor vectors are normalized to fit within a unit hypersphere so that the cosine distance can be used [3].

When a detection cannot be assigned to any track because it is too far from any track according to the distance $d^{(1)}$, or is not visually similar enough to any previous detection according to the distance $d^{(2)}$, a new track is created. The maximum allowed $d^{(1)}$ and $d^{(2)}$ distances when an assignment is still possible is a settable parameter. A new track is also created whenever there are more detected players in a frame than there are existing tracks. A track is abandoned if no assignment has been made to a track for $M \in \mathbb{N}$ consecutive frames, and a new track ID will be generated if the same object re-appears later in the video.

3.3. Activity Detection

The object detector provides the location of all the players present in the scene in bounding boxes, but it has no information regarding the movements or activities of the players. Some information about movement can be obtained by first tracking the players and then by calculating the shift of the bounding box centroid across frames. However, it is expected that strong and sudden changes in both velocity and appearance that cannot be described by the shift of the bounding box centroid alone will characterize various actions in sports videos. Three activity measures are proposed (DT+OF, DT+STIPS, and DT+Y) to better capture the information about the activity of the players, as noted earlier.

Both the optical flow estimation method and STIPs detection by themselves only consider sections of video frames without taking the object identities into account. Thus, the information about player locations obtained with Yolo is combined with the activity measure obtained from either optical flow or STIPs within the player's bounding box to measure the individual player's activity.

3.3.1. Optical Flow-Based Activity Measure—DT+OF

In the proposed DT+OF activity measure, the optical flow estimate from time-varying image intensity is used to capture the information regarding speed and direction of movement of image patches within players bounding boxes B_b, which may correspond to player activity [21]. Movement of any point on the image plane produces a two-dimensional (2D) path $x(t) \equiv (x(t), y(t))^T$ in camera-centered coordinates. The current direction is the velocity $V = dx(t)/dt$. The 2D velocities of all visible surface points make up the 2D motion field.

The optical flow motion field in consecutive video frames is estimated while using the Lucas–Kanade method [34]. In the Lucas–Kanade method, the original images are first divided into smaller sections, and it is assumed that the pixels within each section will have similar velocity vectors. The result is the vector field V of velocities, where, at each point (x, y), the magnitude of the vector represents the movement speed and its angle represents the movement direction.

Figure 5 visualizes an example of the calculated optical flow vectors between correspondent bounding boxes in two video frames from the dataset. The direction and length of blue arrows represent the direction and magnitude of optical flow at each point.

Figure 5. Player bounding box (yellow) and optical flow vectors (blue).

The activity measure (A_b^{OF}) of a player detected with its bounding box (B_b) is calculated in each frame as the maximum optical flow magnitude $V_{x,y}$ within the area P_b of the bounding box:

$$A_b^{OF} = \max_{B_b} |V_{x,y}|; \ x, y \text{ within } B_b. \tag{5}$$

The assumption is that the players performing a sports action make more sudden movements that will result in larger optical flow vectors within the player bounding boxes. The maximum value of the optical flow is used in order to make the comparison between players with different bounding box sizes simple.

3.3.2. STIPs-Based Activity Measure—DT+STIP

The DT+STIP activity measure is based on spatiotemporal interest points (STIPs). STIPs are an extension of the idea of local points of interests in the spatial domain, i.e. of points with a significant local variation of image intensities into both spatial and temporal domains, by requiring that image values have large variation in both spatial and temporal directions around a point.

It is expected that the higher density of STIPs in a certain area will point to a region of higher movement, which might correspond to the higher level of player's activity, since STIPs capture the spatiotemporal "interestingness" at different points in the image, and most actions in sports videos are characterized by strong variations in velocity and appearance over time [16]. Thus, an activity measure A_b^{STIP} that is based on STIPs density is calculated for a player with a bounding box B_b and area P_b. in a certain frame, as:

$$A_b^{STIP} = \frac{\#STIP}{P_b}; \ STIP \text{ within } B_b. \tag{6}$$

Figure 6 shows an example of detected STIPS in a frame of video, as well as the detected STIPs superimposed on the detected players' bounding boxes.

Figure 6. Fusion of bounding box and spatiotemporal interest points.

A number of methods have been proposed to detect the STIPs. The method that is proposed in [35] is based on the detection of the spatiotemporal corners derived from Harris corner operator (Harris3D), in [36] on Dollar's detector that uses a Gaussian filter in the space domain and a Gabor band-pass filter in the time domain and obtains a denser sampling by avoiding scale selection, in [37] on Hessian3D derived from SURF or selective STIPs detector [38] that focuses on detecting the STIPs that likely belong to persons and not to the possibly moving background. The Harris3D detector is used in the experiments described here.

In the Harris3D STIPs detector, a linear scale-space representation L is first computed by convolving the input signal f with an anisotropic Gaussian kernel g, i.e., a kernel with independent spatial and temporal variances [35]:

$$L\left(.;\sigma^2,\tau^2\right) = g\left(.;\sigma^2,\tau^2\right) * f(.),\tag{7}$$

where σ^2 and τ^2 represent the spatial and temporal variance, respectively, and the kernel g is defined as:

$$g\left(x,y,t;\sigma^2,\tau^2\right) = \frac{1}{\sqrt{(2\pi)^3\sigma^4\tau^2}}e^{\frac{-(x^2+y^2)}{2\sigma^2}-\frac{t^2}{2\tau^2}}.\tag{8}$$

A spatiotemporal second-moment matrix μ, which is composed of first order spatial and temporal derivatives of L averaged using a Gaussian weighting function g, is computed:

$$\mu = g\left(.;\sigma_i^2,\tau_i^2\right) * \begin{pmatrix} L_x^2 & L_xL_y & L_xL_t \\ L_xL_y & L_y^2 & L_yL_t \\ L_xL_t & L_yL_t & L_t^2 \end{pmatrix},\tag{9}$$

where $L_x, L_y,$ and L_t are the first-order derivatives of $g * f$ in the x, y, and t directions.

The spatiotemporal interest points are obtained from the local maxima of the corner function H:

$$H = \det(\mu) - k \cdot \text{trace}^3(\mu),\tag{10}$$

where k is a parameter that was experimentally set to be $k \approx 0.005$ [35].

The STIPs are extracted while using the with default parameters from the whole video.

3.3.3. CNN-Based Activity Measure—DT+Y

The DT+Y activity measure uses the YOLOv3 to recognize the level of activity of detected player, or to determine whether the player is active or not.

The YOLOv3 network that was, in the previous cases, only used for player detection, was additionally trained on custom data to classify the detected player as either active or inactive so that the class person was replaced with two classes: active player and inactive player. Aside from the number of detected classes, the architecture of the network is the same as the network used for player detection in DT+OF and DT+STIPS.

The training was completed for 70,000 iterations on 3232 frames that were extracted from 61 videos in our training dataset. The frames were manually labeled either as active or as inactive player. The learning rate was 0.001, momentum 0.9, and decay was 0.0005. The input image size was 608 × 608 pixels without using YOLO's multiscale training. Data augmentation in the form of random image saturation and exposure transformations, with maximum factors of 1.5, and with random hue shifts by, at most, 0.1 was used.

The activity measure A_b^Y for each bounding box B_b corresponds to confidence value c of the active player class, such as:

$$A_b^Y = c(active\ player\ class(B_b)). \tag{11}$$

Additionally, a threshold value can be defined for the classification accuracy of the *active player class*, above which the player will be considered active, and inactive otherwise.

3.4. Determining the Most Active Player

Throughout the sequence, different players will have the highest activity scores at different times. The activity scores for each player in individual frames are first aggregated into the track scores to select the most active player in the whole sequence. In Figure 7, a thick white solid bounding box line indicates the most active player for each frame (here player with ID 6). As shown in the Figure 7, in the fourth frame, in the case where multiple players have the same activity level, none is marked as being active.

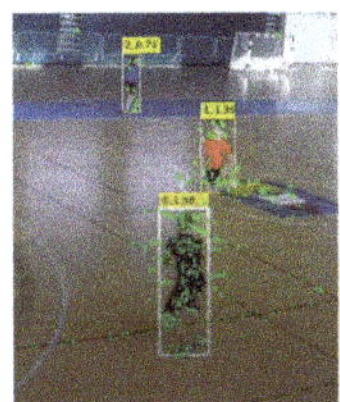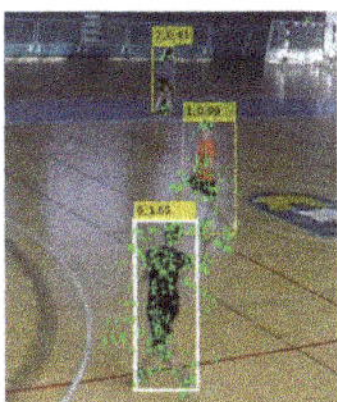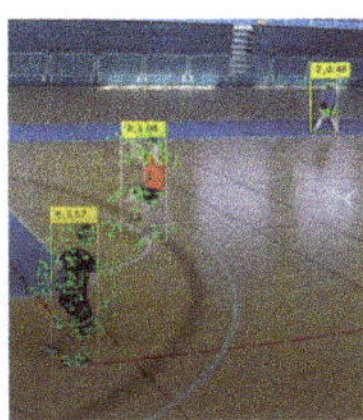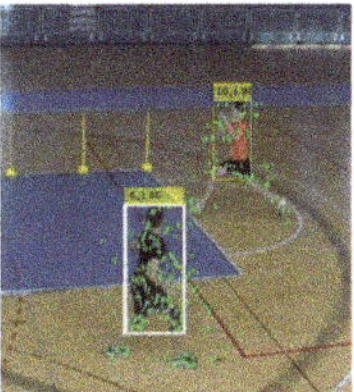

Figure 7. Indicating the most active player on each frame (marked with thick white bounding box).

The track scores are represented by a vector whose elements represent the ranking by activity measure in a frame. For example, if at the first frame the track with ID 6 (track 6) has the highest activity measure, and track 1 has the second largest activity measure, the track score at frame 1 for track 6 will be 1, and for the track 2 it will be 2. Table 1 provides an example of the ranking of players in each individual frame by their activity level and determining the most active player in a given time frame.

Finally, the active player's track is the one that is most commonly ranked first by the activity level in the whole sequence, i.e. whose track score vector contains the most ones among all track score vectors.

Table 1. The track score vectors for each detected player on video sequence is presented in Figure 7. Number of wins for each player in a sequence and final rankings are highlighted.

	Player ID	F1 Ranking	F2 Ranking	F3 Ranking	F4 Ranking	F5 Ranking	Number of Wins	Final Ranking
	1	2	2				1	0
Track score vectors	2	3	3	2	3		0	0
	6	1	1	1	1	1	5	1
	9				2	2	0	0

The result of the proposed method for determining the most active player is a series of active player bounding box representations $B_b(C_b, P_b)$ that are stored in the player track $\mathcal{T}_b = (B_1,\ B_1,\ \ldots . B_b)$ from the beginning to the end of performing an action. The graphical representation of the centroids C_b of each bounding box $B_b(C_b, P_b)$ through the motion sequence corresponds to the trajectory of his movement that is shown with the yellow line (Figure 8).

Figure 8. Detected most active player (white thick bounding box) and his trajectory through the whole sequence (yellow line).

The active player track can ultimately be represented by a collage of thumbnails that correspond to the bounding boxes of the most active player, showing the stages of an action (Figure 9).

Figure 9. Extracted collage of the most active player's action.

4. Performance Evaluation of Active Player Detection and Discussion

The proposed method was tested on a custom dataset that consists of scenes that were recorded during a handball training session. The handball training was organized in a sports hall or sometimes in an outdoor terrain, with a variable number of players without uniform jerseys, with challenging artificial lighting and strong sunshine, with cluttered background and other adverse conditions.

The dataset consists of 751 videos, each containing one of the handball actions, such as passing, shooting, jump-shot, or dribbling. Multiple players, on average, 12 of them, appear in each video, as shown in Figure 10. Table 2 provides the statistics of the number of players per frame.

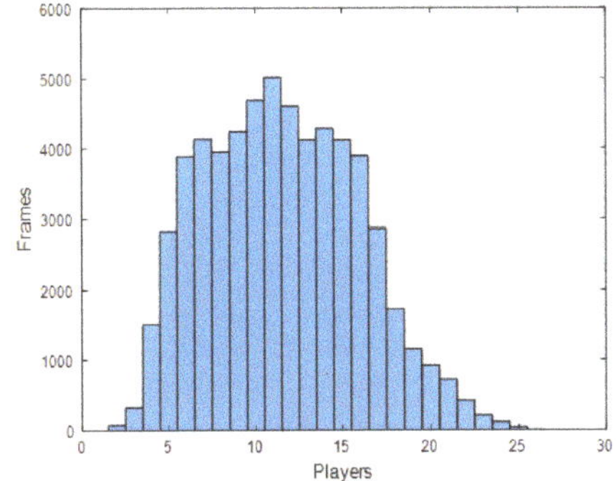

Figure 10. Distribution of number of players per frames.

Table 2. Number of detected players per frame.

	Min	Mean	Max	Median
No. players	2	11.55	26	11

Each player can move in any direction, but mostly one player performs an action of interest, so he is considered as the active player. Therefore, each file is only marked with one action of interest. The scenes were shot with stationary GoPro cameras that were mounted on the left or right side of the playground, from different angles and in different lighting conditions (indoor and outdoor). In the internal scenes, the camera is mounted at a height of 3.5 m and in the outside scenes at 1.5 m. The videos were recorded in full HD (1920 × 1080) at 30 frames per second. The total duration of the recordings used for the experiment was 1990 s.

In the experiment, the individual steps of the proposed method and the method as a whole were tested, from person detection to selecting the trajectory of an active player.

For the player detection task, the Yolo detector was tested on full-resolution videos and without frame skipping. The performance of the Yolo detector was evaluated in terms of recall, precision, and F1 score [39]. The detections are considered to be true positive when the intersection over union of the detected bounding box and the ground truth box exceeded the threshold of 0.5. The intersection over union measure (IoU) is defined as the ratio of the intersection of the detected bounding box and the ground truth (GT) bounding box and their union, as in Figure 11.

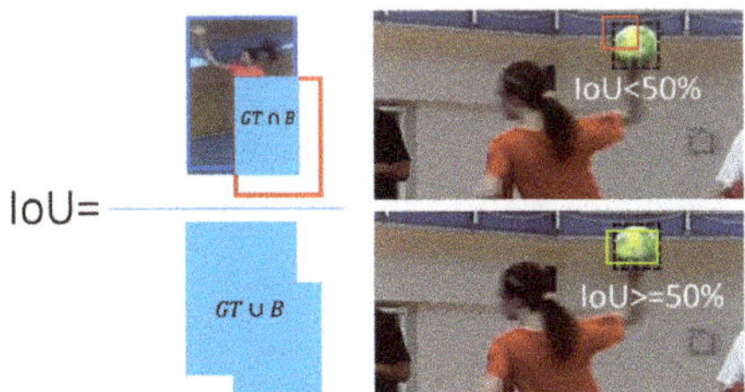

Figure 11. Visual representation of intersection over union (IoU) criteria equal to or greater than 50%.

The results of the detector greatly depend on the complexity of the scene, which in the used dataset depends on the number of players on the scene and their size, as well as on the distance from the camera and the number of occlusions. For example, if there are up to eight players on the scene, close to the camera, the results are significantly better than in the case when there are nine or more players on scene that are far from the camera and occlusions exist, as shown in Figure 12. The background is similar in both cases, and it includes the ground surface and its boundaries, the advertisements along the edge of the terrain, and chairs in the auditorium.

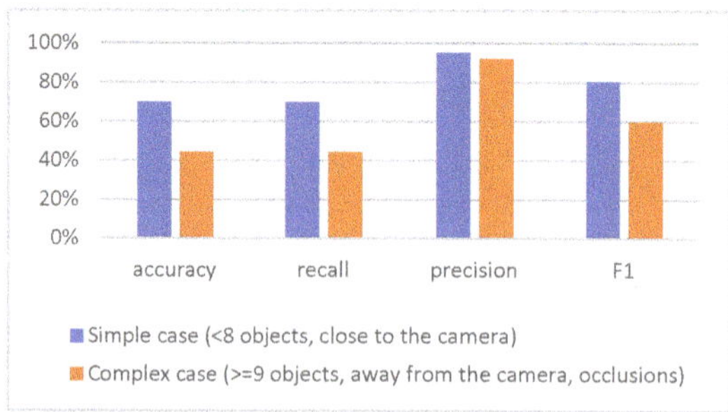

Figure 12. Results of player detection with Yolo in simple and complex scenarios.

The obtained results confirm that player detection, which is the first step of the active player detection method, provides a good starting point for the next phases of the algorithm.

In the next step, we tested the performance of each activity measure in terms of precision, recall, and F1. Activity measures for each detected player in a frame are computed based on motion features that are extracted within each player's bounding box in that video frame in the case of DT+OF and DT+STIP and according to the confidence value of active player class detection in the case of DT+Y. The performance of all measures is evaluated on each frame by comparing the bounding box with the highest activity measure with the ground truth active player bounding box, as the DT+OF and DT+STIP measures are used in combination with the bounding boxes of detected players proposed by Yolo and DT+Y measure, as one of the outputs gives the bounding box of the active player.

A detection was considered to be true positive if it was correctly labeled as active and had the largest IoU with the ground truth active player bounding box among the detected boxes. If the goal is to evaluate the performance of an activity measure, that is, how many times the measure has selected the right player as the most active, the exact match of the detected box with the ground truth is not crucial, but it is important to match GT as closely as possible (large IoU) if one wants to monitor player postures while performing an action. The minimum IoU overlap between the detected bounding box and ground truth was set to 10%, since, in this part, the focus was on the ability of the measure to distinguish between activity or inactivity of the player and not on the correct localization.

Figure 13 shows the evaluation results for all activity measures and indicates significantly higher values for the DT+Y activity measure. The DT+Y achieves the best F1 score of 73%, which is significantly better than the results that were achieved by other measures. DT+OF has a precision of 51%, a DT+STIP of 67%, and DT+Y of 87%. All of the measures achieve lower recall scores than precision scores, DT+OF achieves only 20%, DT+STIP 23%, and DT+Y 63%; however, this is an even more significant difference in DT+Y performance when compared to other measures.

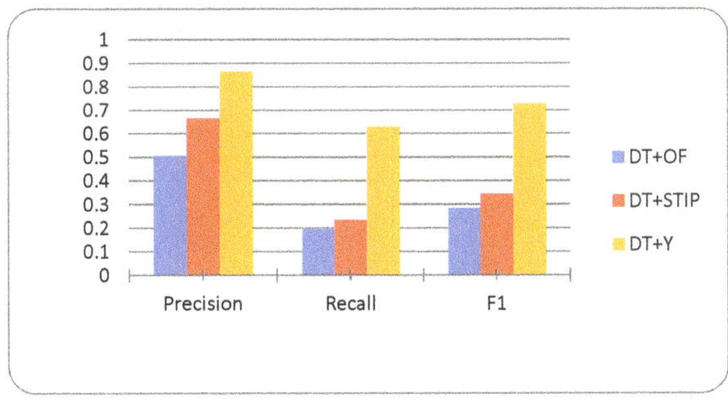

Figure 13. Evaluation results for DT+OF, DT+STIP, and DT+Y activity measures.

Figure 14 shows examples of positive and negative detection of an active player. Negative detection occurred because the other player on the field had a greater activity measure than the player performing the required action who is supposed to be the most active.

(a)   (b)

Figure 14. Marking of the leading player (white thick bounding box); (**a**) Correct detections; (**b**) Wrong detections. Examples were taken in different handball halls and under different indoor and outdoor lighting conditions.

It is necessary to know their activity measure in each frame, but also to uniquely identify that player throughout the period of interest, i.e., to track them through the sequence, to determine a player's overall activity in a time period. The track of the active player $\mathcal{T}_b = (B_1, B_1, \ldots B_b)$ on sequential frames is compared with the ground truth in order to evaluate the results of the proposed method for detecting an active player over a period of time.

A true positive rate was calculated as a number of true positives divided by the number of tested sequences to quantify the performance. An active player track is considered to be true positive if the

IoU of the detected active player is equal to or greater than a threshold value $\alpha = 50\%$ in more than θ of frames (here set to 50%). Formally, the measure for active player track evaluation is defined, as follows:

Let be $\mathcal{T}$ a sequence of bounding box representations B, $\mathcal{T}_{GT}$ a ground truth sequence, $IoU : \mathcal{T} x \mathcal{T}_{GT} \rightarrow [0,1]$ given function and $f : \mathcal{T} \rightarrow \{0,1\}$, being defined as:

$$f(x) = \begin{cases} 1 & IoU(x,g) \geq \alpha \\ 0 & IoU(x,g) < \alpha \end{cases} \tag{12}$$

then a true positive $(TP_{\mathcal{T}})$ for active player track $\mathcal{T}$. is defined as:

$$TPR_{\mathcal{T}} = \begin{cases} 1 & \frac{\sum_{B_i \in \mathcal{T}} f(B_i)}{|\mathcal{T}|} \geq \theta \\ 0 & \text{else} \end{cases}. \tag{13}$$

The $TP_{\mathcal{T}}$ is computed for every track in a test set, so TPR% is the ratio of the true positive tracks to the total number of tracks in the test set.

This means that the measure is strictly defined, because only those results that have a true positive detection of the player in θ consecutive frames are considered to be positive, as long as the action is completed, with the additional condition that it should be the given handball action. Thus, the case when a player is detected and properly tracked during his movement, but does not perform an action of interest in more that θ frames is not counted as the true result. Additionally, when the player is correctly identified and followed only in the part of the sequence (less than θ) is not counted as a positive result.

Table 3 provides the evaluation results for active player detection method along the entire motion trajectory for all activity measures where IoU is set to $\alpha = 50\%$ and at least $\theta = 50\%$ of frames per track correctly detected.

Table 3. True positive rates (TPR %) for active player detection.

Tracking	TPR % for Activity Measures		
	DT+OF	DT+ STIP	DT+Y
Hungarian	42.9	41.9	32.8
Deep Sort	50.0	53.2	67.2

The best overall score was achieved with the DT+Y activity measure that was used in conjunction with the Deep Sort tracker with 67% correct active player sequences, while the second-best result was achieved using the DT+STIP activity measure and Deep Sort tracker, with 53% correct sequences. The gap in performance between the DT+STIP and DT+OF methods from the DT+Y is smaller here, where whole sequences are taken into account, than when being examined for their performance of activity detection in isolation. When considering whole sequences, if player detection and tracking performs well, some mislabeling of players as inactive in certain frames will not affect the label of the whole sequence, so the result will not degrade either. This is important to note, because even though the DT+Y activity measure performs the best for determining the activity of the player, to use it is necessary to have annotated training data, which, if not available, can be a time consuming and tedious job to acquire, and this is not required for the DT+OF and DT+STIP measures. An even larger set of learning data and data balancing is needed to achieve even better results of DT+Y measure, because the number of inactive athletes on the scene is much higher than the inactive ones.

With all activity measures, better results were obtained while using the Deep Sort tracker than with Hungarian algorithm-based tracking, so it seems a clear choice, since it does not require much more effort to use.

We tested the proposed methods with different α threshold parameter and Figure 15 presents the results. In the case where only the information that the player is active for a certain period is relevant, the accuracy of detection is not as important (low overlap with the ground truth is acceptable) as in the case when we want to observe the player's movements and postures when performing an action (large overlap with the ground truth is desirable).

Similar rankings hold for most values of the minimum required IoU (parameter α), as can be seen in Figure 16.

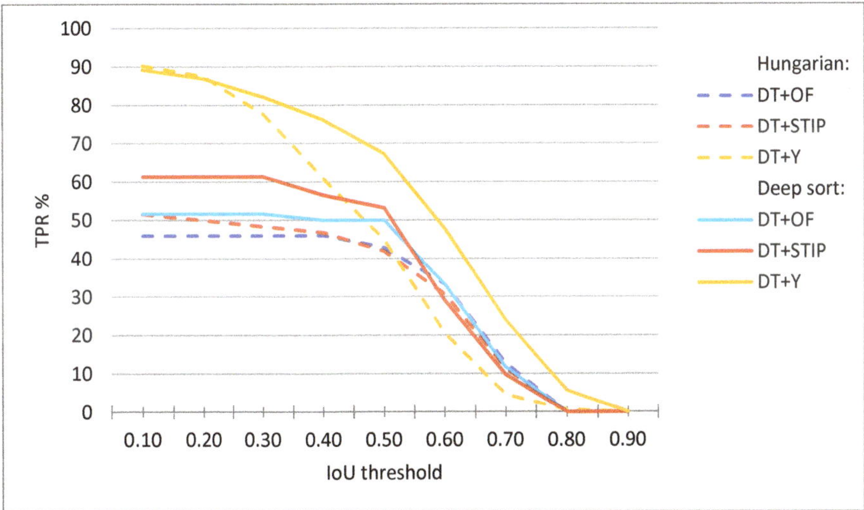

Figure 15. Evaluation results for player tracks obtained with DT+OF, DT+STIP, and DT+Y activity measures using Deep sort or Hungarian algorithm for tracking. The results at minimum $\theta = 50\%$ correct frames per sequence.

When considering different values of minimum correct frames per sequence (parameter θ.), again similar rankings can be observed, but, as expected, with different achieved TPR%. For example, Figure 16 shows that the TPR% is lower for various values of α when θ is increased to 70% than in the case shown in Figure 15. Here, the TPR score at lowest values of α is mostly limited by the performance of the tracking and activity detection, while at higher values of α. with the precision of player detection.

In the following, some typical cases and issues are shown and commented on. Figure 17 shows a collage of thumbnails that present good examples of player tracks that include all correctly detected bounding boxes on the frames in the sequence. As each player performs a particular action in a specific way, the number of frames is variable, even when performing the same handball action or technique according to the same rules.

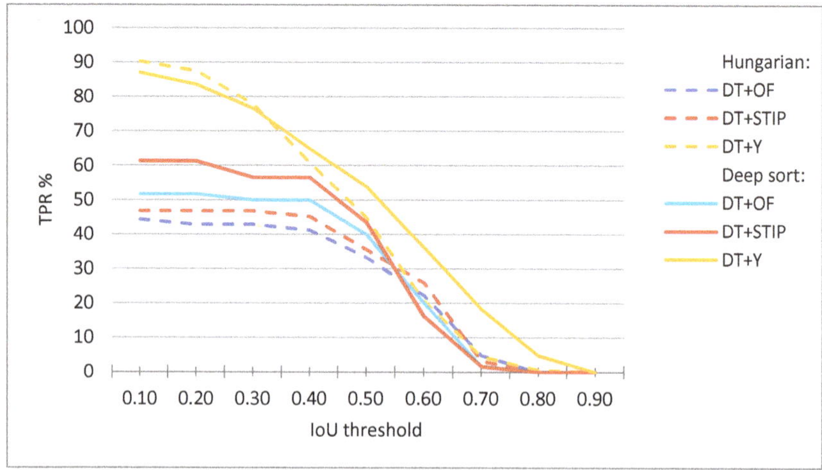

Figure 16. Evaluation results for player tracks obtained with DT+OF, DT+STIP, and DT+Y activity measures using Deep sort or Hungarian algorithm for tracking. The results at minimum $\theta = 70\%$ correct frames per sequence.

Figure 17. Examples of shooting action with and without jump performed by different players in different positions with respect to the camera in both outdoor and indoor fields.

The number of frames in player track depends on detector performance and tracking at all stages of the action. Figure 18 shows the problem of imprecise detection in some parts of the player tracking, where the detector has correctly detected the player, but the IoU is less than the threshold because his legs are outside of the bounding box. Figure 19 shows a negative example of an active player track that is caused by poor player detection that has propagated to other stages of the algorithm. The thumbnails show that the player is active and well tracked as long as it was detected. Additional training of the person class on samples from the handball domain can reduce the problem of imprecise detection.

Figure 18. Examples of imprecise player detection.

Figure 19. Tracking errors due to poor player detection.

Figure 20 shows a negative example of an active player's track due to improper tracking of players occurring due to cluttered scenes, players overlapping, or changing position.

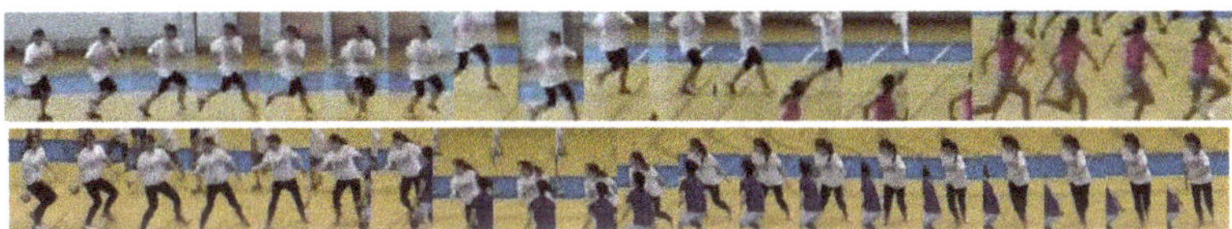

Figure 20. Examples of tracking problems.

A detector detects all players on each frame and the activity level is calculated for each one and their movement is monitored and recorded. The proposed method will select the player with the highest level of activity in most frames as the most active player; however, it is possible that the most active player is not the one performing the given action, so, in this case, there is no match to the GT, even though everything is done correctly. Figure 21 shows the cases when players who were preparing to perform an action or ran in the queue had a higher level of activity then player who was performing the default action and so were wrongly selected as active players.

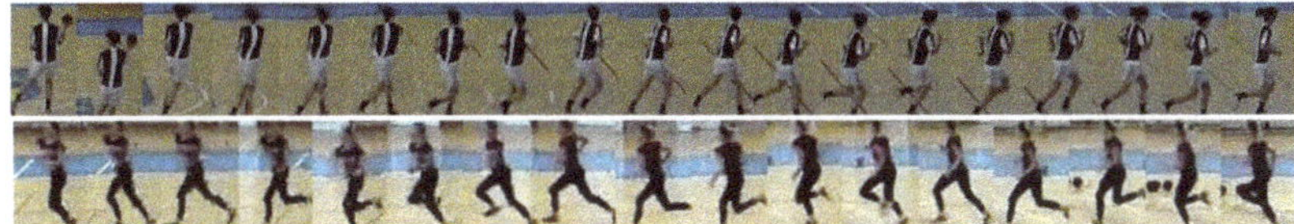

Figure 21. Tracking of players that perform actions outside of the scope.

As the two measures of activities DT + OF and DT + STIPS do not require any learning, we have tested them in completely different sports scenes from other team sports to test how general the proposed activity measures are.

We used the basketball event detection dataset [26] for the experiment. The dataset contains automatically obtained player detections and tracks for basketball events in 11 classes, such as three-point success, three-point failure, free-throw success, free-throw failure, and slam-dunk. Scenes were taken from a basketball game so that only one ball was present on the field. The clips are four seconds long and subsampled to 6fps from the original frame rate. The clips are tailored to include the entire event, from throwing the ball to the outcome of the shot. A subset of 850 clips also have the known ball location in the frame where the ball leaves the shooter's hands.

We have used two activity measures, DT+OF and DT+STIPS, on the basketball event dataset without any modification. The DT+YOLO method was not tested, because it needs to be trained before use, and it was not trained on the basketball domain. The player detections and tracks provided with the dataset were used to evaluate only the impact of the activity measures on active player detection, Figure 22.

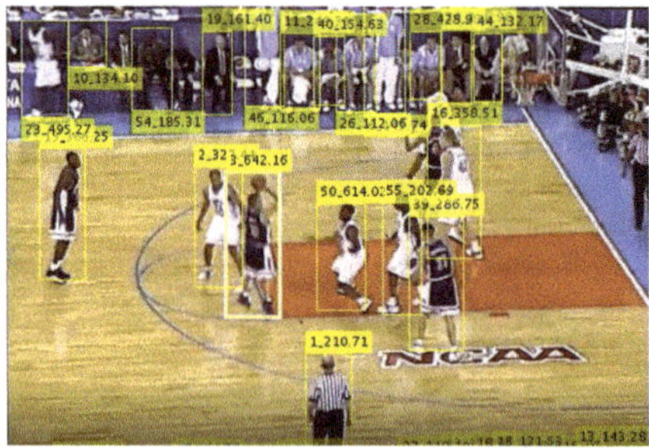

Figure 22. Marking of the leading player in basketball scenes (white thick bounding box).

In [26], the authors selected the player(s) who was closest in image space to the ball as the leading player ("shooter") to evaluate the attention mechanism. On the other hand, the handball scenes correspond to individual handball techniques and they were taken during training so that each player has his own ball, therefore the possession of the ball could not be taken as a measure of player activity and was not included in the design of proposed activity measures. For this reason, we selected a part of the basketball test set and determined the active player for the entire duration of the given action and then evaluated the active player detection metrics against these criteria.

The evaluation results for active basketball players detection along the entire motion trajectory for DT+OF and DT+SIFT activity measures, where at least $\theta = 50\%$ of frames per track are correctly detected, as given in Table 4.

Table 4. True positive rates for active player detection in basketball event detection scenes.

Activity Measure	DT+OF	DT+ STIP	Shooter Detection [26]
TPR %	43.87	45.26	43.8

The obtained results are lower than in the handball scenes (up to 8%), primarily because the video sequences in basketball dataset are event-driven rather than determined by the action for which the method was designed. This is most noticeable on actions, such as three-point throw, where it happens that the player who threw the ball is not even present in most of the frames as the camera zooms in on the ball's trajectory and the surrounding players and basket to track the outcome of the throw. Contrary to the method of detecting events in the basketball game, the proposed methods are defined from the perspective of the player and they correspond to some handball techniques, for example, a shot jump takes from the moment the player receives the ball, takes up to three steps, and throws the ball towards the goal, regardless of the outcome of the action (the ball enters/does not enter the goal).

The results are significant and promising, because they are comparable and even slightly better than the results that were reported in [26], although the method in [26] was purposefully trained to detect an active player on the basketball dataset while using 20 GPUs.

The track results exhibit similar properties as on the handball dataset, with problems stemming from the same sources—player detection, tracking, or activity measure. The difference is that basketball has fewer players on the field, smaller terrain, bigger ball, and the player's occlusion in basketball is even greater than in handball scenes because of the speed and difference in the rules of the game. Additionally, in our handball dataset, the actions are defined from the player's perspective and they correspond to individual handball techniques.

Finally, Figure 23 presents a visual sample of results on the basketball set.

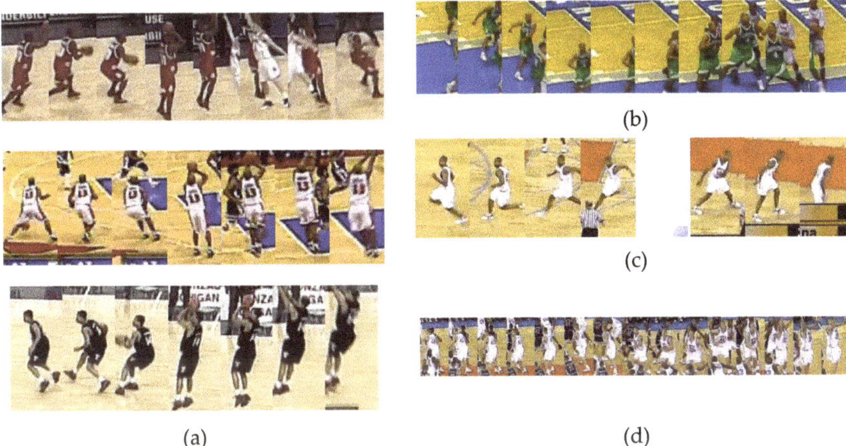

Figure 23. Example of results on the basketball dataset. (**a**) Correct tracks for the three-point failure event; (**b**) Wrong result due to imprecise player detection; (**c**) Incorrect result due to higher activity of player who is not performing the action of interest; (**d**) Incorrect results due to bias to longer tracks.

For this reason, for the final track selection, our method assumes that the player should be tracked throughout the whole video sequence, so it is, by design, biased towards longer tracks. It often misses choosing the shooter in cases, such as three-point shot, where the player who throws the ball might not even be present in the majority of the clips (Figure 23d). Track fragmentation, due to poor tracking, where the player that should be labeled as active is actually labeled with more than one track ID, impairs the ability to select the correct track for the same reason. The wrong selection of active track can happen even though the correct player might have been the most active in the frame the ball leaves the shooter's hands. A similar bias toward longer tracks is reported in [26] for the tracking-based attention model. In future work, the method should be able to switch the leading player track from player to player to handle this problem and extend the application to longer, temporally unsegmented clips.

5. Conclusions

There is great interest in the automatic interpretation of sports videos, for the purposes of analysis of game tactics and individual athlete's performance, for retrieval, sports analysis and monitoring and refinement of player action techniques and playing style. To facilitate automatic analysis in team sports, such as handball, the detection and tracking of players who are currently performing a relevant action is of particular interest, as the interpretation of the whole scene often depends on that player's action. Additionally, during training sessions, the time should be efficiently utilized, so that all players are active in sufficient proportion.

In this paper, an active player detection method is proposed, which entails frame-by-frame object detection, player activity measures, and tracking methods to detect and track active players in video sequences. The CNN-based Yolo object detector was used for player detection, Hungarian assignment algorithm based method, and Deep SORT were considered for tracking the detected players along the time dimension, and three different measures are proposed for the task of player activity detection in a certain frame, DT+OF, DT+STIP, and DT+Y. The DT+STIP measure is calculated from the spatiotemporal interest points (STIPs) that were detected within the area of a player, the simpler DT+OF method depends on the computed optical flow (OF) motion field and the DT+Y method uses convolutional neural network (CNN) for determining player activity and classifying active and inactive players.

The proposed method and its components were tested on the test set of handball videos from several training sessions that were acquired in the wild. A sequence-based evaluation measure was proposed for overall evaluation, and the best result was achieved with the Deep Sort tracking method with the CNN-based DT+Y activity measure, achieving the rate of 67% correct sequences on the test set. The DT+Y method requires training data, and if it is not available, the DT+STIP or DT+OF method can be used, which achieved 53% TPR and 50% TPR on the test set, respectively.

To test how general the proposed activity measures are, we have chosen completely different sport scenes from basketball sports domain and evaluated the performance of DT+OF and DT+STIPS measures, since they do not require any leaning. The obtained results are lower than in the handball scenes (up to 8%), primarily because the video sequences are event-driven rather than determined by the action, as in our case, and there is an even greater occlusion of the players, as basketball is played faster and on a smaller field. The results are promising, because they are comparable to the results of a method that was trained on that basketball dataset to detect an active player in a video sequence.

The achieved results confirm that active player detection is a very challenging task and it could be further improved by improving all three components (player detection, tracking, and activity measure). Particularly, player detection should be improved first, because both other components directly depend on its performance. Activity measure should be calculated inside the detected bounding box, so imprecise detection will negatively affect this measure, since parts of player body may be outside the detected box and irrelevant parts of background within. Similarly, in addition to the inability to track objects that are not detected at all, the Deep Sort tracking method uses visual history of previous appearances of players, so their imprecise detection impacts the ability of the tracker to correctly re-identify them after occlusion.

For application in match setting, where a single ball exists in play, ball detection could be incorporated as an additional significant clue for detecting the leading player. It is indisputable that the information that the ball has in the game is important for determining the action and the player, but it is a small object that is hidden in the player's hand most of the time, and, when thrown quickly, moves and changes direction, so it is often blurred and tracking is still a challenge that is being intensively tried to tackle.

In the future, the method should be extended to better deal with multi-player actions, such as crossing or defense, and the player activity could be monitored in relation to the type of action the player is performing. Additionally, the method should be extended to handle long-term video clips, where the leading player changes in different time intervals.

Author Contributions: Conceptualization, M.P. and M.I.-K.; Data curation, M.P.; Formal analysis, M.I.-K.; Investigation, M.P. and M.I.-K.; Methodology, M.P. and M.I.-K.; Validation, M.P. and M.I.-K.; Visualization, M.P.; Writing—original draft, M.P. and M.I.-K.; Writing—review & editing, M.I.-K. All authors have read and agreed to the published version of the manuscript.

Funding: This research was funded by the Croatian Science Foundation, grant number HRZZ-IP-2016-06-8345 "Automatic recognition of actions and activities in multimedia content from the sports domain" (RAASS).

Conflicts of Interest: The authors declare no conflict of interest

References

1. Buric, M.; Pobar, M.; Ivasic-Kos, M. An overview of action recognition in videos. In Proceedings of the 40th International Convention on Information and Communication Technology, Electronics and Microelectronics (MIPRO), Opatija, Croatia, 22–26 May 2017; pp. 1098–1103.
2. Redmon, J.; Farhadi, A. Yolov3: An incremental improvement. *arXiv* **2018**, arXiv:1804.02767 2018.
3. Wojke, N.; Bewley, A.; Paulus, D. Simple online and realtime tracking with a deep association metric. In Proceedings of the 2017 IEEE International Conference on Image Processing (ICIP), Beijing, China, 17–20 September 2017; pp. 3645–3649.
4. Kristan, M.; Leonardis, A.; Matas, J.; Felsberg, M.; Pflugfelder, R.; Cehovin Zajc, L.; Vojir, T.; Hager, G.; Lukezic, A.; Eldesokey, A. The visual object tracking vot2017 challenge results. In Proceedings of the IEEE International Conference on Computer Vision Workshops, Venice, Italy, 22–29 October 2017; pp. 1949–1972.

5. Dendorfer, P.; Rezatofighi, H.; Milan, A.; Shi, J.; Cremers, D.; Reid, I.; Roth, S.; Schindler, K.; Leal-Taixe, L. CVPR19 Tracking and Detection Challenge: How crowded can it get? *arXiv* **2019**, arXiv:1906.04567 2019.

6. Okuma, K.; Taleghani, A.; De Freitas, N.; Little, J.J.; Lowe, D.G. A boosted Particle Filter: Multitarget Detection and Tracking. In *European Conference on Computer VISION*; Springer: Berlin/Heidelberg, Germany, 2004; pp. 28–39.

7. Pers, J.; Kovacic, S. Computer vision system for tracking players in sports games. In Proceedings of the First International Workshop on Image and Signal Processing and Analysis. Conjunction with 22nd International Conference on Information Technology Interfaces, IWISPA 2000, Pula, Croatia, 14–15 June 2000; pp. 177–182.

8. Needham, C.J.; Roger, D.B. Tracking multiple sports players through occlusion, congestion and scale. *BMVC* **2001**, *1*, 93–102.

9. Erikson, M.; Ferreira, A.; Cunha, S.A.; Barros, R.M.L.; Rocha, A.; Goldenstein, S. A multiple camera methodology for automatic localization and tracking of futsal players. *Pattern Recognit. Lett.* **2014**, *39*, 21–30.

10. Bebie, T.; Bieri, H. SoccerMan-reconstructing soccer games from video sequences. In Proceedings of the 1998 International Conference on Image Processing. ICIP98 (Cat. No.98CB36269), Chicago, IL, USA, 7 October 1998; Volume 1, pp. 898–902. [CrossRef]

11. Xu, M.; Orwell, J.; Jones, G. Tracking football players with multiple cameras. In Proceedings of the 2004 International Conference on Image Processing, ICIP'04, Singapore, 24–27 October 2004; Volume 5, pp. 2909–2912.

12. Jia, L.; Tong, X.; Li, W.; Wang, T.; Zhang, Y.; Wang, H. Automatic player detection, labeling and tracking in broadcast soccer video. *Pattern Recognit. Lett.* **2009**, *30*, 103–113.

13. Zhu, G.; Xu, C.; Huang, Q.; Gao, W. Automatic multi-player detection and tracking in broadcast sports video using support vector machine and particle filter. In Proceedings of the 2006 IEEE International Conference on Multimedia and Expo, Toronto, ON, Canada, 9–12 July 2006; pp. 1629–1632.

14. Lehuger, A.; Duffner, S.; Garcia, C. *A Robust Method for Automatic Player Detection in Sport Videos*; Orange Labs: Paris, France, 2007; Volume 4.

15. Bewley, A.; Ge, Z.; Ott, L.; Ramos, F.; Upcroft, B. Simple online and realtime tracking. In Proceedings of the 2016 IEEE International Conference on Image Processing (ICIP), Phoenix, AZ, USA, 25–28 September 2016; pp. 3464–3468.

16. Pobar, M.; Ivašić-Kos, M. Detection of the leading player in handball scenes using Mask R-CNN and STIPS. In Proceedings of the Eleventh International Conference on Machine Vision (ICMV 2018), Munich, Germany, 1–3 November 2018; International Society for Optics and Photonics: Bellingham, WA, USA, 2019; Volume 11041, p. 110411V.

17. Ali, B.; Itti, L. State-of-the-art in visual attention modeling. *IEEE Trans. Pattern Anal. Mach. Intell.* **2012**, *35*, 185–207.

18. Laurent, I.; Koch, C.; Niebur, E. A model of saliency-based visual attention for rapid scene analysis. *IEEE Trans. Pattern Anal. Mach. Intell.* **1998**, *11*, 1254–1259.

19. Junwei, H.; Zhang, D.; Cheng, G.; Liu, N.; Xu, D. Advanced deep-learning techniques for salient and category-specific object detection: A survey. *IEEE Signal Process. Mag.* **2018**, *35*, 84–100.

20. Marat, S.; Phuoc, T.H.; Granjon, L.; Guyader, N.; Pellerin, D.; Guérin-Dugué, A. Modelling spatio-temporal saliency to predict gaze direction for short videos. *Int. J. Comput. Vis.* **2009**, *82*, 231. [CrossRef]

21. Pobar, M.; Ivasic-Kos, M. Mask R-CNN and Optical flow based method for detection and marking of handball actions. In Proceedings of the 11th International Congress on Image and Signal Processing, BioMedical Engineering and Informatics (CISP-BMEI), Beijing, China, 13–15 October 2018; pp. 1–6.

22. Mihir, J.; Jan van, G.; Jégou, H.; Bouthemy, P.; Snoek, C.G.M. Action localization with tubelets from motion. In Proceedings of the IEEE Conference on Computer Vision and Pattern Recognition, Columbus, OH, USA, 24–27 June 2014; pp. 740–747.

23. Georgia, G.; Malik, J. Finding action tubes. In Proceedings of the IEEE Conference on Computer Vision and Pattern Recognition, Boston, MA, USA, 7–12 June 2015; pp. 759–768.

24. Alexander, K.; Marszałek, M.; Schmid, C.; Zisserman, A. Human focused action localization in video. In Proceedings of the European Conference on Computer Vision, Heraklion, Crete, Greece, 5–11 September 2010; Springer: Berlin/Heidelberg, Germany, 2010; pp. 219–233.

25. Dalal, N.; Triggs, B. Histograms of oriented gradients for human detection. In Proceedings of the IEEE Computer Society Conference on Computer Vision and Pattern Recognition (CVPR'05), San Diego, CA, USA, 20–25 June 2005; Volume 1, pp. 886–893.

26. Ramanathan, V.; Huang, J.; Abu-El-Haija, S.; Gorban, A.; Murphy, K.; Fei-Fei, L. Detecting events and key actors in multi-person videos. In Proceedings of the IEEE Conference on Computer Vision and Pattern Recognition, Las Vegas, NV, USA, 26 June–1 July 2016; pp. 3043–3053.

27. Ivasic-Kos, M.; Iosifidis, A.; Tefas, A.; Pitas, I. Person de-identification in activity videos. In Proceedings of the 37th International Convention on Information and Communication Technology, Electronics and Microelectronics (MIPRO), Opatija, Croatia, 26–30 May 2014; pp. 1294–1299.

28. Buric, M.; Pobar, M.; Ivašić-Kos, M.I. Object Detection in Sports Videos. In Proceedings of the 41st International Convention on Information and Communication Technology, Electronics and Microelectronics (MIPRO), Opatija, Croatia, 21–25 May 2018.

29. Burić, M.; Pobar, M.; Ivašić-Kos, M. Ball detection using YOLO and Mask R-CNN. In Proceedings of the International Conference on Computational Science and Computational Intelligence (CSCI), Las Vegas, NV, USA, 13–15 December 2018.

30. He, K.; Gkioxari, G.; Dollár, P.; Girshick, R. Mask r-cnn. In Proceedings of the IEEE International Conference on Computer Vision, Piscataway, NJ, USA, 22–29 October 2017; pp. 2961–2969.

31. Burić, M.; Pobar, M.; Ivašić-Kos, M. Adapting YOLO network for Ball and Player Detection. In Proceedings of the 8th International Conference on Pattern Recognition Applications and Methods (ICPRAM 2019), Prague, Czech Republic, 19–21 February 2019; pp. 845–851.

32. Redmon, J.; Divvala, S.; Girshick, R.; Farhadi, A. You only look once: Unified, real-time object detection. In Proceedings of the IEEE Conference on Computer Vision and Pattern Recognition, Las Vegas, NV, USA, 26 June–1 July 2016; pp. 779–788.

33. Munkres, J. Algorithms for the assignment and transportation problems. *J. Soc. Ind. Appl. Math.* **1957**, *5*, 32–38. [CrossRef]

34. Barron, J.L.; Fleet, D.J.; Beauchemin, S.S. Performance of optical flow techniques. *Int. J. Comput. Vis.* **1994**, *12*, 43–77. [CrossRef]

35. Laptev, I. On space-time interest points. *Int. J. Comput. Vis.* **2005**, *64*, 107–123. [CrossRef]

36. Dollár, P.; Rabaud, V.; Cottrell, G.; Belongie, S. Behavior recognition via sparse spatio-temporal features. In Proceedings of the IEEE International Workshop on Visual Surveillance and Performance Evaluation of Tracking and Surveillance, Beijing, China, 15–16 October 2005; pp. 65–72.

37. Klaser, A.; Marszalek, M.; Schmid, C. A Spatio-Temporal Descriptor Based on 3D-Gradients. In Proceedings of the BMVC 2008-19th British Machine Vision Conference, British Machine Vision Association, Leeds, UK, 1–4 September 2008.

38. Chakraborty, B.; Holte, M.B.; Moeslund, T.B.; Gonzàlez, J. Selective spatio-temporal interest points. *Comput. Vis. Image Underst.* **2012**, *116*, 396–410. [CrossRef]

39. Ivasic-Kos, M.; Pobar, M.; Ribaric, S. Two-tier image annotation model based on a multi-label classifier and fuzzy-knowledge representation scheme. *Pattern Recognit.* **2016**, *52*, 287–305. [CrossRef]

Article

Hand Gesture Recognition Using Compact CNN via Surface Electromyography Signals

Lin Chen [1,2,†], **Jianting Fu** [1,2,†], **Yuheng Wu** [1,3], **Haochen Li** [1,2] **and Bin Zheng** [1,*]

[1] Chongqing Institute of Green and Intelligent Technology, Chinese Academy of Sciences, Chongqing 400700, China; chenlin18@cigit.ac.cn (L.C.); jiantingfu@cigit.ac.cn (J.F.); wuyuheng21@163.com (Y.W.); lihaochen@csu.ac.cn (H.L.)

[2] School of Computer Science and Technology, University of Chinese Academy of Sciences, Beijing 100049, China

[3] School of Mechatronical Engineering, Changchun University of Science and Technology, Changchun 130022, China

[*] Correspondence: zhengbin@cigit.ac.cn

[†] These authors contributed equally.

Received: 25 December 2019; Accepted: 22 January 2020; Published: 26 January 2020

Abstract: By training the deep neural network model, the hidden features in Surface Electromyography(sEMG) signals can be extracted. The motion intention of the human can be predicted by analysis of sEMG. However, the models recently proposed by researchers often have a large number of parameters. Therefore, we designed a compact Convolution Neural Network (CNN) model, which not only improves the classification accuracy but also reduces the number of parameters in the model. Our proposed model was validated on the Ninapro DB5 Dataset and the Myo Dataset. The classification accuracy of gesture recognition achieved good results.

Keywords: surface electromyography (sEMG); convolution neural networks (CNNs); hand gesture recognition

---

1. Introduction

In recent years, Surface Electromyography(sEMG) signals have been widely used in artificial limb control, medical devices, human-computer interaction, and other fields. With the development of artificial intelligence and robotics technology, the intention of human hand movements can be obtained by using an artificial intelligence algorithm to analyze the sEMG signals collected from the residual limb. Robotics and artificial intelligence can be leveraged to better help the disabled people to independently complete some basic interactions in their daily life. The sEMG signals, which are non-stationary, represent the sum of subcutaneous athletic action potentials generated through muscular contraction [1]. Also, it is one of the main physical signals of an intelligent algorithm to identify motion intention.

Distinguishing sEMG signals collected from different gestures is the core part of the related applications using sEMG signals as intermediate media. At present, the literature on gesture recognition or artificial limb control by sEMG signals primarily focuses on the time and frequency domain feature extraction of sEMG signals, which aims to distinguish sEMG signals by feature recognition [1–3]. After years of exploration by researchers, some effective feature combinations have been proposed in both the time domain and frequency domain [4–6], and some fruitful results have been achieved with their respective datasets. Choosing feature extraction is particularly important in that different gestures can be distinguished by traditional methods. However, it is difficult to improve the performance of gesture recognition based on sEMG by traditional methods. Nevertheless, the process of designing and selecting features can be complicated and the combinations of features are diverse, leading to increasing of workload and dissatisfied results [7].

Using deep neural networks to distinguish sEMG signals has been proposed by researchers. Wu et al. [7] proposed LCNN and CNN_LSTM models (These models can be thought of as autoencoders for automatic feature extraction.), which do not require the process of traditional feature extraction. In recent years, deep learning has achieved great success in the field of image recognition. An important idea was put forward in [8,9] that the signals of a channel can form a graph, after the short- time Fourier transform or wavelet transform of sEMG signals. It was a good idea to convert the sEMG signal into an image and inspired us with the transform of the sEMG signal. Researchers such as Côté-Allard et al. [8], who regarded the original sEMG signals as an image, constructed the ConvNet model to further improve the classification accuracy of sEMG signals. However, the LCNN and CNN_LSTM models proposed by Wu et al. [7], and the ConvNet model used by Côté-Allard et al. [8], contain a large number of parameters.

For deep learning algorithms, the final test accuracy is directly affected by the size of the training data, but one participant cannot be expected to generate tens of thousands of examples in one experiment during the data collection process. However, a large amount of data can be obtained by aggregating the records of multiple participants, so that the model can be pre-trained to reduce the amount of data required by new participants. On the other hand, designing a compact network structure to reduce the number of parameters can also reduce the demand for data size.

In order to reduce the number of model parameters and improve the accuracy of model classification, we present a new compact deep convolutional neural network model for gesture recognition, called as EMGNet. It was validated on the Myo Dataset that the average recognition accuracy of EMGNet can achieve 98.81%. The NinaPro DB5 dataset has often been used to test classical machine learning methods. The accuracy of the EMGNet on these datasets was higher than that of the traditional machine learning methods. Figure 1 shows the overall flow chart of sEMG signal acquisition and identification.

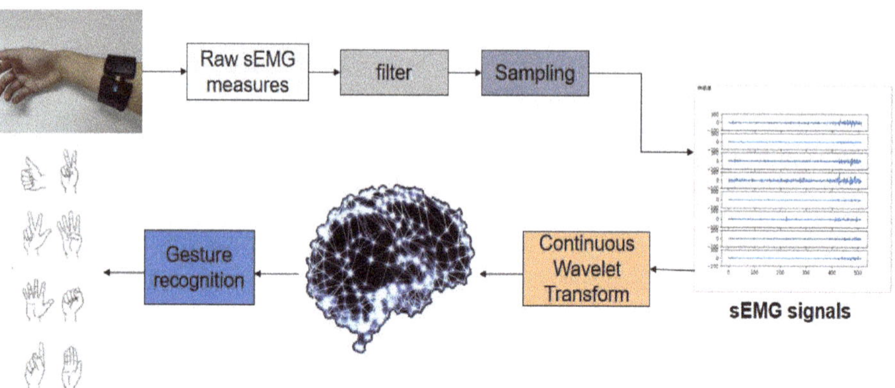

Figure 1. Surface Electromyography(sEMG) signals collection and classification process. Myo armband was used to collect the original signal, and then the collected signal was filtered and sampled to get sEMG signals. Continuous wavelet transform was selected to obtain the signal spectrum, and the neural network model was applied to classify the spectrum to achieve gesture recognition.

The rest of this paper is organized as follows. The related work of gesture recognition through deep learning is outlined in Section 2. The data processing and network architecture are described in detail in Section 3. The proposed network model is compared with the current excellent deep network framework and the classical machine learning classification method in Section 4. Finally, we present the conclusion in Section 5.

2. Related Work

People will produce different signals when completing the same action even lots of precise control electrodes are used to sense them [10]. Therefore, it is difficult to recognize sEMG signals. Since the AlexNet network proposed by Krizhevksy et al. [11] won the ImageNet challenge in 2012, deep learning has achieved great success in image classification, speech recognition, and other fields. Images can be accurately classified by training the neural network model to learn the characteristics of images. Nowadays, exploring network architecture has become part of deep learning research.

Currently, some researchers have successfully applied deep learning to sEMG signal classification and explored several effective network frameworks [8,12–16]. Using CNN to classify sEMG signals, literature [12,17,18] took the raw signals as input space. The spectrograms of raw sEMG signals were obtained by Short-Time Fourier Transform (STFT) and fed into the convolutional network (ConvNets) [13,19]. Literature [8] used the ConvNets to classify the characterizations of the sEMG signals extracted by short-time Fourier transform-based spectrogram and Continuous Wavelet Transform (CWT). Since sEMG signals correspond to the timing signal, we proposed to classify the sEMG signal by combining Long Short-Term Memory (LSTM) and CNN from our previous work [7]. The temporal information in the signal is retained and the ability of CNN to extract features is utilized. We take advantage of the complementarity of CNNs and LSTMs by combing them into one unified architecture. Meanwhile, we analyze the effect of adding CNN before the LSTM. We propose LCNN and CNN-LSTM models, which can directly input pre- processed EMG signals into the network [7]. In practical work, we verified that the performance of the LCNN model is better than CNN-LSTM. Figure 2 simply depicts the architecture of the LCNN model. We use PReLU [20] as the non-linear activation function. ADAM [21] is utilized for the optimization of the model.

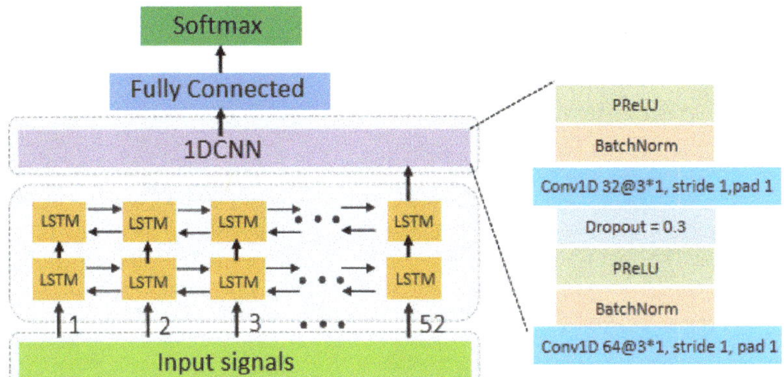

Figure 2. LCNN Architecture diagram, the LCNN consists of 2 LSTM layers, 2 one- dimensional convolution layers and 1 output layer. We use 2 LSTM layers, and each LSTM layer has 52 cells, and every cell has 64 hidden layers.

However, the ConvNets model (Shown in Figure 3) in [8] was too complicated, and the LSTM model was introduced in [7], which led to expensive computation in gesture recognition. Therefore, a new network model was proposed in this paper, and it was proved by experiment that this model not only improves the accuracy of recognition, but also reduces the complexity of the network model.

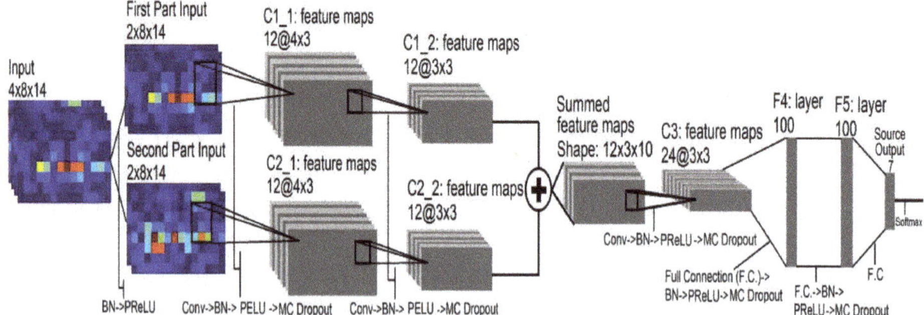

Figure 3. Schematic diagram of ConvNet architecture. In this figure, Conv refer to Convolution and F.C. to Fully Connected layers.

3. sEMG Signals Recognition Algorithm

3.1. sEMG Signals Feature Extraction

Designing a sEMG signal feature is one of the main tasks of the algorithm. It can not only easily distinguish the sEMG signals produced by different movements, but also maintain a small variance between the sEMG signals produced by the same movements. Four feature sets—Time domain characteristics (TD) [5], Enhanced TD [4], Nina Pro Features [6,22], and SampEn Pipeline [23]—were selected as features of the classical machine learning algorithm compared to the proposed method. The TD feature set includes four features: Mean Absolute Value (MAV), Zero Crossing (ZC), Slope Sign Changes (SSC), and Waveform Length (WL) [5]. The Enhanced TD includes features such as Root Mean Square (RMS) and Autoregressive Coefficient (AC) [3]. Nina Pro Features includes Root Mean Square, TD features, etc. [6,22]. SampEn Pipeline includes features such as Sample Entropy (SampEn) [24], Root Mean Square, and Waveform Length [23].

Since the sEMG signals is a non-stationary signal [25], the analysis of these signals using the Fourier transform is limited. One technique to solve this problem is the Short Time Fourier Transform (STFT), which separates the signal into smaller segments by applying a sliding window and separately calculates the Fourier transform for each of the segments. The spectrogram of the signal can be calculated from the square of the signal after STFT transformation. When the signal $x(t)$ and the window function $w(t)$ are given, the spectrogram is calculated as follows:

$$spectrogram\ (x(t), w(t)) = \left|STFT_x(t, f)\right|^2 \tag{1}$$

$$STFT_x(t, f) = \int_{-\infty}^{+\infty} [x(u)w(u-t)]e^{-j2\pi fu}du \tag{2}$$

where f represents frequency. The wavelet transform(WT) is similar to the STFT, but it overcomes the shortcomings of the window in the STFT which does not change with frequency. The WT adapts to the change in frequency in the signal by adjusting the width of the window. When the frequency in the signal climbs high, the WT increases the resolution by narrowing the time window. In addition, WT, which is an ideal signal analysis tool, has the ability to obtain the amplitude and frequency of mutations in a signal.

$$X(a, b) = \frac{1}{\sqrt{b}} \int_{-\infty}^{\infty} x(t)\varphi\left(\frac{t-a}{b}\right)dt \tag{3}$$

$$\int_{-\infty}^{+\infty} \frac{\left|\varphi(\omega)\right|^2}{\omega}d\omega < \infty \tag{4}$$

where the Fourier Transform $\varphi(\omega)$ of $\varphi(t)$ must satisfy Equation (4). $\varphi(t)$, also known as the mother wavelet function, is a signal with a limited duration, varying frequency and a mean of zero. The scaling factor b controls the scaling of the wavelet function. The translation amount a controls the translation of the wavelet function.

After the continuous wavelet transform of the sEMG signal, the corresponding spectrum information will be obtained, which is similar to the image in scale and also contains the frequency-domain information of the timing sequence data. The datasets tested in this article were all collected by Myo armband, as shown in Figure 4. Myo is an 8-channel, dry-electrodes, low-sampling rate (200 Hz), low-cost consumer grade sEMG armband, which is convenient to wear and easy to use [7]. The data of one channel was separated by applying sliding windows of 52 samples (260 ms). The mother wavelet of continuous wavelet transform adopts the Mexican Hat wavelet function. The CWTs were calculated with 32 scales obtaining a 32×52 matrix. The scale (32×52) sampled down 0.5 in spectrum information is taken as the input of EMGNet model. Thus the input of the EMGNet model has 8 channels, each of which consists of a matrix of size 15×25. Figure 5b is the spectrum of the signal shown in Figure 5a after wavelet transformation.

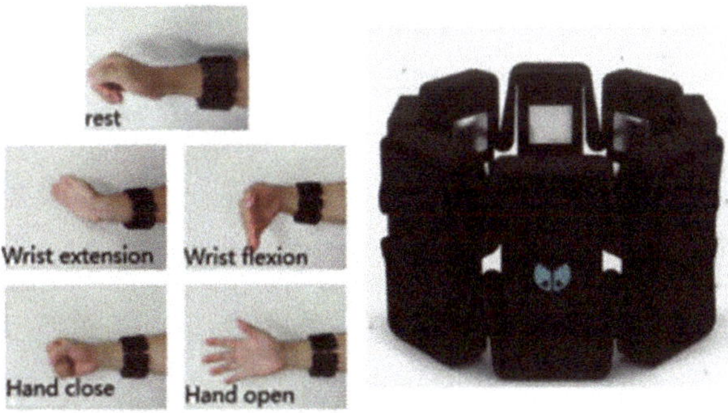

Figure 4. The 5 hand/wrist gestures and Myo armband. In this figure, the left is a schematic diagram of five gestures, and the right is the Myo armband.

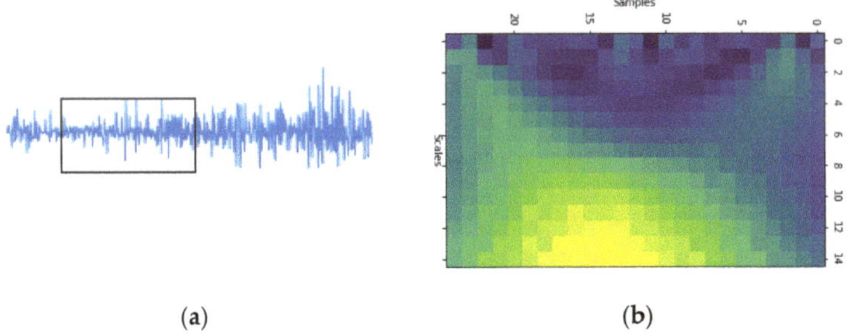

Figure 5. Part (**a**) shows the waveform of sEMG signals and Part (**b**) the spectrum of the sEMG signals shown in (**a**) after wavelet transformation.

3.2. EMGNet Architecture

A new network model called EMGNet is proposed in this paper, and shown in Figure 6. It consists of four convolutional layers and a max pooling layer without using the full connection layer as the final output. ConvNet architecture (Shown in Figure 3) contains 67179 learnable parameters used in the CWT+TL method proposed in Literature [8]. As shown in Table 1, the model proposed in this paper contains fewer parameters than the models used in current advanced methods. In view of the better performance of the EMGNet model in the actual measurement, this section mainly introduces the EMGNet model.

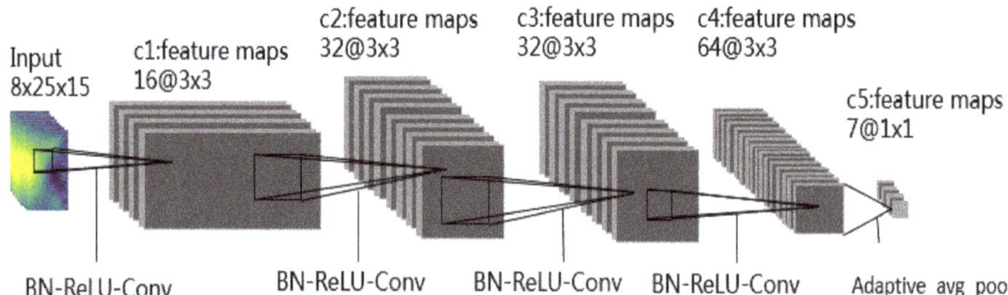

Figure 6. EMGNet Architecture contains four convolutional layers and a max pooling layer without using the full connection layer as the final output. In this figure, Conv refer to Convolution and avg_pool to a max pooling layer.

Table 1. Number of parameters used by various models.

Model	CNN_LSTM	LCNN	ConvNet	EMGNet
Number of parameters	56645	154353	67179	34311

The cost function Loss can be computed as follows:

$$Loss = -\sum_{i=1}^{n} y_i \log(y_i')$$

(5)

where y_i is the truth value of the ith category, n is the number of categories, and y_i' is the ith category predicted value of the output. Because we adopted One-Hot Encoding, the true value of the one category is 1 while the others are 0.

The advanced optimization method is used for the backpropagation of the EMGNet, and our final target is to minimize the cost function Loss. In the field of image recognition, the size of convolution kernels commonly used by researchers is 3×3 and 5×5 [11,26,27]. We found that the experimental results obtained by setting the size of the convolution kernel to 3×3 are better. Therefore, the size of all convolution kernels in the EMGNet model is set to 3×3. Meanwhile, in order to reduce the parameters of the model as much as possible, the feature map of each layer of the EMGNet model is also set smaller. When the feature map of the network model increases, the step length in the convolution process is set to 2, so as to achieve the effect of halving. In other cases, the step length and the padding are set to 1, so that the feature remains unchanged. In order to further reduce the number of network parameters, the output of the model does not use the full connection layer, but carries out adaptive mean sampling first, and then uses the convolution layer for classification.

4. Experiment and Results

We evaluated our mode on two publicly available hand gesture recognition datasets composed of Myo Dataset [8,9] and NinaPro DB5 Dataset [6]. First, we compared the method proposed with the methods of classical machine learning and three other methods (CNN-LSTM, LCNN, CWT+TL) on the Myo Dataset [8,9]. Then, it was compared with the methods of classical machine learning and the methods (CNN-LSTM, LCNN) proposed earlier on the NinaPro DB5 Dataset [6].

4.1. Evaluation Dataset

Containing two different sub-datasets, this gesture dataset (Myo Dataset [8,9]) was collected using the Myo armband. The first Dataset was used as the pre-training dataset in [8] and the second includes the part of training and testing. The former is mainly used to establish, verify, and optimize the classification model, which consists of 19 subjects. The latter, used only for final training and validation, consists of 17 participants. The second Dataset contains a training section and two test sections in the literature [8,9], which is an unreasonable arrangement and leads to a decrease in the amount of training data. In order to facilitate the comparative experiment, this article uses the same settings. The Myo Dataset contains 7 types of gestures, and there are significant differences between gestures. It provides sufficient amount of data, and its gestures are shown in Figure 7.

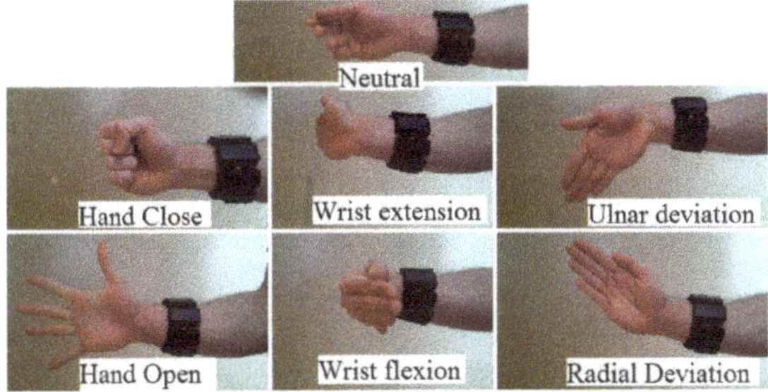

Figure 7. The 7 hand/wrist gestures in the Myo Dataset. In the Myo dataset, seven gestures are included: Netral, Hand Close, Wrist Extension, Ulnar Deviation, Hand Open, Wrist Flexion, and Radial Deviation.

This dataset (NinaPro DB5 [6]) is based on benchmark sEMG-based gesture recognition algorithms [6] containing data from 10 able-bodied participants divided into three exercise sets—Exercise A, B, and C contain 12, 17, and 23 different movements (including neutral) respectively. It is recorded by two Myo armbands, and only one of them is used in this work. One of the characteristics is that there are some similar gestures and the training data is not enough. Therefore, the model is prone to overfitting in the training process. Figure 8 shows the gesture categories in the dataset of Exercise A.

Exercise A								
1	Index flexion		5	Ring flexion		9	Thumb adduction	
2	Index extension		6	Ring extension		10	Thumb adbuction	
3	Middle flexion		7	Little finger flexion		11	Thumb flexion	
4	Middle extension		8	Little finger extension		12	Thumb extension	

Figure 8. The gesture categories in the Exercise A dataset.

4.2. Method of Training

Adam [11] is used as the optimization method of network model training in this work. The length of data collected by each gesture in the two datasets was the same. After the same segmentation method, the data amount of each gesture was the same, and the samples are balanced. In the Myo dataset, each person has 2280 samples for each gesture, with a total of 19 participants, while in the Nina Pro dataset, each person has 1140 samples for each gesture, with a total of 10 participants.

At the same time, after the samples are segmented, we used the shuffle algorithm to shuffle the samples of each gesture, and then took 60% as a training sample set, 10% as the verification sample set, and the last 30% of each gesture as the test sample set.

The amount of data fed into the network by each training batch was 128 samples, and a total of 50 rounds of iterative training were conducted. We initially set the learning rate at 0.01 and shrank the learning rate by 10 times at epoch 20 and 40. In order to prevent over-fitting of the network, L2 regularization is used in this paper. The parameter settings in the training process are shown in Table 2.

Table 2. Training policy parameter setting (lr represents the learning rate).

Dataset	Initial lr	lr in 20 Epoch	lr in 40 Epoch	L2 Regularization
Myo Dataset	0.01	0.001	0.0001	1e-2
DB5	0.01	0.001	1e-3	1e-3

4.3. Myo Dataset Classification Results

The loss and accuracy curves during training and testing on the Myo Dataset using the EMGNet model are shown in Figure 9.

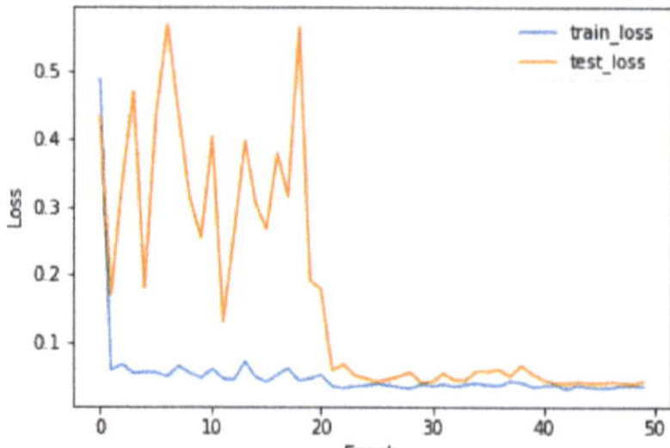
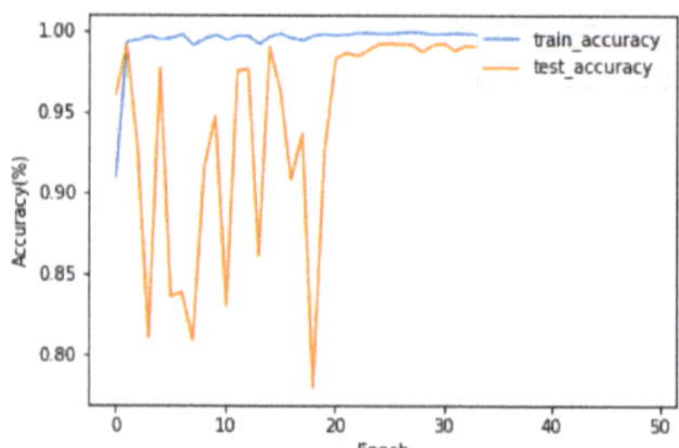

Figure 9. The loss and accuracy curves during training and testing on the Myo Dataset. when the training reaches convergence, there is no phenomenon where the accuracy of the training set is high and the accuracy of the test set is low, which is over-fitting.

As shown in Figure 9, the EMGNet network successfully completed the classification task in the Myo Dataset, and the over-fitting phenomenon did not appear in the training and testing. We tested the accuracy of our model and compared it with the current three most advanced methods. Table 3 shows the accuracy of each method. According to the results shown in Table 3, the accuracy of our proposed model is better than the current advanced methods.

Table 3. Classification accuracy on the Myo Dataset. (We used the same samples to test these methods).

Model	TD+LDA	TD+SVM	Enhanced TD+LDA	Enhanced TD+SVM	NinaPro+LDA	NinaPro+SVM	SampEn+LDA	SampEn+SVM	LCNN	CNN-LSTM	CWT+TL	CWT+EMGNet
Accuracy (%)	97.54	96.61	98.16	96.40	97.54	94.73	98.02	95.95	97.88	96.77	98.31	98.81
STD (%)	2.60	5.50	2.30	4.70	2.60	6.70	2.10	5.70	2.20	2.80	2.16	1.51

[1] The STD represents the standard deviation in accuracy for the 20 runs over the 17 participants.

4.4. NinaPro DB5 Dataset Classification Results

Figure 10 shows that the loss and accuracy curves during training and testing on exercise A of the NinaPro DB5 Dataset. During training of the EMGNet model, the phenomenon of overfitting appears. Reducing the number of layers in the EMGNet does not solve this problem.

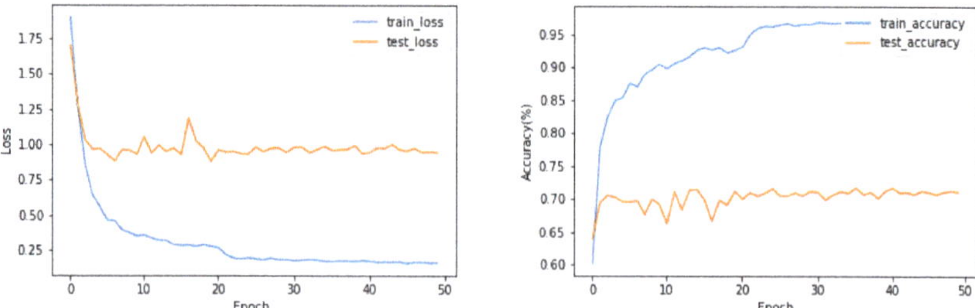

Figure 10. The loss and accuracy curves during training and testing on the exercise A of the NinaPro DB5 Dataset. During training and testing on the exercise A of the NinaPro DB5 Dataset, the phenomenon of overfitting appears.

Table 4 shows the accuracy of three subsets of DB5 Dataset. Time domain characteristics (TD) [5], Enhanced TD [4], Nina Pro Features [6,22] and SampEn Pipeline [23] were selected as the classification features of the classical machine learning algorithm (LDA and SVM). The LCNN and CNN_LSTM we proposed previously did not perform feature extraction on the data and directly processed the sEMG signal. The proposed method uses continuous wavelet transform (CWT) to process the data as the input of EMGNet. From the experimental results, we can conclude that the classification accuracy of our proposed EMGNet model is higher than that of the classical machine learning algorithm.

With the increase of the categories of gestures, the classification algorithms decline at different degrees. The degree of decline of EMGNet is lower than that of the classical machine learning algorithms (see Figure 11 and Table 4).

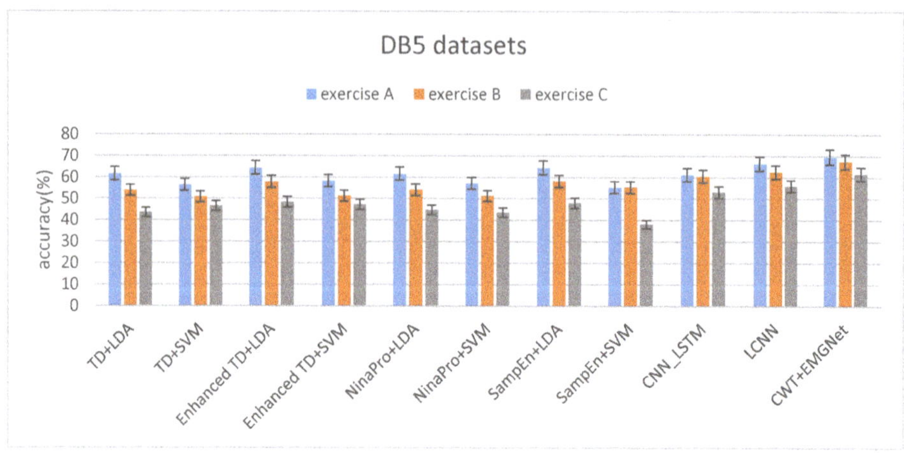

Figure 11. The average accuracy of three subsets on the DB5 Dataset.

Table 4. Classification accuracy on the DB5 Dataset. (We use the same samples to test these methods).

Model	TD+LDA	TD+SVM	Enhanced TD+LDA	Enhanced TD+SVM	NinaPro+LDA	NinaPro+SVM	SampEn+LDA	SampEn+SVM	CNN_LSTM	LCNN	CWT+EMGNet
EA(%)	61.61	56.40	64.36	58.32	61.66	57.12	64.53	55.28	61.21	66.38	69.62
EB(%)	53.80	50.70	57.80	51.17	54.05	51.18	58.20	55.34	60.46	62.51	67.42
EC(%)	43.48	46.49	48.26	47.18	44.61	43.46	47.89	38.12	52.99	55.87	61.63
$\frac{EA - EC}{EA}$	29.4%	17.5%	25.0%	19.1%	29.5%	23.9%	25.8%	31.0%	13.4%	15.8%	11.5%

[1] EA, EB and EC represent the accuracy of exercise A, exercise B and exercise C respectively.

The classification accuracy of both the classical machine learning method and EMGNet is not as high as that tested on the Myo Dataset (see Table 4). Two reasons are as follows:

(1) The DB5 dataset has a relatively small amount of data per gesture and a relatively large number of gesture categories. For example, there are at most 7 gestures in the Myo dataset, while the smallest exercise A in DB5 dataset has 12 gestures.

(2) There are a large number of similar gestures in the DB5 dataset. Figure 12 shows the sEMG signal spectrum of two processed samples in exercise A. The two samples shown in Figure 12c,d belong to different gesture categories, and it can be seen from Figure 12 that the channels with large fluctuations of the two gestures are almost the same. When the sample number of each gesture is insufficient, the model can easily misidentify it as the same gesture.

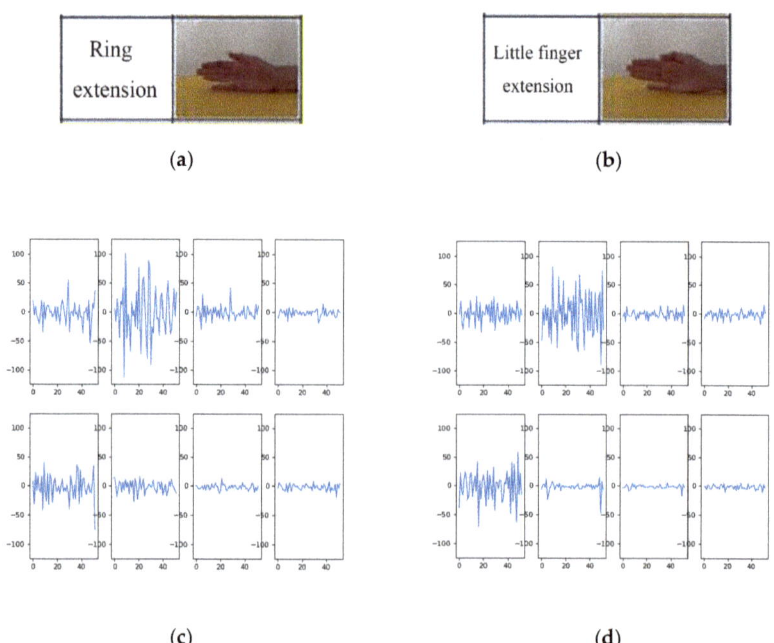

Figure 12. Parts (**a**) and (**b**) represent pictures of two different gestures in exercise A, Parts (**c**) and (**d**) show the spectrum diagrams of the sEMG signal generated by (**a**) and (**b**) respectively.

5. Conclusions

This paper presents a novel CNN architecture consisting of four convolutional layers and a max pooling layer with a compact structure and fewer parameters. The experimental results show that the proposed EMGNet not only reduces the complexity of the model but also improves the accuracy of the sEMG signal classification. It is highly competitive with the classical classifiers and deep learning frameworks currently used to classify sEMG signals, which also shows the potential of deep learning in classifying sEMG signals. (The elimination of transition time between successive gestures is explained in Appendix A).

Author Contributions: Writing—original draft, L.C.; Investigation, L.C. and Y.W.; Conceptualization and methodology, J.F.; Software, Y.W. and H.L.; Visualilization, H.L.; Experiment design, Y.W.; Data curation, Y.W.; Writing—review and editing, B.Z. All authors have read and agreed to the published version of the manuscript.

Funding: This work was supported by the Priority research & design program of Chongqing technology innovation and application demonstration, Grant No. cstc2017zdcy-zdyfX0036.

Conflicts of Interest: The authors declare no conflict of interest.

Abbreviations

The following abbreviations are used in this manuscript:

EMG	Electromyography
SEMG	surface Electromyography
Adam	adaptive moment estimation
CNN	convolutional neural network
TD	time domain
MAV	mean absolute value
SSC	slop sign changes
ZC	zero crossing
WL	waveform length
WT	wavelet transform
RMS	root mean square
AR	autoregressive coefficient
STFT	short time Fourier transform
CWT	continuous wavelet transform
LSTM	Long Short-Term Memory

Appendix A

We assume that each gesture can be held for a certain time. As shown in the figure below, if three sliding windows among four consecutive sliding windows are judged as gesture A, then we judge this signal as gesture A, if not, we judge this signal as the previous gesture remains unchanged. We use the same method for other gestures. (The numbers 4 and 3 used above are only examples, not actual numbers.) Our proposed method can identify the signals generated by different gestures during the signal acquisition process.

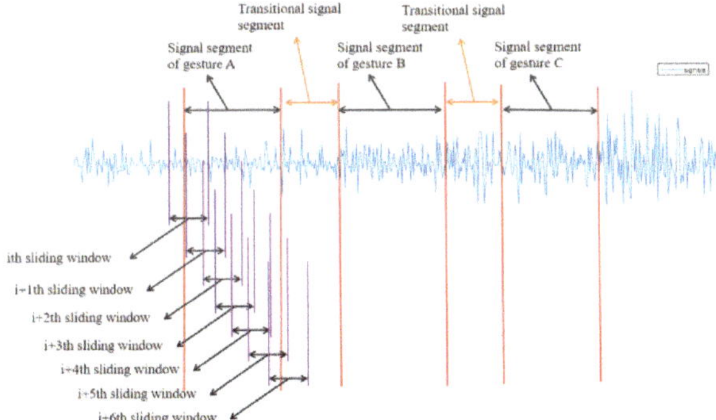

Figure A1. Schematic of continuous signal recognition.

References

1. Oskoei, M.A.; Hu, H. Myoelectric control systems—A survey. *Biomed. Signal Process. Control* **2007**, *2*, 275–294. [CrossRef]
2. Phinyomark, A.; Hirunviriya, S.; Limsakul, C.; Phukpattaranont, P. Evaluation of EMG feature extraction for hand movement recognition based on Euclidean distance and standard deviation. In Proceedings of the ECTI-CON2010: The 2010 ECTI International Confernce on Electrical Engineering/Electronics, Computer, Telecommunications and Information Technology, Chiang Mai, Thailand, 19–21 May 2010; pp. 856–860.

3. Phinyomark, A.; Phukpattaranont, P.; Limsakul, C. Feature reduction and selection for EMG signal classification. *Expert Syst. Appl.* **2012**, *39*, 7420–7431. [CrossRef]

4. Khushaba, R.N.; Kodagoda, S. Electromyogram (EMG) Feature Reduction Using Mutual Components Analysis for Multifunction Prosthetic Fingers Control. In Proceedings of the 2012 12th International Conference on Control Automation Robotics Vision (ICARCV), Guangzhou, China, 5–7 December 2012; pp. 1534–1539.

5. Englehart, K.; Hudgins, B. A robust, real-time control scheme for multifunction myoelectric control. *IEEE Trans. Biomed. Eng.* **2003**, *50*, 848–854. [CrossRef] [PubMed]

6. Pizzolato, S.; Tagliapietra, L.; Cognolato, M.; Reggiani, M.; Müller, H.; Atzori, M. Comparison of six electromyography acquisition setups on hand movement classification tasks. *PLoS ONE* **2017**, *12*, e0186132. [CrossRef] [PubMed]

7. Wu, Y.; Zheng, B.; Zhao, Y. Dynamic Gesture Recognition Based on LSTM-CNN. In Proceedings of the Chinese Automation Congress (CAC), Xi'an, China, 30 November—2 December 2018; pp. 2446–2450.

8. Côté-Allard, U.; Fall, C.L.; Drouin, A.; Campeau-Lecours, A.; Gosselin, C.; Glette, K.; Laviolette, F.; Gosselin, B. Deep Learning for Electromyographic Hand Gesture Signal Classification Using Transfer Learning. *IEEE Trans. Neural Syst. Rehabil. Eng.* **2019**, *27*, 760–771. [CrossRef] [PubMed]

9. Cote-Allard, U.; Fall, C.L.; Campeau-Lecours, A.; Gosselin, C.; Laviolette, F.; Gosselin, B. Transfer learning for sEMG hand gestures recognition using convolutional neural networks. In Proceedings of the 2017 IEEE International Conference on Systems, Man, and Cybernetics (SMC), Banff, AB, Canada, 5–8 October 2017; pp. 1663–1668.

10. Castellini, C.; Fiorilla, A.E.; Sandini, G. Multi-subject/dailylife activity EMG-based control of mechanical hands. *Neuroeng. Rehabil.* **2009**, *6*, 41. [CrossRef] [PubMed]

11. Krizhevsky, A.; Sutskever, I.; Hinton, G.E. Imagenet classification with deep convolutional neural networks. In *Proceedings of the Advances in Neural Information Processing Systems 25 (NIPS 2012), Lake Tahoe, NV, USA, 3–6 December 2012*; Neural Information Processing Systems Foundation Inc.: La Jolla, CA, USA, 2012.

12. Zia ur Rehman, M.; Waris, A.; Gilani, S.O.; Jochumsen, M.; Niazi, I.K.; Jamil, M.; Farina, D.; Kamavuako, E.N. Multiday EMG-Based Classification of Hand Motions with Deep Learning Techniques. *Sensors* **2018**, *18*, 2497. [CrossRef] [PubMed]

13. Allard, U.C.; Nougarou, F.; Fall, C.L.; Giguere, P.; Gosselin, C.; LaViolette, F.; Gosselin, B. A convolutional neural network for robotic arm guidance using sEMG based frequency-features. In Proceedings of the 2016 IEEE/RSJ International Conference on Intelligent Robots and Systems (IROS), Daejeon, Korea, 9–14 October 2016; pp. 2464–2470.

14. Tommasi, T.; Orabona, F.; Castellini, C.; Caputo, B. Improving Control of Dexterous Hand Prostheses Using Adaptive Learning. *IEEE Trans. Robot.* **2012**, *29*, 207–219. [CrossRef]

15. Patricia, N.; Tommasit, T.; Caputo, B. Multi-source Adaptive Learning for Fast Control of Prosthetics Hand. In Proceedings of the 2014 22nd International Conference on Pattern Recognition, Stockholm, Sweden, 24–28 August 2014; pp. 2769–2774.

16. Orabona, F.; Castellini, C.; Caputo, B.; Fiorilla, A.E.; Sandini, G. Model adaptation with least-squares SVM for adaptive hand prosthetics. In Proceedings of the IEEE International Conference on Robotics and Automation (ICRA), Kobe, Japan, 12–17 May 2009; pp. 2897–2903.

17. Atzori, M.; Cognolato, M.; Muller, H. Deep learning with convolutional neural networks applied to electromyography data: A resource for the classification of movements for prosthetic hands. *Front. Neurorobotics* **2016**, *10*, 9. [CrossRef] [PubMed]

18. Geng, W.; Du, Y.; Jin, W.; Wei, W.; Hu, Y.; Li, J. Gesture recognition by instantaneous surface EMG images. *Sci. Rep.* **2016**, *6*, 36571. [CrossRef] [PubMed]

19. Du, Y.; Jin, W.; Wei, W.; Hu, Y.; Geng, W. Surface EMG-Based Inter-Session Gesture Recognition Enhanced by Deep Domain Adaptation. *Sensors* **2017**, *17*, 458. [CrossRef] [PubMed]

20. He, K.; Zhang, X.; Ren, S.; Sun, J. Delving Deep into Rectifiers: Surpassing Human-Level Performance on ImageNet Classification. In Proceedings of the 2015 IEEE International Conference on Computer Vision (ICCV), Santiago, Chile, 7–3 December 2015; pp. 1026–1034.

21. Kingma, D.P.; Ba, J. Adam: A method for stochastic optimization. *arXiv* **2014**, arXiv:1412.6980.

22. Atzori, M.; Gijsberts, A.; Castellini, C. Electromyography data for non-invasive naturally-controlled robotic hand prostheses. *Sci. Data* **2014**, *1*, 1–3. [CrossRef] [PubMed]

23. Phinyomark, A.; Quaine, F.; Charbonnier, S.; Serviere, C.; Tarpin-Bernard, F.; Laurillau, Y. EMG feature evaluation for improving myoelectric pattern recognition robustness. *Expert Syst. Appl.* **2013**, *40*, 4832–4840. [CrossRef]
24. Zhang, X.; Zhou, P. Sample entropy analysis of surface EMG for improved muscle activity onset detection against spurious background spikes. *J. Electromyogr. Kinesiol.* **2012**, *22*, 901–907. [CrossRef] [PubMed]
25. Khushaba, R.N.; Kodagoda, S.; Takruri, M.; Dissanayake, G. Toward improved control of prosthetic fingers using surface electromyogram (EMG) signals. *Expert Syst. Appl.* **2012**, *39*, 10731–10738. [CrossRef]
26. LeCun, Y.; Haffner, P.; Bottou, L.; Bengio, Y. Object Recognition with Gradient-Based Learning. In *Shape, Contour and Grouping in Computer Vision*; Forsyth, D.A., Mundy, J.L., di Gesú, V., Cipolla, R., Eds.; Lecture Notes in Computer Science; Springer: Berlin/Heidelberg, Germany, 1999; pp. 319–345.
27. Simonyan, K.; Zisserman, A. Very deep convolutional networks for large-scale image recognition. *arXiv* **2014**, arXiv:1409.1556.

Article

RFI Artefacts Detection in Sentinel-1 Level-1 SLC Data Based On Image Processing Techniques

Agnieszka Chojka [1,*], Piotr Artiemjew [2] and Jacek Rapiński [1]

[1] Faculty of Geoengineering, University of Warmia and Mazury in Olsztyn, 10-719 Olsztyn, Poland; jacek.rapinski@uwm.edu.pl
[2] Faculty of Mathematics and Computer Science, University of Warmia and Mazury in Olsztyn, 10-719 Olsztyn, Poland; artem@matman.uwm.edu.pl
*   Correspondence: agnieszka.chojka@uwm.edu.pl

Received: 31 March 2020; Accepted: 19 May 2020; Published: 21 May 2020

Abstract: Interferometric Synthetic Aperture Radar (InSAR) data are often contaminated by Radio-Frequency Interference (RFI) artefacts that make processing them more challenging. Therefore, easy to implement techniques for artefacts recognition have the potential to support the automatic Permanent Scatterers InSAR (PSInSAR) processing workflow during which faulty input data can lead to misinterpretation of the final outcomes. To address this issue, an efficient methodology was developed to mark images with RFI artefacts and as a consequence remove them from the stack of Synthetic Aperture Radar (SAR) images required in the PSInSAR processing workflow to calculate the ground displacements. Techniques presented in this paper for the purpose of RFI detection are based on image processing methods with the use of feature extraction involving pixel convolution, thresholding and nearest neighbor structure filtering. As the reference classifier, a convolutional neural network was used.

Keywords: RFI; artefacts; InSAR; image processing; pixel convolution; thresholding; nearest neighbor filtering; deep learning

---

1. Introduction

Since NASA launched the first satellite (at that time known as the Earth Resources Technology Satellite) of the United States' Landsat program in 1972 [1], satellite remote sensing has developed strongly. They are currently placing newer satellites in orbit around Earth, equipped with advanced sensors for Earth monitoring. Among them worthy of emphasis are satellites with Synthetic Aperture Radar (SAR) instruments on board.

SAR is an active microwave remote sensing tool [2] with wide spatial coverage, fine resolution, all-weather and day-and-night image acquisition capability [3,4]. These features allow the ability to use SAR images for a multitude of scientific research, commercial and defense applications ranging from geoscience to Earth system monitoring [2,4,5].

Interferometric Synthetic Aperture Radar (InSAR) technology exploits differences in the phase of the waves returning to the satellite of at least two complex SAR images to generate, for instance, surface topography and deformation maps or Digital Elevation Models (DEMs) [6–8].

Among various InSAR techniques, satellite Differential Interferometric Synthetic Aperture Radar (DInSAR) has emerged as a powerful instrument to measure surface deformation associated with ground subsidence on a large scale with centimeter to millimeter accuracy [9–11].

DInSAR exploits the phase information of SAR images to calculate the ground displacements between two different satellite acquisitions [12]. The Permanent Scatterers InSAR (PSInSAR) method, an upgrade of DInSAR, for analytical purposes uses long stacks of co-registered SAR images to identify coherent points that provide consistent and stable response to the radar on board of a satellite [13].

Phase information obtained from these persistent scatterers is used to derive the ground displacement information and its temporal evolution [9].

Thanks to the increasing availability of a large amount of SAR data from such missions like ALOS-2, COSMO-SkyMed, PAZ, RADARSAT-2, Sentinel-1 or TerraSAR-X [3,11] and due to high-quality images covering a wide area [14,15], in the last decades, Earth observation techniques have become a valuable and indispensable remote sensing tool in geophysical monitoring of natural hazards such as earthquakes [16], volcanic activity [17] or landslides [18], mine subsidence monitoring [19] and structural engineering, especially monitoring of subsidence [20] and structural stability of buildings [15] or bridges [10].

DInSAR and PSInSAR are complementary methods and both have essential advantages and some disadvantages [16]. By way of illustration, PSInSAR is considered to be more precise than DSInSAR, because it requires about 15–20 SAR data acquisitions for a successful result, whereas DInSAR needs only 2 [13].

Despite the obvious benefits, InSAR technology has some limitations. InSAR measurements are often affected by various artefacts that not only make interpreting them more challenging, but also affect the reliability and accuracy of its outcomes.

One of the most significant is the effect of the atmosphere. In general, it results from the phenomenon of electromagnetic waves delay when traveling through the troposphere and accelerate when traveling through the ionosphere [3]. Atmospheric artefacts are usually strongly correlated with the topography (elevation) and the proximity of the sea [11,21]. Over the past decades, numerous methods were investigated to identify and mitigate these artefacts e.g., [3,8,11,22,23].

Another type of InSAR data failure described in the literature is a border noise [24–27]. This undesired processing artefact appeared in all of the Sentinel-1 GRD products generated before March 2018 [24]. Although this problem has been solved for the newly generated products, it did not cover the entire range of products and researchers still develop new methods and tools to effectively detect and remove this particular type of noise [24–27].

Contrary to the unwanted artefacts influence on InSAR data, Bouvet [28] proposed a new indicator of deforestation based on geometric artefact, called the shadowing effect. It appears in SAR images in the form of a shadow at the border of the deforested patch.

The primary issue of this paper is one more contamination frequently appearing in SAR images that is called Radio-Frequency Interference (RFI). Since the work is in progress on the development of the automatic monitoring system for high-energy para-seismic events, the urgent matter is to elaborate the effective method supporting the automatic PSInSAR processing workflow by the removal of faulty SAR data (with artefacts) that can lead to misinterpretation of the final results.

Our main goal is to find the easiest to implement technique for marking images with RFI artefacts. The solution presented in this paper for the purpose of RFI detection was based on image processing methods with the use of feature extraction involving pixel convolution, thresholding and nearest neighbor structure filtering techniques. As the reference classifier we used a convolutional neural network.

After a short introduction on the characteristics of RFI, common forms and sources of this artefact, included in Section 2, the used materials are presented in Section 3. The applied methodology and techniques are depicted in Section 4. Their results are presented and comprehensively discussed in Section 5. The main findings and recommendations for future work are summarized in Section 6.

2. RFI Artefact

RFI is defined as 'the effect of unwanted energy due to one or a combination of emissions, radiations or inductions upon reception in a radio communication system, manifested by any performance degradation, misinterpretation or loss of information which could be extracted in the absence of such unwanted energy' according to Article 1.166 of the International Telecommunication Union Radio Regulations [29].

In the SAR images these incoherent electromagnetic interference signals usually appear as various kind of bright linear features [30,31] like bright stripes with curvature or dense raindrops [2,4]. RFI introduces artefacts in image by slight haziness [2,31] that acutely degrade its quality [4,30]. Such affected images may lead to wrong interpretation process and results [2].

The reason for RFI contamination for SAR is that many different radiation sources operate with the same frequency band as the SAR system [2,30]. In general they can be grouped into terrestrial and space-borne sources [2]. Most of these incoherent electromagnetic interference signals are emitted by terrestrial commercial or industrial radio devices [2], e.g., communication systems, television networks, air-traffic surveillance radars, meteorological radars, radiolocation radars, amateur radios and other, mainly military-based, radiation sources [2,4,30,31]. An example of space-borne RFI source are signals broadcasting from other satellites, such as global navigation satellite systems (GNSSs) constellations, communication satellites or other active remote sensing systems [2].

Over the past years, great efforts have been made to better understand RFI effects and to develop robust methods for detecting and mitigating this artefact, e.g., [32–37], in particular from SAR data. See [2] for general review. Although in the majority of cases, research is focused on the recognition and removal of RFI signatures from L-band SAR data, where this artefact is commonly observed [30]. In case of SAR systems, the most susceptible for RFI effects are signals operating in the low band frequency region, such as P, L, S and even C-band [2,30,31]. Studies are usually conducted on raw data [2,4,30,31,38]. To fill this gap in the literature, our research addressed the detection of RFI artefacts in SAR data, especially in recently available Sentinel-1 data. Additionally, our work meets the recommendations, proposed among others by Tao [2] and Itschner [39], concerning the application of artificial intelligent techniques, such as deep learning methods, for RFI recognition.

3. Materials

This section shortly describes the materials we used in our investigation. We depict the source data and datasets used to carry out experiments and tests of our solution.

3.1. Source Data

A set of dedicated satellites, the Sentinel families, are developed by the European Space Agency (ESA) for the operational needs of the Copernicus programme [40]. The European Union's Earth observation programme delivers operational data and information services openly and freely in a wide range of applications in a variety of areas, such as urban area management, agriculture, tourism, civil protection, infrastructure and transport [41].

The Sentinel-1 mission is a polar-orbiting, all-weather, day-and-night C-band synthetic aperture radar imaging mission for land and ocean services. It is based on a constellation of two satellites: Sentinel-1A was launched on 3 April 2014 and Sentinel-1B on 25 April 2016 [40,42].

Sentinel-1 data are intended to be available systematically and free of charge, without limitations, to all data users including the general public, scientific and commercial users [40]; however, the most reachable for the majority of users are Level-1 products, provided as Single Look Complex (SLC) and Ground Range Detected (GRD) [40].

Level-1 Single Look Complex (SLC) products, as other Sentinel-1 data products, are delivered as a package containing, among others, metadata, measurement data sets and previews [40].

The Sentinel-1 SAR instrument mainly operates in the Interferometric Wide (IW) swath mode over land [26,40] that is one of four Sentinel-1 acquisition modes. IW uses the Terrain Observation with Progressive Scans SAR (TOPSAR) technique [43] and provides data with a large swath width of 250 km at 5 m × 20 m (range × azimuth) spatial resolution for single look data.

In this study, we used quick-look images contained in the preview folder, a part of Sentinel-1 data package. As in our case we have dual polarization products, data are represented by a single composite color image in RGB [40]. Because quick-look data has a lower resolution version of the source image, it is easy to detect RFI artefacts and then exclude data contaminated by RFI from further processing, thus improving the whole PSInSAR processing workflow, which in our case required a stack of 25–30 SAR images to calculate the ground displacements. Faulty data included in this dataset would lead to misinterpretation of the processing results. Besides, elimination of failure data does not affect the quality and reliability of the final data processing results as data are obtained continuously and observations can be repeated with a frequency better than 6 days, considering two satellites (Sentinel 1A and 1B) and both ascending and descending passes [40].

3.2. Experimental Datasets

Due to these advantages of Sentinel-1 data products, in summary, high and free of charge availability, wide ground coverage and fine resolution, these data are used in developing the automatic monitoring system for high-energy para-seismic events and its influence on the surface in the study area, the Żelazny Most, with the use of GNSS, seismic and PSInSAR techniques. The Żelazny Most is the largest tailings storage facility in Europe and the second largest in the world [44,45]. It is an integral part of copper production technological chain [46] that since 1977 collects the tailings from the three mines of the KGHM Polska Miedź (formerly - the Polish State Mining and Metallurgical Combine) [45]. This project exploits Level-1 SLC products in IW mode. Images comprising the area of interest are regularly acquired since September 2018 and processed using PSInSAR technique.

RFI characteristic depends strongly on the geographic position of data acquisition [2]. According to the RFI power map with continental coverage over Europe [38] our study area was located in potential RFI-affected zone. On the other hand, the IEEE GRSS RFI Observations Display System [47] did not indicate any irregularities in C-band frequency (ranging from 4–8 GHz) over this area.

In order to carry out experiments and to verify the proposed approach for RFI artefacts detection, we used 3 different quick-look datasets, named RIYADH, MOSCOW and ASMOW, for the needs of our research. This selection was based on above mentioned RFI power map and IEEE GRSS RFI Observations Display System. Each collection includes both correct images and images with various levels of RFI contamination. RIYADH dataset consists of 136 images that were collected between May 2015 and January 2019, and comprise the area of the capital of Saudi Arabia. The MOSCOW set includes 99 images acquired between January and December 2019. These images cover the area surrounding the capital of Russia. The ASMOW collection consists of 53 images covering the study area, located in southwest Poland, in Lower Silesia, east of the town of Polkowice in the municipality of Rudna. These data were acquired from September 2018 to February 2020.

4. Methodology

In this section, we introduce basic information about the techniques we used in the experimental part. Here we start by discussing an example of how to store an image in digital form.

4.1. Digital Representation of the Image

A digital image can be simply represented by binary numbers of pixel color saturation in the relevant system [48,49]. Greyscale image pixels are represented by single Bytes that take decimal values from 0 to 255. In binary format from 00000000 to 11111111. Sample change 01101011 ⇒ 01101010 would not be noticeable to the human eye and in that way we could exploit that shortcoming of a human eye to hide data. RGB image pixels are represented by triples of Bytes which take decimal values from 0 to 255. For example, white color in decimal form is represented by $(255, 255, 255)$ and in binary form by $(11111111, 11111111, 11111111)$. Data in digital systems are often shown in hexadecimal form for reading convenience. The white color is (FF, FF, FF) where F is the letter of the hexadecimal system alphabet to which a decimal value of 15 is assigned.

In the pre-processing step, the images are converted to greyscale representation. This conversion makes it very easy to use feature extraction techniques. Next, we discuss a sample conversion.

4.2. Conversion from RGB to Greyscale

$$new_{pixel} = R_{old_{pixel}} * 0.292 + G_{old_{pixel}} * 0.594 + B_{old_{pixel}} * 0.114 \tag{1}$$

The above formula is a transformation from OpenCV library [50]. An example of a conversion is in Figure 1.

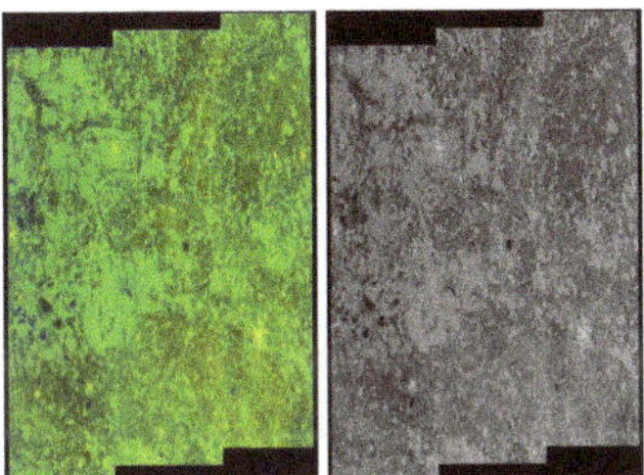

Figure 1. ASMOW quick-look—demonstration of conversion to greyscale using the formula (Section 4.2).

In the next section we discuss the technique of image features extraction based on pixel convolution [51–53].

4.3. Feature Extraction Based on Pixel Convolution

In Figure 2 we have an example of how a pixel convolution works. We have used the 3×3 Gaussian blur mask.

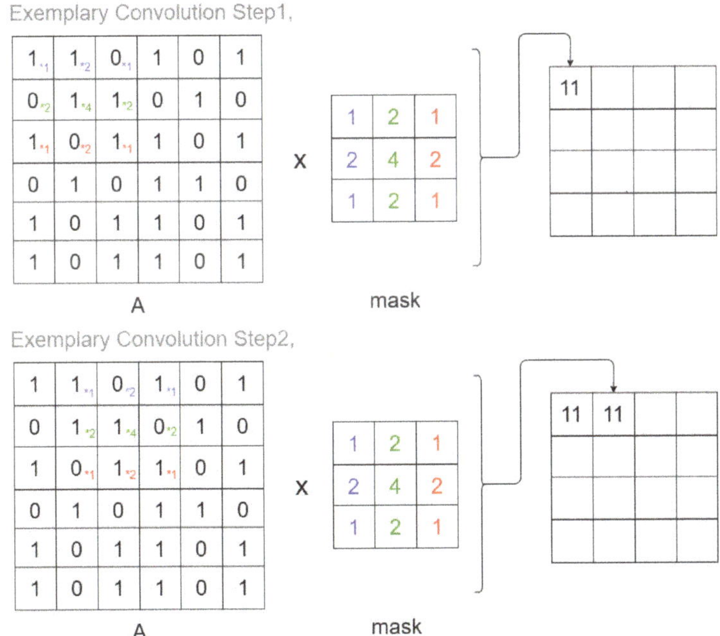

Figure 2. Exemplary convolution—first two steps based on Gaussian blur mask 3×3, $A_{x,y} \times mask = \sum_{i=1}^{mask\_width} \sum_{j=1}^{mask\_height} A_{x+i-1} * mask_{ij}$, (x, y) are the top left coordinates of convolved pixels.

Other useful techniques that were applied are image filtering techniques by means of thresholding [51,52,54]. The method allowed us to focus on areas that are visible with a fixed, properly defined frequency or belonging to a defined range.

4.4. Thresholding

We used two types of thresholding [52]. The first type consists of giving the image pixels that do not exceed a certain fixed normalized threshold of color saturation.

The second method consisted of using a histogram of pixel values and filtering out those which (e.g., giving black color) belong to a fixed frequency range.

The last method we tested for an artefact classifier was the use of structural filtering with the nearest pixel neighbors [51,52,55,56].

4.5. Nearest Neighbor Structure Filtering

We also applied the technique of filtering the image structure by replacing the pixel with its closest neighbor in the sense of Manhattan distance [57,58]. Filtering allows you to eliminate unstructured pixels in the image. This step allowed us to better discriminate against images with artefacts.

As a reference classifier defined in the following section, we used the Convolutional Neural Network (CNN) [59,60].

4.6. Exemplary Deep Neural Network Architecture as Referenced Classifier

To verify the effectiveness of our artefacts extraction method, we used a simple deep neural network [59,61] defined in this section for classification. We used Python related tools such as PyTorch, TorchVision and NumPy. The visualization of results was performed using Seaborn and Matplotlib. A transformation was made when loading images, scaling everything to 400×600 pixels to ensure the same size of the input for the network. The data were divided into train and test sets in 80/20 ration in a random way. A simple network with two convolutional layers, linear transformations and pooling was proposed. The activation function was RELU ($f(colour\ saturation) = max(0, colour\ saturation)$), and loss function took the form of categorical Cross Entropy (thus it can be higher than one).

The Adam optimizer [62] was used. The training was done over 15 epochs. To split the data, the following function was used:

$$train\_dataset,\ test\_dataset = torch.utils.data.random\_split(dataset, [train\_size, test\_size])$$

A detailed definition of the neural network is given in Listing 1.

Listing 1: Neural network configuration.

```python
# Define network
class Net(nn.Module):
def __init__(self):
super(Net, self).__init__()
self.conv1 = nn.Conv2d(3, 6, 5)
self.pool = nn.MaxPool2d(2, 2)
self.conv2 = nn.Conv2d(6, 16, 5)
self.fc1 = nn.Linear(16 * 96 * 146, 120)
self.fc2 = nn.Linear(120, 84)
self.fc3 = nn.Linear(84, 2)

def forward(self, x):
x = self.pool(F.relu(self.conv1(x)))
x = self.pool(F.relu(self.conv2(x)))
x = x.view(-1, 16 * 96 * 146)
x = F.relu(self.fc1(x))
x = F.relu(self.fc2(x))
x = self.fc3(x)
return~x

net = Net()
```

To make it easier to reproduce our research results, here are the versions of the libraries we installed to carry out the experiments (see Listing 2) .

Listing 2: Versions of the libraries used.

```
#version for GPU
conda install -c pytorch torchvision
conda install pytorch seaborn numpy matplotlib
#version for CPU
conda install pytorch==1.2.0 torchvision==0.4.0 cpuonly -c pytorch
conda install seaborn numpy matplotlib
```

5. Experimental Session Details

To carry out the research we have used the bitmap library.hpp—see [63] library, which allows simple image processing. We used RIYADH (37 objects with strong defects, 72 without damages and 27 with weak artefacts), MOSCOW (53 correct, 26 difficult and 20 with strong artefacts) and ASMOW dataset (data from our project, we have only few defects; 50 correct objects and only 3 damaged) to check our methods. With RIYADH and MOSCOW collections, our aim was to find a pre-processing technique that will improve learning by means of a deep neural network. With ASMOW (due to its small class with artefacts), we were looking for a technique to effectively mark artefacts in an image, without classifying them with a network. In an experiment with deep neural network classification—see details in Section 4.6—the image sets were divided into a training subset where the network is taught and the validation test set, 20 percent of the objects on which the final neural network was tested. To estimate the quality of the classification on RIYADH and MOSCOW data we used the Monte Carlo Cross Validation [59,64] 5 technique (MCCV5, i.e., five times train and test), presenting average results. In the tests, we considered two binary classifications and one of three classes. In case of separation of three classes, in the network settings (from Section 4.6) we set the number of classes and we consider three outputs from the network.

5.1. Artefacts Detection in the RIYADH Dataset

Consider the examples of problems with artefacts detection from our experimental groups of quick-look images. The first one is the RIYADH quick-look dataset. In Figure 3 we present sample images with artefacts and undamaged ones.

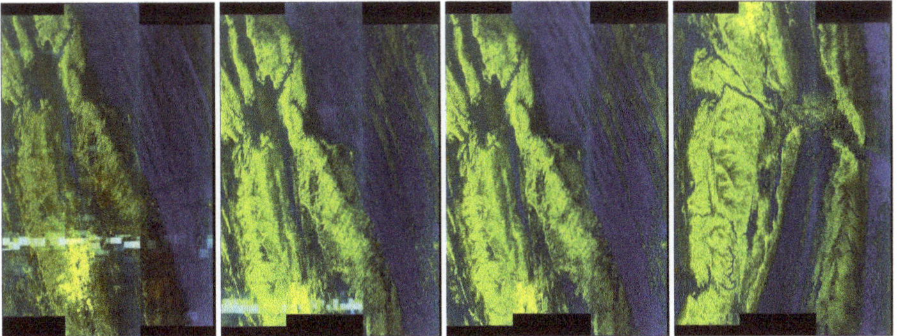

Figure 3. RIYADH quick-looks—exemplary pictures—the two on the left with artefacts, two on the right undamaged.

Next, we describe the steps of searching for our method of extracting artefacts on the RIYADH collection.

The first step was to convert the image into a greyscale form—according to the formula from Section 4.2.

In the next step, we tested the detectors of features based on pixel convolution.

5.2. Overview of Feature Detectors Based on Convolution

In this section, we present the selected feature detectors on the basis of an image with artefacts—see the left picture (Figure 4). We treated this step as a pre-processing stage allowing us to extract specific features of images. For example, masks allowed us to create frames of artefacts potentially useful for shape detection in the image. Some examples of well-functioning masks in this context are seen in Figures 4 and 5. Consider the effect of selected popular filters [49] usage: in the middle picture (Figure 4) we have Sobel's gradient sharpening, in the right picture (Figure 4) we have Laplace filtering (sharpening), in the left picture (Figure 5) we have Gaussian blur mask and finally in the right picture (Figure 5) we have Emboss mask usage. According to our tests, Gaussian blur filter is one the best at extracting artefacts from the RIYADH images. We applied this mask to the hybrid technique together with thresholding.

We have considered the following feature detectors:

```
Sobel's gradient sharpening        Laplace filtering (sharpening)
1   0 -1                              0 -1  0
2   0 -2                             -1  5 -1
1   0 -1                              0 -1~0

Gaussian blur mask                     Emboss mask
1   2   1                            1   0   0
2   4   2                            0   0   0
1   2   1                            0   0  -1
```

Figure 4. On the left side we have sample image for testing feature detectors. In the middle there is convolution based on Sobel's gradient sharpening. On the right convolution based on Laplace filtering (sharpening).

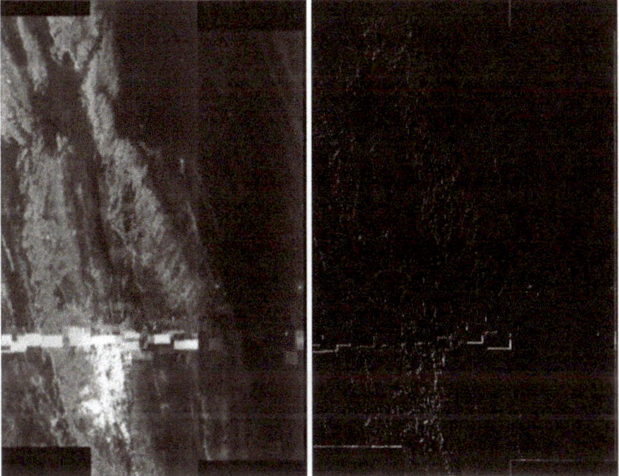

Figure 5. On the left side we have convolution based on Gaussian blur mask. On the right convolution based on Emboss mask.

We provide more extensive testing of the Gaussian blur on the RIYADH data in Figure 6.

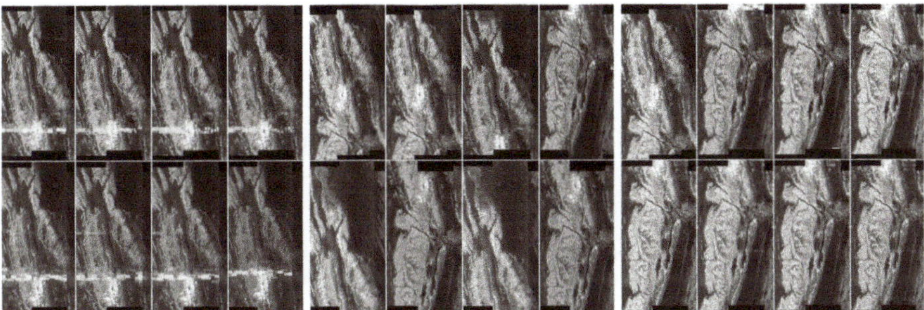

Figure 6. Comparison of the effects of Gaussian blur 3 × 3 convolution in the pictures, from the left: with artefacts, without artefacts and difficult—the discrimination of classes is not explicit.

5.3. Application of Thresholding

Next, we discuss the results of experiments with the application of thresholding techniques from Section 4.4. In Figure 7 we have a demonstration of the first threshold method from Section 4.4. As we can observe, discrimination is not very accurate when using this technique alone. Using the hybrid method, combining Gaussian blur and thresholding—see Figure 8, we get a slightly better artefacts separation effect. The application of the second thresholding method based on the pixel histogram can be seen in Figure 9.

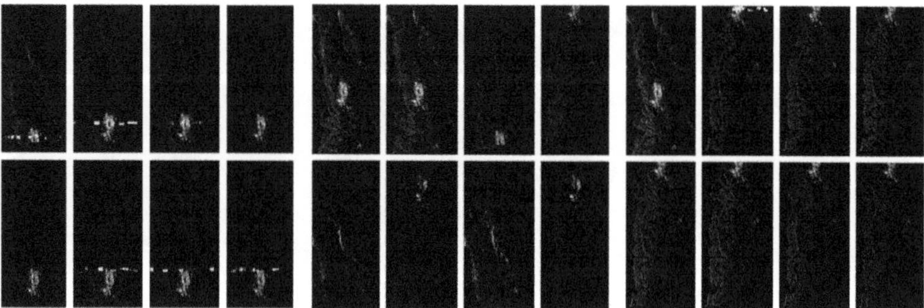

Figure 7. Comparison of the effects of thresholding in the pictures, from the left: with artefacts, without artefacts and difficult—the discrimination of classes is not explicit.

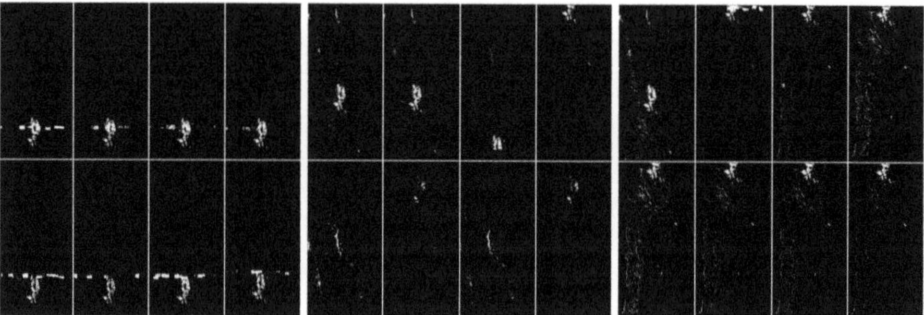

Figure 8. Demonstration of the application of two steps, the convolution with Gaussian blur 3×3 and thresholding 0.1, from the left: with artefacts, without artefacts and difficult. Example of a Python (Jupiter Notebook) thresholding. The level of artefacts separation is higher after these two steps. Classification may consist in calculating the narrowest possible uniform white surface.

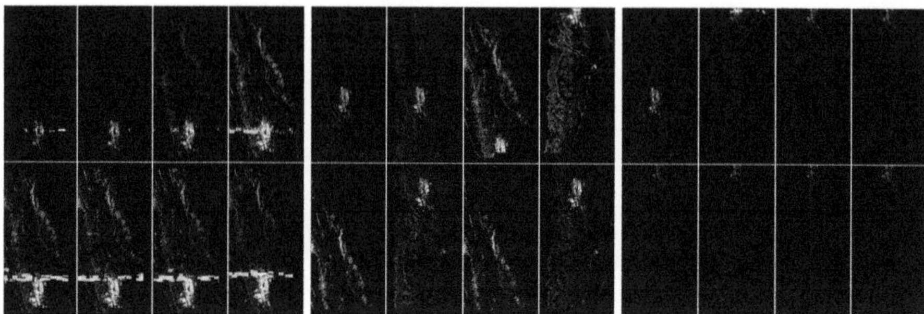

Figure 9. Visualization of the application of pixel frequency threshold. In this case a histogram is created for $[0, 255]$, and the frequency threshold is set to 300. All color values that occur more often are replaced by black. After thresholding on the basis of frequency, it is clear that the artefacts have become exposed; however, there are also many unnecessary white pixels in the picture. From the left: with artefacts, without artefacts and difficult.

5.4. Application of Nearest Neighbor Filtering

In Figures 10 and 11, we present hybrid methods which involved thresholding based on pixel histogram and noise filtering technique using the nearest neighbors of pixels. The use of these methods clearly shows the possibility of separating artefacts.

Figure 10. Visualization of the application of pixel frequency threshold and one nearest neighbor filtering. In this case, a histogram was created for $[0, 255]$, and the frequency threshold was set to 300. All color values that occur more often are replaced by black. The next step of the image processing was to swap pixels with their nearest neighbors. This step filtered out some unstructured pixels. The level of discrimination has clearly increased. From the left: with artefacts, without artefacts and difficult.

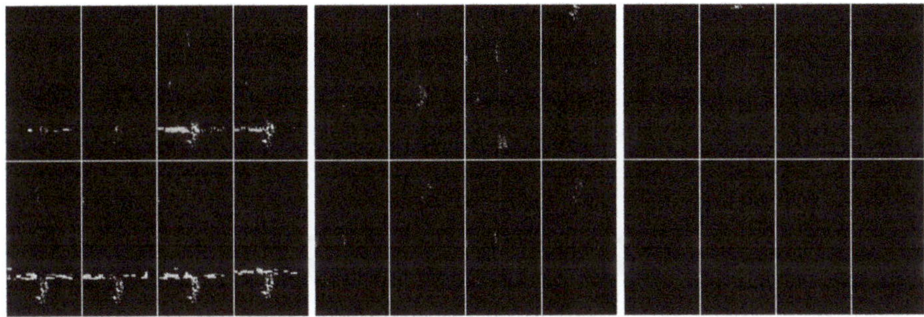

Figure 11. Visualization of the application of pixel frequency threshold and one nearest neighbor filtering. In this case a histogram is created for $[0, 255]$, and the frequency threshold is set to 300. All color values that occur more often are replaced by black. The next step of the image processing was to swap pixels with their nearest neighbors. In the present case, we have done the swap procedure twice. This step filtered out some unstructured pixels. The level of discrimination has clearly increased. From the left: with artefacts, without artefacts and difficult.

5.5. Summary of Results for Detection Artefacts in RIYADH Dataset

Summarizing the results obtained, we can state that the most effective technique of detecting artefacts, from among the ones we have studied, is the use of thresholding based on histograms of pixel values and noise filtering using the closest neighbors method. The result of good performance of this combination was predictable, because artefact colors usually have a low frequency in pixel histograms; therefore, it is quite easy to visualize clear artefacts. Then the filtering method allows us to remove single, unstructured pixels. Let us present the results (before and after the application of our method) on the RIYADH dataset using the neural network described in Section 4.6. The exemplary epochs of learning for class *ok* (without artefacts) and *er* (with strong artefacts) (before and after the application of our method) is available in Figure 12. A similar result for classification of *ok* class and class *difficult* (with weak artefacts) is in Figure 13. In Figure 14 we have learning between the three mentioned classes.

The exact results from the MCCV5 test can be seen in Table 1. The results are promising, considering the separation of the undamaged image class from the heavily damaged ones—the reference classification level was about 74 percent, and after applying our method, the degree of class distinction (artefacts detection) increased to a level close to 92 percent of accuracy. Similarly, the detection accuracy of weak artefacts has increased from about 68 percent to 84 percent. A spectacular increase can be observed in the level of discrimination between undamaged images and weak and strong artefacts from 54 percent to nearly 81 percent on validation set. Seeing Figures 12–14, the process of learning the neural network after applying our method seems to be more stable. Standard deviation (see Table 1) without pre-processing reaches 13 percentage points, after applying our method, it is within 5, 6 percentage points for variants (er vs. ok), and classifying all three classes. In case of the classification of weak artefacts is it similar in both cases within 8 percentage points.

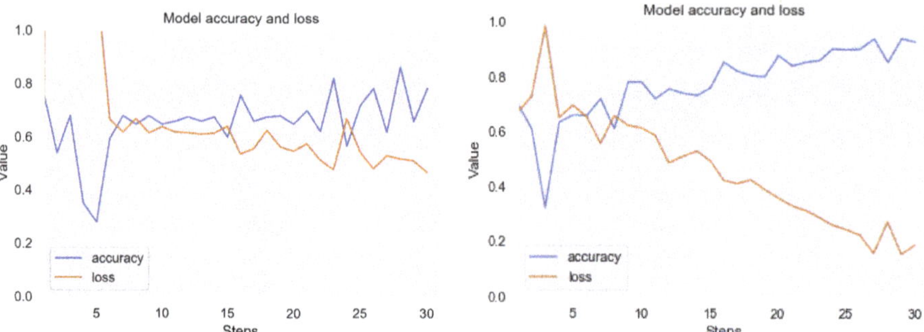

Figure 12. Exemplary learning effect before (left side) and after using our technique (right side). Result for RIYADH dataset. Class ok (without artefacts) vs. class er (with strong artefacts). The line graphs shows accuracy (blue) and loss (orange) for 15 epochs (30 steps from batch). Although the image dataset is relatively small for Convolutional Neural Network (CNN), there can be observed a good trend in increasing accuracy, while minimizing loss.

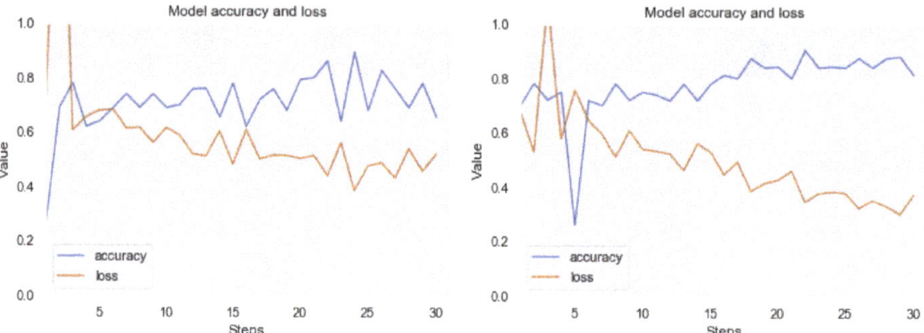

Figure 13. Exemplary learning effect before (left side) and after using our technique (right side). Result for RIYADH dataset. Class ok (without artefacts) vs. class difficult (with weak artefacts). The line graphs shows accuracy (blue) and loss (orange) for 15 epochs (30 steps from batch). Although the image dataset is relatively small for CNN, there can be observed a good trend in increasing accuracy, while minimizing loss.

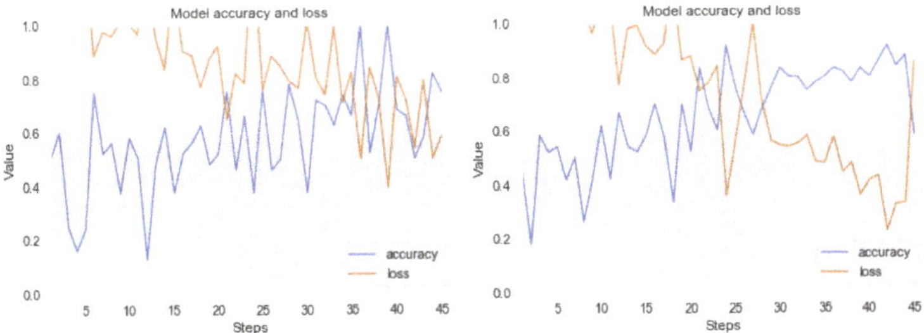

Figure 14. Exemplary learning effect before (left side) and after using our technique (right side). Result for RIYADH dataset. Class ok (without artefacts) vs. class er (with strong artefacts) vs. class difficult (with weak artefacts). The line graphs shows accuracy (blue) and loss (orange) for 15 epochs (30 steps from batch). Although the image dataset is relatively small for CNN, there can be observed a good trend in increasing accuracy, while minimizing loss. The accuracy of classification on the validation set came out nearly 55 percent.

Table 1. Summary of results for RIYADH–MCCV5 technique; nil.ok.er = accuracy of classification of er and ok classes before pre-processing, ok.er = accuracy of classification of er and ok classes after application of our method, nil.ok.tr, ok.tr, nil.all, all = analogous parameters, showing the class separation before and after application of our technique, SD = standard deviation of results, avg = average result.

Test No.	nil.ok.er	ok.er	nil.ok.tr	ok.tr	nil.all	All
1	0.8181	1.0	0.7	0.8571	0.5358	0.7586
2	0.7727	0.8696	0.7	0.9048	0.3928	0.8621
3	0.5	0.9130	0.75	0.7143	0.5714	0.8621
4	0.7727	0.9565	0.55	0.8571	0.6071	0.7586
5	0.8181	0.8696	0.7	0.9048	0.6071	0.8276
avg	0.73632	0.92174	0.68	0.84762	0.54284	0.8138
SD	0.134042986	0.056671051	0.075828754	0.078251307	0.088932463	0.052321841

Next, we discuss the results on artefacts collection from the MOSCOW database.

5.6. Results for MOSCOW Dataset

We considered three classes: *0* (no artefacts), *I* (weak artefacts) and *II* (strong artefacts)—see Figure 15. We conducted experiments with similar network settings as in Section 5.1. The size of classes *0*, *I* and *II* was 53, 26 and 20, respectively.

Figure 15. MOSCOW quick-looks—exemplary pictures, from the left to right, without artefacts (class *0*), with weak artefacts (class *I*) and with strong artefacts (class *II*).

We applied the same steps as for the RIYADH dataset. We used a frequency range of 1000 for thresholding. Samples of data after applying our method on the MOSCOW collection can be seen in Figure 16.

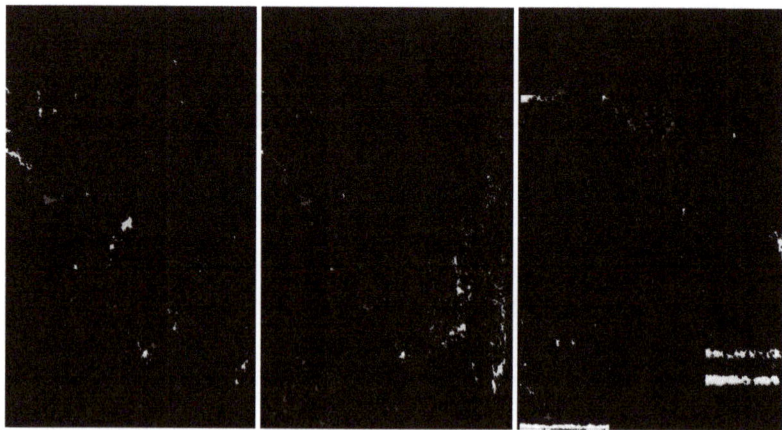

Figure 16. MOSCOW quick-looks—exemplary pictures, from the left to right, without artefacts (class *0*), with weak artefacts (class *I*) and with strong artefacts (class *II*).

Next, we briefly present our results.

5.7. Summary of Results for MOSCOW Dataset

Detailed test results using the MCCV5 method are shown in Table 2. An example of the learning effect on these data before and after the application of our method can be seen in Figures 17–19. The results are not as good as those of RIYADH, but they indicate the positive effects of our method. For large artefacts, efficiency has increased from 61 percent to 80 percent after using our method. The results for the classification of weak artefacts (class *I*) are comparable on the validation set. Despite the fact that the quality of neural network learning on three classes has increased significantly (see Figure 19), the classification on the validation set increased by around 8 percentage points. Standard deviation (see Table 2) without pre-processing reaches around 15 percentage points, after applying our method, it is within 5, 6 percentage points for variants (*II* vs. *0*) and classifying all three classes. In case of the classification of weak artefacts the results are comparatively low with high standard deviation up to 18 percentage points.

Table 2. Summary of results for MOSCOW–MCCV5 technique; nil.ok.er = accuracy of classification of er and ok classes before pre-processing, ok.er = accuracy of classification of er and ok classes after application of our method, nil.ok.tr, ok.tr, nil.all, all = analogous parameters, showing the class separation before and after application of our technique, SD = standard deviation of results, avg = average result.

Test No.	nil.0.II	0.II	nil.0.I	0.I	nil.all	All
1	0.533	0.8	0.563	0.5	0.55	0.7
2	0.467	0.8	0.563	0.5	0.6	0.6
3	0.733	0.866	0.5	0.937	0.45	0.55
4	0.8	0.8	0.563	0.63	0.5	0.6
5	0.533	0.733	0.813	0.75	0.55	0.6
avg	0.613	0.8	0.6	0.663	0.53	0.61
SD	0.145	0.047	0.122	0.185	0.057	0.055

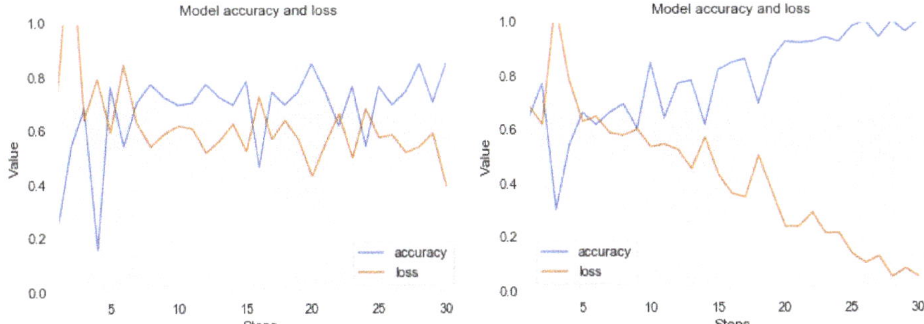

Figure 17. Exemplary learning effect before (left side) and after using our technique (right side). Result for MOSCOW dataset. Class *0* (without artefacts) vs. class *I* (with weak artefacts). The line graphs shows accuracy (blue) and loss (orange) for 15 epochs (30 steps from batch). Although the image dataset is relatively small for CNN, there can be observed a good trend in increasing accuracy, while minimizing loss.

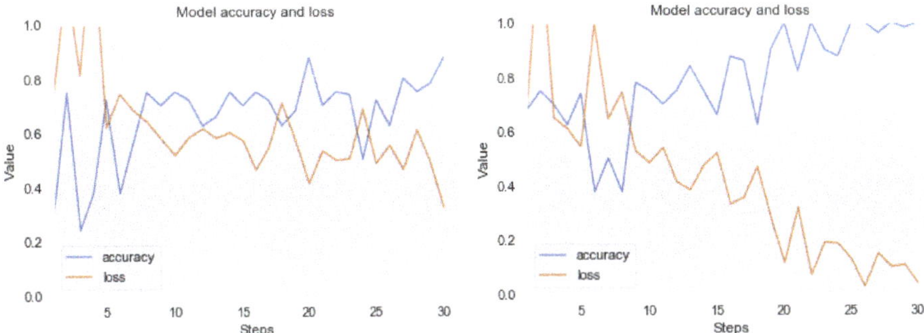

Figure 18. Exemplary learning effect before (left side) and after using our technique (right side). Result for MOSCOW dataset. Class *0* (without artefacts) vs. class *II* (with strong artefacts). The line graphs shows accuracy (blue) and loss (orange) for 15 epochs (30 steps from batch). Although the image dataset is relatively small for CNN, there can be observed a good trend in increasing accuracy, while minimizing loss.

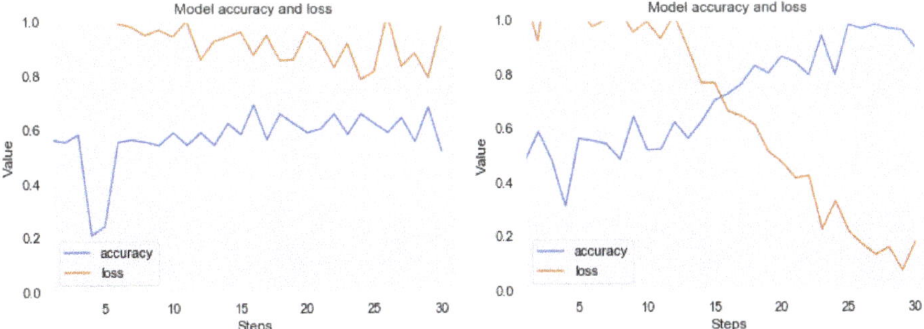

Figure 19. Exemplary learning effect before (left side) and after using our technique (right side). Result for MOSCOW dataset. Class *0* (without artefacts) vs. class *I* (with weak artefacts) vs. class *II* (with strong artefacts). The line graphs shows accuracy (blue) and loss (orange) for 15 epochs (30 steps from batch). Although the image dataset is relatively small for CNN, there can be observed a good trend in increasing accuracy, while minimizing loss.

Let us move on to test the ASMOW collection, an important one for our project.

5.8. Results for ASMOW Dataset

In this section, we describe the detection of artefacts, which appear in the ASMOW quick-look dataset. Performing referenced deep learning classification was not possible due to the availability of RFI-affected images. From here we simply present the procedure for detecting artefacts on this data—with the position of the artefact marked in the picture.

The detection procedure is as follows:

1. Considering sample data where the middle picture contains an artefact—see Figure 20. We convolved the pixels using the Mask14 (Gaussian blur) see Figure 21.
2. We applied a hybrid threshold method based on the histogram (with a fixed frequency threshold of 10,000), and using the 1 *nn* technique for noise reduction, see Figure 22.
3. We have applied pixel binarization [65], setting the threshold of 120, every pixel with saturation less than 120 is black, the rest are white, see Figure 23.
4. Then we detected the area whose pixels are arranged in a thick straight line—see Figure 24. The thickness was set at 7 vertical pixels. We applied the expected linear structure length threshold of 40, so our artefact was detected and distinguished from undamaged images. The area of artefact was automatically marked with a line.

Figure 20. Example of original data from the ASMOW quick-look dataset, there is a sample with an artefact in the middle.

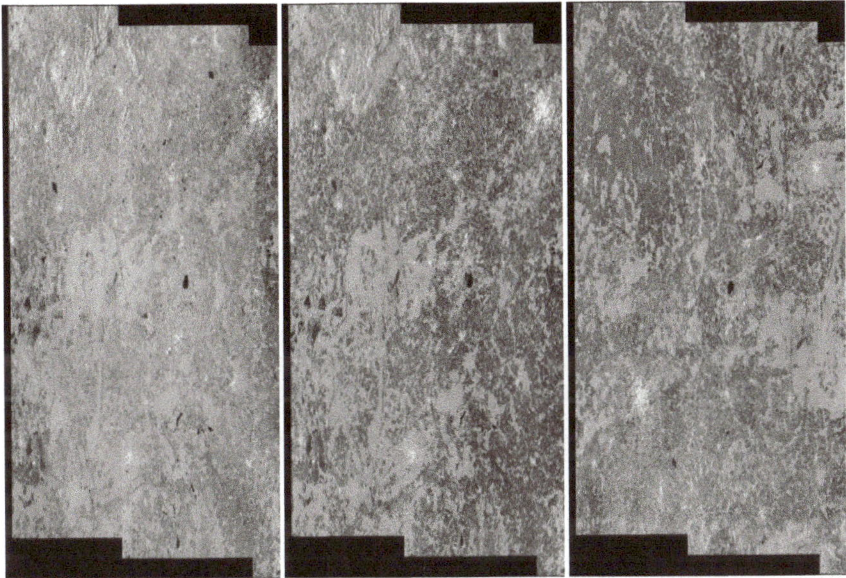

Figure 21. Step1: We carried out the pixel convolution with the Mask14 (Gaussian blur), on the data from Figure 20.

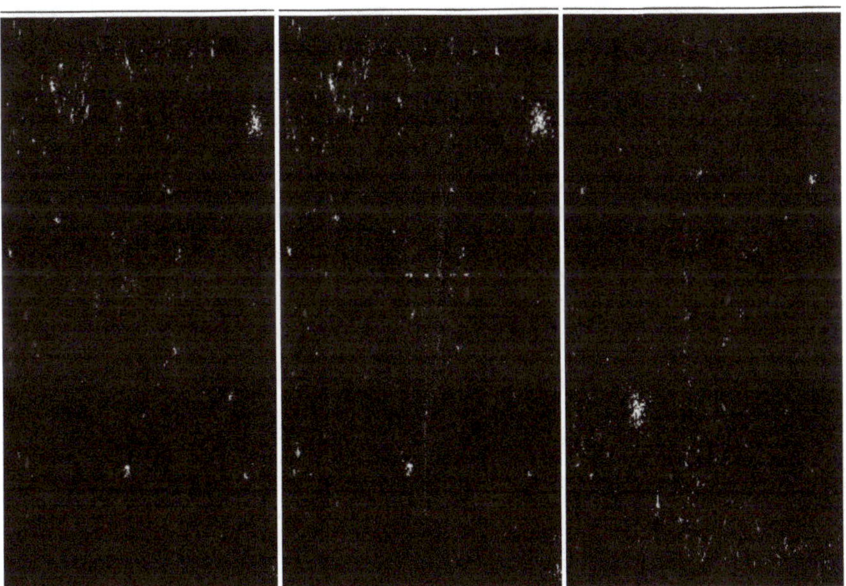

Figure 22. Step2: The thresholding described in Section 4.4 and the noise reduction by method 1 nn described in Section 4.5.

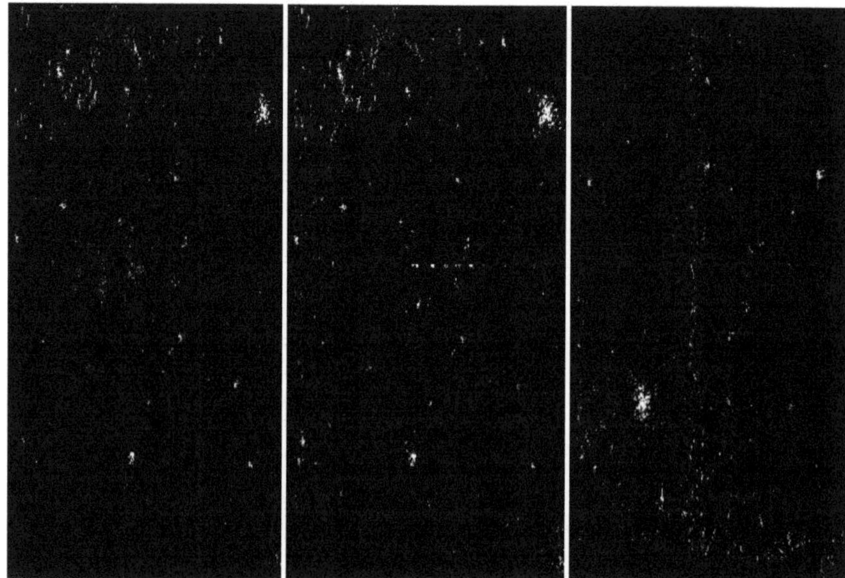

Figure 23. Step3: In this step, we binarized the pixels for better recognition of the artefact.

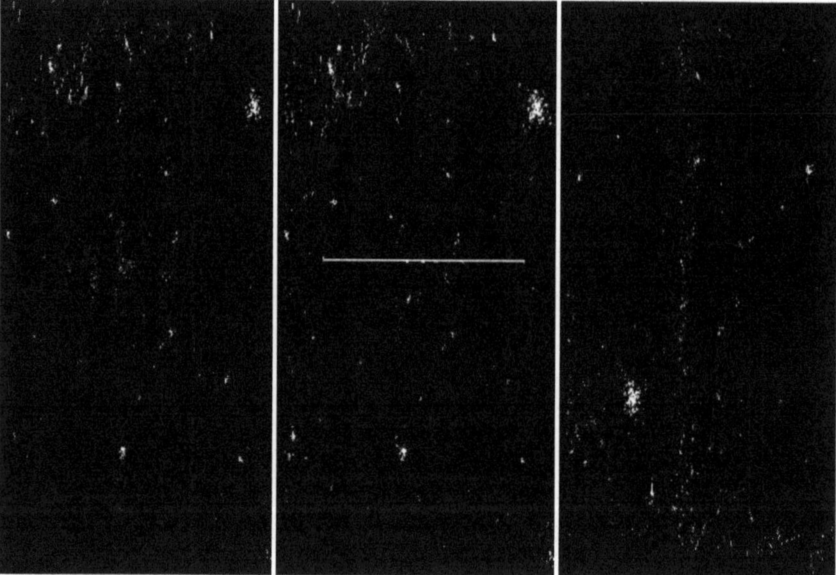

Figure 24. Step4: In the last step, we detected the artefact: a horizontal plane that passes unevenly through the line. We established that we are looking for a sufficiently long plane, consisting of seven-pixel blocks passing horizontally through the image at the same height. Our acceptance threshold for detecting the artefact, resulting from the experiments, is 40. So the artefact consists of a minimum of 40 seven-pixel blocks arranged horizontally.

5.9. Summary of Results for ASMOW Dataset

The experimental results show that the detection of these very little visible artefacts requires the following steps. The first two are analogous to the detection of the artefacts of RIYADH dataset, that is, (1) we use pixel convolutions by means of Gaussian blur, (2) thresholding on the basis of the color occurrence frequency, (3) defrosting using the 1 nn technique, (4) binarization and (5) detection of a plane consisting of pixel blocks. Artefacts can be successfully detected as we can see in Figure 24.

6. Conclusions

In this work, we tested a group of image processing techniques to separate clear RFI artefacts from undamaged images. We reviewed the masks used to select features in the process of convolution. Then we tested methods of thresholding. The first one is based on the selection of pixels in a fixed standardized range. The second one involves filtering pixels that do not meet the established criterion of color frequency (based on color histogram). Then we tested the hybrid solution with thresholding and filtration method based on the nearest neighbor of pixels. We verified the results of our methods of separating artefacts using a convolutional neural network as a reference classifier. The classification was carried out on raw data and on data prepared by our methods—with the Monte Carlo Cross Validation 5 model. In our work, we considered three datasets with artefacts. The first one comes from RIYADH, the second one from MOSCOW and the third one from ASMOW quick-look dataset. In case of RIYADH and MOSCOW datasets we see that for the separation of our RFI artefacts, the best level of their separation is given by the Gaussian blur. The best method, among the tested ones, that gives a clear separation of large artefacts is a hybrid of frequency based on thresholding and filtration with the nearest neighbors method. In case of the ASMOW dataset of RFI artefacts, we applied the same steps as with RIYADH and MOSCOW, additionally applying binarization and detection of the plane arranged horizontally on the image consisting of pixel blocks. The initial goal was achieved, we found an easy to implement method of separating large RFI artefacts—not repairable with image filtering methods. The proposed solution can effectively support the automatic PSInSAR processing workflow by recognizing RFI-affected data, and as a consequence, removing them from the stack of SAR images required to determine the ground displacements. Our method improves (compared to the classification on raw data) the efficiency of artefact detection by up to 27 percentage points depending on the classification context under consideration. The standard deviation of the results after application of our methods is nearly $5, 6$ percentage points (except for the unstable classification of weak artefacts). The obvious conclusion: it is difficult to find a method to generalize the problem of searching for artefacts. Each dataset containing artefacts should be treated individually. The model structure should be selected in a personalized way.

One of the methods of developing our detection system was the use of techniques for the recognition of specific shapes relevant to the appearing RFI artefacts and the use of complex convolutional neural networks—which is the foreground of our future research.

Author Contributions: Conceptualization, P.A., J.R. and A.C.; methodology, software, validation, formal analysis and investigation, P.A.; data acquisition, J.R. and A.C.; resources, A.C. and P.A.; writing—original draft preparation, A.C. and P.A.; writing—review and editing, A.C. and P.A.; funding acquisition, J.R. All authors have read and agreed to the published version of the manuscript.

Funding: This research conducted under the project entitled automatic system for monitoring the influence of high-energy paraseismic tremors on the surface using GNSS/PSInSAR satellite observations and seismic measurements (Project No. POIR.04.01.04-00-0056/17), co-financed from the European Regional Development Fund within the Smart Growth Operational Programme 2014-2020.

Conflicts of Interest: The authors declare no conflict of interest.

References

1. NASA. Landsat 1. Landsat Science. 2020. Available online: https://landsat.gsfc.nasa.gov/landsat-1/ (accessed on 28 February 2020).

2.	Tao, M.; Su, J.; Huang, Y.; Wang, L. Mitigation of Radio Frequency Interference in Synthetic Aperture Radar Data: Current Status and Future Trends. *Remote Sens.* **2019**, *11*, 2438. [CrossRef]

3.	Ding, X.-L.; Li, Z.-W.; Zhu, J.-J.; Feng, G.-C.; Long, J.-P. Atmospheric Effects on InSAR Measurements and Their Mitigation. *Sensors* **2008**, *8*, 5426–5448. [CrossRef] [PubMed]

4.	Yang, L.; Zheng, H.; Feng, J.; Li, N.; Chen, J. Detection and suppression of narrow band RFI for synthetic aperture radar imaging. *Chin. J. Aeronaut.* **2015**, *28*, 1189–1198. [CrossRef]

5.	Wang, J.; Yu, W.; Deng, Y.; Wang, R.; Wang, Y.; Zhang, H.; Zheng, M. Demonstration of Time-Series InSAR Processing in Beijing Using a Small Stack of Gaofen-3 Differential Interferograms. *J. Sens.* **2019**. [CrossRef]

6.	Massonnet, D.; Feigl, K.L. Radar interferometry and its application to changes in the earth's surface. *Rev. Geophys.* **1998**, *36*, 441–500. [CrossRef]

7.	Burgmann, R.; Rosen, P.A.; Fielding, E.J. Synthetic aperture radar interferometry to measure Earth's surface topography and its deformation. *Annu. Rev. Earth Planet. Sci.* **2000**, *28*, 169–209. [CrossRef]

8.	Liu, Z.; Zhou, C.; Fu, H.; Zhu, J.; Zuo, T. A Framework for Correcting Ionospheric Artifacts and Atmospheric Effects to Generate High Accuracy InSAR DEM. *Remote Sens.* **2020**, *12*, 318. [CrossRef]

9.	Solari, L.; Ciampalini, A.; Raspini, F.; Bianchini, S.; Moretti, S. PSInSAR Analysis in the Pisa Urban Area (Italy): A Case Study of Subsidence Related to Stratigraphical Factors and Urbanization. *Remote Sens.* **2016**, *8*, 120. [CrossRef]

10.	Qin, X.; Ding, X.; Liao, M.; Zhang, L.; Wang, C. A bridge-tailored multi-temporal DInSAR approach for remote exploration of deformation characteristics and mechanisms of complexly structured bridges. *ISPRS J. Photogramm. Remote Sens.* **2019**, *156*, 27–50. [CrossRef]

11.	Hu, Z.; Mallorquí, J.J. An Accurate Method to Correct Atmospheric Phase Delay for InSAR with the ERA5 Global Atmospheric Model. *Remote Sens.* **2019**, *11*, 1969. [CrossRef]

12.	Massonnet, D.; Feigl, K.L. Discrimination of geophysical phenomena in satellite radar interferograms. *Geophys. Res. Lett.* **1995**, *22*, 1537–1540. [CrossRef]

13.	Ferretti, A.; Prati, C.; Rocca, F. Permanent Scatterers in SAR Interferometry. *IEEE Trans. Geosci. Remote Sens.* **2001**, *39*, 8–20. [CrossRef]

14.	Dai, K.; Li, Z.; Tomás, R.; Liu, G.; Yu, B.; Wang, X.; Cheng, H.; Chen, J.; Stockamp, J. Monitoring activity at the Daguangbao mega-landslide (China) using Sentinel-1 TOPS time series interferometry. *Remote Sens. Environ.* **2016**, *186*, 501–513. [CrossRef]

15.	Yang, K.; Yan, L.; Huang, G.; Chen, C.; Wu, Z. Monitoring Building Deformation with InSAR: Experiments and Validation. *Sensors* **2016**, *16*, 2182. [CrossRef] [PubMed]

16.	Oštir, K.; Komac, M. PSInSAR and DInSAR methodology comparison and their applicability in the field of surface deformations—A case of NW Slovenia. *Geologija* **2007**, *50*, 77–96. [CrossRef]

17.	Babu, A.; Kumar, S. PSInSAR Processing for Volcanic Ground Deformation Monitoring Over Fogo Island. In Proceedings of the 2nd International Electronic Conference on Geosciences, Online, 8–15 June 2019. [CrossRef]

18.	Tofani, V.; Raspini, F.; Catani, F.; Casagli, N. Persistent Scatterer Interferometry (PSI) Technique for Landslide Characterization and Monitoring. *Remote Sens.* **2013**, *5*, 1045–1065. [CrossRef]

19.	Ge, L.; Chang, H.C.; Rizos, C. Mine subsidence monitoring using multi-source satellite SAR images. *Photogramm. Eng. Remote Sens.* **2007**, *73*, 259–266. [CrossRef]

20.	Benattou, M.M.; Balz, T.; Lia, M. Measuring surface subsidence in Wuhan, China with Sentinel-1 data using PSInSAR. *Int. Arch. Photogramm. Remote Sens. Spat. Inf. Sci.* **2018**, *XLII-3*, 73–77. [CrossRef]

21.	Liao, M.; Jiang, H.; Wang, Y.; Wang, T.; Zhang, L. Improved topographic mapping through high-resolution SAR interferometry with atmospheric removal. *ISPRS J. Photogram. Remote Sens.* **2013**, *80*, 72–79. [CrossRef]

22.	Zebker, H.A.; Rosen, P.A.; Hensley, S. Atmospheric effects in interferometric synthetic aperture radar surface deformation and topographic maps. *J. Geophys. Res.* **1997**, *102*, 7547–7563. [CrossRef]

23.	Emardson, T.R.; Simons, M.; Webb, F.H. Neutral atmospheric delay in interferometric synthetic aperture radar applications: Statistical description and mitigation. *J. Geophys. Res.* **2003**, *108*, 2231–2238. [CrossRef]

24.	Stasolla, M.; Neyt, X. An Operational Tool for the Automatic Detection and Removal of Border Noise in Sentinel-1 GRD Products. *Sensors* **2018**, *18*, 3454. [CrossRef] [PubMed]

25.	Hajduch, G.; Miranda, N. *Masking "No-Value" Pixels on GRD Products Generated by the Sentinel-1 ESA IPF*; Document Reference MPC-0243; S-1 Mission Performance Centre, ESA: Paris, France, 2018.

26. Ali, I.; Cao, S.; Naeimi, V.; Paulik, C.; Wagner, W. Methods to Remove the Border Noise From Sentinel-1 Synthetic Aperture Radar Data: Implications and Importance For Time-Series Analysis. *IEEE J. Sel. Top. Appl. Earth Obs. Remote Sens.* **2018**, *11*, 777–786. [CrossRef]

27. Luo, Y.; Flett, D. Sentinel-1 Data Border Noise Removal and Seamless Synthetic Aperture Radar Mosaic Generation. In Proceedings of the 2nd International Electronic Conference on Remote Sensing, Online, 22 March–5 April 2018. [CrossRef]

28. Bouvet, A.; Mermoz, S.; Ballère, M.; Koleck, T.; Le Toan, T. Use of the SAR Shadowing Effect for Deforestation Detection with Sentinel-1 Time Series. *Remote Sens.* **2018**, *10*, 1250. [CrossRef]

29. International Telecommunication Union. Radio Regulations Articles, Section VII—Frequency sharing, article 1.166, definition: Interference. 2016. Available online: http://search.itu.int/history/HistoryDigitalCollectionDocLibrary/1.43.48.en.101.pdf (accessed on 28 February 2020).

30. Meyer, F.J.; Nicoll, J.B.; Doulgeris, A.P. Correction and Characterization of Radio Frequency Interference Signatures in L-Band Synthetic Aperture Radar Data. *IEEE Trans. Geosci. Remote Sens.* **2013**, *51*, 4961–4972. [CrossRef]

31. Parasher, P.; Aggarwal, K.M.; Ramanujam, V.M. RFI detection and mitigation in SAR data. In Proceedings of the Conference: 2019 URSI Asia-Pacific Radio Science Conference (AP-RASC), New Delhi, India, 9–15 March 2019. [CrossRef]

32. Cucho-Padin, G.; Wang, Y.; Li, E.; Waldrop, L.; Tian, Z.; Kamalabadi, F.; Perillat, P. Radio Frequency Interference Detection and Mitigation Using Compressive Statistical Sensing. *Radio Sci.* **2019**, *54*, 11. [CrossRef]

33. Querol, J.; Perez, A.; Camps, A. A Review of RFI Mitigation Techniques in Microwave Radiometry. *Remote Sens.* **2019**, *11*, 3042. [CrossRef]

34. Shen, W.; Qin, Z.; Lin, Z. A New Restoration Method for Radio Frequency Interference Effects on AMSR-2 over North America. *Remote Sens.* **2019**, *11*, 2917. [CrossRef]

35. Soldo, Y.; Le Vine, D.; de Matthaeis, P. Detection of Residual "Hot Spots" in RFI-Filtered SMAP Data. *Remote Sens.* **2019**, *11*, 2935. [CrossRef]

36. Johnson, J.T.; Ball, C.; Chen, C.; McKelvey, C.; Smith, G.E.; Andrews, M.; O'Brien, A.; Garry, J.L.; Misra, S.; Bendig, R.; et al. Real-Time Detection and Filtering of Radio Frequency Interference Onboard a Spaceborne Microwave Radiometer: The CubeRRT Mission. *IEEE J. Sel. Top. Appl. Earth Obs. Remote Sens.* **2020**, *13*. [CrossRef]

37. Yang, Z.; Yu, C.; Xiao, J.; Zhang, B. Deep residual detection of radio frequency interference for FAST. *Mon. Not. R. Astron. Soc.* **2020**, *492*, 1. [CrossRef]

38. Monti-Guarnieri, A.; Giudici, D.; Recchia, A. Identification of C-Band Radio Frequency Interferences from Sentinel-1 Data. *Remote Sens.* **2017**, *9*, 1183. [CrossRef]

39. Itschner, I.; Li, X. Radio Frequency Interference (RFI) Detection in Instrumentation Radar Systems: A Deep Learning Approach. In Proceedings of the IEEE Radar Conference (RadarConf) 2019, Boston, MA, USA, 22–26 April 2019. [CrossRef]

40. ESA. Sentinel Online Technical Website. Sentinel-1. 2020. Available online: https://sentinel.esa.int/web/sentinel/missions/sentinel-1 (accessed on 2 February 2020).

41. Copernicus. The European Union's Earth Observation Programme. 2020. Available online: https://www.copernicus.eu/en/about-copernicus/copernicus-brief (accessed on 28 February 2020).

42. Torres, R.; Snoeij, P.; Geudtner, D.; Bibby, D.; Davidson, M.; Attema, E.; Potin, P.; Rommen, B.; Floury, N.; Brown, M.; et al. GMES Sentinel-1 mission. *Remote Sens. Environ.* **2012**, *120*, 9–24. [CrossRef]

43. De Zan, F.; Monti-Guarnieri, A. TOPSAR: Terrain Observation by Progressive Scans. *IEEE Trans. Geosci. Remote Sens.* **2006**, *44*, 2352–2360. [CrossRef]

44. Lasocki, S; Antoniuk, J; Moscicki, J. Environmental Protection Problems in the Vicinity of the Żelazny Most Flotation Wastes Depository in Poland. *J. Environ. Sci. Health Part A* **2003**, *38*, 1435–1443. [CrossRef] [PubMed]

45. KGHM Polska Miedź. Experts discuss „Żelazny Most". 2019. Available online: https://media.kghm.com/en/news-and-press-releases/experts-discuss-zelazny-most (accessed on 27 February 2020).

46. Major, K. Jest Największy w Europie i Rośnie. Kluczowa Inwestycja KGHM. [The largest in Europe and Continues to Raise. Crucial investment for the KGHM.]WP Polska Miedź. 2019. Available online: http://polskamiedz.wp.pl/artykul/jest-najwiekszy-w-europie-i-rosnie-kluczowa-inwestycja-kghm (accessed on 27 February 2020).

47. IEEE FARS Technical Committee. Database of Frequency Allocations for Microwave Remote Sensing and Observed Radio Frequency Interference. 2020. Available online: http://grss-ieee.org/microwave-interferers/ (accessed on 28 February 2020).
48. Trussell, H.J.; Saber, E.; Vrhel, M. Color image processing [basics and special issue overview]. *IEEE Signal Process. Mag.* **2005**, *22*, 14–22. [CrossRef]
49. Gonzalez, R. *Digital Image Processing*; Pearson: New York, NY, USA, 2018; ISBN 978-0-13-335672-4.
50. Bradski, G.; Kaehler, A. *Learning OpenCV: Computer vision with the OpenCV library*; O'Reilly Media: Sebastopol, CA, USA, 2008; ISBN 978-0596516130.
51. Shapiro, L.G.; Stockman, G.C. *Computer Vision*; Pearson: London, UK, 2001; ISBN 978-0130307965.
52. Nixon, M.S.; Aguado, A.S. *Feature Extraction and Image Processing*; Newnes: Oxford, UK, 2002; ISBN 978-0750650786.
53. Mo, J.; Wang, B.; Zhang, Z.; Chen, Z.; Huang, Z.; Zhang, J.; Ni, X. A convolution-based approach for fixed-pattern noise removal in OCR. In Proceedings of the International Conference on Artificial Intelligence and Big Data (ICAIBD) 2018, Chengdu, China, 26–28 May 2018; pp. 134–138. [CrossRef]
54. Bradley, D.; Roth, G. Adaptive Thresholding using the Integral Image. *J. Graph. Tools* **2007**, *12*, 13–21. [CrossRef]
55. Cover, T.; Hart, P. Nearest neighbor pattern classification. *IEEE Trans. Inf. Theory* **1967**, *13*, 21–27. [CrossRef]
56. Novotny, J.; Bilokon, P.A.; Galiotos, A.; Délèze, F. Nearest Neighbours. In *Machine Learning and Big Data with kdb+/q*; John Wiley & Sons: Hoboken, NJ, USA, 2019. [CrossRef]
57. Lensch, H. Computer Graphics: Texture Filtering & Sampling Theory. Max Planck Institute for Informatics 2007. Available online: http://resources.mpi-inf.mpg.de/departments/d4/teaching/ws200708/cg/slides/CG09-Textures+Filtering.pdf (accessed on 14 January 2018).
58. Craw, S. Manhattan Distance. In *Encyclopedia of Machine Learning and Data Mining*; Sammut, C., Webb, G.I., Eds.; Springer: Basel, Switzerland, 2017. [CrossRef]
59. Goodfellow, I.; Bengio, Y.; Courville, A. *Deep Learning*; The MIT Press: London, UK, 2016; ISBN 78-0262035613.
60. Lou, G.; Shi, H. Face image recognition based on convolutional neural network. *China Commun.* **2020**, *17*, 2. [CrossRef]
61. Almakky, I.; Palade, V.; Ruiz-Garcia, A. Deep Convolutional Neural Networks for Text Localisation in Figures From Biomedical Literature. In Proceedings of the International Joint Conference on Neural Networks (IJCNN) 2019, Budapest, Hungary, 14–19 July 2019. [CrossRef]
62. Kingma, D.P.; Ba, J. Adam: A method for stochastic optimization. In Proceedings of the 3rd International Conference for Learning Representations, San Diego, CA, USA, 7–9 May 2015. arXiv:1412.698.
63. Partow, A. C++ Bitmap Library. Available online: http://partow.net/programming/bitmap/index.html (accessed on 24 February 2020).
64. Xu, Q.-S.; Liang, Y.-Z. Monte Carlo cross validation. *Chemom. Intell. Lab. Syst.* **2001**, *56*, 1. [CrossRef]
65. Singh, P.; Singh, B. A Review of Document Image Binarization Techniques. *Int. J. Comput. Sci. Eng.* **2019**, *7*, 746–749. [CrossRef]

Article

Waterfall Atrous Spatial Pooling Architecture for Efficient Semantic Segmentation

Bruno Artacho and Andreas Savakis *

Department of Computer Engineering, Rochester Institute of Technology, Rochester, NY 14623, USA; bmartacho@mail.rit.edu
* Correspondence: andreas.savakis@rit.edu

Received: 25 October 2019; Accepted: 29 November 2019; Published: 5 December 2019

Abstract: We propose a new efficient architecture for semantic segmentation, based on a "Waterfall" Atrous Spatial Pooling architecture, that achieves a considerable accuracy increase while decreasing the number of network parameters and memory footprint. The proposed Waterfall architecture leverages the efficiency of progressive filtering in the cascade architecture while maintaining multiscale fields-of-view comparable to spatial pyramid configurations. Additionally, our method does not rely on a postprocessing stage with Conditional Random Fields, which further reduces complexity and required training time. We demonstrate that the Waterfall approach with a ResNet backbone is a robust and efficient architecture for semantic segmentation obtaining state-of-the-art results with significant reduction in the number of parameters for the Pascal VOC dataset and the Cityscapes dataset.

Keywords: semantic segmentation; computer vision; atrous convolution; spatial pooling

---

1. Introduction

Semantic segmentation is an important computer vision task [1–3] with applications in autonomous driving [4], human–machine interaction [5], computational photography [6], and image search engines [7]. The significance of semantic segmentation, in both the development of novel architectures and its practical use, has motivated the development of several approaches that aim to improve the encouraging initial results of Fully Convolutional Networks (FCN) [8]. One important challenge to address is the decrease of the feature map size due to pooling, which requires unpooling to perform pixel-wise labeling of the image for segmentation.

DeepLab [9], for instance, used dilated or Atrous Convolutions to tackle the limitations posed by the loss of resolution inherited from unpooling operations. The advantage of Atrous Convolution is that it maintains the Field-of-View (FOV) at each layer of the network. DeepLab implemented Atrous Spatial Pyramid Pooling (ASPP) blocks in the segmentation network, allowing the utilization of several Atrous Convolutions at different dilation rates for a larger FOV.

A limitation of the ASPP architecture is that the network experiences a significant increase in size and memory required. This limitation was addressed in [10], by replacing ASPP modules with the application of Atrous Convolutions in series, or cascade, with progressive rates of dilation. However, although this approach successfully decreased the size of the network, it presented the setback of decreasing the size of the FOV.

Motivated by the success achieved by a network architecture with parallel branches introduced by the Res2Net module [11], we incorporate Res2Net blocks in a semantic segmentation network. Then, we propose a novel architecture named the Waterfall Atrous Spatial Pooling (WASP) and use it in a semantic segmentation network we refer to as WASPnet (see segmentation examples in Figure 1). Our WASP module combines the cascaded approach used in [10] for Atrous Convolutions with the larger FOV obtained from traditional ASPP in DeepLab for the deconvolutional stages of semantic segmentation.

Figure 1. Semantic segmentation examples using WASPnet.

The WASP approach leverages the progressive extraction of larger FOV from cascade methods, and is able to achieve parallelism of branches with different FOV rates while maintaining reduced parameter size. The resulting architecture has a flow that resembles a waterfall, which is how it gets its name.

The main contributions of this paper are as follows.

- We propose the Waterfall method for Atrous Spatial Pooling that achieves significant reduction in the number of parameters in our semantic segmentation network compared to current methods based on the spatial pyramid architecture.
- Our approach increases the receptive field of the network by combining the benefits of cascade Atrous Convolutions with multiple fields-of-view in a parallel architecture inspired by the spatial pyramid approach.
- Our results show that the Waterfall approach achieves state-of-the-art accuracy with a significant reduction in the number of network parameters.
- Due to the superior performance of the WASP architecture, our network does not require postprocessing of the semantic segmentation result with a CRF module, making it even more efficient in terms of computational complexity.

2. Related Work

The innovations in Convolutional Neural Networks (CNNs) by the authors of [12–15] form the core of image classification and serve as the structural backbone for state-of-the-art methods in semantic segmentation. However, an important challenge with incorporating CNN layers in segmentation is the significant reduction of resolution caused by pooling.

The breakthrough work of Long et al. [8] introduced Fully Convolutional Networks (FCN) by replacing the final fully connected layers with deconvolutional stages. FCN [8] addressed the resolution reduction problem by deploying upsampling strategies across deconvolution layers. These deconvolution stages attempt to reverse the convolution operation and increase the feature map size back to the dimensions of the original image. The contributions of FCN [8] triggered research in semantic segmentation that led to a variety of different approaches that are visually illustrated in Figure 2.

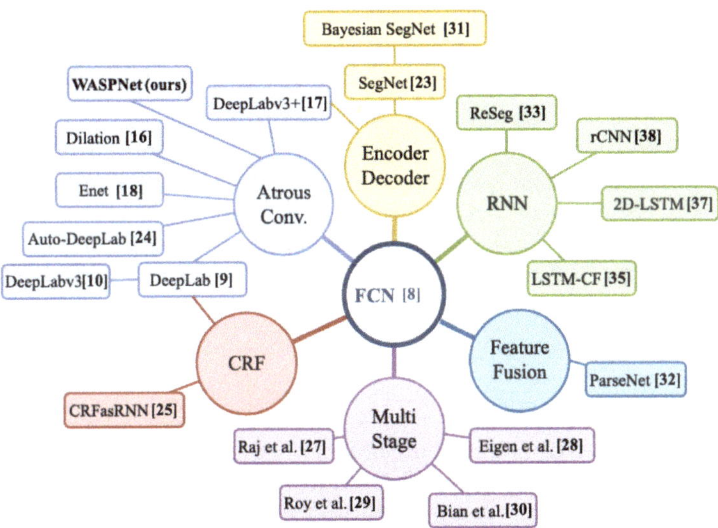

Figure 2. Semantic segmentation research overview.

2.1. Atrous Convolution

The most popular technique shared among semantic segmentation architectures is the use of dilated or Atrous Convolutions. An early work by Yu et al. [16] highlighted the uses of dilation. Atrous convolutions were further explored by the authors of [9,10,17,18]. The main objectives of Atrous Convolutions are to increase the size of the receptive fields in the network, avoid downsampling, and generate a multiscale framework for segmentation.

The name Atrous is derived from the French expression "algorithm à trous", or translated to English "Algorithm with holes". As alluded by its name, Atrous Convolutions alter the convolutional filters by the insertion of "holes", or zero values in the filter, resulting in the increased size of the receptive field, resembling a hybrid of convolution and pooling layers. The use of Atrous Convolutions in the network is shown in Figure 3.

In the simpler case of a one-dimensional convolution, the output of the signal is defined as follows [9],

$$y[i] = \sum_{k=1}^{K} x[i + rk] \cdot w[k] \tag{1}$$

where r is the rate at which the Atrous Convolution is dilated, $w[k]$ is the filter of length K, $x[i]$ is the input, and $y[i]$ is the output of a pixel. As pointed out in [9], a rate value of the unit results in a regular convolution operation.

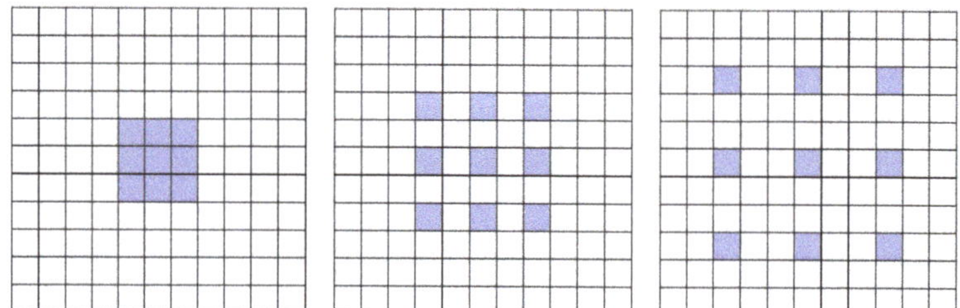

Figure 3. Input pixels using a 3 × 3 Atrous Convolutios with different dilation rates of 1, 2, and 3, respectively.

Leveraging the success of the Spatial Pyramid Pooling (SPP) structure by He et al. [19], the ASPP architecture was introduced in DeepLab [9]. The special configuration of ASPP assembles dilated convolutions in four parallel branches with different rates. The resulting feature maps are combined by fast bilinear interpolation with an additional factor of eight to recover the feature maps in the original resolution.

2.2. DeepLabv3

The application of Atrous Convolution followed the ASPP approach in [9] was later extended in [10] to the cascade approach, that is, the use of several Atrous Convolutions in sequence with rates increasing through its flux. This approach, named Deeplabv3 [10], allows the architecture to perform deeper analysis and increment its performance using approaches similar to those in [20].

Contributions in [10] included module realization in a cascade fashion, investigation of different multi-grid configurations for dilation in the cascade of convolutions, training with different output stride scales for the Atrous Convolutions, and techniques to improve the results when testing and fine-tuning for segmentation challenges. Another addition presented by [10] is the inclusion of a ResNet101 model, pretrained on both ImageNet [21] and JFT-300M [22] datasets.

More recently, DeepLabv3+ [17] proposed the incorporation of ASPP modules with the encoder–decoder structure adopted by [23], reporting a better refinement in the border of the objects being segmented. This novel approach represented a significant improvement in accuracy from previous methods. In a separate development, Auto-DeepLab [24] uses an Auto-ML approach to learn a semantic segmentation architecture by searching both the network level and the cell level of the structure. It achieves results comparable to current methods without requiring ImageNet [21] pre-training or hierarchical architecture search.

2.3. CRF

A complication resulting of the lack of pooling layers is a reduction of spatial invariance. Thus, additional techniques are used to recover spatial definition, namely, Conditional Random Fields (CRF) and Atrous Convolutions. One popular method relying on CRF is CRFasRNN [25]. Aiming to better delineate objects in the image, CRFasRNN combines CNN and CRF in a single network to incorporate the probabilistic method of the Gaussian pairwise potentials during inference. That enables end-to-end training, avoiding the need of postprocessing with a separate CRF module, as done in [9]. A limitation of architectures using CRF is that CRF has difficulty capturing delicate boundaries, as they have low confidence in the unary term of the CRF energy function.

The postprocessing module of CRF performs refining of the prediction by Gaussian filters and iterative comparisons of pixels in the output image. The iteration process aims to minimize the "energy" $E(x)$ below.

$$E(x) = \sum_{i} \theta_i(x_i) + \sum_{ij} \theta_{ij}(x_i, x_j) \tag{2}$$

The energy consists of the summations of the unary potentials $\theta_i(x_i) = -logP(x_i)$, where $P(x_i)$ is the probability (softmax) that pixel i is correctly computed by the CNN, and the pairwise potential energy $\theta_{ij}(x_i, x_j)$, which is determined by the relationship between two pixels. Following the authors of [26], $\theta_{ij}(x_i, x_j)$ is defined as

$$\theta_{ij}(x_i, x_j) = \mu(x_i, x_j) \left[\omega_1 \cdot exp\left(- \frac{||p_i - p_j||^2}{2\sigma_\alpha^2} - \frac{||I_i - I_j||^2}{2\sigma_\beta^2} \right) + \omega_2 \cdot exp\left(- \frac{||p_i - p_j||^2}{2\sigma_\gamma^2} \right) \right] \qquad (3)$$

where the function $\mu(x_i, x_j)$ is defined to be equal to 1 in the case of $x_i \neq x_j$ and zero otherwise, that is, the CRF only accounts for energy that needs to be minimized when the labels differ. The pairwise potential function utilizes two Gaussian kernels: the first depends on pixel positions p and the RGB color I; the second depends only on pixel positions. The Gaussian kernels are controlled by the hyperparameters σ_α, σ_β, and σ_γ, which are determined through the iterations of the CRF, as well as the weights ω_1 and ω_2.

2.4. Other Methods

In contrast to the large scale of segmentation networks using Atrous Convolutions, the Efficient Neural Network (ENet) [18] produces a real-time segmentation by trading-off some of its accuracy for a significant reduction in processing time, ENet is up to $18\times$ faster than other architectures.

During learning, CNN architectures have the tendency to learn information that is specific to the scale of the input image dataset. In an attempt to deal with this issue, a multiscale approach is used. For instance, the authors of [27] proposed a network with two paths containing the original resolution image and another with double the resolution. The former is processed through a short CNN and the latter through a fully convolutional VGG-16. The first path is then combined with the upsampled version of the second resulting in a network that can deal with larger variations in scale. A similar approach is applied in [28–30], expanding the structure to include a larger amount of networks and scales.

Other architectures achieved good results in semantic segmentation by using an encoder–decoder variant. For instance, SegNet [23] utilizes both an encoder and decoder phase, while relying on pooling indices from the encoder phase to aid the decoder phase. The Softmax classifier generates the final segmentation prediction map. The architecture presented by SegNet was further developed to include Bayesian techniques to model uncertainty in the network [31].

Contrasting with the work in [8], ParseNet [32] completes an early fusion in the network, by performing an early merge of the global features from previous layers with the current map of the posterior layer. In ParseNet, the previous layer is unpooled and concatenated to the following layers to generate the final classifier prediction with both having the same size. This approach differs from FCN where the skip connection concatenates maps of different sizes.

Recurrent Neural Networks (RNN) have been used to successfully combine pixel-level information with local region information, enabling the RNN to include global context in the construction of the segmented image. A limitation of RNN, when used for Semantic Segmentation, is that it has difficulty constructing a sequence based on the structure of natural images. ReSeg [33] is a network based on previous work by ReNet [34]. ReSeg presents an approach where RNN blocks from ReNet are applied after a few layers of a VGG structure, generating the final segmentation map by the use of upsampling by transposed convolutions. However, RNN-based architectures suffer from the vanishing gradient problem.

Networks using Long Short-Term Memory (LSTM) aim to tackle the issue of vanishing gradients. For instance, LSTM Context Fusion (LSTM-CF) [35] utilizes the concatenation of an architecture similar to DeepLab to process RGB and depth information. It uses three different scales for the RGB feature response and depth, similar to the work in [36]. Likewise, the authors of [37] used four different LSTM cells, each receiving distinct parts of the image. Recurrent Convolutional Neural Networks (rCNN) [38] recurrently train the network using different input window sizes fed into the RNN. This approach achieves better segmentation and avoids the loss of resolution encountered with fixed window fitting in RNN methods.

3. Methodology

We propose an efficient architecture for Semantic Segmentation making use of the large FOV generated by Atrous Convolutions combined with cascade of convolutions in a "Waterfall" configuration. Our WASP architecture provides benefits due to its multiscale representations as well as efficiency in the reduced size of the network.

The processing pipeline is shown in Figure 4. The input image is initially fed into a deep CNN (namely a ResNet-101 architecture) with the final layers replaced by a WASP module. The resultant score map with the probability distributions obtained from Softmax is processed by a decoder network that performs bilinear interpolation and generates a more efficient segmentation without the use of postprocessing with CRF. We provide a comparison of our WASP architecture with DeepLab's original ASPP architecture and with a modified architecture based on the Res2Net module.

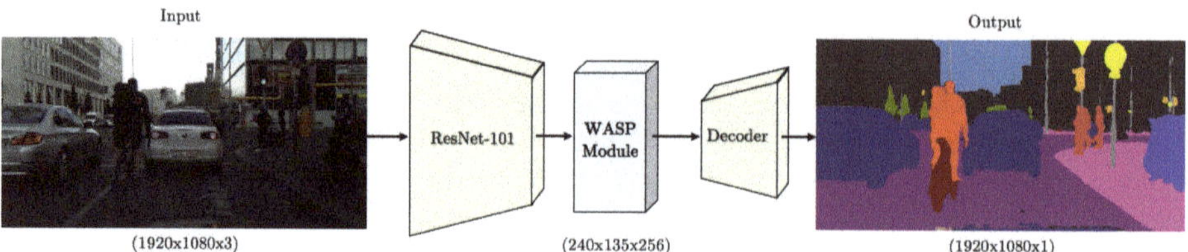

Figure 4. WASPnet architecture for semantic segmentation.

3.1. Res2Net-Seg Module

Res2Net [11] is a recently developed architecture designed to improve upon ResNet [15]. Res2Net incorporates multiscale features with a Squeeze-and-Excitation (SE) block [39] to obtain better representations and achieves promising results. The Res2Net module divides the original bottleneck block into four parallel streams, each containing 25% of the layers that are fed to 4 different 3×3 convolutions. Simultaneously, it incorporates the output of the parallel convolution. The SE block is an adaptable architecture that can recalibrate the responses in the feature map channel by modeling the interdependencies between channels. This allows improvements in performance by exploiting the dependencies between feature maps without increase in the network size.

Inspired by the work in [11], we present a modified version of the Res2Net module that is suitable for segmentation, named Res2Net-Seg. The Res2Net-Seg module, shown in Figure 5, includes the main structure of Res2Net and, additionally, utilizes Atrous Convolutions for each scale for increased FOV and a fifth parallel branch that performs average pooling of all features, which incorporates the original scale in the feature map. The Res2Net-Seg module is utilized in the WASPnet architecture of Figure 4 in place of the WASP module. We next propose the WASP module, inspired by multiscale representations, which an improvement over both the Res2Net-Seg and the ASPP configuration.

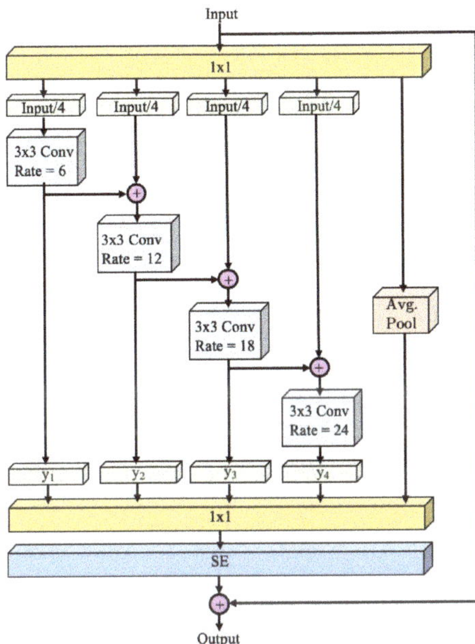

Figure 5. Res2Net-Seg block.

3.2. WASP Module

We propose the "Waterfall Atrous Spatial Pyramid" module, shown in Figure 6. WASP is a novel architecture with Atrous Convolutions that is able to leverage both the larger FOV of the ASPP configuration and the reduced size of the cascade approach.

An important drawback of Atrous Convolution, applied in either the cascade fashion or the ASPP (parallel design), is that it requires a larger number of parameters and more memory for its implementation, compared to standard convolution. In [9], there was experimentation to replace convolutional layers of the network backbone architecture, namely, VGG-16 or ResNet-101, with Atrous Convolution modules, but it was too costly in terms of memory requirements. A compromise solution is to apply the cascade of Atrous Convolutions and ASPP modules starting after block 5 when ResNet-101 was utilized.

We overcome these limitations with our Waterfall architecture for improved performance and efficiency. The Waterfall approach is inspired by multiscale approaches [28,29], the parallel structures of ASPP [9], and Res2Net modules [11], as well as the cascade configuration [10]. It is designed with the goal of reducing the number of parameters and memory required, which are the main limitation of Atrous Convolutions. The WASP module is utilized in the WASPnet architecture shown in Figure 4.

A comparison between the ASPP module, cascade configuration, and the proposed WASP module is visually highlighted in Figures 6 and 7, for the ASPP and cascade modules. The WASP configuration consists of four branches of a Large-FOV being fed forward in a waterfall-like fashion. In contrast, the ASPP module uses parallel branches that use more parameter and are less efficient, while the cascade architecture uses sequential filtering operations lacking the larger FOV.

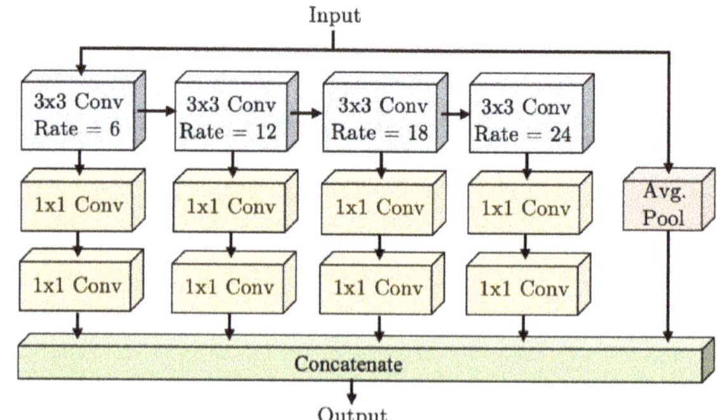

Figure 6. Proposed Waterfall Atrous Spatial Pooling (WASP) module.

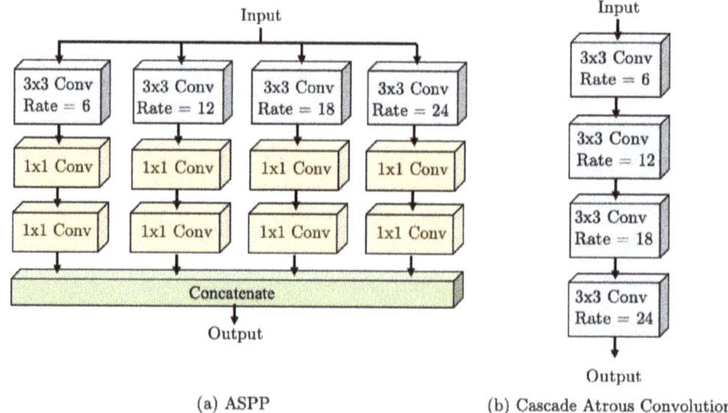

(a) ASPP

(b) Cascade Atrous Convolutions

Figure 7. Comparison for Atrous Spatial Pyramid Pooling (ASPP) [9] and Cascade configuration [10].

3.3. Decoder

To process the score maps resulting from the WASP module, a short decoder stage was implemented containing the concatenation with low level features from the first block of the ResNet backbone, convolutional layers, dropout layers, and bilinear interpolations to generate output maps in the same resolution of the input image.

Figure 8 shows the decoder and the respective stage dimensions and number of layers. The representation considers an input image with dimensions of $1920 \times 1080 \times 3$ for width, height, and RGB color, respectively. In this case, the decoder receives 256 maps of dimensions 240×135 and 256 low level features of dimension 480×270. After matching the dimensions for inputs of the decoder, the layers are concatenated and processed through convolutional layers, dropout, and a final bilinear interpolation to reach the original input size.

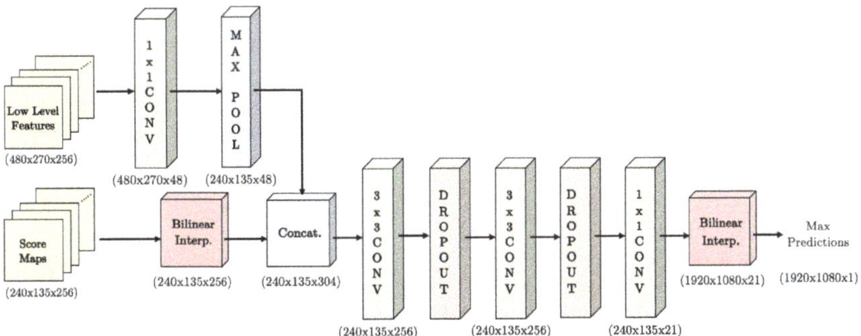

Figure 8. Decoder used in the WASPnet method.

4. Experiments

4.1. Datasets

We performed experiments on three datasets used for pre-training, training, validation, and testing. Microsoft Common Objects in Context (COCO) dataset [40] was used by [9] as pre-training as it includes a large amount of data, allowing a good balance of starting weights when training with different datasets, and consequently allowing the increase in precision of the segmentation.

Pascal Visual Object Class (VOC) 2012 [41] is a dataset containing objects in different scenarios including people, animals, vehicles, and indoor objects. It contains three different types of challenges: classification, detection, and segmentation; the latter was utilized in this paper. For the segmentation benchmark, the dataset contains 1464 images for training, 1449 images for validation, and 1456 images for testing annotated for 21 classes. Data augmentation was used to increase the training set size to 10,582.

Cityscapes [42] is a larger dataset containing urban scene images recorded in street scenes of 50 different cities with pixel annotations of 25,000 frames. In the Cityscapes dataset, 5000 images are finely annotated at pixel level divided into 2975 images for training, 500 for validation, and 1525 for testing. Cityscapes is annotated in 19 semantic classes divided into 7 categories (construction, ground, human, nature, object, sky, and vehicle).

4.2. Evaluation Metrics

We based our comparison of performance to other methods using Mean Intersection over Union (mIOU), considered the most important and more widely used metric for semantic segmentation. A pixel-level analysis of detection is conducted, reporting the intersection of true positive (TP) pixels detection as a percentage of the union of TP with false negative (FN) and false positive (FP) pixels.

4.3. Simulation Parameters

We calculate the learning rate based on the polynomial method ("poly") [32], also adopted in [9]. The poly learning rate LR_{poly} results in more effective updating of the weights when compared to the traditional "step" learning rate, given as

$$LR_{poly} = (1 - \frac{iter}{max\_iter})^{power} \tag{4}$$

where $power = 0.9$ was employed. We utilized a batch size of eight due to physical memory constraints in the hardware available, lower than the batch size of ten used by DeepLab. A subtle improvement in training with a larger batch size is expected for the architectures proposed.

We experimented with different rates of dilation on WASP. We found that larger rates result in better mIOU. A set rate of $r = \{6, 12, 18, 24\}$ was selected for the WASP module. In addition, we performed pre-training using the MS-COCO dataset [40], and data augmentation in randomly selected images scaled between (0.5,1.5).

5. Results

Following training, validation, and testing procedures, the WASPnet architecture was implemented utilizing WASP module, Res2Net-Seg module, or ASPP module. The validation mIOU results are presented in Table 1 for the Pascal VOC dataset. When following similar guidelines as in [9] for training and hyperparameters, and using the WASP module, an mIOU of 80.22% is achieved without the need for CRF postprocessing. Our WASPnet resulted in a gain of 5.07% on the validation set and reduced the number of parameters by 20.69%.

Table 1. Pascal Pascal Visual Object Class (VOC) validation set results.

Architecture	Number of Parameters	Parameter Reduction	mIOU
WASPnet-CRF (ours)	47.482 M	20.69%	80.41%
WASPnet (ours)	47.482 M	20.69%	80.22%
Res2Net-Seg-CRF	50.896 M	14.99%	80.12%
Res2Net-Seg	50.896 M	14.99%	78.53%
Deeplab-CRF [9]	59.869 M	-	77.69%
Deeplab [9]	59.869 M	-	76.35%

The Res2Net-Seg approach results in an mIOU of 78.53% without CRF, achieves mIOU of 80.12% with CRF, and reduces the number of parameters by 14.99%. The Res2Net-Seg approach still shows benefits with the incorporation of CRF as postprocessing, similar to the cascade and ASPP methods.

Overall, the WASP architecture provides the best result and the highest reduction in parameters. Sample results for the WASPnet architecture are shown in Figure 9 for validation images from the Pascal VOC dataset [41]. Note, from the generated segmentation, that our method presents a better definition in the detection shape, being closer to the ground-truth when compared to previous methods utilizing ASPP (DeepLab).

We tested the effects of different dilation rates (in our WASP module) on the final segmentation. In our tests, all kernel sizes were set to 3 following procedures as in [9]. Table 2 reports the accuracy, in mIOU, for the Pascal VOC dataset for different dilation rates in the WASP module. The configuration with dilation rates of $\{6, 12, 18, 24\}$ resulted in the best accuracy for the Pascal VOC dataset, therefore, the following tests were conducted using this dilation rate.

Table 2. Pascal VOC validation set results for different sets of dilation in the WASP module.

WASP Dilation Rates	mIOU
$\{2, 4, 6, 8\}$	79.61%
$\{4, 8, 12, 16\}$	79.72%
$\{6, 12, 18, 24\}$	80.22%
$\{8, 16, 24, 32\}$	79.92%

We also experimented with postprocessing using CRF. The application of CRF has the benefit of better defining the shapes of the segmented areas. Similarly to the procedures followed in [9], we performed parameter tuning, for the parameters of Equation (3), by varying ω_1 between 3 and 6, σ_α from 30 to 100, and σ_β from 3 to 6, while fixing both ω_2 and σ_γ to 3.

(a) Original   (b) Ground-Truth   (c) DeepLab [9]   (d) Ours

Figure 9. Results sample for Pascal VOC dataset [41].

The addition of CRF postprocessing to our WASPnet method resulted in a modest increase of 0.2% in the mIOU for both the validation and test sets of the Pascal VOC dataset. The gains from using CRF are less significant than those in [9], due to more efficient use of FOV by WASPnet. The effects of CRF on accuracy were not consistent across different classes. Classes with objects that do not have extremities, such as bottle, car, bus, and train, benefited most, whereas there was a decrease in accuracy for classes with more delicate boundaries such as bicycle, plant, and motorcycle.

Results on the testing Pascal VOC dataset are shown in Table 3. The additional training dataset column refers to DeepLabv3 types of models where a ResNet-101 model was pretrained on both ImageNet [21] and JFT-300M [22] when performing the test challenge for Pascal VOC. JFT-300M consists of Google's internal dataset of 300 million images labeled in 18,291 categories, and therefore these results cannot be compared directly to other external architectures including this work. The addition of the JFT dataset for training allows the architecture to achieve performance improvements that are not possible without the such a large number of training samples. Note that training of the WASPnet network was performed only on the training dataset provided by the challenge, consisting of 1464 images. Based on these results, WASPnet outperforms all of the other methods that are trained on the same dataset.

Table 3. Pascal VOC test set results.

Architecture	Additional Training Dataset Used	mIOU
DeepLabv3+ [17]	JFT-300M [22]	87.8%
Deeplabv3 [10]	JFT-300M [22]	85.7%
Auto-DeepLab-L [24]	JFT-300M [22]	85.6%
Deeplab [9]	JFT-300M [22]	79.7%
WASPnet-CRF (ours)	-	79.6%
WASPnet (ours)	-	79.4%
Dilation [16]	-	75.3%
CRFasRNN [25]	-	74.7%
ParseNet [32]	-	69.8%
FCN 8s [8]	-	67.2%
Bayesian SegNet [31]	-	60.5%

WASPnet was also used with the Cityscapes dataset [42] following similar procedures. Table 4 shows the results obtained for Cityscapes, resulting in an mIOU of 74.0%, a gain of 4.2% from [9]. The Res2Net-Seg version of the network achieved 72.1% mIOU.

Table 4. Cityscapes validation set results.

Architecture	Number of Parameters	Parameter Reduction	mIOU
WASPnet (ours)	47.482 M	20.69%	74.0%
WASPnet-CRF (ours)	47.482 M	20.69%	73.2%
Res2Net-Seg (ours)	50.896 M	14.99%	72.1%
Deeplab-CRF [9]	59.869 M	-	71.4%
Deeplab [9]	59.869 M	-	71.0%

For both WASP and Res2Net-Seg architectures tested on the Cityscapes dataset, the CRF postprocessing did not have much benefit. A similar result was found with DeepLab where CRF resulted in a small improvement of the mIOU. The higher resolution and shape of detected instances in the Cityscapes dataset likely affected the effectiveness of the CRF. With Cityscapes, we used a batch size of 4 due to hardware constraints during training; other architectures have used batch sizes of up to ten.

Table 5 shows the results of WASPnet on the Cityscapes testing dataset. WASPnet achieved mIOU of 70.5% and outperformed other architectures trained on the dame dataset. We only performed training on the fine annotation images from the Cityscapes dataset, containing 2975 images, whereas the DeepLabv3 style architectures used larger datasets for training, such as JFT-300M containing 300 million images for pre-training and and coarser dataset from Cityscapes containing 20,000 images.

Table 5. Pascal Cityscapes test set results.

Architecture	Additional Training Dataset Used	mIOU
Auto-DeepLab-L [24]	Coarse Cityscapes [42]	82.1%
DeepLabv3+ [17]	Coarse Cityscapes [42]	82.1%
WASPnet (ours)	-	70.5%
Deeplab [9]	-	70.4%
Dilation [16]	-	67.1%
FCN 8s [8]	-	65.3%
CRFasRNN [25]	-	62.5%
ENet [18]	-	58.3%
SegNet [23]	-	55.6%
Mask-RCNN [43]	-	49.9%

Figure 10 shows examples of Cityscapes image segmentations with the WASPnet method. Like our observations from the Pascal VOC dataset, our method produces better defined shapes for the segmentation compared to DeepLab. Our results are closer to the ground-truth data, and show better segmentation of smaller objects that are further away from the camera.

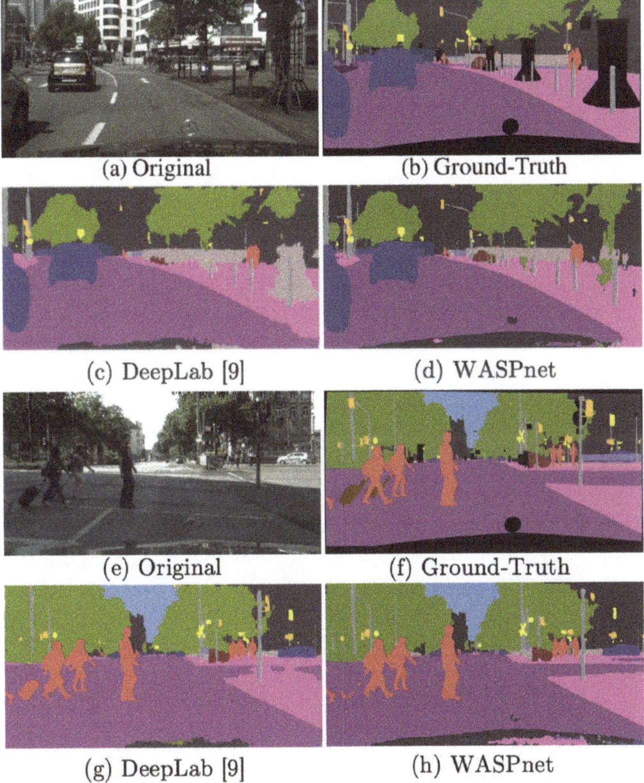

Figure 10. Results sample for Cityscapes dataset [42].

Our results in Table 4 illustrate that postprocessing with CRF slightly decreased the mIOU by 0.8% in the Cityscapes dataset: CRF has difficulty dealing with delicate boundaries, which are common in the Cityscapes dataset. With WASPnet, the presence of larger FOV due to the WASP module is able to offset the potential gains of the CRF module from previous networks. An additional limitation is that CRF requires substantial extra time for processing. For these reasons, we conclude that WASPnet can be used without CRF postprocessing.

Fail Cases

Classes that contain more delicate, and consequently harder to accurately detect, shapes contribute the most to segmentation errors. Particularly, tables, chairs, leaves, and bicycles present a bigger challenge to segmentation networks. These classes also resulted in a lower accuracy when applying CRF. Representative examples of fail cases are shown in Figure 11 for classes chair and bicycle, which are the most difficult to segment. Even in these cases, WASPnet (without CRF) is able to better detect the general shape compared to DeepLab.

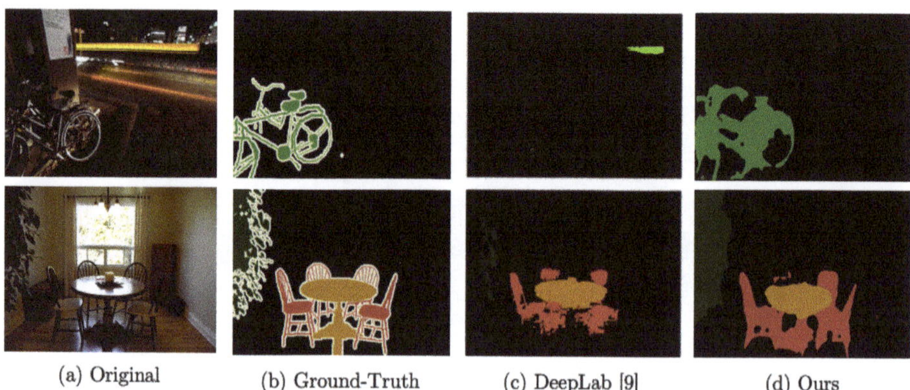

(a) Original   (b) Ground-Truth   (c) DeepLab [9]   (d) Ours

Figure 11. Occurrence of fail cases to detect more delicate boundaries

6. Conclusions

We propose a "Waterfall" architecture based on the WASP module for efficient semantic segmentation that achieves high mIOU scores on the Pascal VOC and Cityscapes datasets. The smaller size of this efficient architecture improves its functionality and reduces the risk of overfitting without the need for postprocessing with the time consuming CRF. The results of WASPnet segmentation demonstrated superior performance compared to Res2Net-Seg and Deeplab. This work provides the foundation for further application of WASP in a broader range of applications for more efficient multiscale analysis.

Author Contributions: Conceptualization, B.A. and A.S.; methodology, B.A.; algorithm and experiments, B.A. and A.S.; original draft preparation, B.A. and A.S.; review and editing, B.A. and A.S.; supervision, A.S.; project administration, A.S.; funding acquisition, A.S.

Funding: This research was funded in part by National Science Foundation grant number 1749376.

Conflicts of Interest: The authors declare no conflict of interest.

Abbreviations

The following abbreviations are used in this manuscript:

ASPP	Atrous Spatial Pyramid Pooling
COCO	Common Objects in Context
CNN	Convolutional Neural Networks
CRF	Conditional Random Fields
ENet	Efficient Neural Network
FCN	Fully Convolutional Networks

FN	False Negative
FOV	Field-of-View
FP	False Positive
LSTM	Long Short-Term Memory
LSTM-CF	Long Short-Term Memory Context Fusion
rCNN	Recurrent Convolutional Neural Networks
mIOU	Mean Intersection over Union
RGB	Red, Green, and Blue
RNN	Recurrent Neural Networks
SE	Squeeze-and-Excitation
TP	True Positive
VOC	Visual Object Class
WASP	Waterfall Atrous Spatial Pooling

References

1. Garcia-Garcia, A.; Orts-Escolano, S.; Oprea, S.; Villena-Martinez, V.; Rodríguez, J.G. A Review on Deep Learning Techniques Applied to Semantic Segmentation. *arXiv* **2017**, arXiv:1704.06857.
2. Zhu, H.; Meng, F.; Cai, J.; Lu, S. A comprehensive survey from bottom-up to semantic image segmentation and cosegmentation. *J. Vis. Commun. Image Represent.* **2016**, *34*, 12–27. [CrossRef]
3. Thoma, M. A Survey of Semantic Segmentation. *arXiv* **2016**, arXiv:1602.0654.
4. Ess, A.; Müller, T.; Grabner, H.; Goo, L.J.V. Segmentation-based urban traffic scene understanding. *BMVC* **2009**, *1*, 2.
5. Oberweger, M.; Wohlhart, P.; Lepetit, V. Hands Deep in Deep Learning for Hand Pose Estimation. *arXiv* **2015**, arXiv:1502.06807.
6. Fan, H.; Liu, D.; Xiong, Z.; Wu, F. Two-stage convolutional neural network for light field super-resolution. In Proceedings of the Image Processing (ICIP) 2017 IEEE International Conference, Beijing, China, 17–20 September 2017; pp. 1167–1171.
7. Tzelepi, M.; Tefas, A. Deep convolutional learning for content based image retrieval. *Neurocomputing* **2018**, *275*, 2467–2478. [CrossRef]
8. Long, J.; Shelhamer, E.; Darrel, T. Fully Convolutional Networks for Semantic Segmentation. In Proceedings of the 2015 IEEE Conference on Computer Vision and Pattern Recognition (CVPR), Boston, MA, USA, 7–12 June 2015.
9. Chen, L.C.; Papandreou, G.; Kokkinos, I.; Murphy, K.; Yuille, L. DeepLab: Semantic Image Segmentation with Deep Convolutional Nets, Atrous Convolution and Fully Connected CFRs. *IEEE Trans. Pattern Anal. Mach. Intell.* **2018**, *40*, 834–845. [CrossRef]
10. Chen, L.C.; Papandreou, G.; Schroff, F.; Adam, H. Rethinking Atrous Convolution for Semantic Image Segmentation. *arXiv* **2017**, arXiv:1706.05587.
11. Gao, S.H.; Cheng, M.M.; Zhao, K.; Zhang, X.Y.; Yang, M.H.; Torr, P. Res2Net: A New Multi-Scale Backbone Architecture. *IEEE Trans. Pattern Anal. Mach. Intell.* **2019**. [CrossRef]
12. Krizhevsky, A.; Sutskever, I.; Hinton, G.E. ImageNet Classification with Deep Convolutional Neural Networks. In Proceedings of the Advances in Neural Information Processing Systems 25 (NIPS), Lake Tahoe, NV, USA, 3–6 December 2012.
13. Simonyan, K.; Zisserman, A. Very Deep Convolutional Networks for Large-Scale Image Recognition. *arXiv* **2015**, arXiv:1409.1556.
14. Szegedy, C.; Liu, W.; Jia, Y.; Sermanet, P.; Reed, S.E.; Anguelov, D.; Erhan, D.; Vanhoucke, V.; Rabinovich, A. Going Deeper with Convolutions. *arXiv* **2014**, arXiv:1409.4842.
15. He, K.; Zhang, X.; Ren, S.; Sun, J. Deep Residual Learning for Image Recognition. *arXiv* **2015**, arXiv:1512.03385.
16. Yu, F.; Koltun, V. Multi-Scale Context Aggregation by Dilated Convolutions. In Proceedings of the ICLR, San Juan, PR, USA, 2–4 May 2016.
17. Chen, L.; Zhu, Y.; Papandreou, G.; Schroff, F.; Adam, H. Encoder-Decoder with Atrous Separable Convolution for Semantic Image Segmentation. *arXiv* **2018**, arXiv:1802.02611.

18. Paszke, A.; Chaurasia, A.; Kim, S.; Culurciello, E. ENet: A Deep Neural Network Architecture for Real-Time Semantic Segmentation. *arXiv* **2016**, arXiv:1606.02147.

19. He, K.; Zhang, X.; Ren, S.; Sun, J. Spatial Pyramid Pooling in Deep Convolutional Networks for Visual Recognition. *arXiv* **2014**, arXiv:1406.4729.

20. Dai, J.; Qi, H.; Xiong, Y.; Li, Y.; Zhang, G.; Hu, H.; Wei, Y. Deformable convolutional networks. In Proceedings of the IEEE International Conference on Computer Vision, Venice, Italy, 22–29 October 2017; pp. 764–773.

21. Deng, J.; Dong, W.; Socher, R.; Li, L.J.; Li, K.; Fei-Fei, L. ImageNet: A Large-Scale Hierarchical Image Database. In Proceedings of the Conference of Computer Vision and Pattern Recognition (CVPR), Miami, FL, USA, 20–25 June 2009.

22. Sun, C.; Shrivastava, A.; Singh, S.; Gupta, A. Revisiting Unreasonable Effectiveness of Data in Deep Learning Era. *arXiv* **2017**, arXiv:1707.02968.

23. Badrinarayanan, V.; Kendall, A.; Cipolla, R. SegNet: A Deep Convolutional Encoder-Decoder Architecture for Image Segmentation. *arXiv* **2015**, arXiv:1511.00561.

24. Liu, C.; Chen, L.C.; Schroff, F.; Adam, H.; Hua, W.; Yuille, A.L.; Fei-Fei, L. Auto-DeepLab: Hierarchical Neural Architecture Search for Semantic Image Segmentationx. In Proceedings of the IEEE Conference on Computer Vision and Pattern Recognition (CVPR), Long Beach, CA, USA, 16–20 June 2019; pp. 82–92.

25. Zheng, S.; Jayasumana, S.; Romera-Paredes, B.; Vineet, V.; Su, Z.; Du, D.; Huang, C.; Torr, P.H.S. Conditional Random Fields as Recurrent Neural Networks. *arXiv* **2015**, arXiv:1502.03240.

26. Krähenühl, P.; Koltun, V. Efficient Inference in Fully Connected CRFs with Gaussian Edge Potentials. In Proceedings of the NIPS, Granada, Spain, 12–17 December 2011.

27. Raj, A.; Maturana, D.; Scherer, S. *Multi-Scale Convolutional Architecture for Semantic Segmentation*; Robotics Institute, Carnegie Mellon University, Tech.: Pittsburgh, PA, USA, 2015.

28. Eigen, D.; Fergus, R. Predicting Depth, Surface Normals and Semantic Labels with a Common Multi-Scale Convolutional Architecture. *arXiv* **2014**, arXiv:1411.4734.

29. Roy, A.; Todorovic, S. A Multi-Scale CNN for Affordance Segmentation in RGB Images. In Proceedings of the IEEE European Conference on Computer Vision (ECCV), Amsterdam, the Netherlands, 11–14 October 2016; pp. 186–201.

30. Bian, X.; Lim, S.N.; Zhou, N. Multiscale fully convolutional network with application to industrial inspection. In Proceedings of the IEEE Winter Conference on Applications of Computer Vision (WACV), Lake Placid, NY, USA, 7–10 March 2016; pp. 1–8.

31. Kendall, A.; Badrinarayanan, V.; Cipolla, R. Bayesian SegNet: Model Uncertainty in Deep Convolutional Encoder-Decoder Architectures for Scene Understanding. *arXiv* **2015**, arXiv:1511.02680.

32. Liu, W.; Rabinovich, A.; Berg, A.C. ParseNet: Looking Wider to See Better. *arXiv* **2015**, arXiv:1506.04579.

33. Visin, F.; Kastner, K.; Courville, A.C.; Bengio, Y.; Matteucci, M.; Cho, K. ReSeg: A Recurrent Neural Network for Object Segmentation. *arXiv* **2015**, arXiv:1511.07053.

34. Visin, F.; Kastner, K.; Cho, K.; Matteucci, M.; Courville, A.C.; Bengio, Y. ReNet: A Recurrent Neural Network Based Alternative to Convolutional Networks. *arXiv* **2015**, arXiv:1505.00393.

35. Li, Z.; Gan, Y.; Liang, X.; Yu, Y.; Cheng, H.; Lin, L. RGB-D Scene Labeling with Long Short-Term Memorized Fusion Model. *arXiv* **2016**, arXiv:1604.05000.

36. Li, G.; Yu, Y. Deep Contrast Learning for Salient Object Detection. *arXiv* **2016**, arXiv:1603.01976.

37. Byeon, W.; Breuel, T.M.; Raue, F.; Liwicki, M. Scene labelingwith lstm recurrent neural networks. In Proceedings of the IEEE Conference on Computer Vision and Pattern Recognition, Boston, MA, USA, 7–12 June 2015; pp. 3547–3555.

38. Pinheiro, P.H.O.; Collobert, R. Recurrent Convolutional Neural Networks for Scene Parsing. *arXiv* **2013**, arXiv:1306.2795.

39. Hu, J.; Shen, L.; Sun, G. Squeeze-and-Excitation Networks. *arXiv* **2017**, arXiv:1709.01507.

40. Lin, T.; Maire, M.; Belongie, S.J.; Bourdev, L.D.; Girshick, R.B.; Hays, J.; Perona, P.; Ramanan, D.; Dollár, P.; Zitnick, C.L. Microsoft COCO: Common Objects in Context. *arXiv* **2014**, arXiv:1405.0312.

41. Everingham, M.; Van Gool, L.; Williams, C.K.I.; Winn, J.; Zisserman, A. The Pascal Visual Object Classes (VOC) Challenge. *Int. J. Comput. Vis.* **2010**, *88*, 303–338. [CrossRef]

42. Cordts, M.; Omran, M.; Ramos, S.; Rehfeld, T.; Enzweiler, M.; Benenson, R.; Franke, U.; Roth, S.; Schiele, B. The Cityscapes Dataset for Semantic Urban Scene Understanding. In Proceedings of the IEEE Conference on Computer Vision and Pattern Recognition (CVPR), Las Vegas, NV, USA, 26 June–1 July 2016; pp. 3213–3223.
43. He, K.; Gkioxari, G.; Dollár, P.; Girshick, R.B. Mask R-CNN. *arXiv* **2017**, arXiv:1703.06870.

Review

Vision-Based Traffic Sign Detection and Recognition Systems: Current Trends and Challenges

Safat B. Wali [1], Majid A. Abdullah [2], Mahammad A. Hannan [2,*], Aini Hussain [1], Salina A. Samad [1], Pin J. Ker [2]and Muhamad Bin Mansor [2]

[1] Centre for Integrated Systems Engineering and Advanced Technologies, Universiti Kebangsaan Malaysia, Bangi 43600, Malaysia; safat.wali@siswa.ukm.edu.my (S.B.W.); draini@ukm.edu.my (A.H.); salinasamad@ukm.edu.my (S.A.S.)

[2] Institute of Power Engineering, Universiti Tenaga Nasional, Kajang 43000, Malaysia; aaamajid2@gmail.com (M.A.A.); pinjern@uniten.edu.my (P.J.K.); muhamadm@uniten.edu.my (M.B.M.)

* Correspondence: hannan@uniten.edu.my

Received: 2 April 2019; Accepted: 26 April 2019; Published: 6 May 2019

Abstract: The automatic traffic sign detection and recognition (TSDR) system is very important research in the development of advanced driver assistance systems (ADAS). Investigations on vision-based TSDR have received substantial interest in the research community, which is mainly motivated by three factors, which are detection, tracking and classification. During the last decade, a substantial number of techniques have been reported for TSDR. This paper provides a comprehensive survey on traffic sign detection, tracking and classification. The details of algorithms, methods and their specifications on detection, tracking and classification are investigated and summarized in the tables along with the corresponding key references. A comparative study on each section has been provided to evaluate the TSDR data, performance metrics and their availability. Current issues and challenges of the existing technologies are illustrated with brief suggestions and a discussion on the progress of driver assistance system research in the future. This review will hopefully lead to increasing efforts towards the development of future vision-based TSDR system.

Keywords: Traffic sign detection and tracking (TSDR); advanced driver assistance system (ADAS); computer vision

---

1. Introduction

In all countries of the world, the important information about the road limitation and condition is presented to drivers as visual signals, such as traffic signs and traffic lanes. Traffic signs are an important part of road infrastructure to provide information about the current state of the road, restrictions, prohibitions, warnings, and other helpful information for navigation [1,2]. This information is encoded in the traffic signs visual traits: Shape, color and pictogram [1]. Disregarding or failing to notice these traffic signs may directly or indirectly contribute to a traffic accident. However, in adverse traffic conditions, the driver may accidentally or deliberately not notice traffic signs [3]. In these circumstances, if there is an automatic detection and recognition system for traffic signs, it can compensate for a driver's possible inattention, decreasing a driver's tiredness by helping him follow the traffic sign, and thus, making driving safer and easier. Traffic sign detection and recognition (TSDR) is an important application in the more recent technology referred to as advanced driver assistance systems (ADAS) [4], which is designed to provide drivers with vital information that would be difficult or impossible to come by through any other means [5]. The TSDR system has received an increasing interest in recent years due to its potential use in various applications. Some of these applications have been well defined and summarized in [6] as checking the presence and condition of the signs on highways, sign inventory in towns and cities, re-localization of autonomous vehicles; as well as its use in the application relevant to this research, as a driver support system. However, a number of challenges remain for a successful TSDR systems; as the performance of these systems is greatly affected by the surrounding conditions that affect road signs visibility [4]. Circumstances that affect road signs visibility are either temporal because of illumination factors and bad weather conditions or permanent because of vandalism and bad postage of signs [7]. Figure 1 shows an example of some non-ideal invariant traffic signs. These non-identical traffic signs cause difficulties for TSDR.

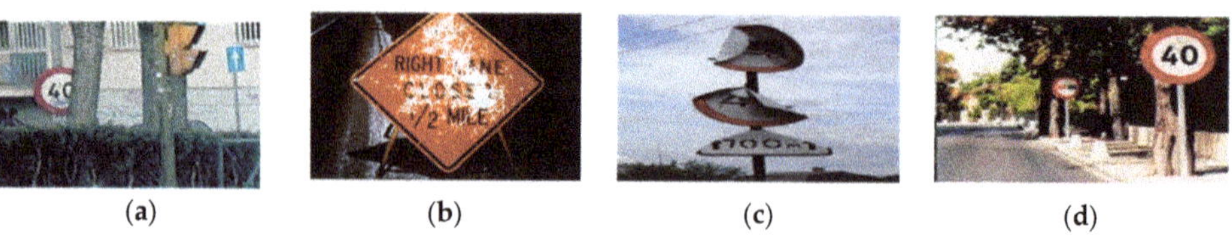

(a)   (b)   (c)   (d)

Figure 1. Non-identical traffic signs: (**a**) Partially occluded traffic sign, (**b**) faded traffic sign, (**c**) damaged traffic sign, (**d**) multiple traffic signs appearing at a time.

This paper provides a comprehensive survey on traffic sign detection, tracking and classification. The details of algorithms, methods and their specifications on detection, tracking and classification are investigated and summarized in the tables along with the corresponding key references. A comparative study on each section has been provided to evaluate the TSDR methods, performance metrics and their availability. Current issues and challenges of the existing technologies are illustrated with brief suggestions and a discussion on the progress of driver assistance system research in the future. The rest of this paper is organized as follows: In Section 2, an overview on traffic signs and recent trends of the research in this field is presented. This is followed by providing a brief review on the available traffic sign databases in Section 3. The methods of detection, tracking, and classification are categorized, reviewed, and compared in Section 4. Section 5 revises current issues and challenges facing the researchers in TSDR. Section 5 summarizes the paper, draws the conclusion and suggestions.

2. Traffic Signs and Research Trends

Aiming at standardizing traffic signs across different countries, an international treaty, commonly known as the Vienna Convention on Road Signs and Signals [8], was agreed upon in 1968. To date, 52 countries have signed this treaty, among which 31 are in Europe. The Vienna convention classified the traffic signs into eight categories, designated with letters A–H: Danger/warning signs (A), priority signs (B), prohibitory or restrictive signs (C), mandatory signs (D), special regulation signs (E), information, facilities or service signs (F), direction, position or indication signs (G), and additional panels (H). Examples of traffic signs in the United Kingdom for each of the categories are shown in Figure 2.

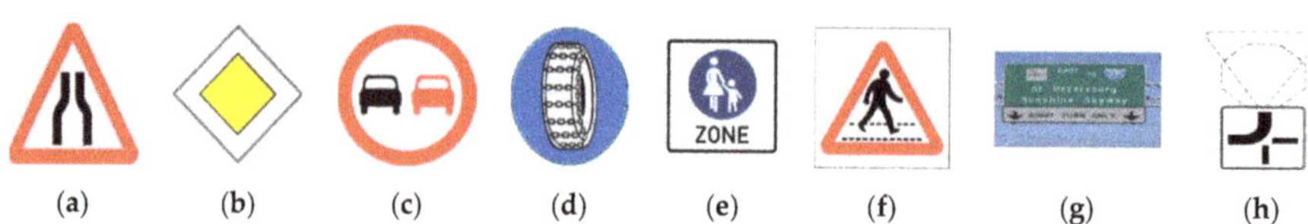

(a)   (b)   (c)   (d)   (e)   (f)   (g)   (h)

Figure 2. Examples of traffic signs: (**a**) A danger warning sign, (**b**) a priority sign, (**c**) a prohibitory sign, (**d**) a mandatory sign, (**e**) a special regulation sign, (**f**) an information sign, (**g**) a direction sign and (**h**) an additional panel.

Despite the well-defined laws in the Vienna Treaty, variations in traffic sign designs still exist among the countries' signatories to the treaty, and in some cases considerable variation within traffic sign designs can exist within the nation itself. These variations are easier to be detected by humans, nevertheless, they may pose a major challenge to an automatic detection system. As an example, different designs of stop signs in different countries are shown in Table 1.

Table 1. Example of stop signs in different countries.

Country	US	Japan	Pakistan	Ethiopia	Libya	New Guinea
Sign						

In terms of research, recently there has been a growing interest in developing efficient and reliable TSDR systems. To show the current state of scientific research regarding this development, a simple search of the term "traffic sign detection and recognition" in the Scopus database has been carried out, with the aim of locating articles published in journals indexed in this database. To focus on the recent and most relevant research, the search has been restricted to the past decade (2009–2018) and only in the subjects of computer science and engineering. In this way, a set of 674 articles and 5414 citations were obtained. The publication and citation trends are shown in Figures 3 and 4, respectively. Generally, the figures indicate a relatively fast growth rate in publications and a rapid increase in citation impact. More importantly, it is clear from the figures that TSDR research has grown remarkably in the last three years (2016–2018), with the highest number of publications and citations representing 41.69% and 60.34%, respectively.

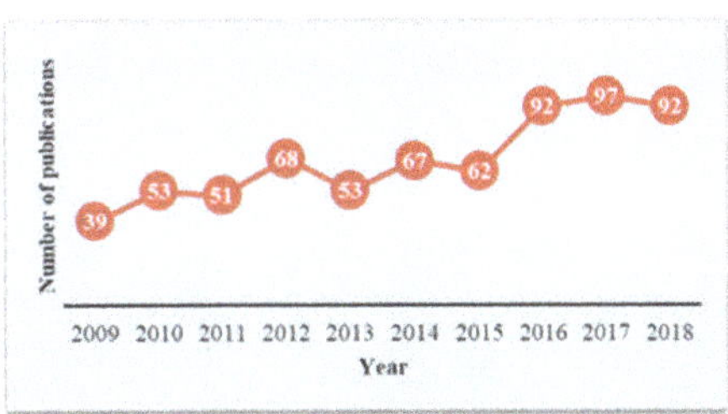

Figure 3. Trends of research for a traffic sign detection and recognition (TSDR) topic based on Scopus analysis tools.

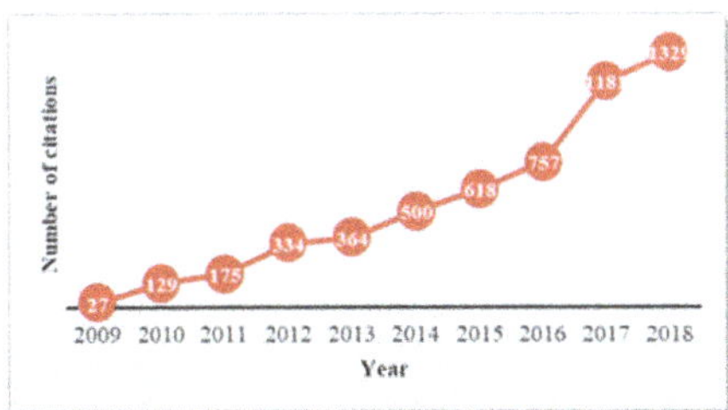

Figure 4. Trends of citations for a TSDR topic based on Scopus analysis tools.

3. Traffic Sign Database

A traffic sign database is an essential requirement in developing any TSDR system. It is used for training and testing the detection and recognition techniques. A traffic sign database contains a large number of traffic sign scenes and images representing samples of all available types of traffic signs: Guide, regulatory, temporary and warning signs. During the past few years, a number of research groups have worked on creating traffic sign datasets for the task of detection, recognition and tracking. Some of these datasets are publicly available for use by the research community. The detailed information regarding the publicly available databases are summarized Table 2. According to [1,9], the first and most widely used dataset is the German traffic sign dataset, which has two datasets: The German Traffic Signs Detection Benchmark (GTSDB) [10] and German Traffic Signs Recognition Benchmark (GTSRB) [11]. This dataset collects three important categories of road signs (prohibitory, danger and mandatory) from various traffic scenes. All traffic signs have been fully annotated with the rectangular regions of interest (ROIs). Examples of traffic scenes in the GTSDB database are shown in Figure 5 [12].

Table 2. Publicly available traffic sign databases [13].

Dataset	Country	Classes	TS Scenes	TS Images	Image Size (px)	Sign Size (px)	Include Videos
GTSDRB (2012 and 2013)	Germany	43	9000	39,209 (training), 12,630 (testing)	15×15 to 250×250	15×15 to 250×250	No
KULD (2009)	Belgium	100+	9006	13,444	1628×1236	100×100 to 1628×1236	Yes, 4 tracks
STSD (2011)	Sweden	7	20,000	3488	1280×960	3×5 to 263×248	No
RUGD (2003)	The Netherlands	3	48	48	360×270	N/A	No
Stereopolis (2010)	France	10	847	251	1920×1080	25×25 to 204×159	No
LISAD (2012)	US	49	6610	7855	640×480 to 1024×52	6×6 to 167×168	All annotations
UKOD (2012)	UK	100+	43,509	1200 (synthetic)	648×480	24×24	No
RTSD (2013)	Russia	140	N/A	80,000+ (synthetic)	1280×720	30×30	No

Figure 5. Examples of traffic scenes in the German Traffic Signs Detection Benchmark (GTSDB) database [12].

4. Traffic Sign Detection, Tracking and Classification Methods

As aforementioned, a TSDR is a driver supportive system that can be used to notify and warn the driver in adverse conditions. This system is a vision-based system that usually has the capability to detect and recognize all traffic signs, even those signs that may be partially occluded or somewhat distorted [14]. Its main tasks are locating the sign, identifying it and distinguishing one sign from another [15,16]. Thus, the procedure of the TSDR system can be divided into three stages, the detection, tracking and classification stages. Detection is concerned with locating traffic signs in the input scene images, whereas classification is about determining what type of sign the system is looking at [17,18]. In other words, traffic sign detection involves generating candidate region of interests (ROIs) that are likely to contain regions of traffic signs, while traffic sign classification gets each candidate ROI and tries to identify the exact type of sign or rejects the identified ROI as a false detection [4,19]. Detection and classification usually constitute recognition in the scientific literature. Figure 6 illustrates the main stages of the traffic sign recognition system. As indicated in the figure, the system is able to work in two modes, the training mode in which a database can be built by collecting a set of traffic signs for training and validation, and a testing mode in which the system can recognize a traffic sign which has not been seen before. In the training mode, a traffic sign image is collected by the camera and stored in the raw image database to be classified and used for training the system. The collected image is then sent to color segmentation process where all background objects and unimportant information in the image are eliminated. The generated image from this step is a binary image containing the traffic sign and any other objects similar to the color of the traffic sign. The noise and small objects in the binary image are cleaned by the object selector process and the generated image is then used to create or update the training image database. According to [20], feature selection has two functions in enhancing the performances of learning tasks. The first function is to eliminate noisy and redundant information, thus getting a better representation and facilitating the classification task. The second function is to make the subsequent computation more efficient through lowering the feature space. In the block diagram, the features are then extracted from the image and used to train the classifier in the subsequent step. In testing mode, the same procedure is followed, but the extracted features are used to directly classify the traffic sign using a pre-trained classifier.

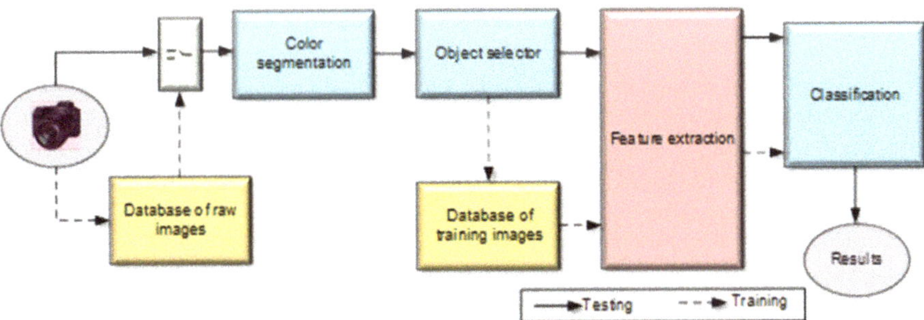

Figure 6. Block diagram of the traffic sign recognition system.

Tracking is used by some research in order to improve the recognition performance [21]. The three stages of a TSDR system are shown in Figure 7, and further discussed in the subsequent sections.

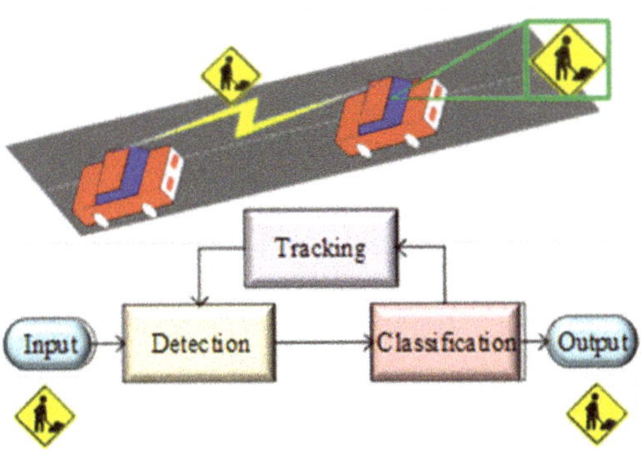

Figure 7. General procedure of TSDR system [22].

4.1. Detection Phase

The initial stage in any TSDR system is locating potential sign image regions from a natural scene image input. This initial stage is called the detection stage, in which a ROI-containing traffic sign ise actually localized [17,23,24]. Traffic signs usually have a strict color scheme (red, blue, and white) and specific shapes (round, square, and triangular). These inherent characteristics distinguish them from other outdoor objects making them suitable to be processed by a computer vision system automatically, thus, allow the TSDR system to distinguish traffic signs from the background scene [21,25]. Therefore, traffic sign detection methods have been traditionally classified into color-based, shape-based and hybrid (color–shape-based) methods [23,26]. Detection methods are outlined in Figure 8 and compared in the following subsections.

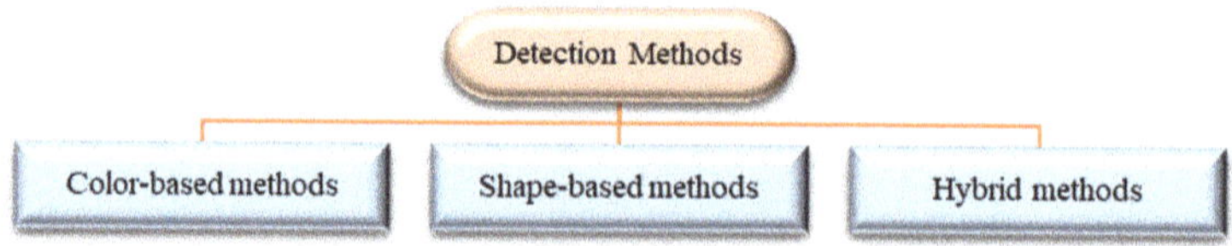

Figure 8. Different methods applied for traffic sign detection.

4.1.1. Color-Based Methods

Color-based methods take advantage of the fact that traffic signs are designed to be easily distinguished from their surroundings, often colored in highly visible contrasting colors [17]. These colors are extracted to detect ROI within an input image based on different image-processing methods. Detection methods based on the color characteristics have low computing, good robustness and other characteristics, which can improve the detection performance to a certain extent [25]. However, methods based on color information can be used with a high-resolution dataset but not with grayscale images [23]. In addition, the main problem with using the color parameter is its sensitivity to various factors such as the distance of the target, weather conditions, time of the day, as well as reflection, age and condition of the signs [17,23].

In color-based approaches, the captured images are partitioned into subsets of connected pixels that share similar color properties [26]. Then the traffic signs are extracted by color thresholding segmentation based on smart data processing. The choice of color space is important during the detection phase, hence, the captured images are usually transformed into a specific color space where the signs are more distinct [9]. According to [27], the developed color-based detection methods are based on the red, green, blue (RGB) color space [28–30], the hue, saturation, and value (HSV) color space [31,32], the hue, saturation, and intensity (HSI) color space [33] and various other color spaces [34,35]. The most common color-based detection methods are represented in Figure 9 and reviewed respectively in Table 3.

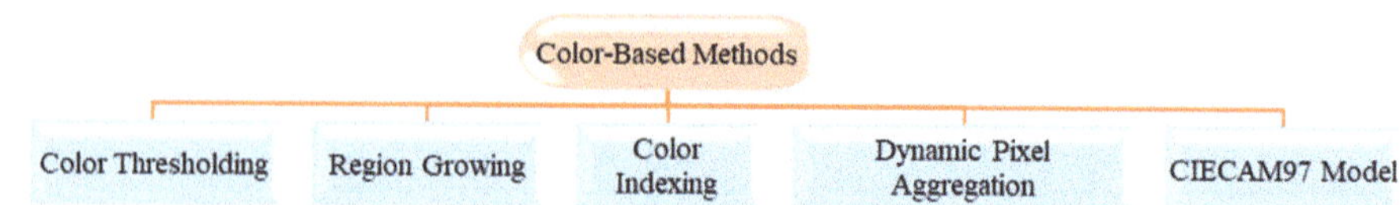

Figure 9. Most popular color-based detection methods.

Color thresholding segmentation is one of the earliest techniques used to segment digital images [26]. Generally, it is based on the assumption that adjacent pixels whose value (grey level, color value, texture, etc.) lies within a certain range belong to the same class [36]. Normal color segmentation was used for traffic sign detection by Varun et al. [37] with their own created dataset, containing 2000 test images, resulting in an accuracy level of 82%. The efficiency was improved in [38] by using color segmentation followed by a color enhancement method. In recent research, color thresholding has commonly been used for pre-processing purposes [39,40]. In [39], pre-filtering was used to train a color classifier, which created a regression problem, whose core was to find a linear function, as shown in (1).

$$f(x) = (w, x) + b, x = (v_1, v_2, v_3)^i \tag{1}$$

where v_i is the intensity value of i_{th} channel (i = 1, 2, 3 for a three-channel RGB image), $(w, b) \in \mathcal{R} \times \mathcal{R}$ are parameters that control the function and the decision rule is given by $\mathrm{sgn}(f(x))$. In [40], Vazquez-Reina et al. used RGB to HSI color space conversion with the additional feature of white sign detection. The main advantage of this feature is its illuminated sign detection. In Refs. [33,41–45], HSI/HSV transformation approach was used for the purpose of detection. The major advantages of the HSI color space over the RGB color space are that it has only two components, hue and saturation, both are very similar to human perception and it is more immune to lighting conditions. In [33], a simple RGB to HSI color space transformation is used for the TSDR purpose. In [44], the HSI color space was used for detection, and then, the detected signal was passed to the distance to borders (DtBs) feature for shape detection to increase the accuracy level. The average accuracy was approximately 88.4% on GRAM database. The main limitation of using HSV transformation is the strong hue dependency of brightness. Hue is only a measurement of the physical lightness of a color, not the perceived brightness. Thus, the value of a fully saturated yellow and blue is the same.

Region growing is another simple and popular technique used for detection in TSDR systems. Region growing is a pixel-based image segmentation method that starts by selecting a starting point or seed pixel. Then, the region develops by adding neighboring pixels that are uniform, according to a certain match criterion, increasing step-by-step the size of the region [46]. This method was used by Nicchiotti et al. [47] and Priese et al. [48] for TSDR. Its efficiency was not very high, approximately 84%. Because this method is dependent on seed values, problems can occur when the seed points lie on edges, and, if the growth process is dominated by the regions, uncertainty around edges of adjacent regions may not be resolved correctly.

The color indexing method is another simple method that identifies objects entirely on the basis of color [49]. It was developed by Swain and Ballard [50] and was used by researchers in the early 1990s. In this method, a comparison of any two colored images is done by comparing their color histogram. For a given pair of histograms, I and M, each containing n bins, the histogram intersections are defined as [50]:

$$\sum_{j=1}^{n} \min\left(I_j, M_j\right). \tag{2}$$

The match value is then,

$$H(I, M) = \frac{\sum_{j=1}^{n} \min\left(I_j, M_j\right)}{\sum_{j=1}^{n} M_j}. \tag{3}$$

The advantage of using color histograms is their robustness with respect to geometric changes of projected objects [51]. However, color indexing is segmentation dependent, and complete, efficient and reliable segmentation cannot be performed prior to recognition. Thus, color indexing is negatively characterized as being an unreliable method.

Another approach to color segmentation is called a dynamic pixel aggregation [52]. In this method, the segmentation process is accomplished by introducing a dynamic thresholding to the pixel aggregation process in the HSV color space. The applied threshold is independent in terms of linearity and its value is defined as [52],

$$a = k - \sin(s_{seed}) \tag{4}$$

where, k is the normalization parameter and S_{seed} is the seed pixel saturation. The main advantage of this approach is hue instability reduction. However, it fails to reduce other segmentation-based problems, such as fading and illumination. This method was tested in [52] on their own created database with 620 outdoor images, resulting in an accuracy level approximately 86.3 to 95.7%.

The International Commission on Illumination 1997 Interim Color Appearance Model (CIECAM97) appearance model is another method has been used to detect and extract color information and to segment and classify traffic signs. Generally, color appearance models are capable of predicting color appearance under a variety of viewing conditions, including different light sources, luminance levels, surrounds, and lightness of backgrounds [53]. This model was used by Gao et al. [54] to transform the image from RGB to (International Commission on Illumination) CIE XYZ values. The main drawback of this model is its chromatic-adaptation transform, which is called the Bradford transform, where chromatic blues appear purple as the chroma is reduced at a constant hue angle.

Table 3. Colors based approaches for TSDR system.

Techniques	Paper	Segmentation Methods	Advantages	Sign Type	No. of Test Images	Test Image Type
Color Thresholding Segmentation	[37]	RGB color segmentation	Simple	Any color	2000	N/A
	[38]	RGB color segmentation with enhancement of color	Fast and high detection rate	Red, blue, yellow	135	Video data
HSI/HSV Transform	[40]	HSI thresholding with addition for white signs	Segments adversely illuminated signs	Any color	N/A	High-res
	[33]	HSI color-based segmentation	Simple and fast	Any color	N/A	N/A
	[41]	RGB to HSI transformation	Segments adversely illuminated signs	Any color	N/A	Low-res
	[42]	RGB to HSI transformation	N/A	Red	N/A	Low-res
	[43]	RGB to HSI transformation	N/A	Any color	3028	Low-res
	[44]	HSI color-based segmentation	Simple and high accuracy rate	Red, blue	N/A	Video data
	[45]	HSI color-based segmentation	Simple and real time application	Any color	632	High-res
Region Growing	[48]	Started with seed and expand to group pixels with similar affinity	N/A	N/A	N/A	N/A
	[47]			N/A	N/A	High-res
Color Indexing	[50]	Comparison of two any-color images is done by comparing their color histogram	Straightforward, fast method	Any color	N/A	Low-res
	[49]			Any color	N/A	N/A
Dynamic Pixel Aggregation	[52]	Dynamic threshold in pixel aggregation on HSV color space	Hue instability reduced	Any color	620	Low-res
CIECAM97 Model	[54]	RGB to CIE XYZ transformation, then to LCH space using CIECAM97 model	Invariant in different lighting conditions	Red, blue	N/A	N/A
YCbCr Color Space	[55]	RGB to YCbCr transformation then dynamic thresholding is performed in Cr component to extract red object	Simple and high accuracy	Red	193	N/A
	[56]		High accuracy less processing time	Any color	N/A	Low-res

The Green (Y), Blue (Cb), Red (Cr) (YCbCr) color space has been considered in recent approaches. Different from the most common color space RGB, which represents color as red, green and blue, YCbCr represents color as brightness and two-color difference signals. It was used for detection in [55], showing an accuracy level over 93% on their own collected database. The efficiency was improved to approximately 97.6% in [56] by first transforming RGB color space to YCbCr color space, then segmenting the image and performing shape-based analysis.

4.1.2. Shape-Based Methods

Just as traffic signs have specific colors, they also have very well-defined shapes that can be searched for. Shape-based methods ignore the color in favor of the characteristic shape of signs [17]. Detection of a traffic sign via its shape follows the defining algorithm of shape detection i.e., to finding the contours and approximating it to reach a final decision based on the number of contours [15,23]. Shape detection is preferred for traffic signs recognition as the colors found on traffic signs changes according to illumination. In addition, shape detection reduces the search for a road sign regions from the whole image to a small number of pixels [57]. However, for this method the memory and computational requirement is quite high for large images [58]. In addition, damaged, partially obscured, faded and blurred traffic signs may cause difficulties in detecting traffic signs accurately, leading to a low accuracy rate. Detection of the traffic signs in these methods is made from the edges of the image analyzed by structural or comprehensive approaches [23]. Many shape-based methods are popular in TSDR systems. These methods are represented in Figure 10 and reviewed respectively in Table 4.

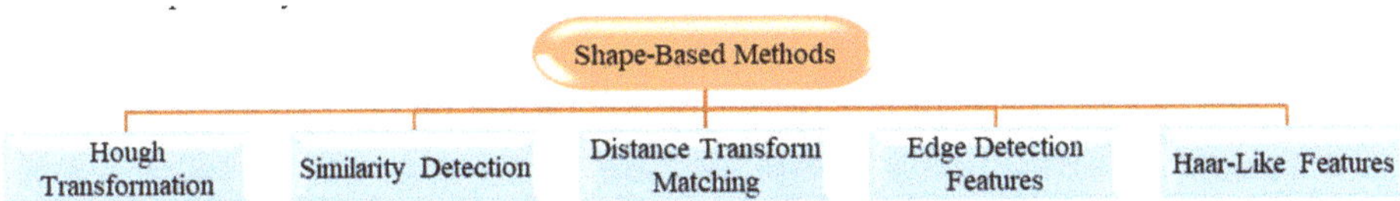

Figure 10. Most popular shape-based detection methods.

The most common shape-based approach is the Hough transformation. The Hough transformation usually isolates features of a particular shape within a given frame/images [15]. It was applied by Zaklouta et al. in [59] to detect triangular and circular signs. Their own test datasets contained 14,763 and 1584 signs, and the accuracy rate was approximately 90%. The main advantage of the Hough transformation technique is that it is tolerant of gaps in feature boundary descriptions and is relatively unaffected by image noise [60]. However, its main disadvantage is the dependency on input data. In addition, it is only efficient for a high number of votes that fall in the correct bin. When the parameters are large, the average number of votes cast for a single bin becomes low, and thus, the detection rate is decreased.

Another shape-based detection method is the similarity detection. In this method the detection is performed by computing a similarity factor between a segmented region and set of binary image samples representing each road sign shape [57]. This method was used by Vitabile et al. [52] on their collected dataset with an accuracy level over 86.3%. The main advantage of this method is its straightforwardness, whilst its main drawback is that the input image should be perfectly segmented and the dimensions have to be same. In [52], the images were initially converted from RGB to HSV, then they were segmented and resized into 36 × 36 pixels. The similarity detection equation is,

$$x' = \frac{x - x_{\min}}{x_{\max} - x_{\min}}.n \quad y' = \frac{y - y_{\min}}{y_{\max} - y_{\min}}.n \tag{5}$$

where, $x_{\max}$, $y_{\max}$, $x_{\min}$ and $y_{\min}$ are the coordinates of the rectangle vertices.

Distance transform matching (DTM) is also another type of shape-based detection method. In this method, the distance transform of the image is formed by assigning each non-edge pixel a value that is a measure of distance to the nearest edge pixel. It was used by Gavrila [61] to capture large variations in object shape by identifying the template features to the nearest feature image from a distribution of distances. This distance is inversely proportional to the matching of the image and the templates of the images. The chamfer distance equation is:

$$D_{chamfer}(T, I) \equiv \frac{1}{T} \sum_{t \in T} d_I(t) \tag{6}$$

where $|T|$ and $d_I(t)$ denote the number of features and the distance between features t in T and the closest feature in I, respectively. In his experiment, Gavrila [61] used DTM to examine 1000 collected test images, and the accuracy was approximately 95%. The DTM technique is efficient for detecting arbitrary shapes within images. However, its main disadvantage is the vulnerability of detecting cluttered images.

Another popular two colorless traffic sign detection methods are edge detection features and Haar-like features. Edge detection refers to the process of identifying and locating sharp discontinuities in an image [62]. By using this method, image data is simplified for the purpose of minimizing the amount of data to be processed. This method was used in [63–67] for indicating the boundaries of objects within the image through finding a set of connected curves. The Haar-like features method was proposed by Paul Viola and Michael Jones [68] based on the Haar wavelet to recognize the target objects. As indicated in Table 4, the Haar-like features based detection method was used in [69,70] for traffic sign detection. The main advantage is its calculating speed, where any size of images can be calculated in a constant time. However, its weakness is the requirement of a large number of training images and high false positive rates [23].

4.1.3. Hybrid Methods

As previously discussed, both color-based and shape-based methods have some advantages and disadvantages. Therefore, researchers recently have tried to improve the efficiency of the TSDR system using a combination of color- and shape-based features. In the hybrid methods, either color-based approaches take shape into account after having looked at colors, or shape detection is used as the main method but integrate some color aspects as well. In color-based approaches a two-stage strategy is usually employed. First, segmentation is done to narrow the search space. Subsequently, shape detection is implemented and is applied only to the segmented regions [58]. Color and shape features were combined into traffic sign detection algorithms in studies [71–76]. In these studies, different signs with various colors and shapes were detected using different datasets.

Table 4. Shape-based methods for TSDR system.

Technique	Paper	Overall Process	Recognition Feature	Advantages	Sign Type	No. of Test Image	Test Image Type
Hough Transform	[77]	Each pixel of edge image votes for the object center at object boundary	N/A	Invariant to in-plane rotation and viewing angle	Octagon, square, triangle	45	Low-res
	[78]		AdaBoost	High accuracy	Any sign	N/A	Low-res
	[79]		N/A	Robustness to illumination, scale, pose, viewpoint change and even partial occlusion	Red (circular), blue (square)	500+	Low-res
	[80]		N/A	Reducing memory consumption and increasing utilization Hough-based SVM	Any sign	3000	High-res
	[81]		N/A	Robustness	Red (circular)	N/A	768 × 580
	[59]		Random Forest	Improve efficiency of K-d tree, random forest and SVM	Triangular and circular	14,763	752 × 480 px
	[82]		SIFT and SURF based MLP	Applying another state refinement	Red circular	N/A	Video data
Similarity Detection	[52]	Computes a region and sets binary samples for representing each traffic sign shape.	NN	Straight forward method	Any color	620	Low-res
DTM	[61]	Capturing object shape by template hierarchy.	RBF Network	Detects objects of arbitrary shape	Circular and triangular	1000	360 × 288 px
Edge Detection Feature	[63]	A set of connected curves is found which indicates the boundaries of objects within the image.	Geometric matching	Invariant in translation, rotation and scaling	Any color	1000	640 × 480
	[64]		Normalized cross correlation	Reliability and high accuracy in real time	Speed limit sign	N/A	320 × 240 px video data
	[65]		N/A	Improved accuracy by training negative sample	Red (circular)	3907	Low-res
	[66]		N/A	Invariant in noise and lighting	Triangle, circular	847	High-res
	[67]		CDT	Invariant in noise and illumination	Red, blue, yellow		
Edges with Haar-like Features	[69]	Sums three pixel intensities and calculates the difference of sums by Haar-like features	CDT	Smoother and noise invariant	Rectangular, any color		Video data
	[70]		SVM	Fast method	Circular, triangular upside-down, rectangle and diamond		640 × 480 px video data

4.2. Tracking Phase

For robust detection and in order to increase the accuracy of the information used in identifying traffic signs, the signs are tracked using a simple motion model and temporal information propagation. This tracking process is very important for real-time applications, by which the TSDR system verifies correctness of the traffic sign and keeps tracking the sign to avoid handling the same detected sign more than once [21,83]. The tracking process is performed by feeding the TSDR system with a video recorded by a camera fixed on the vehicle and monitoring the sign candidates on a number of consecutive frames. The accepted sign candidates are only those shown up more than once. If the object is not a traffic sign or a sign that only shows up once, it can be eliminated as soon as possible, and thus, the computation time of the detection task can be reduced [84]. According to [85] and as shown in Table 5, the most common tracker adapted is the Kalman filter, as in [82,85–88]. The block diagram of a TSDR system with a tracking process based on the Kalman filter as proposed in [82] is shown in Figure 11. In the figure, SIFT, CCD and MLP are abbreviations of scale-invariant feature transform, contracting curve density and multi-layer perceptrons, respectively.

Table 5. Sign tracking based on Kalman Filter approaches.

Technique	Paper	Advantages	Performance
Kalman Filter	[82]	For avoiding incorrect assignment, rule-based approach utilizing combined distance direction difference is used.	N/A
	[89]	Takes less time in tracking and verifying	Using 320 × 240 pixel images, takes 0.1 s to 0.2 s.
	[88]	Used stereo parameters to reduce the error of stereo measurement	N/A
Advanced Kalman Filter	[85]	Fast and advanced method, high detection and tracking rate	Using 400 × 300 pixel images, can process 3.26 frames per second.

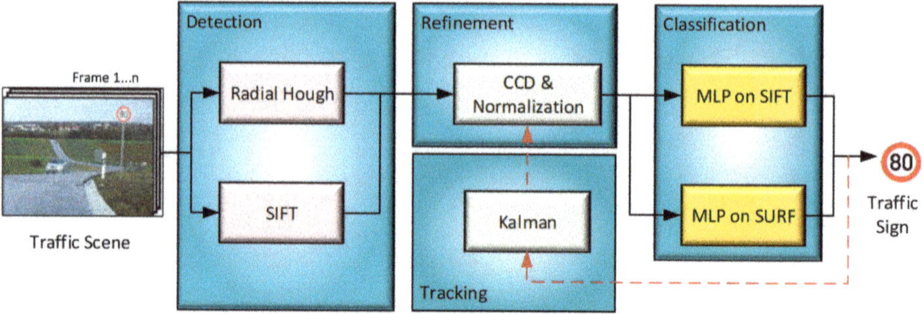

Figure 11. An example of a TSDR system includes tracking process based on Kalman filter [81].

4.3. Classification Phase

After the localization of ROIs, classification techniques are employed to determine the content of the detected traffic signs [1]. Understanding the traffic rule enforced by the sign is achieved by reading the inner part of the detected traffic sign using a classifier method. Classification algorithms are neither color-based nor shape-based. The classifier usually takes a certain set of features as the input, which distinguishes the candidates from each other. Different algorithms are used to classify the traffic signs swiftly and accurately. Some conventional methods used for classification of traffic signs are outlined in Figure 12 and reviewed respectively in Tables 6–13.

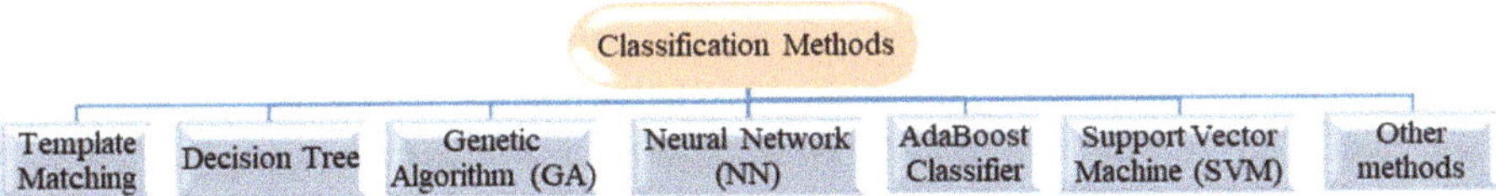

Figure 12. Most popular classification methods.

Table 6. Examples of TSDR systems using a template matching method.

Ref	Detection Feature	Advantages	True Positive Rate	False Positive Rate	No. of Test Images	Overall Accuracy	Time
[90]	RGB to HSV then contrast stretching	Fast and straight forward method	N/A	N/A	N/A	<95%	N/A
[91]	N/A		N/A	N/A	100	90.9%	N/A

Template matching is a common method in image processing and pattern recognition. It is a low-level approach which uses pre-defined templates to search the whole image pixel by pixel or to perform the small window matching [15]. It was used for TSDR by Ohara et al. [90] and Torresen et al. [91]. It has the advantages of being fast, straightforward and accurate (with a hit rate of approximately 90% on their own pictured images dataset). However, the drawback of this method is that it is very sensitive to noise and occlusions. In addition, it requires a separate template for each scale and orientation. Examples of TSDR systems using a template matching method are shown in Table 6.

Another common classification method is the random forest. It is a machine learning method that operates by constructing a multitude of decision trees during the training time and outputting the class that is the mode of the output of the class of individual trees. This method was compared in [92,93] with SVM, MLP, Histogram of Oriented Gradient (HOG)-based classifiers, showing the highest accuracy rate and the lowest computational time. Based on their own dataset, the accuracy was approximately 94.2%, whereas the accuracy of the SVM is 87.8% and that of MLP is 89.2%. In terms of computational time for a single classification, the SVM takes 115.87 ms, MLP takes 1.45 ms, and a decision tree takes 0.15 ms. Despite its high accuracy and low computation time, the main limitation of a random forest is that a large number of trees can make the algorithm slow and ineffective for real-time predictions. Examples of TSDR systems using a decision tree method are shown in Table 7.

Table 7. Examples of TSDR systems using a decision tree method.

Ref	Detection Feature	Advantages	True Positive Rate	False Positive Rate	No. of Test Images	Overall Accuracy	Time	Dataset
[94]	HOG based SVM	Used GTSRB and ETH 80 dataset and compared	90.9%	N/A	12,569	90.46%	17.9 ms	GTSRB and ETH 80
[95,96]		Used Gaussian weighting in HOG to improve performance by 15%	90%	N/A	12,569	97.2%	17.9 ms	Own created
[92]	MSER based HOG	Eliminating hand labeled database, robust to various lighting and illumination	83.3%	0.85	640 × 480 px video data	87.72%	N/A	Own created
[97]	HOG	Remove false alarm up to 94%	N/A	N/A	12,569	92.7%	17.9 ms	Own created

Table 8. Examples of TSDR systems using a genetic algorithm.

Ref	Detection Feature	Advantages	True Positive Rate	False Positive Rate	No. of Test Images	Overall Accuracy	Time
[98,99]	Genetic Algorithm	Unaffected by illumination problem	N/A	N/A	Video data	N/A	N/A

Genetic algorithm is another classification method. It is based on a natural selection process that mimics biological evolution, which was used early in this century. This method was used for traffic sign recognition by Aoyagi et al. [98] and Eccalera et al. [99]. It was proved in these studies that this method is effective in detection of the traffic sign even if the traffic sign has some shape loss or illumination problem. The disadvantage of the genetic algorithm is non-deterministic work time and non-guarantee finding of the best solution [57]. Examples of TSDR systems using a genetic algorithm method are shown in Table 8.

The other most common method for classification is using an artificial neural network (ANN). This method has gained an increasing popularity in recent years due to the advancement in general-purpose computing on graphics processing units (GPGPU) technologies [2]. In addition, it is popular due to its robustness, greater adaptability to changes, flexibility and high accuracy rate [100]. Another key advantage of this method is its ability to recognize and classify objects at the same time, while maintaining high speed and accuracy [2]. ANN-based classifiers were used in [56,99,101–108] for TSDR. In the experiment conducted in [56], the hit rate was 97.6%, and the computational time was 0.2 s. However, in [107] ANN-based methods were described to have some limitations, such as their slowness and the instability in the NN training due to too large a step. This method was compared with a template matching method in [108], concluding that NNs require a large number of training samples for real world applications. Examples of TSDR systems using an ANN method are shown Table 9.

Table 9. Examples of TSDR systems using an ANN method.

Ref	Detection Feature	Advantages	True Positive Rate	False Positive Rate	No. of Test Images	Overall Accuracy	Time	Dataset
[56]	YCbCr and normalized cross correlation	Robustness and adaptability	0.96	0.08	640 × 480 px video data	97.6%	0.2 s	Own created
[101]	N/A	Flexibility and high accuracy	N/A	N/A	N/A	98.52–99.46%	N/A	Own created
[106]	Adaptive shape analysis	Invariant in illumination	N/A	N/A	220	95.4%	0.6 s	Own created
[107]	NN	Robustness	N/A	N/A	467	N/A	N/A	Own created
[108]	Bimodal binarization and thresholding	Compared TM and NN elaborately	0.96	0.08	640 × 480 px video data	97.6%	0.2 s	Own created

Another increasingly popular method in vision-based object recognition is the deep learning method. This method has acquired general interest in recent years owing to its high performance of classification and the power of representational learning from raw data [109,110]. Deep learning is a part of a broader family of machine learning methods. In contrary to task specific methods, deep learning focuses on data representations with supervised, weakly supervised or unsupervised learning. Deep learning methods use a cascade of many layers of nonlinear processing units for feature extraction and transformation. Each successive layer uses the output from the previous layer as input. Higher level features are derived from lower level features to form a hierarchical representation [110]. Among the deep learning models, the convolutional neural networks (CNN) have acquired unique noteworthiness from their repeatedly confirmed superiorities [111]. According to [112], CNN models are the most widely used deep learning algorithms for traffic sign classification to date. Of the examples applied to traffic sign classification are committee CNN [113], multi-scale CNN [114], multi-column CNN [102], multi-task CNN [111,115], hinge-loss CNN [116], deep CNN [46,117], a CNN with diluted convolutions [118], a CNN with a generative adversarial network (GAN) [119], and a CNN with SVM [120]. Based on these studies, a simultaneous detection and classification can be achieved using deep learning-based methods. This simultaneousness results in improved performance, boosted training and testing speeds. Examples of TSDR systems using a deep learning method are shown in Table 10.

Table 10. Examples of TSDR systems using a deep learning method.

Ref	Detection Feature	Advantages	True Positive Rate	False Positive Rate	No. of Test Images	Overall Accuracy	Time	Dataset
[115]	Object bounding box prediction	Predicting position and precise boundary simultaneously	>0.88 mPA	<3 pixels	3,719	91.95%	N/A	GTSDB
[120]	YCbCr model	High accuracy and speed	N/A	N/A	Video data	98.6%	N/A	Own created
[111]	Color space thresholding	Implementing detection and classification	90.2%	2.4%	20,000	95%	N/A	GTSRB
[121]	SVM	Robust against illumination changes	N/A	N/A	Video data	97.9%	N/A	Own created
[117]	Scanning window with a Haar cascade detector	Enhanced detection capability with good time performance	N/A	N/A	16,630	99.36%	N/A	GTSRB

Adaptive boosting or AdaBoost is a combination of multiple learning algorithms that can be utilized for regression or classification [15]. It is a cascade algorithm, which was introduced by Freund and R. Schapire [122]. Its working concept is based on constructing multiple weak classifiers and assembling them into a single strong classifier for the overall task. As indicated in Table 11, the AdaBoost method was used for TSDR in [123–127]. Based on these studies, it can be concluded that the main advantage of the AdaBoost is its simplicity, high prediction power and capability to cascade an architecture for improving the computational efficiency. However, its main disadvantage is that if the input data have wide variations or abrupt changes in the background, then the training time increases and classifier accuracy decreases [121]. In addition, the AdaBoost trained classifier cannot be dynamically adjusted with new coming samples unless retrained from the beginning, which is time consuming and demands storing all historical samples [128]. Examples of TSDR systems using an AdaBoost method are shown in Table 11.

Table 11. Examples of TSDR systems using an AdaBoost method.

Ref	Detection Feature	Advantages	True Positive Rate	False Positive Rate	No. of Test Images	Overall Accuracy	Time	Dataset
[123]	Sobel edge detection	Comparison of SVM and AdaBoost	N/A	0.25	N/A	92%	N/A	Own created
[124]	AdaBoost	Fast	N/A	N/A	200	>90%	50 ms	Own created
[125]	AdaBoost	Invariant in speed, illumination and viewing angle	92.47%	0%	350	94%	51.86 ms	Own created
[126]	AdaBoost and CHT	Real-time and robust system with efficient SLS detection and recognition	0.97	0.26	1850	94.5%	30–40 ms	Own created
[127]	Haar-like method	Reliability and accuracy	0.9	0.4	200	92.7%	50 ms	Own created

Support vector machine (SVM) is another classification method that contracts an N-dimensional hyper plane that optimally separates the data into two categories. More precisely, SVM is a binary classifier that separates two different classes by a subset of data samples called support vectors. It was implemented as a classifier for traffic sign recognition in [44,55,88,129–136]. This classification method is robust, highly accurate and extremely fast which is a good choice for large amounts of training data. In [129], a SVM-based classifier was applied for detecting speed limit signs and it was compared with the artificial neural network multilayer perceptron (MLP), k-nearest neighbors (kNN), least mean squares (LMS), least squares (LS) and extreme learning machine (ELM) based classifiers. Results of the comparison demonstrated that the SVM-based classifier obtained the highest accuracy and lowest standard deviation amongst all other classifiers. Similarly, in a recent study [3], a cascaded linear SVM classifier was used for detecting speed limit signs, and the result was a recall of 99.81% with a precision of 99.08% on the GTSRB dataset. In [55], a SVM-based classifier was used to detect and classify red road signs in 1000 test images, and the accuracy rate was over 95%. In [88,131], SVM was used with Gaussian kernels for the recognition of traffic signs, and the success rate was 92.3% and 92.6%, respectively. In [136], an advanced SVM method was proposed and tested with binary pictogram and gray scale images; the result was achieving high accuracy rates of approximately 99.2% and 95.9%, respectively. SVM has also shown great effectiveness in extracting the most relevant shots of an event of interest in a video, where a new SVM-based classifier called nearly-isotonic SVM classifier (NI-SVM) was proposed in [137] for prioritizing the video shots using a novel notion of semantic saliency. The proposed classifier exhibited higher discriminative power in event analysis tasks. The main disadvantage of SVM is lack of transparency of results. Transparency means how the results were obtained by the kernel and how the results should be interpreted. In SVM such things are unknown and cannot be known due to the high dimensional vector space. Examples of TSDR systems using a SVM method are shown in Table 12.

Table 12. Examples of TSDR systems using a SVM method.

Ref	Detection Feature	Advantages	True Positive Rate	False Positive Rate	No. of Test Images	Overall Accuracy	Time	Dataset
[44]	DtBs and SVM	Fast, high accuracy	N/A	N/A	Video data	92.3%	N/A	GRAM
[55]	Gabor Filter	Simple and high accuracy	N/A	N/A	58	93.1%	N/A	Own created
[130]	CIELab and Ramer–Douglas–Peucker algorithm	Illumination proof and high accuracy	N/A	N/A	405	97%	N/A	Own created
[131]	RGB to HSI then shape analysis	Less processing time	N/A	N/A		92.6%	Avg. 5.67 s	Own created
[88]	Hough transform	Reliability and accuracy	N/A	N/A	Video data	Avg. 92.3%	35 ms	Own created
[132]	RGB to HIS then shape localization	Reduce the memory space and time for testing new sample	N/A	N/A	N/A	95%	N/A	Own created
[133]	MSER	Invariant in illumination and lighting condition	0.97	0.85	43,509	89.2%	N/A	Own created
[134]	HSI and edge detection	Less processing time	N/A	N/A	Video data	N/A	N/A	Own created
[135]	RGB to HSI	Identify the optimal image attributes	0.867	0.12	650	86.7%	0.125 s	Own created
[136]	Edge Adaptive Gabor Filtering	Reliability and Robustness	85.93%	11.62%	387	95.8%.	3.5–5 ms	Own created

Table 13. Examples of TSDR systems using the other methods.

Ref	Method	Detection Feature	Advantages	True Positive Rate	False Positive Rate	No. of Test Images	Overall Accuracy	Time	Dataset
[138]	SIFT matching	N/A	Effective in recognizing low light and damaged signs	N/A	N/A	60	N/A	N/A	Own created
[34]	Fringe-adjusted joint Transform Correlation	Color Feature Extraction using Gabor Filter	Excellent discrimination ability between object and non-object	783	217	587	N/A	N/A	Own created
[139]	Principal Component Analysis	HSV, CIECAM97 and PCA	High accuracy rate	N/A	N/A	N/A	99.2%	2.5 s	Own created
[140]	Improved Fast Radial Symmetry and Pictogram Distribution Histogram based SVM	RGB to LaB color space then IFRS detection	High accuracy rate	N/A	N/A	300	96.93%	N/A	Own created
[144]	Infrastructures of vehicles	N/A	Eliminating possibility of false positive rate because of ID coding	N/A	N/A	Video data	N/A.	N/A	Own created
[145]	FCM and Content Based Image Recorder	Fuzzy c means (FCM)	Effective in real time application	N/A	N/A	Video data	<80%	N/A	Own created
[141]	Template matching and 3D reconstruction algorithm	N/A	Very effective in recognizing damaged or occulted road signs	In 3D, 54 out of 63	In 3D, 6 out of 63 and 3 signs were missing	4800	N/A	N/A	Own created
[142]	Low Rank Matrix Recovery (LRMR)	N/A	Fast computation and parallel execution	N/A	N/A	40,000	97.51%	>0.2	GTSRB
[143]	Karhunen–Loeve Transform and MLP	Oriented gradient maps	Invariant in illumination an different lighting condition	N/A	N/A	12,600	95.9%	0.0054 s/image	GTSRB
[35]	Self-Organizing Map	N/A	Fast and accurate	N/A	N/A	N/A	<99%	N/A	Own created

In addition to these conventional methods, researchers have used other methods for recognition. In [138], the SIFT matching method was used for recognizing broken areas of a traffic sign. This method adjusts the traffic sign to a standard camera axis and then compares it with a reference image. Sebanja et al. in [139] used principal component analysis (PCA) for both TSDR and the accuracy rate was approximately 99.2%. In [140], the researchers used improved fast radial symmetry (IFRS) for detection and a pictogram distribution histogram (PDH) for recognition. Soheilian et al. in [141] used template matching followed by a three dimensional (3D) reconstruction algorithm to reconstruct the traffic signs obtained from video data and to improve the visual angle for detecting traffic signs. In [142], Pei et al. used low rank matrix recovery (LRMR) to recover the correlation for classification with a hit rate of 97.51% in less than 0.2 s. Gonzalez-Reyna et al. [143] used oriented gradient maps for feature extraction, which is invariant in illumination and variable lighting. For classification, they used Karhunen–Loeve transform and MLP. They reported an accuracy of 95.9% and processing time of 0.0054 s per image. In [35], Miguel et al. used a self-organizing map (SOM) for recognition, where in every level, a pre-processor extracts a feature vector characterizing the ROI and passes it to the SOM. The accuracy rate was very high, approximately 99%. Examples of TSDR systems using the other methods are shown in Table 13.

5. Current Issues and Challenges

TSDR is the essential part of the ADAS. It is mainly designed to operate in a real-time environment for enhancing driver safety through the fast acquisition and interpretation of traffic signs. However, there are a number of external non-technical challenges that may face this system in the real environment degrading its performance significantly. Among the many issues that needed to be addressed while developing a TSDR system are the following issues outlined in Figure 13.

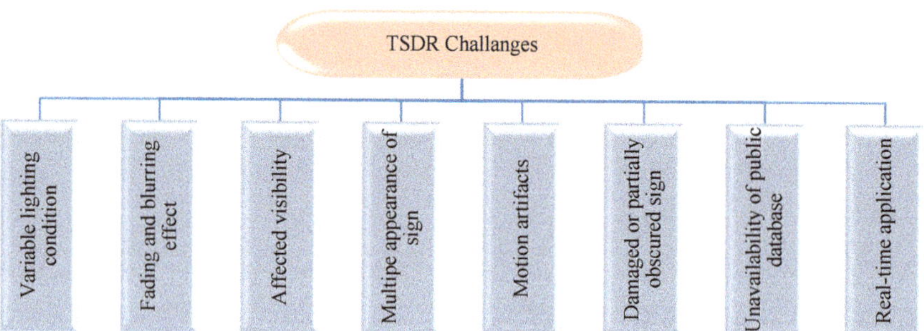

Figure 13. Some of TSDR challenges.

Variable lighting condition: Variable lighting condition is one of the key issues to be considered during TSDR system development. As aforementioned, one of the main distinguishing features of traffic sign is its unique colors which discriminate it from the background information, thus facilitating its detection. However, in outdoor environments illumination changes greatly affects the color of traffic sign, making the color information become completely unreliable as a main feature for traffic sign detection. To cope with such challenge, a method based on adaptive color threshold segmentation and high efficient shape symmetry algorithms has been recently proposed by Xu et al. [26]. This method is claimed to be robust for a complex illumination environment, exceeding a detection rate of 94% on GTSDB dataset.

Fading and blurring effect: Another important difficulty for a TSDR system is the fading and blurring of traffic signs caused by illumination through rain or snow. These conditions can lead to increase in false detections and reduce the effectiveness of a TSDR system. Using a hybrid shape-based detection and recognition method in such conditions can be very useful and may give more superior performance [146].

Affected visibility: Light emitted by the headlamps of the incoming vehicles, shadows, and other weather-related factors such as rains, clouds, snow and fog can lead to poor visibility. Recognizing traffic signs from a road image taken in such cases is a challenging task, and a simple detector may fail to detect these traffic signs. To resolve this problem, it is necessary to enhance the quality of taken images and make them clear by using an image pre-processing technique. A pre-processing makes image filtration and converts input information into usable format for further analysis and detection [147].

Multiple appearances of sign: While detecting traffic signs mainly in city areas, which are more crowded by signs, multiple traffic sign appearing at a time and similar shape man-made objects can cause overlapping of signs and lead to a false detection. The detection process can also be affected by rotation, translation, scaling and partial occlusion. Li et al. in [33], used HSI transform and fuzzy shape recognizer which is robust and unaffected by these problems and its accuracy rate in different weather condition is; sunny 94.66%, cloudy 92.05%, rainy 90.72%.

Motion artifacts: In the ADAS application, the images are captured from a moving vehicle and sometimes using a low resolution camera, thus, these images usually appear blurry. Recognition of blurred images is a challenging task and may lead to false results. In this respect, a TSDR system that integrates color, shape, and motion information could be a possible solution. In such a system, the robustness of recognition is improved through incorporating the detection and classification with tracking using temporal information fusion [73]. The detected traffic signs are tracked, and individual detections from sequential frames (t−t0, . . . , t) are temporally fused for a robust overall recognition.

Damaged or partially obscured sign: The other distinctive feature of traffic sign is its unique shape. However, traffic signs could appear in various conditions including damaged, partly occluded and/or clustered. These conditions can be very problematic for the detection systems, particularly shape-based detection systems. In order to overcome these problems, hybrid color segmentation and shape analysis based methods are recommended [15].

Unavailability of public database: A database is a crucial requirement for developing any TSDR system. It is used for training and testing the detection and recognition methods. One of the obstacles facing this research area is the lack of large, properly organized, and free available public image databases. According to [12], for example, the most commonly used database (GTSDB database) contains only 600 training images and 300 evaluation images. Of the seven categories classified in the Vienna convention, GTSDB covers only three categories of traffic signs for detection: prohibitory, mandatory and danger. All included images are only German traffic signs, which are substantially different from other parts of the world. To resolve the database scarcity problem, perhaps one of the ideas is to create a unified global database containing a large number of images and videos for road scenes in various countries around the world. These scenes must contain all categories of traffic signs under all possible weather conditions and physical states of the signs.

Real-time application: The detection and recognition of traffic signs are caught up with the performance of a system in real-time. Accuracy and speed are surely the two main requirements in practical applications. Achieving these requirements requires a system with efficient algorithms and powerful hardware. A good choice is convolutional neural networks-based learning methods with GPGPU technologies [2].

In brief, although lots of relevant approaches have been presented in the literature, no one can solve the traffic sign recognition problem very well in conditions of different illumination, motion blur, occlusion and so on. Therefore, more effective and more robust approaches need to be developed [12].

6. Conclusions and Suggestion

The major objective of the paper was to analyze the main direction of the research in the field of automatic TSDR and to categorize the main approaches into particular sections to make the topics easy to understand and to visualize the overall research for future directions. Unlike most of the available review papers, the scope of this paper has been broadened to cover all recognition phases: Detection, tracking and classification. In addition, this paper has tried to discuss as many studies as possible, in an attempt to provide a comprehensive review of the various alternative methods available for traffic sign detection and recognition; including along with methods categorization, current trends and research challenges associated with TSDR systems. The overall summary is presented in Figure 14.

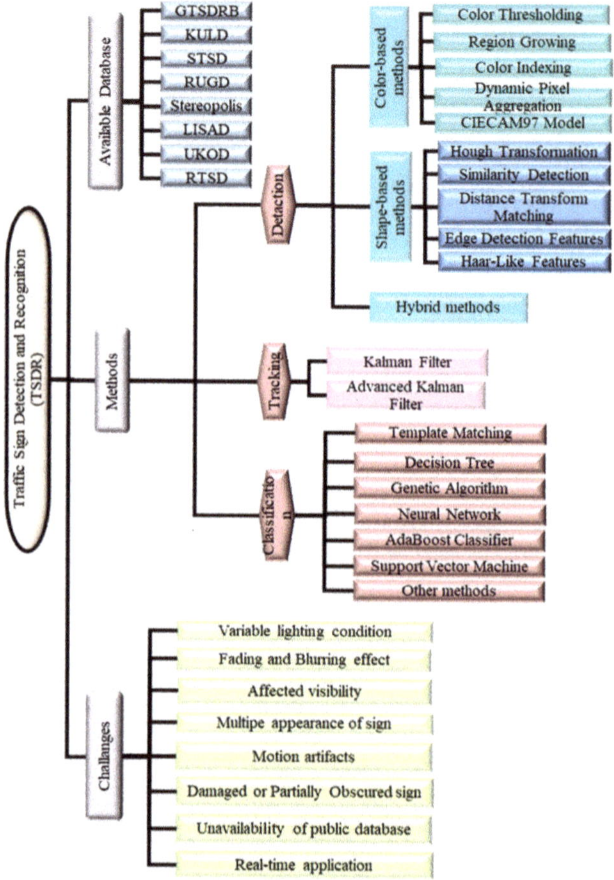

Figure 14. Summary of the paper.

The conducted review reveals that research in traffic sign detection and recognition has grown rapidly, where the number of papers published during the last three years was approximately 280 papers, which represents about 41.69% of the total number of papers published during the last decade as a whole. With regard to the methods used, it was observed that the subject of traffic sign detection and recognition incorporates three main steps: Detection, tracking and classification; and in each step, many methods and algorithms were applied, each has its own merits and demerits. In general, the methods applied in detection and recognition consider either color or shape information of the traffic sign. However, it is well known that the image quality in real-world traffic scenarios is usually poor; due to low resolution, weather condition, varying lighting, motion blur, occlusion, scale and rotation and so on. In addition, traffic signs are usually in a variety of appearances, with high inter-class similarity, and complicated backgrounds. Thus, proper integration of color and shape information in both detection and classification phases is a very promising and exciting task that is in need of much more attention. For tracking, the Kalman filter and its variations are the most common methods. For classification, artificial neural network and support vector machine-based methods were found to be the most popular methods, with a high detection rate, high flexibility and easy adoptability. Despite the recent improvements in the overall performance of TSDR systems, more research is still needed to achieve a rigorous, robust and reliable TSDR system. It is believed that TSDR system performance can be enhanced by merging the detection and classification tasks into one step rather than performing them separately. By doing so, classification can improve detection and vice versa. Another idea for further improvement of TSDR is by using standard, sufficient and large databases for learning, testing and evaluation of the proposed algorithms. In this way, the TSDR system will be able to recognize the eight different categories of the traffic signs in the real environment with different conditions. This paper will be a useful reference for researchers looking for an understanding of the current status of research in the field of TSDR and finding the related research problems in need of solutions.

Author Contributions: Conceptualization, M.A.H. and A.H., S.A.S.; methodology, M.A.H., M.A.A. and S.B.W.; software, M.A.A.; validation, S.B.W.; formal analysis, M.A.A.; investigation, M.A.H., and M.B.M.; resources, M.A.H., S.B.W.; data curation, M.A.A. and S.B.W.; writing—original draft preparation, M.A.H., M.A.A., S.B.W., P.J.K.; writing—review and editing, M.A.H., P.J.K., A.H., S.A.S. and M.B.M.; visualization, M.A.A.; supervision, M.A.H., S.A.S.; project administration, M.A.H.; funding acquisition, M.A.H.

Funding: This research was funded by the TNB Bold Strategic Grant J510050797 under the Universiti Tenaga Nasional and the Universiti Kebangsaan Malaysia under Grant DIP-2018-020.

Acknowledgments: This work is supported by the collaboration between Universiti Tenaga Nasional and Universiti Kebangsaan Malaysia.

Conflicts of Interest: The authors declare no conflict of interest. The funders had no role in the design of the study; in the collection, analyses, or interpretation of data; in the writing of the manuscript, or in the decision to publish the results.

References

1. Alturki, A.S. Traffic Sign Detection and Recognition Using Adaptive Threshold Segmentation with Fuzzy Neural Network Classification. In Proceedings of the 2018 International Symposium on Networks, Computers and Communications (ISNCC), Rome, Italy, 19–21 June 2018; pp. 1–7.
2. Satılmış, Y.; Tufan, F.; Şara, M.; Karslı, M.; Eken, S.; Sayar, A. CNN Based Traffic Sign Recognition for Mini Autonomous Vehicles. In Proceedings of the International Conference on Information Systems Architecture and Technology, Nysa, Poland, 16–18 September 2018; pp. 85–94.
3. Saadna, Y.; Behloul, A.; Mezzoudj, S. Speed limit sign detection and recognition system using SVM and MNIST datasets. *Neural Comput. Appl.* **2019**, 1–11. [CrossRef]
4. Guo, J.; Lu, J.; Qu, Y.; Li, C. Traffic-Sign Spotting in the Wild via Deep Features. In Proceedings of the 2018 IEEE Intelligent Vehicles Symposium (IV), Changshu, China, 26–30 June 2018; pp. 120–125.
5. Vokhidov, H.; Hong, H.; Kang, J.; Hoang, T.; Park, K. Recognition of damaged arrow-road markings by visible light camera sensor based on convolutional neural network. *Sensors* **2016**, *16*, 2160. [CrossRef] [PubMed]
6. De la Escalera, A.; Armingol, J.M.; Mata, M. Traffic sign recognition and analysis for intelligent vehicles. *Image Vis. Comput.* **2003**, *21*, 247–258. [CrossRef]
7. Hoang, T.M.; Baek, N.R.; Cho, S.W.; Kim, K.W.; Park, K.R. Road lane detection robust to shadows based on a fuzzy system using a visible light camera sensor. *Sensors* **2017**, *17*, 2475. [CrossRef] [PubMed]
8. Economic Commission for Europe. *Convention on Traffic Signs and Signals*; Vienna Convention: Vienna, Austria, 1968.
9. Luo, H.; Yang, Y.; Tong, B.; Wu, F.; Fan, B. Traffic sign recognition using a multi-task convolutional neural network. *IEEE Trans. Intell. Transp. Syst.* **2018**, *19*, 1100–1111. [CrossRef]
10. Houben, S.; Stallkamp, J.; Salmen, J.; Schlipsing, M.; Igel, C. Detection of traffic signs in real-world images: The German Traffic Sign Detection Benchmark. In Proceedings of the 2013 International Joint Conference on Neural Networks (IJCNN), Dallas, TX, USA, 4–9 August 2013; pp. 1–8.
11. Stallkamp, J.; Schlipsing, M.; Salmen, J.; Igel, C. The German traffic sign recognition benchmark: A multi-class classification competition. In Proceedings of the 2011 International Joint Conference on Neural Networks (IJCNN), San Jose, CA, USA, 31 July–5 August 2011; pp. 1453–1460.
12. Li, J.; Wang, Z. Real-time traffic sign recognition based on efficient CNNs in the wild. *IEEE Trans. Intell. Transp. Syst.* **2018**, *20*, 975–984. [CrossRef]
13. Madani, A.; Yusof, R. Malaysian Traffic Sign Dataset for Traffic Sign Detection and Recognition Systems. *J. Telecommun. Electron. Comput. Eng. (JTEC)* **2016**, *8*, 137–143.

14. Liu, H.; Ran, B. Vision-based stop sign detection and recognition system for intelligent vehicles. *Transp. Res. Rec. J. Transp. Res. Board* **2001**, *1748*, 161–166. [CrossRef]

15. Nandi, D.; Saif, A.S.; Prottoy, P.; Zubair, K.M.; Shubho, S.A. Traffic Sign Detection based on Color Segmentation of Obscure Image Candidates: A Comprehensive Study. *Int. J. Mod. Educ. Comput. Sci.* **2018**, *10*, 35. [CrossRef]

16. Hannan, M.; Hussain, A.; Mohamed, A.; Samad, S.A.; Wahab, D.A. Decision fusion of a multi-sensing embedded system for occupant safety measures. *Int. J. Automot. Technol.* **2010**, *11*, 57–65. [CrossRef]

17. Møgelmose, A.; Trivedi, M.M.; Moeslund, T.B. Vision-Based Traffic Sign Detection and Analysis for Intelligent Driver Assistance Systems: Perspectives and Survey. *IEEE Trans. Intell. Transp. Syst.* **2012**, *13*, 1484–1497. [CrossRef]

18. Shao, F.; Wang, X.; Meng, F.; Rui, T.; Wang, D.; Tang, J. Real-time traffic sign detection and recognition method based on simplified Gabor wavelets and CNNs. *Sensors* **2018**, *18*, 3192. [CrossRef] [PubMed]

19. Zabihi, S.J.; Zabihi, S.M.; Beauchemin, S.S.; Bauer, M.A. Detection and recognition of traffic signs inside the attentional visual field of drivers. In Proceedings of the 2017 IEEE Intelligent Vehicles Symposium (IV), Los Angeles, CA, USA, 11–14 June 2017.

20. Chang, X.; Yang, Y. Semisupervised feature analysis by mining correlations among multiple tasks. *IEEE Trans. Neural Netw. Learn. Syst.* **2017**, *28*, 2294–2305. [CrossRef] [PubMed]

21. Yuan, Y.; Xiong, Z.; Wang, Q. An incremental framework for video-based traffic sign detection, tracking, and recognition. *IEEE Trans. Intell. Transp. Syst.* **2017**, *18*, 1918–1929. [CrossRef]

22. The Clemson University Vehicular Electronics Laboratory (CVEL). Traffic Sign Recognition Systems. Available online: https://cecas.clemson.edu/cvel/auto/systems/sign-recognition.html (accessed on 5 March 2019).

23. Saadna, Y.; Behloul, A. An overview of traffic sign detection and classification methods. *Int. J. Multimed. Inf. Retr.* **2017**, *6*, 193–210. [CrossRef]

24. Gündüz, H.; Kaplan, S.; Günal, S.; Akınlar, C. Circular traffic sign recognition empowered by circle detection algorithm. In Proceedings of the 2013 21st Signal Processing and Communications Applications Conference (SIU), Haspolat, Turkey, 24–26 April 2013; pp. 1–4.

25. Kuang, X.; Fu, W.; Yang, L. Real-Time Detection and Recognition of Road Traffic Signs using MSER and Random Forests. *Int. J. Online Eng. (IJOE)* **2018**, *14*, 34–51. [CrossRef]

26. Xu, X.; Jin, J.; Zhang, S.; Zhang, L.; Pu, S.; Chen, Z. Smart data driven traffic sign detection method based on adaptive color threshold and shape symmetry. *Future Gener. Comput. Syst.* **2019**, *94*, 381–391. [CrossRef]

27. Liu, C.; Chang, F.; Chen, Z.; Liu, D. Fast traffic sign recognition via high-contrast region extraction and extended sparse representation. *IEEE Trans. Intell. Transp. Syst.* **2016**, *17*, 79–92. [CrossRef]

28. De La Escalera, A.; Armingol, J.M.; Pastor, J.M.; Rodríguez, F.J. Visual sign information extraction and identification by deformable models for intelligent vehicles. *IEEE Trans. Intell. Transp. Syst.* **2004**, *5*, 57–68. [CrossRef]

29. Ruta, A.; Li, Y.; Liu, X. Real-time traffic sign recognition from video by class-specific discriminative features. *Pattern Recognit.* **2010**, *43*, 416–430. [CrossRef]

30. Gómez-Moreno, H.; Maldonado-Bascón, S.; Gil-Jiménez, P.; Lafuente-Arroyo, S. Goal evaluation of segmentation algorithms for traffic sign recognition. *IEEE Trans. Intell. Transp. Syst.* **2010**, *11*, 917–930. [CrossRef]

31. Ren, F.; Huang, J.; Jiang, R.; Klette, R. General traffic sign recognition by feature matching. In Proceedings of the 24th International Conference on Image and Vision Computing New Zealand, IVCNZ'09, Wellington, New Zealand, 23–25 November 2009; pp. 409–414.

32. Fleyeh, H. Shadow and highlight invariant colour segmentation algorithm for traffic signs. In Proceedings of the 2006 IEEE Conference on Cybernetics and Intelligent Systems, Bangkok, Thailand, 7–9 June 2006; pp. 1–7.

33. Maldonado-Bascón, S.; Lafuente-Arroyo, S.; Gil-Jimenez, P.; Gómez-Moreno, H.; López-Ferreras, F. Road-sign detection and recognition based on support vector machines. *IEEE Trans. Intell. Transp. Syst.* **2007**, *8*, 264–278. [CrossRef]

34. Khan, J.F.; Bhuiyan, S.M.; Adhami, R.R. Image segmentation and shape analysis for road-sign detection. *IEEE Trans. Intell. Transp. Syst.* **2011**, *12*, 83–96. [CrossRef]

35. Prieto, M.S.; Allen, A.R. Using self-organising maps in the detection and recognition of road signs. *Image Vis. Comput.* **2009**, *27*, 673–683. [CrossRef]

36. Fan, J.; Yau, D.K.; Elmagarmid, A.K.; Aref, W.G. Automatic image segmentation by integrating color-edge extraction and seeded region growing. *IEEE Trans. Image Process.* **2001**, *10*, 1454–1466.
37. Varun, S.; Singh, S.; Kunte, R.S.; Samuel, R.S.; Philip, B. A road traffic signal recognition system based on template matching employing tree classifier. In Proceedings of the International Conference on Computational Intelligence and Multimedia Applications ICCIMA, Sivakasi, India, 13–15 December 2007; pp. 360–365.
38. Ruta, A.; Li, Y.; Liu, X. Detection, tracking and recognition of traffic signs from video input. In Proceedings of the 11th International IEEE Conference on Intelligent Transportation Systems, ITSC 2008, Beijing, China, 12–15 October 2008; pp. 55–60.
39. Jiang, Y.; Zhou, S.; Jiang, Y.; Gong, J.; Xiong, G.; Chen, H. Traffic sign recognition using ridge regression and Otsu method. In Proceedings of the 2011 IEEE Intelligent Vehicles Symposium (IV), Baden-Baden, Germany, 5–9 June 2011; pp. 613–618.
40. Vázquez-Reina, A.; Lafuente-Arroyo, S.; Siegmann, P.; Maldonado-Bascón, S.; Acevedo-Rodríguez, F. Traffic sign shape classification based on correlation techniques. In Proceedings of the 5th WSEAS International Conference on Signal Processing, Computational Geometry & Artificial Vision, Alcalá de Henares, Malta, 15–17 September 2005; pp. 149–154.
41. Jiménez, P.G.; Bascón, S.M.; Moreno, H.G.; Arroyo, S.L.; Ferreras, F.L. Traffic sign shape classification and localization based on the normalized FFT of the signature of blobs and 2D homographies. *Signal Process.* **2008**, *88*, 2943–2955. [CrossRef]
42. Lafuente-Arroyo, S.; Salcedo-Sanz, S.; Maldonado-Bascón, S.; Portilla-Figueras, J.A.; López-Sastre, R.J. A decision support system for the automatic management of keep-clear signs based on support vector machines and geographic information systems. *Expert Syst. Appl.* **2010**, *37*, 767–773. [CrossRef]
43. Li, L.; Li, J.; Sun, J. Robust traffic sign detection using fuzzy shape recognizer. In Proceedings of the MIPPR 2009: Pattern Recognition and Computer Vision, Yichang, China, 30 October–1 November 2009; p. 74960Z.
44. Wu, J.-Y.; Tseng, C.-C.; Chang, C.-H.; Lien, J.-J.J.; Chen, J.C.; Tu, C.T. Road sign recognition system based on GentleBoost with sharing features. In Proceedings of the 2011 International Conference on System Science and Engineering (ICSSE), Macao, China, 8–10 June 2011; pp. 410–415.
45. Tagunde, G.A.; Uke, N.; Banchhor, C. Detection, classification and recognition of road traffic signs using color and shape features. *Int. J. Adv. Technol. Eng. Res.* **2012**, *2*, 202–206.
46. Deshmukh, V.R.; Patnaik, G.; Patil, M. Real-time traffic sign recognition system based on colour image segmentation. *Int. J. Comput. Appl.* **2013**, *83*, 30–35.
47. Nicchiotti, G.; Ottaviani, E.; Castello, P.; Piccioli, G. Automatic road sign detection and classification from color image sequences. In Proceedings of the 7th International Conference on Image Analysis and Processing, 1994; pp. 623–626.
48. Priese, L.; Rehrmann, V. On hierarchical color segmentation and applications. In Proceedings of the 1993 IEEE Computer Society Conference on Computer Vision and Pattern Recognition, New York, NY, USA, 15–17 June 1993; pp. 633–634.
49. Funt, B.V.; Finlayson, G.D. Color constant color indexing. *IEEE Trans. Pattern Anal. Mach. Intell.* **1995**, *17*, 522–529. [CrossRef]
50. Swain, M.J.; Ballard, D.H. Color indexing. *Int. J. Comput. Vis.* **1991**, *7*, 11–32. [CrossRef]
51. Park, D.-S.; Park, J.-S.; Kim, T.Y.; Han, J.H. Image indexing using weighted color histogram. In Proceedings of the International Conference on Image Analysis and Processing, Venice, Italy, 27–29 September 1999; pp. 909–914.
52. Vitabile, S.; Pollaccia, G.; Pilato, G.; Sorbello, F. Road signs recognition using a dynamic pixel aggregation technique in the HSV color space. In Proceedings of the International Conference on Image Analysis and Processing ICIAP, Palermo, Italy, 26–28 September 2001; p. 0572.
53. Li, C.; Luo, M.R.; Hunt, R.W.; Moroney, N.; Fairchild, M.D.; Newman, T. The performance of CIECAM02. In Proceedings of the Color and Imaging Conference, Scottland, AN, USA, 1 January 2002; pp. 28–32.
54. Gao, X.W.; Podladchikova, L.; Shaposhnikov, D.; Hong, K.; Shevtsova, N. Recognition of traffic signs based on their colour and shape features extracted using human vision models. *J. Vis. Commun. Image Represent.* **2006**, *17*, 675–685. [CrossRef]
55. Fatmehsari, Y.R.; Ghahari, A.; Zoroofi, R.A. Gabor wavelet for road sign detection and recognition using a hybrid classifier. In Proceedings of the 2010 International Conference on Multimedia Computing and Information Technology (MCIT), Sharjah, UAE, 2–4 March 2010; pp. 25–28.

56. Hechri, A.; Mtibaa, A. Automatic detection and recognition of road sign for driver assistance system. In Proceedings of the 2012 16th IEEE Mediterranean Electrotechnical Conference (MELECON), Yasmine Hammamet, Tunisia, 25–28 March 2012; pp. 888–891.

57. Saxena, P.; Gupta, N.; Laskar, S.Y.; Borah, P.P. A study on automatic detection and recognition techniques for road signs. *Int. J. Comput. Eng. Res.* **2015**, *5*, 24–28.

58. Hu, Q.; Paisitkriangkrai, S.; Shen, C.; van den Hengel, A.; Porikli, F. Fast detection of multiple objects in traffic scenes with a common detection framework. *IEEE Trans. Intell. Transp. Syst.* **2016**, *17*, 1002–1014. [CrossRef]

59. Zaklouta, F.; Stanciulescu, B. Real-time traffic-sign recognition using tree classifiers. *IEEE Trans. Intell. Transp. Syst.* **2012**, *13*, 1507–1514. [CrossRef]

60. Yin, S.; Ouyang, P.; Liu, L.; Guo, Y.; Wei, S. Fast traffic sign recognition with a rotation invariant binary pattern based feature. *Sensors* **2015**, *15*, 2161–2180. [CrossRef] [PubMed]

61. Gavrila, D.M. Traffic sign recognition revisited. In *Mustererkennung 1999*; Springer: Berlin, Germany, 1999; pp. 86–93.

62. Kaur, S.; Singh, I. Comparison between edge detection techniques. *Int. J. Comput. Appl.* **2016**, *145*, 15–18. [CrossRef]

63. Xu, S. Robust traffic sign shape recognition using geometric matching. *IET Intell. Transp. Syst.* **2009**, *3*, 10–18. [CrossRef]

64. Barnes, N.; Zelinsky, A.; Fletcher, L.S. Real-time speed sign detection using the radial symmetry detector. *IEEE Trans. Intell. Transp. Syst.* **2008**, *9*, 322–332. [CrossRef]

65. Deguchi, D.; Shirasuna, M.; Doman, K.; Ide, I.; Murase, H. Intelligent traffic sign detector: Adaptive learning based on online gathering of training samples. In Proceedings of the 2011 IEEE Intelligent Vehicles Symposium (IV), Baden-Baden, Germany, 5–9 June 2011; pp. 72–77.

66. Houben, S. A single target voting scheme for traffic sign detection. In Proceedings of the 2011 IEEE Intelligent Vehicles Symposium (IV), Baden-Baden, Germany, 5–9 June 2011; pp. 124–129.

67. Soetedjo, A.; Yamada, K. Improving the performance of traffic sign detection using blob tracking. *IEICE Electron. Express* **2007**, *4*, 684–689. [CrossRef]

68. Viola, P.; Jones, M. Rapid object detection using a boosted cascade of simple features. In Proceedings of the 2001 IEEE Computer Society Conference on Computer Vision and Pattern Recognition, CVPR 2001, Kauai, HI, USA, 8–14 December 2001; pp. 511–518.

69. Ruta, A.; Li, Y.; Liu, X. Towards Real-Time Traffic Sign Recognition by Class-Specific Discriminative Features. In Proceedings of the British Machine Vision Conference, Warwick, UK, 10–13 September 2007.

70. Prisacariu, V.A.; Timofte, R.; Zimmermann, K.; Reid, I.; Van Gool, L. Integrating object detection with 3D tracking towards a better driver assistance system. In Proceedings of the 2010 20th International Conference on Pattern Recognition (ICPR), Istanbul, Turkey, 23–26 August 2010; pp. 3344–3347.

71. Zhu, Z.; Lu, J.; Martin, R.R.; Hu, S. An optimization approach for localization refinement of candidate traffic signs. *IEEE Trans. Intell. Transp. Syst.* **2017**, *18*, 3006–3016. [CrossRef]

72. Bahlmann, C.; Zhu, Y.; Ramesh, V.; Pellkofer, M.; Koehler, T. A system for traffic sign detection, tracking, and recognition using color, shape, and motion information. In Proceedings of the Intelligent Vehicles Symposium, Las Vegas, NY, USA, 6–8 June 2005; pp. 255–260.

73. Chiang, H.-H.; Chen, Y.-L.; Wang, W.-Q.; Lee, T.-T. Road speed sign recognition using edge-voting principle and learning vector quantization network. In Proceedings of the 2010 International Computer Symposium (ICS), Tainan, Taiwan, 16–18 December 2010; pp. 246–251.

74. Gu, Y.; Yendo, T.; Tehrani, M.P.; Fujii, T.; Tanimoto, M. Traffic sign detection in dual-focal active camera system. In Proceedings of the 2011 IEEE Intelligent Vehicles Symposium (IV), Baden-Baden, Germany, 5–9 June 2011; pp. 1054–1059.

75. Wang, G.; Ren, G.; Wu, Z.; Zhao, Y.; Jiang, L. A robust, coarse-to-fine traffic sign detection method. In Proceedings of the 2013 International Joint Conference on Neural Networks (IJCNN), Dallas, TX, USA, 4–9 August 2013; pp. 1–5.

76. Loy, G.; Barnes, N. Fast shape-based road sign detection for a driver assistance system. In Proceedings of the 2004 IEEE/RSJ International Conference on Intelligent Robots and Systems, (IROS 2004), Sendai, Japan, 28 September–2 October 2004; pp. 70–75.

77. Pettersson, N.; Petersson, L.; Andersson, L. The histogram feature-a resource-efficient weak classifier. In Proceedings of the 2008 IEEE Intelligent Vehicles Symposium, Eindhoven, The Netherlands, 4–6 June 2008; pp. 678–683.

78. Xie, Y.; Liu, L.-F.; Li, C.-H.; Qu, Y.-Y. Unifying visual saliency with HOG feature learning for traffic sign detection. In Proceedings of the 2009 IEEE Intelligent Vehicles Symposium, Xi'an, China, 3–5 June 2009; pp. 24–29.

79. Creusen, I.M.; Wijnhoven, R.G.; Herbschleb, E.; de With, P. Color exploitation in hog-based traffic sign detection. In Proceedings of the 2010 IEEE International Conference on Image Processing, Hong Kong, China, 26–29 September 2010; pp. 2669–2672.

80. Overett, G.; Petersson, L. Large scale sign detection using HOG feature variants. In Proceedings of the 2011 IEEE Intelligent Vehicles Symposium (IV), Baden-Baden, Germany, 5–9 June 2011; pp. 326–331.

81. Hoferlin, B.; Zimmermann, K. Towards reliable traffic sign recognition. In Proceedings of the Intelligent Vehicles Symposium, Xi'an, China, 3–5 June 2009; pp. 324–329.

82. Hannan, M.A.; Hussain, A.; Samad, S.A. Decision fusion via integrated sensing system for a smart airbag deployment scheme. *Sens. Mater.* **2011**, *23*, 179–193.

83. Fu, M.-Y.; Huang, Y.-S. A survey of traffic sign recognition. In Proceedings of the 2010 International Conference on Wavelet Analysis and Pattern Recognition (ICWAPR), Qingdao, China, 11–14 July 2010; pp. 119–124.

84. Lafuente-Arroyo, S.; Maldonado-Bascon, S.; Gil-Jimenez, P.; Acevedo-Rodriguez, J.; Lopez-Sastre, R. A tracking system for automated inventory of road signs. In Proceedings of the 2007 IEEE Intelligent Vehicles Symposium, Istanbul, Turkey, 13–15 June 2007; pp. 166–171.

85. Lafuente-Arroyo, S.; Maldonado-Bascon, S.; Gil-Jimenez, P.; Gomez-Moreno, H.; Lopez-Ferreras, F. Road sign tracking with a predictive filter solution. In Proceedings of the IECON 2006—32nd Annual Conference on IEEE Industrial Electronics, Paris, France, 6–10 November 2006; pp. 3314–3319.

86. Wali, S.B.; Hannan, M.A.; Hussain, A.; Samad, S.A. An automatic traffic sign detection and recognition system based on colour segmentation, shape matching, and svm. *Math. Probl. Eng.* **2015**, *2015*, 250461. [CrossRef]

87. Garcia-Garrido, M.; Ocaña, M.; Llorca, D.F.; Sotelo, M.; Arroyo, E.; Llamazares, A. Robust traffic signs detection by means of vision and V2I communications. In Proceedings of the 2011 14th International IEEE Conference on Intelligent Transportation Systems (ITSC), Washington, DC, USA, 5–7 October 2011; pp. 1003–1008.

88. Fang, C.-Y.; Chen, S.-W.; Fuh, C.-S. Road-sign detection and tracking. *IEEE Trans. Veh. Technol.* **2003**, *52*, 1329–1341. [CrossRef]

89. Ohara, H.; Nishikawa, I.; Miki, S.; Yabuki, N. Detection and recognition of road signs using simple layered neural networks. In Proceedings of the 9th International Conference on Neural Information Processing, ICONIP'02, Singapore, 18–22 November 2002; pp. 626–630.

90. Torresen, J.; Bakke, J.W.; Sekanina, L. Efficient recognition of speed limit signs. In Proceedings of the 7th International IEEE Conference on Intelligent Transportation Systems, Washington, WA, USA, 3–6 October 2004; pp. 652–656.

91. Greenhalgh, J.; Mirmehdi, M. Traffic sign recognition using MSER and random forests. In Proceedings of the 2012 20th European Signal Processing Conference (EUSIPCO), Bucharest, Romania, 27–31 August 2012; pp. 1935–1939.

92. Greenhalgh, J.; Mirmehdi, M. Real-time detection and recognition of road traffic signs. *IEEE Trans. Intell. Transp. Syst.* **2012**, *13*, 1498–1506. [CrossRef]

93. Zaklouta, F.; Stanciulescu, B. Real-time traffic sign recognition using spatially weighted HOG trees. In Proceedings of the 2011 15th International Conference on Advanced Robotics (ICAR), Tallinn, Estonia, 20–23 June 2011; pp. 61–66.

94. Zaklouta, F.; Stanciulescu, B. Segmentation masks for real-time traffic sign recognition using weighted HOG-based trees. In Proceedings of the 2011 14th International IEEE Conference on Intelligent Transportation Systems (ITSC), Washington, DC, USA, 5–7 October 2011; pp. 1954–1959.

95. Zaklouta, F.; Stanciulescu, B. Real-time traffic sign recognition in three stages. *Robot. Auton. Syst.* **2014**, *62*, 16–24. [CrossRef]

96. Zaklouta, F.; Stanciulescu, B. Warning traffic sign recognition using a HOG-based Kd tree. In Proceedings of the 2011 IEEE Intelligent Vehicles Symposium (IV), Baden-Baden, Germany, 5–9 June 2011; pp. 1019–1024.

97. Aoyagi, Y.; Asakura, T. A study on traffic sign recognition in scene image using genetic algorithms and neural networks. In Proceedings of the 1996 IEEE IECON 22nd International Conference on Industrial Electronics, Control, and Instrumentation, Taipei, Taiwan, 9 August 1996; pp. 1838–1843.

98. De La Escalera, A.; Armingol, J.M.; Salichs, M. Traffic sign detection for driver support systems. In Proceedings of the International Conference on Field and Service Robotics, Helsinki, Finland, 6–8 June 2001.

99. Hannan, M.; Hussain, A.; Samad, S.; Ishak, K.; Mohamed, A. A Unified Robust Algorithm for Detection of Human and Non-human Object in Intelligent Safety Application. *World Acad. Sci. Eng. Technol. Int. J. Comput. Electr. Autom. Control Inf. Eng.* **2008**, *2*, 3838–3845.

100. Ugolotti, R.; Nashed, Y.S.G.; Cagnoni, S. Real-Time GPU Based Road Sign Detection and Classification. In *Parallel Problem Solving from Nature—PPSN XII*; Coello, C.A.C., Cutello, V., Deb, K., Forrest, S., Nicosia, G., Pavone, M., Eds.; Springer: Berlin/Heidelberg, Germany, 2012; pp. 153–162.

101. CireşAn, D.; Meier, U.; Masci, J.; Schmidhuber, J. Multi-column deep neural network for traffic sign classification. *Neural Netw.* **2012**, *32*, 333–338. [CrossRef]

102. Fang, C.-Y.; Fuh, C.-S.; Yen, P.; Cherng, S.; Chen, S.-W. An automatic road sign recognition system based on a computational model of human recognition processing. *Comput. Vis. Image Underst.* **2004**, *96*, 237–268. [CrossRef]

103. Broggi, A.; Cerri, P.; Medici, P.; Porta, P.P.; Ghisio, G. Real time road signs recognition. In Proceedings of the 2007 IEEE Intelligent Vehicles Symposium, Istanbul, Turkey, 13–15 June 2007; pp. 981–986.

104. Li, L.-B.; Ma, G.-F. Detection and classification of traffic signs in natural environments. *J. Harbin Inst. Technol.* **2009**, *41*, 682–687.

105. Sheng, Y.; Zhang, K.; Ye, C.; Liang, C.; Li, J. Automatic detection and recognition of traffic signs in stereo images based on features and probabilistic neural networks. In Proceedings of the Optical and Digital Image Processing, Strasbourg, France, 25 April 2008; p. 70001I.

106. Fištrek, T.; Lončarić, S. Traffic sign detection and recognition using neural networks and histogram based selection of segmentation method. In Proceedings of the 2011 ELMAR, Zadar, Croatia, 14–16 September 2011; pp. 51–54.

107. Carrasco, J.-P.; de la Escalera, A.D.L.E.; Armingol, J.M. Recognition stage for a speed supervisor based on road sign detection. *Sensors* **2012**, *12*, 12153–12168. [CrossRef]

108. Kim, H.-K.; Park, J.H.; Jung, H.-Y. An Efficient Color Space for Deep-Learning Based Traffic Light Recognition. *J. Adv. Transp.* **2018**, *2018*, 2365414. [CrossRef]

109. Zhu, Y.; Liao, M.; Yang, M.; Liu, W. Cascaded Segmentation-Detection Networks for Text-Based Traffic Sign Detection. *IEEE Trans. Intell. Transp. Syst.* **2018**, *19*, 209–219. [CrossRef]

110. Qian, R.; Zhang, B.; Yue, Y.; Wang, Z.; Coenen, F. Robust Chinese traffic sign detection and recognition with deep convolutional neural network. In Proceedings of the 2015 11th International Conference on Natural Computation (ICNC), Zhangjiajie, China, 15–17 August 2015; pp. 791–796.

111. Kumar, A.D. Novel deep learning model for traffic sign detection using capsule networks. *arXiv* **2018**, arXiv:1805.04424.

112. Ciresan, D.C.; Meier, U.; Masci, J.; Schmidhuber, J. A committee of neural networks for traffic sign classification. In Proceedings of the IJCNN, San Jose, CA, USA, 31 July–5 August 2011; pp. 1918–1921.

113. Sermanet, P.; LeCun, Y. Traffic sign recognition with multi-scale Convolutional Networks. In Proceedings of the IJCNN, San Jose, CA, USA, 31 July–5 August 2011; pp. 2809–2813.

114. Lee, H.S.; Kim, K. Simultaneous traffic sign detection and boundary estimation using convolutional neural network. *IEEE Trans. Intell. Transp. Syst.* **2018**, *19*, 1652–1663. [CrossRef]

115. Jin, J.; Fu, K.; Zhang, C. Traffic sign recognition with hinge loss trained convolutional neural networks. *IEEE Trans. Intell. Transp. Syst.* **2014**, *15*, 1991–2000. [CrossRef]

116. Abdi, L.; Meddeb, A. Deep learning traffic sign detection, recognition and augmentation. In Proceedings of the Symposium on Applied Computing, Marrakech, Morocco, 3–7 April 2017; pp. 131–136.

117. Aghdam, H.H.; Heravi, E.J.; Puig, D. A practical approach for detection and classification of traffic signs using convolutional neural networks. *Robot. Auton. Syst.* **2016**, *84*, 97–112. [CrossRef]

118. Li, J.; Liang, X.; Wei, Y.; Xu, T.; Feng, J.; Yan, S. Perceptual generative adversarial networks for small object detection. In Proceedings of the IEEE Conference on Computer Vision and Pattern Recognition, Honolulu, HI, USA, 21–26 July 2017; pp. 1222–1230.

119. Lai, Y.; Wang, N.; Yang, Y.; Lin, L. Traffic Signs Recognition and Classification based on Deep Feature Learning. In Proceedings of the ICPRAM, Medea, Algeria, 24–25 November 2018; pp. 622–629.

120. Lim, K.; Hong, Y.; Choi, Y.; Byun, H. Real-time traffic sign recognition based on a general purpose GPU and deep-learning. *PLoS ONE* **2017**, *12*, e0173317. [CrossRef] [PubMed]

121. Freund, Y.; Schapire, R.E. A decision-theoretic generalization of on-line learning and an application to boosting. *J. Comput. Syst. Sci.* **1997**, *55*, 119–139. [CrossRef]

122. Li, Y.; Pankanti, S.; Guan, W. Real-time traffic sign detection: An evaluation study. In Proceedings of the 2010 20th International Conference on Pattern Recognition (ICPR), Istanbul, Turkey, 23–26 August 2010; pp. 3033–3036.

123. Chen, L.; Li, Q.; Li, M.; Mao, Q. Traffic sign detection and recognition for intelligent vehicle. In Proceedings of the 2011 IEEE Intelligent Vehicles Symposium (IV), Baden-Baden, Germany, 5–9 June 2011; pp. 908–913.

124. Lin, C.-C.; Wang, M.-S. Road sign recognition with fuzzy adaptive pre-processing models. *Sensors* **2012**, *12*, 6415–6433. [CrossRef]

125. Huang, Y.-S.; Le, Y.-S.; Cheng, F.-H. A method of detecting and recognizing speed-limit signs. In Proceedings of the 2012 Eighth International Conference on Intelligent Information Hiding and Multimedia Signal Processing (IIH-MSP), Piraeus, Greece, 18–20 July 2012; pp. 371–374.

126. Chen, L.; Li, Q.; Li, M.; Zhang, L.; Mao, Q. Design of a multi-sensor cooperation travel environment perception system for autonomous vehicle. *Sensors* **2012**, *12*, 12386–12404. [CrossRef]

127. Liu, C.; Li, S.; Chang, F.; Dong, W. Supplemental Boosting and Cascaded ConvNet Based Transfer Learning Structure for Fast Traffic Sign Detection in Unknown Application Scenes. *Sensors* **2018**, *18*, 2386. [CrossRef]

128. Gomes, S.L.; Rebouças, E.D.S.; Neto, E.C.; Papa, J.P.; de Albuquerque, V.H.; Rebouças Filho, P.P.; Tavares, J.M.R. Embedded real-time speed limit sign recognition using image processing and machine learning techniques. *Neural Comput. Appl.* **2017**, *28*, 573–584. [CrossRef]

129. Soendoro, D.; Supriana, I. Traffic sign recognition with Color-based Method, shape-arc estimation and SVM. In Proceedings of the 2011 International Conference on Electrical Engineering and Informatics (ICEEI), Bandung, Indonesia, 17–19 July 2011; pp. 1–6.

130. Siyan, Y.; Xiaoying, W.; Qiguang, M. Road-sign segmentation and recognition in natural scenes. In Proceedings of the 2011 IEEE International Conference on Signal Processing, Communications and Computing (ICSPCC), Xi'an, China, 14–16 September 2011; pp. 1–4.

131. Bascón, S.M.; Rodríguez, J.A.; Arroyo, S.L.; Caballero, A.F.; López-Ferreras, F. An optimization on pictogram identification for the road-sign recognition task using SVMs. *Comput. Vis. Image Underst.* **2010**, *114*, 373–383. [CrossRef]

132. Martinović, A.; Glavaš, G.; Juribašić, M.; Sutić, D.; Kalafatić, Z. Real-time detection and recognition of traffic signs. In Proceedings of the 2010 Proceedings of the 33rd International Convention (MIPRO), Opatija, Croatia, 24–28 May 2010; pp. 760–765.

133. Min, K.-I.; Oh, J.-S.; Kim, B.-W. Traffic sign extract and recognition on unmanned vehicle using image processing based on support vector machine. In Proceedings of the 2011 11th International Conference on Control, Automation and Systems (ICCAS), Gyeonggi-do, Korea, 26–29 October 2011; pp. 750–753.

134. Bui-Minh, T.; Ghita, O.; Whelan, P.F.; Hoang, T. A robust algorithm for detection and classification of traffic signs in video data. In Proceedings of the 2012 International Conference on Control, Automation and Information Sciences (ICCAIS), Ho Chi Minh City, Vietnam, 26–29 November 2012; pp. 108–113.

135. Park, J.-G.; Kim, K.-J. Design of a visual perception model with edge-adaptive Gabor filter and support vector machine for traffic sign detection. *Expert Syst. Appl.* **2013**, *40*, 3679–3687. [CrossRef]

136. Chang, X.; Yu, Y.-L.; Yang, Y.; Xing, E.P. Semantic pooling for complex event analysis in untrimmed videos. *IEEE Trans. Pattern Anal. Mach. Intell.* **2017**, *39*, 1617–1632. [CrossRef]

137. Yang, L.; Kwon, K.-R.; Moon, K.; Lee, S.-H.; Kwon, S.-G. Broken traffic sign recognition based on local histogram matching. In Proceedings of the Computing, Communications and Applications Conference (ComComAp), Hong Kong, China, 11–13 January 2012; pp. 415–419.

138. Sebanja, I.; Megherbi, D. Automatic detection and recognition of traffic road signs for intelligent autonomous unmanned vehicles for urban surveillance and rescue. In Proceedings of the 2010 IEEE International Conference on Technologies for Homeland Security (HST), Waltham, MA, USA, 8–10 November 2010; pp. 132–138.

139. Huang, Y.-S.; Fu, M.-Y.; Ma, H.-B. A combined method for traffic sign detection and classification. In Proceedings of the 2010 Chinese Conference on Pattern Recognition (CCPR), Chongqing, China, 21–23 October 2010; pp. 1–5.

140. Soheilian, B.; Paparoditis, N.; Vallet, B. Detection and 3D reconstruction of traffic signs from multiple view color images. *ISPRS J. Photogramm. Remote Sens.* **2013**, *77*, 1–20. [CrossRef]

141. Pei, D.; Sun, F.; Liu, H. Supervised low-rank matrix recovery for traffic sign recognition in image sequences. *IEEE Signal Process. Lett.* **2013**, *20*, 241–244. [CrossRef]

142. Gonzalez-Reyna, S.E.; Avina-Cervantes, J.G.; Ledesma-Orozco, S.E.; Cruz-Aceves, I. Eigen-gradients for traffic sign recognition. *Math. Probl. Eng.* **2013**, *2013*, 364305. [CrossRef]

143. Măriuţ, F.; Foşalău, C.; Zet, C.; Petrişor, D. Experimental traffic sign detection using I2V communication. In Proceedings of the 2012 35th International Conference on Telecommunications and Signal Processing (TSP), Prague, Czech Republic, 3–4 July 2012; pp. 141–145.

144. Wang, W.; Wei, C.-H.; Zhang, L.; Wang, X. Traffic-signs recognition system based on multi-features. In Proceedings of the 2012 IEEE International Conference on Computational Intelligence for Measurement Systems and Applications (CIMSA), Tianjin, China, 2–4 July 2012; pp. 120–123.

145. Islam, K.T.; Raj, R.G. Real-time (vision-based) road sign recognition using an artificial neural network. *Sensors* **2017**, *17*, 853. [CrossRef]

146. Khan, J.; Yeo, D.; Shin, H. New dark area sensitive tone mapping for deep learning based traffic sign recognition. *Sensors* **2018**, *18*, 3776. [CrossRef] [PubMed]

147. Laguna, R.; Barrientos, R.; Blázquez, L.F.; Miguel, L.J. Traffic sign recognition application based on image processing techniques. *IFAC Proc. Vol.* **2014**, *47*, 104–109. [CrossRef]

Article

Liver Tumor Segmentation in CT Scans Using Modified SegNet

Sultan Almotairi [1], Ghada Kareem [2], Mohamed Aouf [2], Badr Almutairi [3] and Mohammed A.-M. Salem [4,5,*]

1   Department of Natural and Applied Sciences, Faculty of Community College, Majmaah University, Majmaah 11952, Saudi Arabia; almotairi@mu.edu.sa
2   Department of Biomedical Engineering, Higher Technological Institute, 10th Ramadan City 44629, Egypt; ghadakareem5@gmail.com (G.K.); maoufmedical@yahoo.com (M.A.)
3   Department of Information Technology, College of Computer Sciences and Information Technology College, Majmaah University, Al-Majmaah 11952, Saudi Arabia; b.algoian@mu.edu.sa
4   Faculty of Computer and Information Sciences, Ain Shams University, Cairo 11566, Egypt
5   Faculty of Media Engineering & amp; Technology, German University in Cairo, Cairo 11835, Egypt
*   Correspondence: mohammed.salem@guc.edu.eg

Received: 5 December 2019; Accepted: 4 March 2020; Published: 10 March 2020

Abstract: The main cause of death related to cancer worldwide is from hepatic cancer. Detection of hepatic cancer early using computed tomography (CT) could prevent millions of patients' death every year. However, reading hundreds or even tens of those CT scans is an enormous burden for radiologists. Therefore, there is an immediate need is to read, detect, and evaluate CT scans automatically, quickly, and accurately. However, liver segmentation and extraction from the CT scans is a bottleneck for any system, and is still a challenging problem. In this work, a deep learning-based technique that was proposed for semantic pixel-wise classification of road scenes is adopted and modified to fit liver CT segmentation and classification. The architecture of the deep convolutional encoder–decoder is named SegNet, and consists of a hierarchical correspondence of encode–decoder layers. The proposed architecture was tested on a standard dataset for liver CT scans and achieved tumor accuracy of up to 99.9% in the training phase.

Keywords: deep learning; CT images; convolutional neural networks; hepatic cancer

---

1. Introduction

The liver is the largest organ, located underneath the right ribs and below the lung base. It has a role in digesting food [1,2]. It is responsible for filtering blood cells, processing and storing nutrients, and converting some these nutrients into energy; it also breaks down toxic agents [3,4]. There are two main hepatic lobes, the left and right lobes. When the liver is viewed from the undersurface, there are two more lobes, the quadrate and caudate lobes [5].

Hepatocellular carcinoma (HCC) [6] may occur when the liver cells begin to grow out of control and can spread to other areas in the body. Primary hepatic malignancies develop when there is abnormal behavior of the cells [7].

Liver cancer has been reported to be the second most frequent cancer to cause death in men, and sixth for women. About 750,000 people got diagnosed with liver cancer in 2008, 696,000 of which died from it. Globally, the rate of infection of males is twice than that of females [8]. Liver cancer can be developed from viral hepatitis, which is much more problematic. According to the World Health Organization, WHO, about 1.45 million deaths a year occur because of this infection [9]. In 2015, Egypt was named as the country with the highest rate of adults infected by viral hepatitis C (HCV), at 7% [9]. Because the treatment could not be reached by all infected people, the Egypt's government launched a

"100 Million *Seha*" (*seha* is an Arabic word meaning "health") national campaign between October 2018 and April 2019. At the end of March 2019, around 35 million people had been examined for HCV [10].

Primary hepatic malignancy is more prevalent in Southeast Asia and Africa than in the United States [11,12]. The reported survival rate is generally 18%. However, survival rates rely on the stage of disease at the time of diagnosis [13].

Primary hepatic malignancy is diagnosed by clinical, laboratory, and imaging tests, including ultrasound scans, magnetic resonance imaging (MRI) scans, and computed tomography (CT) scans [14]. A CT scan utilizes radiation to capture detailed images around the body from different angles, including sagittal, coronal, and axial images. It shows organs, bones, and soft tissues; the information is then processed by the computer to create images, usually in DICOM format [15]. Quite often, the examination requires intravenous injection of contrast material. The scans can help to differentiate malignant lesions from acute infection, chronic inflammation, fibrosis, and cirrhosis [16].

Staging of hepatic malignancies depends on the size and location of the malignancy [16,17]. Hence, it is important to develop an automatic procedure to detect and extract the cancer region from the CT scan accurately. Image segmentation is the process of partitioning the liver region in the CT scan into regions, where each region represents a semantic part of the liver [18,19]. This is a fundamental step to support the diagnosis by radiologists, and a fundamental step to create automatic computer-aided diagnosis (CAD) systems [20–22]. CT scans of the liver are usually interpreted by manual or semi-manual techniques, but these techniques are subjective, expensive, time-consuming, and highly error prone. Figure 1 shows an example where the gray level intensities of the liver and the spleen are too similar to be differentiated by the naked eye. To overcome these obstacles and improve the quality of liver tumors' diagnosis, multiple computer-aided methods have been developed. However, these systems have not been that great at the segmentation of the liver and lesions due to multiple challenges, such as the low contrast between the liver and neighboring organs and between the liver and tumors, different contrast levels in tumors, variation in the numbers and sizes of tumors, tissues' abnormalities, and irregular tumor growth in response to medical treatment. Therefore, a new approach must be used to overcome these obstacles [23].

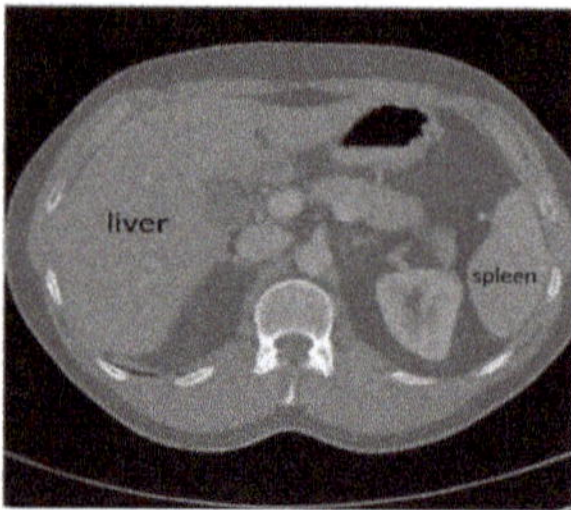

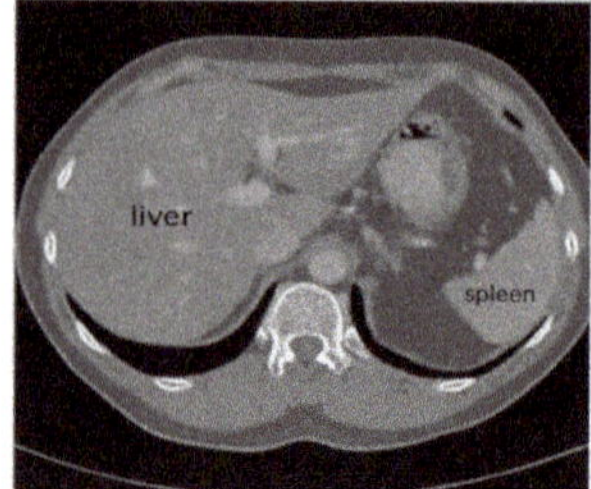

Figure 1. Example of the similarity in gray levels between the liver and the spleen in computed tomography (CT) images. Imported from the Medical Image Computing and Computer Assisted Intervention (MICCAI) SLIVER07 workshop datasets [24,25].

In this work, a review of wide variety of recent publications of image analysis for liver malignancy segmentation is introduced. In recent years, extensive research has depended on supervised learning methods. The supervised method use inputs labeled to train a model for a specific task—liver or tumor segmentation, in this case. On top of these learning methods are the deep learning methods [26,27]. There are many different models of deep learning that have been introduced, such as stacked auto-encoder (SAE), deep belief nets (DBN), convolutional neural networks (CNNs), and Deep Boltzmann Machines (DBM) [28–31]. The superiority of the deep learning models in terms of accuracy has been established. However, it is still a challenge to find proper training dataset, which should be huge in size and prepared by experts.

CNNs are considered the best of deep learning methods used. Elshaer et al. [13] reduced the computation time of a large number of slices by using two trained deep CNN models. The first model was used to get the liver region, and the second model was used for avoiding fogginess from image re-sampling and for avoiding missed small lesions.

Wen Li et al. [28] utilized a convolutional neural network (CNN) that uses image patches. It considers an image patch for each pixel, such that the pixel of interest is in the center of that patch. The patches are divided into normal or tumor liver tissue. If the patch contains at least 50 percent or more of tumor tissue, the patch is labeled as a positive sample. The reported accuracy reached about 80.6%. The work presented in [12,13] reported s more than 94% accuracy rate for classifying the images either as normal or abnormal if the image showed a liver with tumor regions. The CNN model has different architectures—i.e., Alex Net, VGG-Net, ResNet, etc. [32–34]. The work presented by Bellver et al. [5] used VGG-16 architecture as the base network in their work. Other work [11,16,29,32,35,36] has used two-dimensional (2D) U-Net, which is designed mainly for medical image segmentation.

The main objective of this work is to present a novel segmentation technique for liver cross-sectional CT scans based on a deep learning model that has proven successful in image segmentation for scene understanding, namely SegNet [37]. Memory and performance efficiency are the main advantages of this architecture over the other models. The model has been modified to fit two-class classification tasks.

The paper is organized as follows. In the next section, a review is presented on recent segmentation approaches for the liver and lesions in CT images, as well as a short introduction to the basic concepts addressed in this work. Section 3 presents the proposed method and the experimental dataset. Experimental results are presented in Section 4. Finally, conclusions are presented and discussed in Section 5.

2. Basic Concepts

Convolutional neural networks are similar to traditional neural networks [20,38,39]. A convolutional neural network (CNN) includes one or more layers of convolutional, fully connected, pooling, or fully connected and rectified linear unit (ReLU) layers. Generally, as the network becomes deeper with many more parameters, the accuracy of the results increases, but it also becomes more computationally complex.

Recently, CNN models have been used widely in image classification for different applications [20,34,40–42] or to extract features from the convolutional layers before or after the down sampling layers [41,43]. However, the architectures discussed above are not suitable for image segmentation or pixel-wise classifications. VGG-16 network architecture [44] is a type of CNN model. The network includes 41 layers. There are 16 layers with learnable weights: there are 13 convolutional layers and three fully connected layers. Figure 2 shows the architecture of VGG-16 as introduced by Simonyan and Zisserman [44].

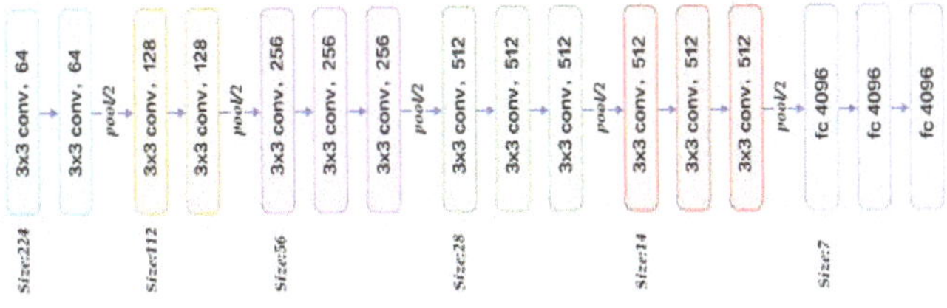

Figure 2. VGG-16 network architecture [44].

Most pixel-wise classification network architectures are of encoder–decoder architecture, where the encoder part is the VGG-16 model. The encoder gradually decreases the spatial dimension of the images with pooling layers; however, the decoder retrieves the details of the object and spatial dimensions for fast and precise segmentation of images. U-Net [45,46] is a convolutional encoder-decoder network used widely for semantic image segmentation. It is interesting because it applies a fully convolutional network architecture for medical images. However, it is very time- and memory-consuming.

The semantic image segmentation approach uses the predetermined weights of the pertained VGG-16 network [45]. Badrinarayanan et al. [37], have proposed an encoder–decoder deep network, named SegNet, for scene understanding applications tested on road and indoor scenes. The main parts of the core trainable segmentation engine are an encoder network, a decoder network, and a pixel-wise classification layer. The architecture of the encoder network is similar to the 13 convolutional layers in the VGG-16 network. The function of the decoder network is mapping the features of encoder with low to full-input resolution feature maps for pixel-wise classification. Figure 3 shows a simple illustration of the SegNet model during the down sampling (max-pooling or subsampling layers) of the encoder part. Instead of transferring the pixel values to the decoder, the indices of the chosen pixel are saved and synchronized with the decoder for the up-sampling process. In SegNet, more shortcut connections are presented. The indices are copied from max pooling instead of copying the features of encoder, such as in FCN [47], so the memory and performance of SegNet is much more efficient than FCN and U-Net.

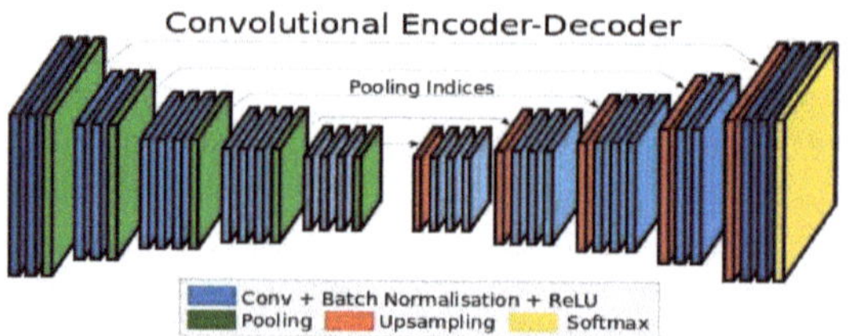

Figure 3. SegNet network architecture [37].

3. Materials and Method

This section discusses the steps and the implementation of the proposed method for segmentation of a liver tumor. The proposed method follows the conventional pattern recognition scheme: preprocessing, feature extraction and classification, and post-processing.

3.1. Dataset

The 3D-IRCADb-01 database is composed of three-dimensional (3D) CT-scans of 20 different patients (10 females and 10 males), with hepatic tumors in 15 of those cases. Each image has a resolution of 512×512 width and height. The depth or the number of slices per patient ranges between 74 and 260. Along with patient images in DICOM format, labeled images and mask images are given that could be used as ground truth for the segmentation process. The place of tumors is exposed by Couinaud segmentation [48]. This shows the main difficulties in segmentation of the liver via software [49].

3.2. Image Preprocessing

In the preprocessing steps, the DICOM CT images were subject to file format conversion to portable network graphics (PNG). The PNG file format was chosen to preserve the image quality, as it is a lossless format. In DICOM format, the pixel values are in Hounsfield, in the range [−1000, 4000]. In this format, the images cannot be displayed, and many image processing operations will fail. Therefore, the color depth conversion, and hence the range of the pixel's values mapping to the positive 1 byte integer, is necessary. The mapping is done according to the following formula:

$$g = \frac{h - m_1}{m_2 - m_1} * 255 \tag{1}$$

where h is the pixel value in Hounsfield, g is the corresponding predicted gray level value, and m_1 and m_2 are the minimum and maximum of the Hounsfield range, respectively.

The second step is to put the images in an acceptable format for the SegNet model [37]. The images have been converted to three channels, similar to the RGB color space, by simply duplicating the slice in each channel and resizing each to be the dimension $360 \times 480 \times 3$. Figure 4 shows three samples of the input images before color depth correction. The images in this format have too low contrast and are not suitable for use by the deep learning model.

Figure 4. Raw CT slices in DICOM format for three different patients imported from the IRCAD dataset [50].

In order to increase the performance of the system, the training images were subject to data augmentation, where the images are transformed by a set of affine transformations, such as flipping, rotation, and mirroring, as well as augmenting the color values [38,51,52]. Perez et al. [53] discuss the effectiveness of data augmentation on the classification results when deep learning is used, and showed that the traditional augmentation techniques can improve the results by about 7%.

3.3. Training and Classification

The goodness of CNN features was compared to other traditional feature extraction methods, such as LBP, GLCM, Wavelet and Spectral. The feature extractors, which give good performance in comparison with the other texture extractor features, are a CNN. CNN training consumes some time; however, features can be extracted from the trained convolutional network, compared to other complex textural methods. CNNs have proven to be effective in classification tasks [26]. The training data and data augmentation are combined by reading batches of training data, applying data augmentation, and sending the augmented data to the training algorithm. The training is started by taking the data source, which contains the training images, pixel labels, and their augmentation forms.

4. Experimental Results

4.1. Evaluation Metrics

The output results of classification were compared against the ground truth given by the dataset. The comparison was done on a pixel-to-pixel basis. To evaluate the results, we applied the evaluation metrics given below. Table 1 represent the confusion matrix for binary class classification.

Table 1. Terms used to define sensitivity, specificity, and accuracy.

		Predicted	
		Positive	**Negative**
Actual	**Positive**	TP	FN
	Negative	FP	TN

Out of the confusion matrix, some important metrics were computed, such as

1. Overall Accuracy: this represents the percentage of correctly classified pixels to the whole number of pixels. This could be formulated as in Equation (2):

$$accuracy = \frac{TN + TP}{TN + TP + FN + FP} \tag{2}$$

while the mean accuracy is the mean of accuracies reported across the different testing folds.

2. Recall (Re) or true positive rate (TPR): this represents the capability of the system to correctly detect tumor pixels relative to the total number of true tumor pixels, as formulated in Equation (3):

$$Re = \frac{TP}{TP + FN} \tag{3}$$

3. Specificity of the true negative rate (TNR): this represents the rate of the correctly detected background or normal tissue, as formulated in Equation (4):

$$Specificity = \frac{TN}{TN + FP} \tag{4}$$

Since most of image is normal or background, the percentage of global accuracy is significantly influenced by the TNR. Therefore, some other measures for the tumor class are computed.

4. Intersection over union (IoU): this is the ratio of correctly classified pixels relative to the union of predicted and actual number of pixels for the same class. Equation (5) shows the formulation of the IoU:

$$IoU = \frac{TP}{TP + FP + FN} \tag{5}$$

5. Precision (Pr): this measures the trust in the predicted positive class, i.e., prediction of a tumor. It is formulated as in Equation (6):

$$Pr = \frac{TP}{TP + FP} \tag{6}$$

6. F1 score (F1): this is a harmonic mean of recall (true positive rate) and precision, as formulated in Equation (7). It measures whether a point on the predicted boundary has a match on the ground truth boundary or not:

$$F1 = \frac{2(Pr * Re)}{(Pr + Re)} \tag{7}$$

4.2. Data Set and Preprocessing

As mentioned before, the dataset used to test the proposed algorithm is 3D-IRCADb. The 3D-IRACDb dataset is offered by the French Research Institute against Digestive Tract, or IRCAD [50]. It has two subsets: the first one, 3DIRACDb-01, is the one appropriate for liver tumor segmentation. This subset consists of publicly available 3D CT scans of 20 patients, half of them for women patients and half for men, with hepatic tumors in 75% of the cases. All the scans are available in DICOM format with axial dimensions of 512 × 512. For each case, tens of 2D images are available, together with labeled images and masked images prepared by radiologists. In this work, we have considered all 15 cases with a total of 2063 images for training and testing. The dataset is used widely and recently, as in [54–57].

All image slices were subject to preprocessing, as discussed above. The labeled images provided by the dataset are preprocessed by the same procedure, except the step of range mapping, since they are given as binary images in the range [0,255]. Figures 5 and 6 show the examples of the preprocessing steps on input images. Associated with the input (patient) images are the labeled images, which are labeled by experts and are fed to the system as ground truth for the segmentation process.

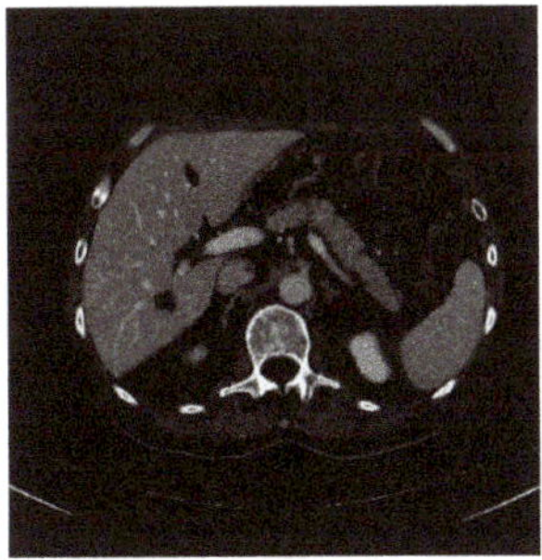 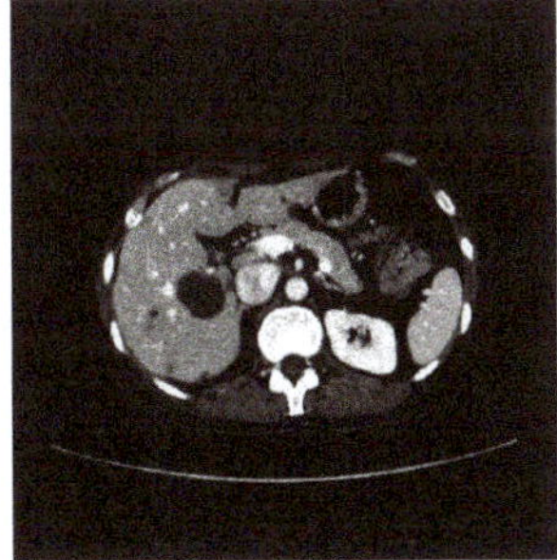 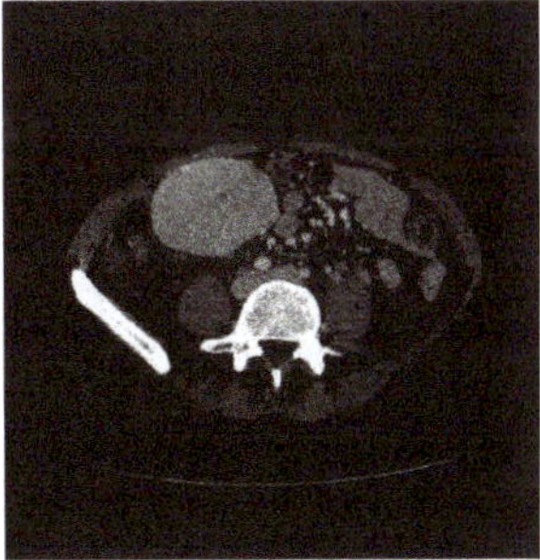

Figure 5. Samples of the slices after color range mapping to [0, 255]. The three images correspond respectively to the images in Figure 4.

 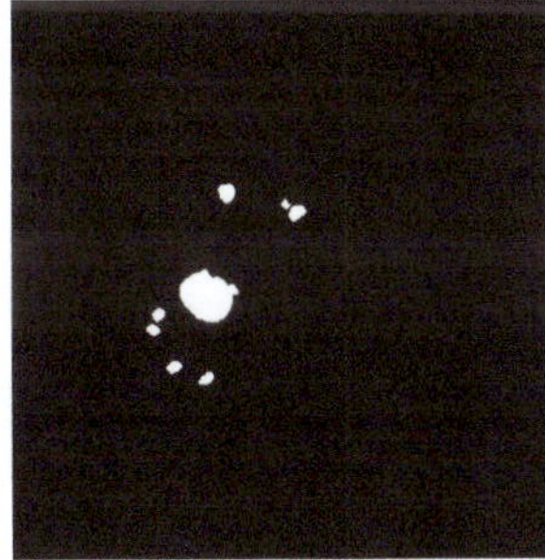 

Figure 6. Liver tumor labeled images from the ground truth given by the dataset. The three images correspond respectively to the images in Figures 4 and 5.

4.3. Training and Classification

Three of the 15 cases of the dataset were used for testing and evaluation, with a total of 475 images. Among these, 454 images were used for training and validation, and 45 images were used for testing.

The first training and testing experiments were carried out using the U-Net model in [45]. The U-Net model is trained to perform semantic segmentation on medical images. It is based on VGG-16, as discussed before. The results were near perfect to extract the liver region. However, it failed completely when tested to extract the tumor regions from the image. In this case, the tumor region was almost missed or predicted as others.

The proposed architecture is based on the SegNet model [37], which is an encoder network, and a corresponding decoder network connected to a 2D multi-classification layer for pixel-based semantic segmentation. However, the final classification layer was replaced by 2D binary classification. The VGG-16 trained model was imported for the encoder part. Figure 7 shows an illustration of the proposed network architecture. To improve the training, class weighting was used to balance the classes and calculate the median frequency class weights.

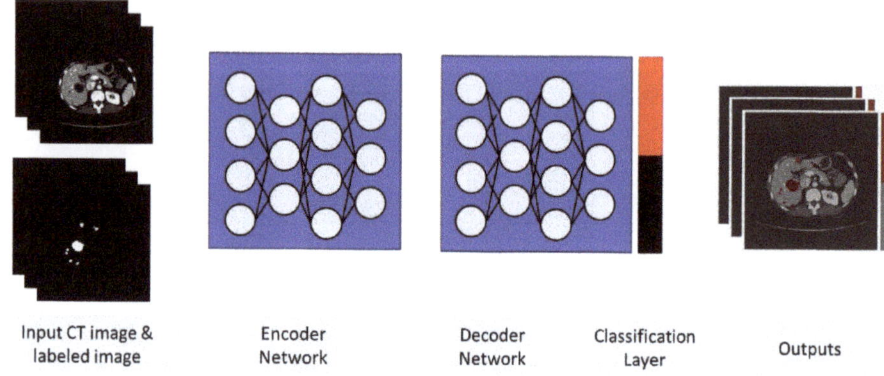

Figure 7. The proposed architecture.

For testing, a semantic segmentation was returned from the input image with the classification scores for each categorical label, in order to run the network for one image from test set.

4.4. Testing and Evaluation

The proposed method was trained on a machine with NVIDIA GTX 1050 4GB RAM GPU on an Intel Core i7-7700HQ 2.20 GHz 16 GB RAM, and developed with MATLAB 2018b software, which offers a Neural Network Toolbox and an Image Processing Toolbox.

The images of the tested cases were divided randomly into two groups for training and testing by the ratio 9:1. The results of the training are normally higher than that achieved by testing. Figure 8 shows three samples of testing output, where the resulted binary segmentation is augmented on the input gray-level images. At this stage, an almost perfect segmentation was achieved. In Table 2 are the evaluation metrics for the three cases. The network training performed by 1000 iterations per epoch for 100 epochs on a single GPU with a constant learning rate was 0.001. It is clear from Table 2 that as the number of training images increases, the segmentation quality increases up to perfect results, as in case 3.

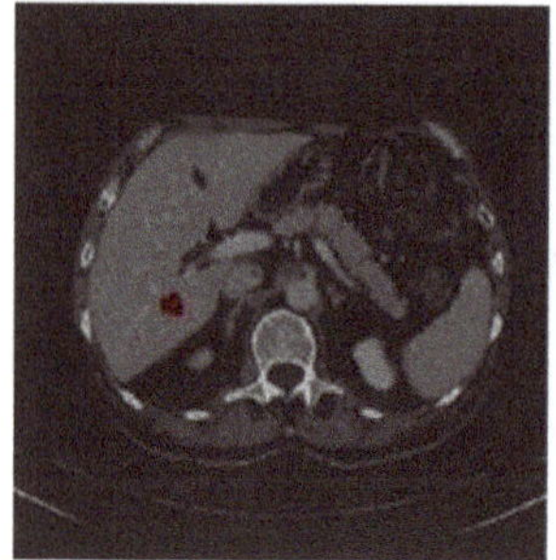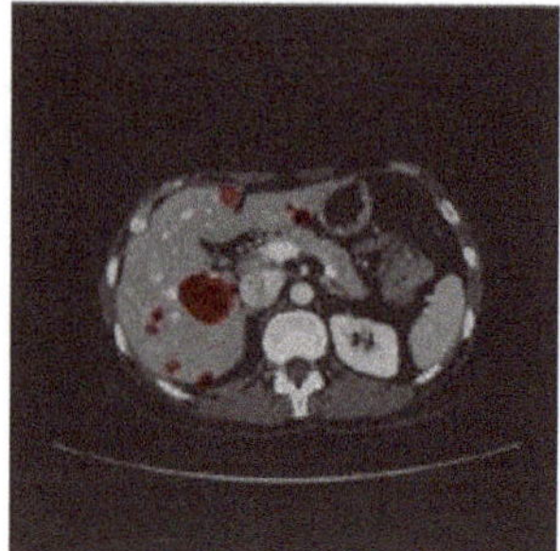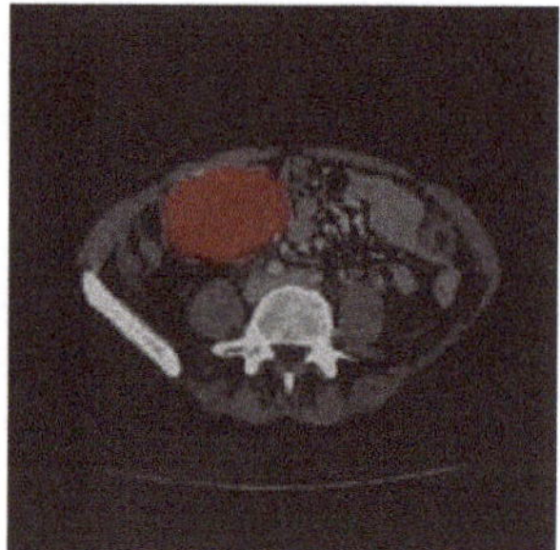

Figure 8. Samples of the results of testing. The predicted tumor position is highlighted with the red color.

Table 2. Evaluation metrics for the training of the three test cases.

	# Training Images	Overall Ac.	Mean Ac.	Tumor Ac. (TPR)	Background Ac. (TNR)	Weighted IoU	F1-Score
Case 1	70	91.27%	93.43%	95.58%	91.27%	91.26%	21 91%
Case 2	124	93.36%	95.18%	97.34%	93.03%	92.99%	46.58%
Case 3	260	94.57%	97.26%	99.99%	94.52%	93.72%	62.16%

For testing, a semantic segmentation is returned for the input image, with the classification scores for each categorical label. Figure 9 shows an illustration of the evaluation method, where the resulted segmented images are superimposed over the ground truth image. The correctly classified tumor pixels, known as true positive, are colored in white. It is clear from this figure that the results of the first are the one with the least accuracy, while the results of case 3 are perfect in terms of tumor detection; however, the tumor appears larger than it actually is.

Figure 9. Samples of the resulting segmented image superimposed over the ground truth image. The correctly classified tumor pixels (known as true positive (TP)) are colored in white. The missed tumor pixels are colored in purple. The pixels that are predicted to belong to the tumor, but actually are pixels representing normal tissue or the background, are colored in green. The black color represents pixels that are correctly classified as normal or background.

The experimental results are presented in confusion matrices in Tables 3–5 for the test cases 1,2 and 3, respectively. The results displayed are normalized.

Table 3. Normalized confusion matrix for test case 1.

Case 1		Predicted	
		Tumor	Others
Actual	Tumor	45.82%	54.28%
	Others	0.31%	99.69%

Table 4. Normalized confusion matrix for test case 2.

Case 1		Predicted	
		Tumor	Others
Actual	Tumor	67.13%	32.87%
	Others	3.04%	96.95%

Table 5. Normalized confusion matrix for test case 3.

Case 1		Predicted	
		Tumor	Others
Actual	Tumor	99.99%	0%
	Others	5.48%	94.52%

In order to increase the insight on the presented results, Table 6 presents a comparison between the overall accuracy of the proposed method compared to some chosen work from the literature, according to the results reported in their papers. From this work, we have achieved higher accuracy than the work in the comparison.

Table 6. Comparison on the overall accuracy for the proposed method against other work in the literature.

Author (s)	Application	Method	Accuracy
Chlebus [11]	Liver tumor candidate classification	Random Forest	90%
Christ [12]	Automatic liver and tumor segmentation of CT and MRI	Cascaded fully convolutional neural networks (CFCNs) with dense 3D conditional random fields (CRFs)	94%
Yang [33]	Liver segmentation of CT volumes	A deep image-to-image network (DI2IN)	95%
Bi [7]	Liver segmentation	Deep residual network (Cascaded ResNet)	95.9%
Li [46]	Liver and tumor segmentation from CT Volumes	H-DenseUNet	96.5%
Yuan [32]	Automatic liver and tumor segmentation	Hierarchical Convolutional—Deconvolutional Neural Networks	96.7%
Wen Li et al. [28]	Patch-based liver tumor classification	Conventional Convolutional Neural Network (CNN)	80.6%
Our method	Liver tumor in CT Scans segmentation	Modified SegNet	98.8%

5. Conclusions

This paper presents experimental work to adopt a deep learning model, used for semantic segmentation of road scene understanding, for tumor segmentation in CT Liver scans in DICOM format.

SegNet is recent encoder–decoder network architecture that employs the trained VGG-16 image classification network as encoder, and employs corresponding decoder architecture to transform the features back into the image domain to reach a pixel-wise classification at the end. The advantage of SegNet over standard auto-encoder architecture is in the simple yet very efficient modification where the max-pooling indices of the feature map are saved, instead of saving the feature maps in full. As a result, the architecture is much more efficient in training time, memory requirements, and accuracy.

To facilitate binary segmentation of medical images, the classification layer was replaced with binary pixel classification layer. For training and testing, the standard 3D-IRCADb-01 dataset was used. The proposed method correctly detects most parts of the tumor, with accuracy above 86% for tumor classification. However, by examining the results, there were few false positives that could be improved by applying false positive filters or by training the model on a larger dataset.

As a future work, we propose using a new deep learning model as an additional level to increase the localization accuracy of the tumor, and hence reduce the FN rate and increase the IoU metric, like the work introduced in [20].

Author Contributions: Conceptualization and methodology, M.A.-M.S. and S.A.; software and validation, G.K. and M.A.; formal analysis, M.A. and M.A.-M.S.; writing—original draft preparation, M.A.-M.S.; writing—review and editing, G.K., M.A., and B.A.; supervision, M.A.-M.S.; project administration, S.A.; funding acquisition, S.A. and B.A. Reviewed by all authors. All authors have read and agreed to the published version of the manuscript.

Funding: This research received external funding for publication by the Deanship of Scientific Research at Majmaah University, which is funding this work under project RGP-2019-29.

Acknowledgments: The authors would like to thank Mamdouh Mahfouz, Professor of Radiology at Kasr-El-Aini School of Medicine, Cairo University, Egypt, for his help in revising the introduction in order to correct any misused medical terms.

Conflicts of Interest: The authors declare no conflict of interest.

References

1. Al-Shaikhli, S.D.S.; Yang, M.Y.; Rosenhahn, B. Automatic 3d liver segmentation using sparse representation of global and local image information via level set formulation. *arXiv* **2015**, arXiv:abs/1508.01521.

2. Whitcomb, E.; Choi, W.T.; Jerome, K.R.; Cook, L.; Landis, C.; Ahn, J.; Westerhoff, M. Biopsy Specimens From Allograft Liver Contain Histologic Features of Hepatitis C Virus Infection After Virus Eradication. *Clin. Gastroenterol. Hepatol.* **2017**, *15*, 1279–1285. [CrossRef] [PubMed]

3. Woźniak, M.; Połap, D. Bio-Inspired methods modeled for respiratory disease detection from medical images. *Swarm Evol. Comput.* **2018**, *41*, 69–96. [CrossRef]

4. Mohamed, A.S.E.; Mohammed Salem, A.-M.; Hegazy, D.; Howida Shedeed, A. Improved Watershed Algorithm for CT Liver Segmentation Using Intraclass Variance Minimization. *ICIST* **2017**. [CrossRef]

5. Bellver, M.; Maninis, K.-K.; Pont-Tuset, J.; Nieto, X.G.; Torres, J.; Gool, L.V. Detection-Aided liver lesion segmentation using deep learning. Machine Learning 4 Health Workshop - NIPS. *arXiv* **2017**, arXiv:abs/1711.11069.

6. Ben-Cohen, A.; Klang, E.; Amitai, M.M.; Goldberger, J.; Greenspan, H. Anatomical data augmentation for CNN based pixel-wise classification. In Proceedings of the IEEE 15th International Symposium on Biomedical Imaging (ISBI 2018), Washington, DC, USA, 4–7 April 2018; pp. 1096–1099.

7. Bi, L.; Kim, J.; Kumar, A.; Feng, D. Automatic liver lesion detection using cascaded deep residual networks. *arXiv* **2017**, arXiv:abs/1704.02703.

8. Jemal, A.; Bray, F.; Center, M.M.; Ferlay, J.; Ward, E.; Forman, D. Global cancer statistics. *CA Cancer J. Clin.* **2017**, *61*, 60–90.

9. World Health Organization. Available online: http://www.emro.who.int/media/news/world-hepatitis-day-in-egypt-focuses-on-hepatitis-b-and-c-prevention (accessed on 24 February 2020).

10. World Health Organization Hepatitis. Available online: https://www.who.int/hepatitis/news-events/egypt-hepatitis-c-testing/en/ (accessed on 24 February 2020).

11. Chlebus, G.; Meine, H.; Moltz, J.H.; Schenk, A. Neural network-Based automatic liver tumor segmentation with random forest-Based candidate filtering. *arXiv* **2017**, arXiv:abs/1706.00842.

12. Christ, P.F.; Elshaer, M.E.A.; Ettlinger, F.; Tatavarty, S.; Bickel, M.; Bilic, P.; Rempfler, M.; Armbruster, M.; Hofmann, F.; D'Anastasi, M.; et al. Automatic liver and lesion segmentation in CT using cascaded fully convolutional neural networks and 3d conditional random fields. *arXiv* **2016**, arXiv:abs/1610.02177.

13. Christ, P.F.; Ettlinger, F.; Grün, F.; Elshaer, M.E.A.; Lipkova, J.; Schlecht, S.; Ahmaddy, F.; Tatavarty, S.; Bickel, M.; Bilic, P.; et al. Automatic liver and tumor segmentation of CT and MRI volumes using cascaded fully convolutional neural networks. *arXiv* **2017**, arXiv:abs/1702.05970.

14. Wei, W.; Zhou, B.; Połap, D.; Woźniak, M. A regional adaptive variational PDE model for computed tomography image reconstruction. *Pattern Recognit.* **2019**, *92*, 64–81. [CrossRef]

15. Varma, D.R. Managing DICOM images: Tips and tricks for the radiologist. *Indian J. Radiol. Imaging* **2012**, *22*, 4–13. [CrossRef] [PubMed]

16. The American Cancer Society medical and editorial content team. Liver Cancer Early Detection, Diagnosis, and Staging. Available online: https://www.cancer.org/content/dam/CRC/PDF/Public/8700.00.pdf (accessed on 22 July 2019).

17. Wozniak, M.; Polap, D.; Capizzi, G.; Sciuto, G.L.; Kosmider, L.; Frankiewicz, K. Small lung nodules detection based on local variance analysis and probabilistic neural network. *Comput. Methods Programs Biomed.* **2018**, *161*, 173–180. [CrossRef]

18. Mohamed, A.S.E.; Salem, M.A.-.M.; Hegazy, D.; Shedeed, H. Probablistic-Based framework for medical CT images segmentation. In Proceedings of the 2015 IEEE Seventh International Conference on Intelligent Computing and Information Systems (ICICIS), Cairo, Egypt, 12–14 December 2015; pp. 149–155. [CrossRef]

19. Salem, M.A.; Meffert, B. Resolution mosaic EM algorithm for medical image segmentation. In Proceedings of the 2009 International Conference on High Performance Computing & Simulation, Leipzig, Germany, 21–24 June 2009; pp. 208–215. [CrossRef]

20. El-Regaily, S.A.; Salem, M.A.M.; Abdel Aziz, M.H.; Roushdy, M.I. Multi-View convolutional neural network for lung nodule false positive reduction. *Expert Syst. Appl.* **2019**. [CrossRef]

21. Atteya, M.A.; Salem, M.A.-.M.; Hegazy, D.; Roushdy, M.I. Image segmentation and particles classification using texture analysis method. *Rev. Bras. de Eng. Biomed.* **2016**, *32*, 243–252. [CrossRef]

22. Han, X. Automatic liver lesion segmentation using A deep convolutional neural network method. *arXiv* **2017**, arXiv:abs/1704.07239.

23. National Institute of Biomedical Imaging and Bioengineering (NIBIB). Computed Tomography (CT), URL. Available online: https://www.nibib.nih.gov/science-education/science-topics/computed-tomography-ct (accessed on 18 May 2019).

24. Heimann, T.; Van Ginneken, B.; Styner, M. Sliver07 Grand Challenge. Available online: https://sliver07.grand-challenge.org/Home/ (accessed on 20 July 2019).

25. Heimann, T.; van Ginneken, B.; Styner, M.A.; Arzhaeva, Y.; Aurich, V.; Bauer, C.; Beck, A.; Becker, C.; Beichel, R.; Bekes, G.; et al. Comparison and evaluation of methods for liver segmentation from ct datasets. *IEEE Trans. Med. Imaging* **2009**, *28*, 1251–1265. [CrossRef]

26. Salem, M.A.-M.; Atef, A.; Salah, A.; Shams, M. Recent survey on medical image segmentation. *Comput. Vis. Concepts Methodol. Tools Appl.* **2018**, 129–169. [CrossRef]

27. Horn, Z.C.; Auret, L.; Mccoy, J.T.; Aldrich, C.; Herbst, B.M. Performance of convolutional neural networks for feature extraction in froth flotation sensing. *IFAC Pap. Line* **2017**, *50*, 13–18. [CrossRef]

28. Li, W.; Jia, F.; Hu, Q. Automatic segmentation of liver tumor in CT images with deep convolutional neural networks. *J. Comput. Commun.* **2015**, *3*, 146–151. [CrossRef]

29. Li, X.; Chen, H.; Qi, X.; Dou, Q.; Fu, C.; Heng, P. H-DenseUNet: Hybrid Densely Connected UNet for Liver and Tumor Segmentation From CT Volumes. *IEEE Trans. Med. Imaging* **2018**, *37*, 2663–2674. [CrossRef]

30. Rajeshwaran, M.; Ahila, A. Segmentation of Liver Cancer Using SVM Techniques. *Int. J. Comput. Commun. Inf. Syst. (IJCCIS)* **2014**, *6*, 78–80.

31. El-Regaily, S.A.; Salem, M.A.M.; Aziz, M.H.A.; Roushdy, M.I. Lung nodule segmentation and detection in computed tomography. In Proceedings of the 8th IEEE International Conference on Intelligent Computing and Information Systems (ICICIS), Cairo, Egypt, 5–7 December 2017; pp. 72–78. [CrossRef]

32. Yuan, Y. Hierarchical convolutional-Deconvolutional neural networks for automatic liver and tumor segmentation. *arXiv* **2017**, arXiv:abs/1710.04540.

33. Yang, D.; Xu, D.; Zhou, S.K.; Georgescu, B.; Chen, M.; Grbic, S.; Metaxas, D.N.; Comaniciu, D. Automatic liver segmentation using an adversarial image-To-Image network. *arXiv* **2017**, arXiv:abs/1707.08037.

34. Shafaey, M.A.; Salem, M.A.-M.; Ebied, H.M.; Al-Berry, M.N.; Tolba, M.F. Deep Learning for Satellite Image Classification. In Proceedings of the 3rd International Conference on Intelligent Systems and Informatics (AISI2018), Cairo, Egypt, 1–3 September 2018.

35. Sutskever, I.; Vinyals, O.; Le, Q.V. Sequence to sequence learning with neural networks. *arXiv* **2014**, arXiv:abs/1409.3215.

36. Ke, Q.; Zhang, J.; Wei, W.; Damaševičius, R.; Wozniak, M. Adaptive Independent Subspace Analysis of Brain Magnetic Resonance Imaging Data. *IEEE Access* **2019**, *7*, 12252–12261. [CrossRef]

37. Badrinarayanan, V.; Kendall, A.; Cipolla, R. SegNet: A Deep Convolutional Encoder-Decoder Architecture for Image Segmentation. *IEEE Trans. Pattern Anal. Mach. Intell.* **2017**, *39*, 2481–2495. [CrossRef]

38. Ahmed, A.H.; Salem, M.A. Mammogram-Based cancer detection using deep convolutional neural networks. In Proceedings of the 2018 13th International Conference on Computer Engineering and Systems (ICCES), Cairo, Egypt, 18–19 December 2018; pp. 694–699. [CrossRef]

39. Lu, F.; Wu, F.; Hu, P.; Peng, Z.; Kong, D. Automatic 3d liver location and segmentation via convolutional neural networks and graph cut. *Int. J. Comput. Assist Radiol. Surg.* **2017**, *12*, 171–182. [CrossRef]

40. Wang, P.; Qian, Y.; Soong, F.K.; He, L.; Zhao, H. Part-Of-Speech tagging with bidirectional long short-Term memory recurrent neural network. *arXiv* **2015**, arXiv:abs/1510.06168.

41. Abdelhalim, A.; Salem, M.A. Intelligent Organization of Multiuser Photo Galleries Using Sub-Event Detection. In Proceedings of the 12th International Conference on Computer Engineering & Systems (ICCES), Cairo, Egypt, 19–20 December 2017; Available online: https://ieeexplore.ieee.org/document/347/ (accessed on 22 January 2020).

42. Elsayed, O.A.; Marzouky, N.A.M.; Atefz, E.; Mohammed, A.-S. Abnormal Action detection in video surveillance. In Proceedings of the 9th IEEE International Conference on Intelligent Computing and Information Systems (ICICIS), Cairo, Egypt, 8–10 December 2019.

43. El-Masry, M.; Fakhr, M.; Salem, M. Action Recognition by Discriminative EdgeBoxes. *IET Comput. Vis. Dec.* **2017**. Available online: http://digital-library.theiet.org/content/journals/10.1049/iet-cvi.2017.0335 (accessed on 22 January 2020). [CrossRef]

44. Simonyan, K.; Zisserman, A. Very deep convolutional networks for largescale image recognition. *arXiv* **2014**, arXiv:abs/1409.1556.

45. Ronneberger, O.; Fischer, P.; Brox, T. U-net: Convolutional networks for biomedical image segmentation. *arXiv* **2015**, arXiv:abs/1505.04597.

46. Ke, Q.; Zhang, J.; Wei, W.; Połap, D.; Woźniak, M.; Kośmider, L.; Damaševǐcius, R. A neuro-Heuristic approach for recognition of lung diseases from X-Ray images. *Expert Syst. Appl.* **2019**, *126*, 218–232. [CrossRef]

47. Noh, H.; Hong, S.; Han, B. Learning Deconvolution Network for Semantic Segmentation. In Proceedings of the 2015 IEEE International Conference on Computer Vision (ICCV), Santiago, Chile, 7–13 December 2015; pp. 1520–1528.

48. Vadera, S.; Jone, J. Couinaud classification of hepatic segments. Available online: https://radiopaedia.org/articles/couinaud-classification-of-hepatic-segments (accessed on 28 November 2019).

49. Zayene, O.; Jouini, B.; Mahjoub, M.A. Automatic liver segmentation method in CT images. *arXiv* **2012**, arXiv:abs/1204.1634.

50. IRCAD. Hôpitaux Universitaires. 1, place de l'Hôpital, 67091 Strasbourg Cedex, FRANCE. Available online: https://www.ircad.fr/contact/ (accessed on 2 February 2020).

51. Golan, R.; Jacob, C.; Denzinger, J. Lung nodule detection in CT images using deep convolutional neural networks. In Proceedings of the 2016 International Joint Conference on Neural Networks (IJCNN), Vancouver, BC, Canada, 24–29 July 2016; pp. 243–250.

52. El-Regaily, S.A.; Salem, M.A.; Abdel Aziz, M.H.A.; Roushdy, M.I. Survey of Computer Aided Detection Systems for Lung Cancer in Computed Tomography. *Current. Med. Imaging* **2018**, *14*, 3–18. [CrossRef]

53. Perez, L.; Wang, J. The Effectiveness of Data Augmentation in Image Classification Using Deep Learning. *arXiv* **2017**, arXiv:abs/1712.04621.

54. Sun, C.; Guo, S.; Zhang, H.; Li, J.; Chen, M.; Ma, S.; Jin, L.; Liu, X.; Li, X.; Qian, X. Automatic segmentation of Liver tumors from multiphase contrast-Enhanced CT images based on FCN. *Artificial Intell. Med.* **2017**, *83*, 58–66. [CrossRef] [PubMed]

55. Badura, P.; Wieclawe, W. Calibrating level set approach by granular computing in computed tomography abdominal organs segmentation. *Appl. Soft Comput.* **2016**, *49*, 887–900. [CrossRef]

56. Shi, C.; Cheng, Y.; Liu, F.; Wang, Y.; Bai, J.; Tamura, S. A hierarchical local region-Based sparse shape composition for liver segmentation in CT scans. *Pattern Recognit.* **2015**. [CrossRef]

57. Hesamian, M.H.; Jia, W.; He, X.; Kennedy, P. Deep Learning Techniques for Medical Image Segmentation: Achievements and challenges. *J. Digit. Imaging* **2019**. [CrossRef] [PubMed]

Article

HEMIGEN: Human Embryo Image Generator Based on Generative Adversarial Networks

Darius Dirvanauskas [1], Rytis Maskeliūnas [1], Vidas Raudonis [2], Robertas Damaševičius [3,4,*] and Rafal Scherer [5]

[1]   Department of Multimedia Engineering, Kaunas University of Technology, 51368 Kaunas, Lithuania
[2]   Department of Control Systems, Kaunas University of Technology, 51367 Kaunas, Lithuania
[3]   Department of Software Engineering, Kaunas University of Technology, 51368 Kaunas, Lithuania
[4]   Institute of Mathematics, Silesian University of Technology, 44-100 Gliwice, Poland
[5]   Institute of Computational Intelligence, Czestochowa University of Technology, 42-200 Czestochowa, Poland
*   Correspondence: robertas.damasevicius@ktu.lt

Received: 26 July 2019; Accepted: 14 August 2019; Published: 16 August 2019

Abstract: We propose a method for generating the synthetic images of human embryo cells that could later be used for classification, analysis, and training, thus resulting in the creation of new synthetic image datasets for research areas lacking real-world data. Our focus was not only to generate the generic image of a cell such, but to make sure that it has all necessary attributes of a real cell image to provide a fully realistic synthetic version. We use human embryo images obtained during cell development processes for training a deep neural network (DNN). The proposed algorithm used generative adversarial network (GAN) to generate one-, two-, and four-cell stage images. We achieved a misclassification rate of 12.3% for the generated images, while the expert evaluation showed the true recognition rate (TRR) of 80.00% (for four-cell images), 86.8% (for two-cell images), and 96.2% (for one-cell images). Texture-based comparison using the Haralick features showed that there is no statistically (using the Student's t-test) significant ($p < 0.01$) differences between the real and synthetic embryo images except for the sum of variance (for one-cell and four-cell images), and variance and sum of average (for two-cell images) features. The obtained synthetic images can be later adapted to facilitate the development, training, and evaluation of new algorithms for embryo image processing tasks.

Keywords: deep learning; neural network; generative adversarial network; synthetic images

1. Introduction

Deep neural networks (DNN) have become one of the most popular modern tools for image analysis and classification [1]. One of the first accurate implementations of DNN, AlexNet [2], was quickly bested by a deep convolutional activation feature (DeCAF) network, which extracted features from AlexNet and evaluated the efficacy of these features on generic vision tasks [3]. VGGNet demonstrated that increasing the depth of convolutional neural network (CNN) is beneficial for the classification accuracy [4]. However, the deeper neural network is more difficult to train. ResNet presented a residual learning framework to ease the training for very deep networks [5]. The ResNet architecture, which combined semantic information from a deep, coarse layer with appearance information from a shallow network, was adapted for image segmentation [6]. Region CNN (R-CNN) was proposed as a combination of high-capacity CNNs with bottom-up region proposals in order to localize and segment images [7].

The breakthrough of DNNs in quality led to numerous adaptations in solving medical image processing problems such as the analysis of cancer cells [8] and cancer type analysis [9]. Deep max-pooling CNNs were used to detect mitosis in breast histology images using a patch centered on the pixel as context [10]. The U-Net architecture with data augmentation and elastic deformations achieved very good performance on different biomedical segmentation applications [11]. A supervised max-pooling CNN was trained to detect cell pixels in regions that are preselected by a support vector machine (SVM) classifier [12]. After a pre-processing step to remove artefacts from the input images, fully CNNs were used to produce the embryo inner cell mass segmentation [13]. A set of Levenberg–Marquardt NNs trained using textural descriptors allowed for predicting the quality of embryos [14].

Image analysis methods were also applied to embryo image analysis. Techniques for extracting, classifying, and grouping properties were used to measure the quality of embryos. Real-time grading techniques for determining the number of embryonic cells from time-lapse microscope images help embryologists to monitor the dividing cells [15]. Conditional random field (CRF) models [16] were used for cell counting and assessing various aspects of the developing embryo and predicting the stage of the embryonic development. In addition to grading, embryonic positioning was achieved using linear chain Markov model [17]. Different solutions were developed for segmenting and calculating the number of cells: ImageJ, MIPAV, VisSeg [18]. Moreover, cell segmentation by marking its center and edges makes it possible to determine their shapes and quantities more quickly and accurately [19]. Two-stage classifier for embryo image classification was proposed in [20].

The rising progress of generative adversarial networks (GANs) applications on medical imaging state that most of the research is focused on synthetic imaging, the reconstruction, segmentation, and classification and proves the importance of the sector. Wasserstein-based GANs were applied for the synthesis of cells imaged by fluorescence microscopy capturing these relationships to be relevant for biological application [21]. The quality of generated artificial images also improved due to the improvements to deep neural networks [22] as well as on progressive growing of GANs for image data augmentation [23]. Variational autoencoders (VAE) and generative adversarial networks (GANs) are currently the most distinguished according to the quality of their results. The VAE models are more often used for image compression and recovery. Both methods have drawbacks, as recovering a part of the information loses some data and often introduces image fading or blurring. This effect could be reduced by matching the data as well as the loss distributions of the real and fake images by a pair of autoencoders used as the generator and the discriminator in the adversarial training [24]. Utilization of both generator and discriminator growing progressively from a low-resolution standpoint and adding new layers for an increase in fine details allowed for achieving a current state-of-the-art inception score of 8.80 on CIFAR10 dataset [25]. Network modifications can be applied to minimize fading of the resulting images [24]. Generative stochastic networks (GSN) were used to learn the transition operator of a Markov chain whose stationary distribution estimates the data distribution [26]. Deep recurrent attentive writer (DRAW) neural network architecture [27] was used to generate highly realistic natural images such as photographs of house numbers or other digits in the MNIST database. Dosovitskiy et al. [28] introduced an algorithm to find relevant information from the existing 3D chair models and to generate new chair images using that information. Denton et al. [29] introduced a generative parametric model capable of producing high-quality samples of natural images based on a cascade of convolutional networks within a Laplacian pyramid framework. Radford and Metz [30] introduced a simplification of training by a modification called deep convolutional GANs (DCGANs). GANs can also recover images from bridging text and image modelling, thus translating visual concepts from characters to pixels [31]. Zhang et al. [32] proposed stacked GANs (StackGAN) with conditioning augmentation for synthesizing photo-realistic images. GANs could also be applied for generating motion images, e.g., the motion and content decomposed GAN (MoCoGAN) framework for video generation by mapping a sequence of random vectors to a sequence of video frames [33]. The GAN for video (VGAN) model was based on a GAN with a spatial-temporal convolutional architecture that untangles the scene's foreground from the background and can be used at predicting plausible futures of static images [34]. Temporal GAN (TGAN) was used to learn a semantic representation of unlabeled videos by using different types of generators via Wasserstein GAN, and a method to train it stably in an end-to-end manner [35]. MIT presented a 3D generative adversarial network (3D-GAN) model to recreate 3D objects from a probabilistic space by leveraging recent advances in volumetric CNNs and GANs [36]. Li et al. [37] used multiscale GAN (DR-Net) to remove rain streaks from a single image. Zhu [38] used GANs for generating synthetic saliency maps for given natural images. Ma et al. [39] used background augmentation GANs (BAGANs) for synthesizing background images for augmented reality (AR) applications. Han et al. presented a system based on a Wasserstein GANs for generating realistic synthetic brain MR images [40].

All these results are achieved by using large expert-annotated and ready-made databases and also exposing the problem of lacking good, core training datasets. In this work, we aimed to develop a method to generate realistic synthetic images that could later be used for classification, analysis, and training, thus resulting in the creation of novel synthetic datasets for research areas lacking data such as human embryo images. Here we used human embryo images obtained during cell development processes for training a DNN. We propose an algorithm for generating one-, two-, and four-cell images (the selection was based on the initial dataset provided by our medical partners) to increase the overall number of unique images available for training. For generating images, we have developed a generative adversarial network for image recovery, filling, and improvement. The significance of our approach would be that the method was applied to embryonic cell images as a type of image. It was very important for the GAN to accurately reproduce the outline of the cell (often poorly visible even in microscopy picture), since the whole image was almost the same (gray) and the cell itself is translucent. Our focus was not only to generate the generic image of a cell as such, but to make sure that it has all necessary attributes of a real cell image to provide a fully realistic synthetic version. We believe that as a large number of real embryo images required for training neural networks are difficult to obtain due to ethics requirements, the synthetic images generated by the GAN can be later adapted to facilitate the development, training, and evaluation of new algorithms for embryo image-processing tasks.

2. Materials and Methods

2.1. Architecture of the Generative Adversarial Network (GAN)

The generative adversarial network (GAN) consists of two parts: A generator *G* and a discriminator *D* (Figure 1).

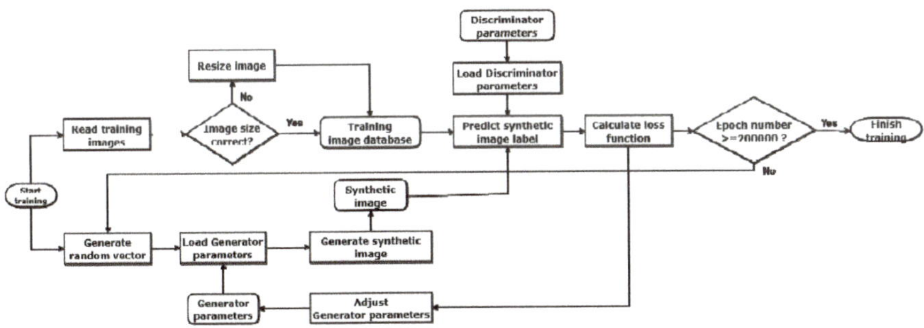

Figure 1. A pipeline of a generative adversarial network (GAN).

The discriminator tries to distinguish true images from synthetic, generator-generated images. The generator tries to create images with which the discriminator could be deceived. The discriminator is implemented as a fully connected neural network with dense layers that represents an image as a probability vector and classifies it into two classes, a real or a fake image. The generator is a reverse model that restores the former image from a random noise vector. During the training, the discriminator is trained to maximize the probability of assigning the correct class to the training images and images generated by the generator. At the same time, the generator is trained to minimize classification error

$\log(1 - D(G(z)))$. This process can be expressed in terms of game theory as a two-player (discriminator and generator) minimax game with cost function $V(G, D)$ [41]:

$$\max_{G} \min_{D} V(D, G) = E_{x \sim P(x)}[\log(D(x))] + E_{z \sim P(z)}[\log(1 - D(G(z)))]. \tag{1}$$

The discriminator evaluates the images created by the training database and the generator. The architecture of the discriminator network is shown in Figure 2. The network consists of six layers. The monochrome 200×200 pixel image in the first layer is transformed and expanded into a single line vector. The dense layer is used in the second, fourth, and sixth layers. The LeakyReLU [42] function, where $\alpha = 0.2$, is used in the third and fifth layers as it allows a small gradient when the unit is not active:

$$\text{LReLu}(x) = \max(x, 0) + \alpha \min(x, 0). \tag{2}$$

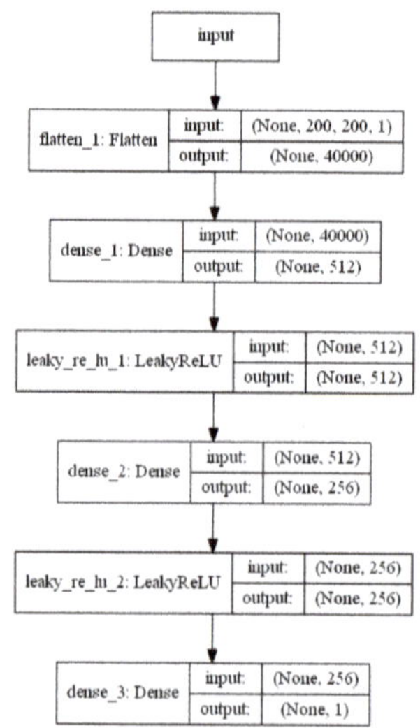

Figure 2. Architecture of the discriminator network.

At the network output, we get a one if the network guesses that the image is real or zero if the network guesses that the image is fake.

The generator is composed of 11 layers (Figure 3). The input layer contains 1×100 vector of randomly generated data. These data are transformed using the dense, LeakyReLU, and batch normalization layers. The ReLU layer uses $\alpha = 0.2$. The batch normalization layer was used to normalize the activations of the previous layer at each batch, by applying a transformation that maintains the mean activation close to 0 and the activation standard deviation close to 1 [43]:

$$\hat{x}_i \leftarrow \frac{x_i - \mu_\beta}{\sqrt{\sigma_\beta^2 + \varepsilon}}. \tag{3}$$

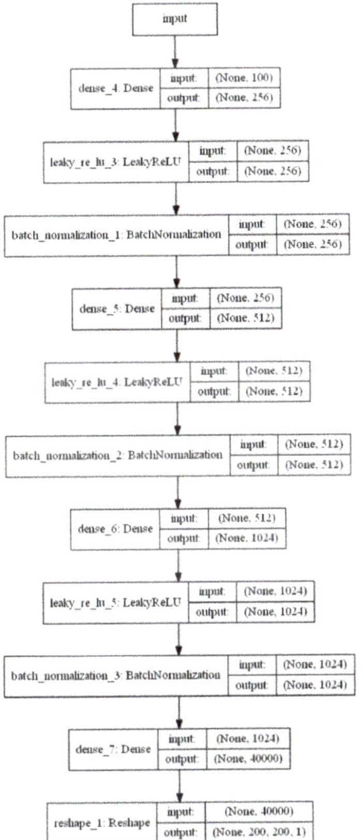

Figure 3. Architecture of the generator network.

The batch normalization momentum value was set to 0.8, while all other values were used as per default Keras model values. The output produces a 200 × 200 pixel sized black-and-white image.

2.2. Training

We trained one (single) network to generate all types of cell images. For the training, we have used the Adam optimization algorithm [44] with a fixed training speed of 0.00002, and $\beta_1 = 0.7$ for training the generator. During training, the Adam algorithm calculates the gradient as follows:

$$g_t \leftarrow \Delta_\theta f_t(\theta_{t-1}). \tag{4}$$

The biased first moment estimate is updated as follows:

$$m_t \leftarrow \beta_1 m_{t-1} + (1 - \beta_1) g_t. \tag{5}$$

The biased second moment estimate is update as follows:

$$v_t \leftarrow \beta_2 v_{t-1} + (1 - \beta_2) g_t^2. \tag{6}$$

Then we compute the bias corrected first moment estimate as follows:

$$\hat{m}_t \leftarrow \frac{m_t}{1 - \beta_1^t}.$$ (7)

The bias corrected second moment estimate is computed as follows:

$$\hat{v}_t \leftarrow \frac{v_t}{1 - \beta_2^t}.$$ (8)

This gives the update parameters as follows:

$$\theta_t \leftarrow \frac{\theta_{t-1} - a\hat{m}_t}{\sqrt{\hat{v}_t''} + \varepsilon}.$$ (9)

The binary cross-entropy function is used to evaluate the error as follows:

$$-\frac{1}{N}\sum_{i=1}^{N}[y_i \log(\hat{y}_i) + (1 - y_i)\log(1 - \hat{y}_i)].$$ (10)

2.3. Evaluation of Image Quality

The problem with using the GAN model is that there are no clear ways to measure it qualitatively. There are few most commonly used criteria to evaluate GAN network: Average log-likelihood, classifier classification [41], and visual fidelity of samples [45]. These methods have both advantages and disadvantages. Log-likelihood and classifiers attribute the image to a certain class with a matching factor. However, they only indicate whether the generated image is like the average of one of the classes but does not value a qualitatively generated image. Such a qualitative assessment requires a human evaluator, but with a large amount of data, such estimates can change independently and be biased [46]. Likewise, a different expert person could evaluate the same data somewhat differently.

We use a combined method to evaluate the results obtained in our study, which includes: (1) Human expert evaluation, (2) histogram comparison, and (3) texture feature comparison. The generated images are evaluated by human experts (three medical research professionals) to determine if they are being reproduced qualitatively. If we can determine its class from the received image, the image is quality-restored. The expert based scoring was calculated using a visual Turing test [47] as this method has already been proven effective in the evaluation of synthetic images generated by GANs [48].

The histogram comparison method checks whether the histogram of the generated images corresponds to the histogram of the images in the training database. For comparison, we used the normalized average histogram of training data H_1, and the normalized average histogram of generated images H_2. Further, we apply four different methods for comparison as it is recommended in [49]. The correlation function determines the likelihood of two image histograms. Its value of two identical histograms equals one:

$$C(H_1, H_2) = \frac{\sum_I \left(H_1(I) - \overline{H}_1\right)\left(H_2(I) - \overline{H}_2\right)}{\sqrt{\sum_I \left(H_1(I) - \overline{H}_1\right)^2 \sum_I \left(H_2(I) - \overline{H}_2\right)^2}}.$$ (11)

The Chi-square function takes the squared difference between two histograms. The squared differences are divided by the number of samples, and the sum of these weighted squared differences is the likelihood value:

$$\chi^2(H_1, H_2) = \sum_I \frac{(H_1(I) - H_2(I))^2}{H_1(I)}.$$ (12)

The histogram intersection calculates the similarity of two discretized histograms, with a possible value of the intersection lying between no overlap and identical distributions. It works well on categorical data and deals with null values by making them part of the distribution.

$$I(H_1, H_2) = \sum_I \min(H_1(I), H_2(I)) \tag{13}$$

The Bhattacharyya distance approximates the normalized distance between the histograms using the maximum likelihood of two object histograms as follows:

$$B(H_1, H_2) = \sqrt{1 - \frac{1}{\sqrt{H_1 H_2 N^2}} \sum_I \sqrt{H_1(I) H_2(I)}}. \tag{14}$$

The texture-based comparison uses the texture analysis features based on grey level co-occurrence matrix (GLCM), which are related to second-order image statistics that were introduced by Haralick [50] as follows: (1) Angular second moment (energy), (2) contrast, (3) correlation, (4) variance, (5) inverse difference moment (homogeneity), (6) sum average, (7) sum variance, (8) sum entropy, (9) entropy, (10) difference variance, (11) difference entropy, (12) information measure of correlation, (13) information measure of correlation II, and (14) maximal correlation coefficient.

2.4. Ethics Declaration

The study was conducted in accordance with the Declaration of Helsinki, and the protocol was approved by the Ethics Committee of Faculty of Informatics, Kaunas University of Technology (No. IFEP201809-1, 2018-09-07).

3. Experiment and Results

3.1. Dataset and Equipment

We have used the embryo photos taken with Esco Global incubator series, called Miri TL (time lapse). The embryo image sets were registered in the German, Chinese, and Singapore clinics. No identity data were ever provided to the authors of this paper. The Esco's embryo database consists of three image classes that together have 5000 images with a resolution of 600×600 pixels. Number of different images in classes: One-cell—1764 images, two-cell—1938 images, four-cell—1298 images. Images are obtained from 22 different growing embryos in up to five stages of cell evolution. The database was then magnified by rotating each photo at 90, 180, and 270 degrees. Resulting number of different images in classes: (one-cell images—$1764 \times 4 = 7056$; two-cell images—$1938 \times 4 = 7752$; four-cell images—$1298 \times 4 = 5192$). The images were taken in the following culturing conditions: Temperature of 37 °C, stable level of 5% $CO2$, and the controllable values of nitrogen and oxygen mixture. The embryos were photographed in a culture coin dish made from polypropylene with a neutral media pH = 7. The inverted microscopy principle was used, with 20x lenses, without zoom, with focusing and with field of view of 350 um. A camera sensor used was IDS UI-3260CP-M-GL (IDS Imaging Development Systems GmbH, Obersulm, Germany). An example of embryo images in the Esco dataset is presented in Figure 4. The evaluation and training of GANs was done on an Intel i5-4570 CPU with a GeForce 1060 GPU and 8 GB of RAM.

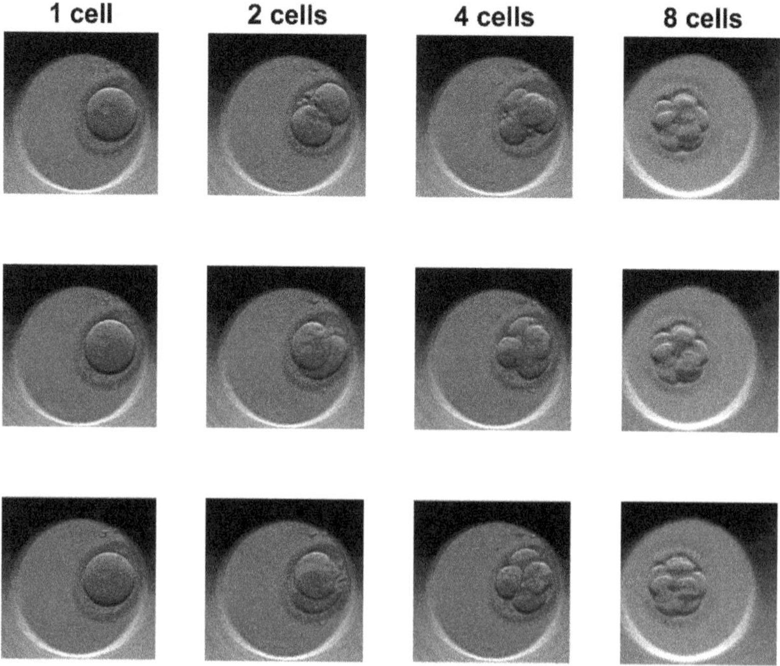

Figure 4. Sample images from Esco embryo image dataset.

3.2. Results

The training of GAN was carried out in 200,000 iterations, with a sample size of 256 images per batch. The duration of training using GeForce 1060 GPU was about 15 h 45 min.

From the final generated images, one can easily count the number of embryonic cells and determine the class (Figure 5). During subsequent iterations, the generator restored the embryo image and reduced the amount of noise in the generated images.

Table 1. Evolution of embryo cell images during GAN training.

Epoch	0	25,000	50,000	75,000	100,000	125,000	150,000	175,000	200,000
one-cell									
two-cells									
four-cells									

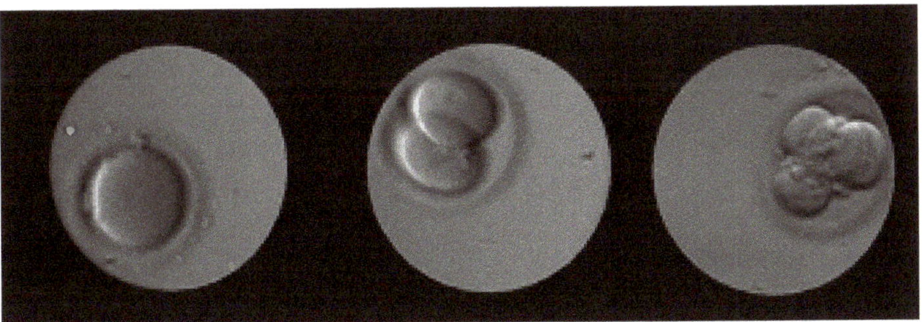

Figure 5. Example of final generated embryo cell images using the proposed algorithm. From left to right: One, two, and four cells. Images were filtered from "salt and pepper" noise using a median filter. See Table 1 for raw outputs.

During the training, embryo cell images were generated at every 25,000 iterations. An example of these images is shown in Table 1.

After 25,000 iterations, we can see that he managed to restore the plate's flair and release the background, but the embryo itself was not restored. After 100,000 iterations, the generator has been able to clearly reproduce the image of the embryo cell.

The error function values for both generator and discriminator are shown in Figures 6 and 7. During an initial training up to 20,000 iterations, the error function value of the generator increases rapidly. Further increase of the error function results in a slower pace. Once the training number reaches 75,000 iterations, the value of the generator error function begins to increase with the training of two-cell and four-cell images. The generator becomes able to produce more complex images, which is also visible from the generated sample images (see Table 1). The one-cell image can be identified from 50,000 iterations, whereas two-cell and four-cell images can be recognized from 75,000–125,000 iterations. From the discriminator error feature, we can see that starting from 20,000 iterations it becomes difficult to separate real images from artificially generated images, and the value of the error function is less than 0.15. A value of discriminator log loss increases, as discriminator is unable to differentiate between a real image and a generated image.

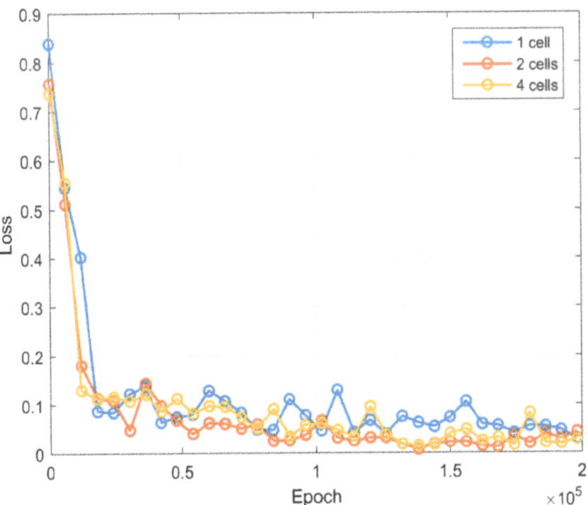

Figure 6. Loss log graph for the generator network.

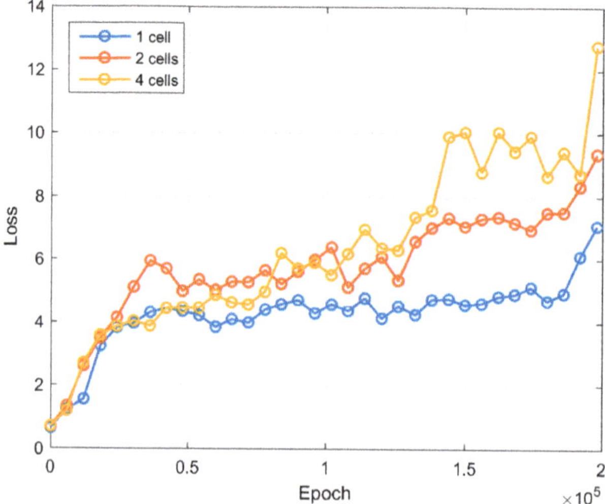

Figure 7. Loss log graph for the discriminator network.

For the expert-based evaluation (see Table 2), 1500 images (500 in each class) were generated. The number of artificial one-cell images, where one cell was clearly recognized, was 96.2%. The quality of two-cell and four-cell image generation of more complex images deteriorated. Of the two-cell images, the number of cells could be clearly determined from 86.8% of the images. When evaluating the four-cell images, 80% accuracy was obtained, i.e., one out of five images was generated inaccurately (as decided by an expert).

Table 2. Evaluation of the generated embryo image cells by human experts.

Class	Test Images (Generated)	Good Images	Accuracy by Expert Selection
one-cell	500	481	96.2%
two-cells	500	434	86.8%
four-cells	500	400	80.00%

Additionally, the image similarity was evaluated by comparing their histograms. In Figure 8, one can see the comparison of image histograms, where the blue curve is the average histogram of the training image dataset and the red is the average histogram of the artificially generated images. In all three classes, we can see that the generated image is brighter than the real one. This could be explained due to the slight "salt-and-pepper"-type noise in the generated image. The highest histogram match was obtained in the single-cell-generated images. This shows that the generator was better able to reproduce images of a simpler structure. Note that Figure 8 shows an evaluation of a generated cell image with a maximum epoch number of 200,000. The particular histogram cannot be identical as a single network generates images of different cells. This comparison was done by evaluating the average histogram of all (training and generated) image histograms.

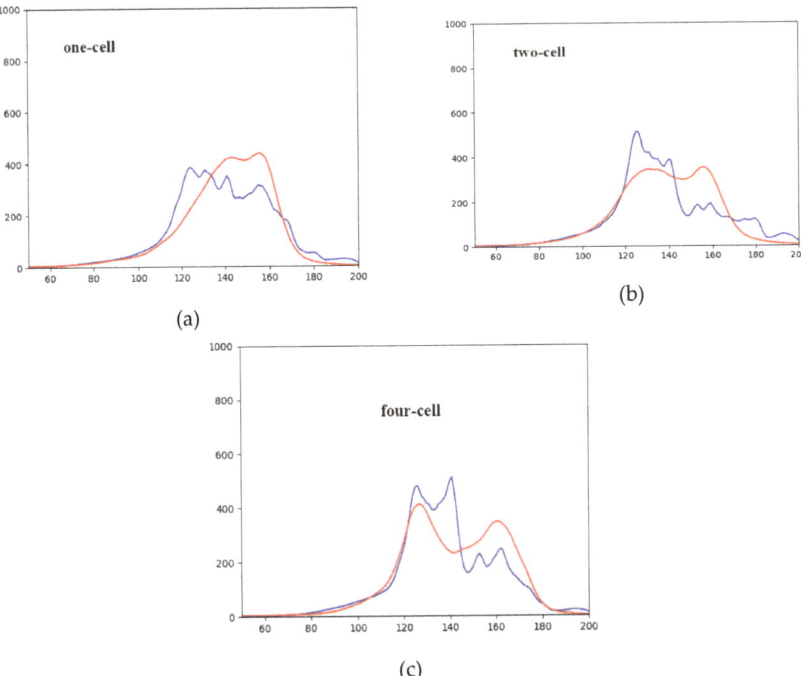

Figure 8. Comparison of histograms of real vs. generated images (real images—blue, generated images—red): (**a**) one-cell images, (**b**) two-cell images, (**c**) four-cell images.

Table 3 shows the results of a comparison of normalized histograms using correlation, Chi-square, intersection, and Bhattacharyya distance formulas. In all four cases, the largest histogram coincidence was shown by the one-cell generated images, while the two-cell and four-cell images performed worse.

Table 3. Evaluation of generated artificial embryo cell images using histogram comparison criteria.

	One-cell	Two-cells	Four-cells
Correlation	0.995	0.990	0.986
Chi-square	0.236	0.455	0.442
Intersection	1.883	1.849	1.873
Bhattacharyya	0.147	0.210	0.208

We also compared the values of the textural (Haralick) features for the original and generated embryo images and evaluated the statistical difference between the values of features using the Student's t-test. The results were significant ($p < 0.01$) for all Haralick features except the sum variance feature (one-cell images), variance and sum average (two-cell images), and the sum variance feature (four-cell images).

To compare the Haralick features from both (original and generated) image datasets, we also used the Pearson correlation between the principal components of feature sets extracted using principal component analysis (PCA). For both feature sets, the first principal component (PC1) explains more than 98% of variance, so we use only PC1 for further comparison. The results are presented in Table 4 and show a very high correlation between the PC vectors.

Table 4. Results of principal component analysis (PCA) of Haralick features from original and synthetic embryo images.

	One-cell	Two-cells	Four-cells
Explained variance of PC1 (original)	99.90%	99.94%	99.86%
Explained variance of PC1 (synthetic)	98.95%	99.60%	99.73%
Correlation between values of PC1 (original) and PC1 (synthetic)	1.0000	0.9997	0.9999

4. Discussion

As the correct comparison with other algorithms not possible due to very different source datasets used, we tried to compare within the misclassification rate, by comparing percentage of the images from all generated groups were assigned to correct class or not at all. Our HEMIGEN method was rated at 12.3%.

Table 5 provides a comparison of the misclassification rate of synthetic images when compared to the results obtained by other authors. Gaussian mixture deep generative network (DGN) demonstrated 36.02% misclassification rate [51]. DGN with auxiliary variables and two stochastic layers and skip connections achieved the 16.61% misclassification rate [52]. Semi-supervised classification and image generation with four-layer generator demonstrated 8.11% misclassification rate on house number image generation [48]. The adversarial learned inference (ALI) model, which jointly learns a generation network and an inference network using an adversarial process, reached a misclassification rate of 7.42% using CIFAR10 test set (tiny images) [53]. The WGAN-based approaches ranged from 6.9% [21] to 50% [40] depending on the application. The methods indicated are only "loosely" comparable, taking into account the differences (importance) of features of synthetic image targeted and scopes of the works by other researchers.

Table 5. Misclassification rate.

Model	Images Analyzed	Overall Misclassification Rate
DGN [51]	Digits 0 to 9 and combination (based on SVHN, and NORB sets) (*realistic images*)	36.02%
Skip Deep Generative Model [52]	Digits 0 to 9 and combination (based on SVHN, and NORB sets) (*realistic images*)	16.61%
GAN (feature matching) [48]	LSVRC2012 dataset with 1,000 categories (*non-realistic images*)	8.11%
ALI [53]	25 types of shapes (*non-realistic images*)	7.42%
WGAN [40]	6 classes of brain MR images (*realistic images*)	50%
WGAN-GP [21]	8 classes of florescent microscopy images (*realistic images*)	6.9%
HEMIGEN (our approach)	Once cell, Two cells, Four cells (*realistic images*)	12.3%

We also checked for the problem of mode collapse where the generator would produce the same image over and over, thus fooling discriminator, while no new image would be generated in real-life. We have not found this case in our approach, possibly due to an adequate number of different embryos used. Some generated images had embryos, which overlap one another, but this was not considered a failure, but a realistic case of cell division.

5. Conclusions

We used generative adversarial networks trained on real human embryo cell images to generate a dataset of synthetic one-, two-, and four-cell embryo images. We have achieved the highest quality of generated images for single-cell embryo images, where 96.2% of the synthetic embryo images were recognized as accurate and usable by human experts. The worst accuracy was achieved for the synthetic four-cell images, of which only 80% could be identified correctly. These results were confirmed by the histogram comparison, which achieved the highest scores for synthetic single-cell images (an average correlation of 0.995 was achieved when comparing histograms of real and synthetic one-cell embryo images), as well as by comparison of image textures analyzed using the Haralick features.

Sensors **2019**, *19*, 3578

As our algorithm allows us to manipulate the size, position, and number of the artificially generated embryo cell images, these images can then be used to train and validate other embryo image processing algorithms, when the real embryo images are not available, or the number of available real embryo images is too small for training neural networks.

Author Contributions: Conceptualization, R.M. and V.R.; data curation, D.D.; funding acquisition, R.S.; investigation, D.D, R.M., and V.R.; methodology, R.M.; software, D.D; supervision, R.M.; validation, D.D and R.M.; visualization, D.D; writing—original draft, D.D, R.M., and R.D.; writing—review and editing, R.M. and R.S.

Funding: The project partially financed under the program of the Polish Minister of Science and Higher Education under the name "Regional Initiative of Excellence" in the years 2019–2022 project number 020/RID/2018/19, the amount of financing 12,000,000.00 PLN.

Acknowledgments: The authors also would like to thank Esco Global for kindly provided embryo image dataset.

Conflicts of Interest: The authors declare no conflict of interest.

References

1. Russakovsky, O.; Deng, J.; Su, H.; Krause, J.; Satheesh, S.; Ma, S.; Huang, Z.; Karpathy, A.; Khosla, A.; Bernstein, M.; et al. ImageNet Large Scale Visual Recognition Challenge. *Int. J. Comput. Vis.* **2015**, *115*, 211. [CrossRef]

2. Krizhevsky, A.; Sutskever, I.; Hinton, G.E. ImageNet Classification with Deep Convolutional Neural Networks. *Commun. ACM* **2017**, *60*, 84–90. [CrossRef]

3. Donahue, J.; Jia, Y.; Vinyals, O.; Hoffman, J.; Zhang, N.; Tzeng, E.; Darrell, T. DeCAF: A Deep Convolutional Activation Feature for Generic Visual Recognition. In Proceedings of the 31st International Conference on Machine Learning, Beijing, China, 21–26 June 2014; Volume 32, pp. 647–655.

4. Simonyan, K.; Zisserman, A. Very Deep Convolutional Networks for Large-Scale Image Recognition. In Proceedings of the International Conference on Learning Representations ICLR, San Diego, CA, USA, 7–9 May 2015; pp. 1–14.

5. He, K.; Zhang, X.; Ren, S.; Sun, J. Deep Residual Learning for Image Recognition. In Proceedings of the 2016 IEEE Conference on Computer Vision and Pattern Recognition (CVPR), Jersey, NJ, USA, 26 June–1 July 2016; IEEE: Piscataway, NJ, USA. [CrossRef]

6. Long, J.; Shelhamer, E.; Darrell, T. Fully convolutional networks for semantic segmentation. In Proceedings of the 2015 IEEE Conference on Computer Vision and Pattern Recognition (CVPR), Boston, MA, USA, 7–12 June 2015; pp. 3431–3440. [CrossRef]

7. Girshick, R.B.; Donahue, J.; Darrell, T.; Malik, J. Rich feature hierarchies for accurate object detection and semantic segmentation. In Proceedings of the 2014 IEEE Conference on Computer Vision and Pattern Recognition (CVPR '14), Washington, DC, USA, 23–28 June 2014; pp. 580–587. [CrossRef]

8. Esteva, A.; Kuprel, B.; Novoa, R.A.; Ko, J.; Swetter, S.M.; Blau, H.M.; Thrun, S. Dermatologist-level classification of skin cancer with deep neural networks. *Nature* **2017**, *542*, 115–118. [CrossRef] [PubMed]

9. Fakoor, R.; Ladhak, F.; Nazi, A.; Huber, M. Using deep learning to enhance cancer diagnosis and classification. In *ICML Workshop on the Role of Machine Learning in Transforming Healthcare Proceedings of the International Conference on Machine Learning, Atlanta, AG, USA, 16–21 June 2013*; ACM: New York, NY, USA, 2013; pp. 1–7.

10. Ciresan, D.C.; Giusti, A.; Gambardella, L.M.; Schmidhuber, J. Mitosis Detection in Breast Cancer Histology Images with Deep Neural Networks. In Proceedings of the Medical Image Computing and Computer-Assisted Intervention, MICCAI 2013, Nagoya, Japan, 22–26 September 2013; pp. 411–418.

11. Ronneberger, O.; Fischer, P.; Brox, T. U-Net: Convolutional Networks for Biomedical Image Segmentation. In Proceedings of the Medical Image Computing and Computer-Assisted Intervention, MICCAI 2015, Munich, Germany, 5–9 October 2015; Lecture Notes in Computer Science. Springer: Cham, Switzerland, 2015; Volume 9351, pp. 234–241.

12. Dong, B.; Shao, L.; Da Costa, M.; Bandmann, O.; Frangi, A.F. Deep learning for automatic cell detection in wide-field microscopy zebrafish images. In Proceedings of the 2015 IEEE 12th International Symposium on Biomedical Imaging (ISBI), Brooklyn, NY, USA, 16–19 April 2015; pp. 772–776.

13. Kheradmand, S.; Singh, A.; Saeedi, P.; Au, J.; Havelock, J. Inner cell mass segmentation in human HMC embryo images using fully convolutional network. In Proceedings of the 2017 IEEE International Conference on Image Processing (ICIP), Beijing, China, 17–20 September 2017; pp. 1752–1756.

14. Manna, C.; Nanni, L.; Lumini, A.; Pappalardo, S. Artificial intelligence techniques for embryo and oocyte classification. *Reprod. Biomed. Online* **2013**, *26*, 42–49. [CrossRef] [PubMed]

15. Wang, Y.; Moussavi, F.; Lorenzen, P. Automated Embryo Stage Classification in Time-Lapse Microscopy Video of Early Human Embryo Development. In Proceedings of the International Conference on Medical Image Computing and Computer-Assisted Intervention, MICCAI (2013), Nagoya, Japan, 22–26 September 2013; pp. 460–467. [CrossRef]

16. Khan, A.; Gould, S.; Salzmann, M. Automated monitoring of human embryonic cells up to the 5-cell stage in time-lapse microscopy images. In Proceedings of the 2015 IEEE 12th International Symposium on Biomedical Imaging (ISBI), New York, NY, USA, 16–19 April 2015; pp. 389–393.

17. Khan, A.; Gould, S.; Salzmann, M. A Linear Chain Markov Model for Detection and Localization of Cells in Early Stage Embryo Development. In Proceedings of the 2015 IEEE Winter Conference on Applications of Computer Vision, Waikoloa, HI, USA, 6–9 January 2015; pp. 526–533.

18. Nandy, K.; Kim, J.; McCullough, D.P.; McAuliffe, M.; Meaburn, K.J.; Yamaguchi, T.P.; Gudla, P.R.; Lockett, S.J. Segmentation and Quantitative Analysis of Individual Cells in Developmental Tissues. In *Mouse Molecular Embryology; Methods in Molecular Biology (Methods and Protocols)*; Lewandoski, M., Ed.; Humana Press: Boston, MA, USA, 2014; Volume 1092, pp. 235–253.

19. Baggett, D.; Nakaya, M.-A.; McAuliffe, M.; Yamaguchi, T.P.; Lockett, S. Whole cell segmentation in solid tissue sections. *Cytometry* **2005**, *67A*, 137–143. [CrossRef] [PubMed]

20. Dirvanauskas, D.; Maskeliunas, R.; Raudonis, V.; Damasevicius, R. Embryo development stage prediction algorithm for automated time lapse incubators. *Comput. Methods Programs Biomed.* **2019**, *177*, 161–174. [CrossRef] [PubMed]

21. Osokin, A.; Chessel, A.; Salas, R.E.C.; Vaggi, F. GANs for Biological Image Synthesis. In Proceedings of the 2017 IEEE International Conference on Computer Vision (ICCV), Venice, Italy, 22–29 October 2017; IEEE: Piscataway, NJ, USA. [CrossRef]

22. Karras, T.; Aila, T.; Laine, S.; Lehtinen, J. Progressive growing of GANs for improved quality, stability, and variation. *arXiv* **2017**, arXiv:1710.10196.

23. Han, C.; Rundo, L.; Araki, R.; Furukawa, Y.; Mauri, G.; Nakayama, H.; Hayashi, H. Infinite Brain MR Images: PGGAN-based Data Augmentation for Tumor Detection. *arXiv* **2019**, arXiv:1903.12564.

24. Khan, S.H.; Hayat, M.; Barnes, N. Adversarial Training of Variational Auto-encoders for High Fidelity Image Generation. In Proceedings of the 2018 IEEE Winter Conference on Applications of Computer Vision (WACV), Waikoloa Village, HI, USA, 12–15 March 2018; IEEE: Piscataway, NJ, USA, 2018. [CrossRef]

25. Yi, X.; Walia, E.; Babyn, P. Generative adversarial network in medical imaging: A review. *arXiv* **2018**, arXiv:1809.07294.

26. Bengio, Y.; Thibodeau-Laufer, É.; Alain, G.; Yosinski, J. Deep Generative Stochastic Networks Trainable by Backprop. In Proceedings of the 31st International Conference on International Conference on Machine Learning, Beijing, China, 21–26 June 2014; Volume 32, pp. II-226–II-234.

27. Gregor, K.; Danihelka, I.; Graves, A.; Jimenez Rezende, D.; Wierstra, D. DRAW: A Recurrent Neural Network for Image Generation. In Proceedings of the 32nd International Conference on Machine Learning, Lille, France, 6–11 July 2015; Volume 37, pp. 1462–1471.

28. Dosovitskiy, A.; Springenberg, J.T.; Brox, T. Learning to generate chairs with convolutional neural networks. In Proceedings of the 2015 IEEE Conference on Computer Vision and Pattern Recognition (CVPR), Boston, MA, USA, 7–12 June 2015; pp. 1538–1546.

29. Denton, E.; Chintala, S.; Szlam, A.; Fergus, R. Deep Generative Image Models using a Laplacian Pyramid of Adversarial Networks. In Proceedings of the 28th International Conference on Neural Information Processing Systems-Volume 1 (NIPS'15), Montreal, QC, Canada, 7–12 December 2015; MIT Press: Cambridge, MA, USA, 2015; Volume 1, pp. 1486–1494.

30. Radford, A.; Metz, L.; Chintala, S. Unsupervised Representation Learning with Deep Convolutional Generative Adversarial Networks. *arXiv* **2015**, arXiv:1511.06434.

31. Reed, S.; Akata, Z.; Yan, X.; Logeswaran, L.; Schiele, B.; Lee, H. Generative Adversarial Text to Image Synthesis. In Proceedings of the 33rd International Conference on Machine Learning, New York, NY, USA, 20–22 June 2016; Volume 48, pp. 1060–1069.

32. Zhang, H.; Xu, T.; Li, H.; Zhang, S.; Wang, X.; Huang, X.; Metaxas, D. StackGAN: Text to Photo-realistic Image Synthesis with Stacked Generative Adversarial Networks. In Proceedings of the 2017 IEEE International Conference on Computer Vision (ICCV), Venice, Italy, 22–29 October 2017; pp. 5908–5916.

33. Tulyakov, S.; Liu, M.-Y.; Yang, X.; Kautz, J. MoCoGAN: Decomposing Motion and Content for Video Generation. In Proceedings of the IEEE Conference on Computer Vision and Pattern Recognition (CVPR), Salt Lake, UT, USA, 18–22 June 2018; pp. 1526–1535.

34. Vondrick, C.; Pirsiavash, H.; Torralba, A. Generating Videos with Scene Dynamics. In Proceedings of the 30th International Conference on Neural Information Processing Systems (NIPS'16), Long Beach, CA, USA, 4–9 December 2017; pp. 613–621.

35. Saito, M.; Matsumoto, E.; Saito, S. Temporal Generative Adversarial Nets with Singular Value Clipping. In Proceedings of the 2017 IEEE International Conference on Computer Vision (ICCV), Venice, Italy, 22–29 October 2017; pp. 2849–2858. [CrossRef]

36. Wu, J.; Zhang, C.; Xue, T.; Freeman, W.T.; Tenenbaum, J.B. Learning a Probabilistic Latent Space of Object Shapes via 3D Generative-Adversarial Modeling. In Proceedings of the 30th International Conference on Neural Information Processing Systems (NIPS'16), Boston, MA, USA, 15–19 September 2016; pp. 82–90.

37. Li, C.; Guo, Y.; Liu, Q.; Liu, X. DR-Net: A Novel Generative Adversarial Network for Single Image Deraining. *Secur. Commun. Netw.* **2018**, *2018*, 7350324. [CrossRef]

38. Zhu, D.; Dai, L.; Luo, Y.; Zhang, G.; Shao, X.; Itti, L.; Lu, J. Multi-Scale Adversarial Feature Learning for Saliency Detection. *Symmetry* **2018**, *10*, 457. [CrossRef]

39. Ma, Y.; Liu, K.; Guan, Z.; Xu, X.; Qian, X.; Bao, H. Background Augmentation Generative Adversarial Networks (BAGANs): Effective Data Generation Based on GAN-Augmented 3D Synthesizing. *Symmetry* **2018**, *10*, 734. [CrossRef]

40. Han, C.; Hayashi, H.; Rundo, L.; Araki, R.; Shimoda, W.; Muramatsu, S.; Nakayama, H. GAN-based synthetic brain MR image generation. In Proceedings of the 2018 IEEE 15th International Symposium on Biomedical Imaging (ISBI 2018), Washington, DC, USA, 4–7 April 2018; IEEE: Piscataway, NJ, USA, 2018. [CrossRef]

41. Goodfellow, I.J.; Pouget-Abadie, J.; Mirza, M.; Xu, B.; Warde-Farley, D.; Ozair, S.; Courville, A.; Bengio, Y. Generative adversarial nets. In Proceedings of the 27th International Conference on Neural Information Processing Systems-Volume 2 (NIPS'14), Montreal, QC, Canada, 8–13 December 2014; MIT Press: Cambridge, MA, USA, 2014; pp. 2672–2680.

42. Maas, A.L.; Hannun, A.Y.; Ng, A.Y. Rectifier nonlinearities improve neural network acoustic model. In Proceedings of the International Conference on Machine ICML 2013, Atlanta, GA, USA, 16–21 June 2013; Volume 30, p. 3.

43. Ioffe, S.; Szegedy, C. Batch Normalization: Accelerating Deep Network Training by Reducing Internal Covariate Shift. In Proceedings of the 32nd International Conference on International Conference on Machine Learning-Volume 37 (ICML'15), Lille, France, 7–9 July 2015; Volume 37, pp. 448–456.

44. Kingma, D.P.; Ba, J. Adam: A Method for Stochastic Optimization. In Proceedings of the 3rd International Conference on Learning Representations, Indianapolis, IN, USA, 24–28 March 2014.

45. Theis, L.; van den Oord, A.; Bethge, M. A note on the evaluation of generative models. In Proceedings of the International Conference on Learning Representations (ICLR), San Juan, Puerto Rico, 19 November 2015.

46. Im, D.J.; Kim, C.D.; Jiang, H.; Memisevic, R. Generating Images with Recurrent Adversarial Networks. *arXiv* **2016**, arXiv:1602.05110.

47. Geman, D.; Geman, S.; Hallonquist, N.; Younes, L. Visual Turing test for computer vision systems. In Proceedings of the National Academy of Sciences, Washington, DC, USA, 22–29 December 2015; p. 201422953. [CrossRef]

48. Salimans, T.; Goodfellow, I.; Zaremba, W.; Cheung, V.; Radford, A.; Chen, X. Improved techniques for training GANs. In Proceedings of the 30th International Conference on Neural Information Processing Systems (NIPS'16), Barcelona, Spain, 5–10 December 2016; Curran Associates Inc.: Red Hook, NY, USA, 2016; pp. 2234–2242.

49. Tazehkandi, A.A. *Computer Vision with OpenCV 3 and Qt5*; Packt Publishing: Birmingham, UK, 2018.

Sensors **2019**, *19*, 3578

50. Haralick, R.M.; Shanmugam, K.; Dinstein, I. Textural Features for Image Classification. *IEEE Trans. Syst. Man Cybern.* **1973**, *SMC-3*, 610–621. [CrossRef]
51. Kingma, D.P.; Rezende, D.J.; Mohamed, S.; Welling, M. Semi-supervised learning with deep generative models. In Proceedings of the 27th International Conference on Neural Information Processing Systems-Volume 2 (NIPS'14), Montreal, QC, Canada, 8–13 December 2014; MIT Press: Cambridge, MA, USA, 2014; pp. 3581–3589.
52. Maaløe, L.; Sønderby, C.K.; Sønderby, S.K.; Winther, O. Auxiliary deep generative models. In Proceedings of the 33rd International Conference on Machine Learning, New York, NY, USA, 19–24 June 2016; Volume 48, pp. 1445–1453.
53. Dumoulin, V.; Belghazi, I.; Poole, B.; Lamb, M.A.; Mastropietro, O.; Courville, A. Adversarially Learned Inference. In Proceedings of the International Conference on Learning Representations (ICLR), Toulon, France, 24–26 April 2017.

Article

Image Thresholding Improves 3-Dimensional Convolutional Neural Network Diagnosis of Different Acute Brain Hemorrhages on Computed Tomography Scans

Justin Ker [1], Satya P. Singh [2], Yeqi Bai [2], Jai Rao [1], Tchoyoson Lim [3] and Lipo Wang [2,*]

[1] Neurosurgery, National Neuroscience Institute, Singapore 308433, Singapore; justinker@gmail.com (J.K.); jai.rao@singhealth.com.sg (J.R.)
[2] School of Electrical and Electronic Engineering, Nanyang Technological University, Singapore 639798, Singapore; satya@ntu.edu.sg (S.P.S.); BA0001QI@e.ntu.edu.sg (Y.B.)
[3] Neuroradiology, National Neuroscience Institute, Singapore 308433, Singapore; tchoyoson.lim@singhealth.com.sg
* Correspondence: ELPWang@ntu.edu.sg

Received: 1 April 2019; Accepted: 4 May 2019; Published: 10 May 2019

Abstract: Intracranial hemorrhage is a medical emergency that requires urgent diagnosis and immediate treatment to improve patient outcome. Machine learning algorithms can be used to perform medical image classification and assist clinicians in diagnosing radiological scans. In this paper, we apply 3-dimensional convolutional neural networks (3D CNN) to classify computed tomography (CT) brain scans into normal scans (N) and abnormal scans containing subarachnoid hemorrhage (SAH), intraparenchymal hemorrhage (IPH), acute subdural hemorrhage (ASDH) and brain polytrauma hemorrhage (BPH). The dataset used consists of 399 volumetric CT brain images representing approximately 12,000 images from the National Neuroscience Institute, Singapore. We used a 3D CNN to perform both 2-class (normal versus a specific abnormal class) and 4-class classification (between normal, SAH, IPH, ASDH). We apply image thresholding at the image pre-processing step, that improves 3D CNN classification accuracy and performance by accentuating the pixel intensities that contribute most to feature discrimination. For 2-class classification, the F1 scores for various pairs of medical diagnoses ranged from 0.706 to 0.902 without thresholding. With thresholding implemented, the F1 scores improved and ranged from 0.919 to 0.952. Our results are comparable to, and in some cases, exceed the results published in other work applying 3D CNN to CT or magnetic resonance imaging (MRI) brain scan classification. This work represents a direct application of a 3D CNN to a real hospital scenario involving a medically emergent CT brain diagnosis.

Keywords: 3D convolutional neural networks; machine learning; CT brain; brain hemorrhage

---

1. Introduction

Intracranial hemorrhage is a medical emergency that can have high morbidity and mortality if not diagnosed and treated immediately. This condition affects 40,000 to 67,000 patients in the United States annually and up to 52% of patients die within one month [1]. Three commonly-encountered sub-types of intracranial hemorrhage are subarachnoid hemorrhage (SAH), intraparenchymal hemorrhage (IPH), acute subdural hemorrhage (ASDH). In a severe brain trauma, various permutations of SAH, IPH, and ASDH can be seen, which we have termed brain polytrauma hemorrhage (BPH) in this work. The common causes of SAH are trauma and cerebral aneurysmal rupture, while IPH can be caused by hypertension, amyloid angiopathy, brain tumor hemorrhage or trauma. ASDH and BPH appear because of head trauma.

When patients with intracranial hemorrhage present to the emergency department, a computed tomography (CT) scan of the brain is done to diagnosis intracranial hemorrhage, so that medical or surgical treatment can follow. CT scans work by exploiting the differential absorptive properties of body tissues placed between an x-ray emitter and detector. Brain tissue, blood, muscle, and bone give rise to different levels of x-ray attenuation, expressed in Hounsfield units. By moving the x-ray emitter and detector circumferentially around a subject, a three-dimensional image of the subject's internal tissues is obtained.

Limitations in the availability or experience of clinicians, especially in rural or resource-strapped health systems, to diagnose CT brains quickly can cause treatment delays. Automating the diagnosis of CT brain scans, or assisting the clinician in triaging critical from normal scans, would help patients by expediting their treatments and improve outcomes. Figure 1 shows the axial slices of five CT brain scans showing a normal brain, SAH, IPH, ASDH, and BPH. The outer rim of uniform white skull bone surrounds the dark grey brain tissue, and areas of acute hemorrhage, appear as patchy, light-grey areas of varying shapes.

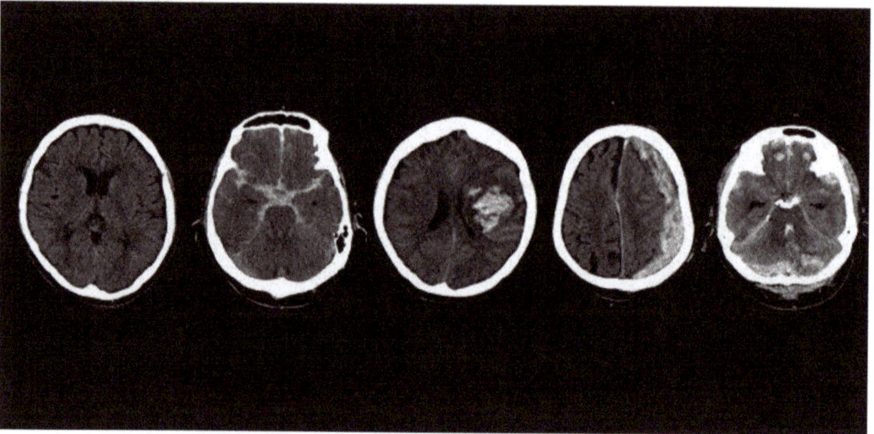

Figure 1. Computed tomography (CT) brain scans. From left, normal (N), subarachnoid hemorrhage (SAH), intraparenchymal hemorrhage (IPH), acute subdural hemorrhage (ASDH), brain polytrauma hemorrhage (BPH). Each image represents an individual image slice. One patient's complete stack of CT images contained between 24 to 34 image slices in our dataset.

The artificial neuron was first described by McCulloch and Pitts in 1943 [2]. This has evolved through the symbolic, rule-based artificial intelligence (AI) paradigms, to manual feature-handcrafting algorithms, and to modern multi-layered or "deep" neural networks, which perform feature detection and classification automatically. Convolutional neural networks (CNN) owe their inception to Fukushima's Neocognitron model in 1982 [3], and their popularity to Lecun et al. [4] and Krizhevsky et al. [5] The latter employed a CNN to win the 2012 Imagenet Large Scale Visual Recognition Challenge, and since then CNNs have been used for many image classification tasks. The advantage of CNNs in image classification is the ability to perform feature-extraction and learn high-level image features automatically without feature-handcrafting, leading CNNs to become the dominant machine learning architecture in image recognition tasks. CNNs have been widely used in machine vision to perform a variety of tasks, such as image classification, object detection, and semantic segmentation. In the medical image analysis space, CNNs have been used for the classification and diagnosis of 2-dimensional medical images, such as chest x-rays, retinal photographs, skin dermoscopic images, and histology images, with performance comparable to or exceeding human clinicians [6–9].

In analyzing volumetric magnetic resonance imaging (MRI) or CT data, various machine learning approaches including 2D CNN have been attempted. These efforts have involved manual slice selection, extensive manual pre-processing, feature hand-crafting, and segmentation, before classification [10–12]. A number of these approaches classify individual images from a volumetric image stack one image at a time, turning a 3D classification problem into a 2D task. Other authors have used employed a combination of 2D CNN in the three planes (axial, coronal, sagittal) that define a 3D volume [13–15]. Roth et al. [14] detected abnormal lymph nodes on thoracic CT scans, by decomposing a 3D volume of interest into re-sampled 2D views, and then training their CNN on augmented variations of these 2D views. Their CNN achieved a satisfactory sensitivity of 90%, at six false positives per patient. The disadvantages of these previous strategies include the manual time and effort in hand-crafting, feature segmentation, and stripping, and the potential loss of spatial contextual information when a 3D volume is analyzed using 2D slices.

Three-dimensional CNNs are an emerging architecture, and have been used mainly in analyzing video or 3-dimensional volumetric medical images. Previously, the use of 3D CNN was limited as it was computationally expensive and lengthy to process 3-dimensional kernels and entire volumes of images. However, more papers on 3D CNN have appeared in the scientific literature with their adoption likely aided by decreasing computational hardware costs. In medical image analysis, 3D CNNs have been applied to detecting abnormalities (tumors, hemorrhage, ischemia) in brain, heart, lung, and liver organs on CT or MRI imaging [16–18]. As a measure of the interest in the clinical problem of automatically detecting intracerebral hemorrhage on MRI or CT, there is also a growing number of publications on this topic using 3D CNNs [19,20]. Dou et al. [20] analyzed cerebral microbleeds on MRI brain scans using a two-stage 3D CNN that screened and then detected microbleeds. Their method achieved 93% sensitivity, at a cost of 44% precision and 2.7 false positives per patient.

We propose a 3D CNN that classifies CT brain scans into normal (N), subarachnoid hemorrhage (SAH), intra-parenchymal hemorrhage (IPH), acute subdural hemorrhage (ASDH), and brain polytrauma hemorrhage (BPH).

In this work, we aim to create a 3D CNN that can automatically detect and diagnose SAH, ASDH and IPH on CT brain scans, and distinguish them from normal scans. This work would have direct clinical application in emergency departments and acute-care hospitals worldwide. We propose a simple and fast 3D CNN that is effective and accurate. It is hoped that the straightforward implementation of this 3D CNN will lead to its widespread adoption. Specifically, we modify well-known 2D CNN architectures into 3D CNN. The novel aspects of our work are as follow:

(1) To our knowledge, this is the first demonstration of a 3D CNN on volumetric CT brain data that classifies patient scans into different acute hemorrhagic variants, which impacts subsequent medical treatment. Previous work has been limited to detecting normal versus abnormal scans. We also demonstrate that our network performance gives state-of-the-art results in classification accuracy.

(2) We present a novel image thresholding method optimized for the detection of acute hemorrhage on CT brain scans, which mimics the visual analysis of radiological images by human radiologists. We demonstrate that the application of our method improves the classifier performance.

This paper is organized into the following sections. Section 2 describes our methods, with details on the dataset, network architectures, and training protocols. Section 3 reports our results and experimentation with various network parameters. Section 4 analyzes the impact, limitations and future work stemming from our results. Section 5 summarizes and concludes this paper.

2. Methods

2.1. 3-Dimensional Convolutional Neural Networks

The feature map of a convolution layer is formed by convolving the feature map of the previous layer with learnable kernels, adding a bias term, and then applying an activation function [21]. Specifically, we can define h_k^p as the kth feature map of the pth layer, and h_j^{p-1} as the jth feature map of the previous layer, $p-1$. $W_{j,k}^p$ is a learnable kernel, and b^p is the bias term. σ is the activation function, commonly a rectified linear activation unit (ReLu). This is written as:

$$h_k^p = \sigma\left(\sum_j h_j^{p-1} \times W_{j,k}^p + b_k^p\right) \tag{1}$$

A 2-dimensional convolution can be defined for convolving an input image I with a kernel K. Extending the equation for a 2-dimensional convolution [22] into a 3-dimensional convolution, we obtain:

$$S(\ell,m,n) = (K \times I)(l,m,n) = \sum_a \sum_b \sum_c I(\ell-a, m-b, n-c) \, K(a,b,c) \tag{2}$$

Equation (2) may be expressed as:

$$S(\ell,m,n) = \sum_{a,b,c} h_j^{p-1}(\ell-a, m-b, n-c) W_{j,k}^p(a,b,c) \tag{3}$$

In Equation (3), h_j^{p-1} is the jth 3-dimensional feature map of the previous layer $p-1$, and $W_{j,k}^p$ is the 3-dimensional kernel. Substituting Equations (1) and (3), we obtain the equation for h_k^p, which is a 3-dimensional feature map:

$$h_k^p = \sigma\left(S_{j,k}^p + b_k^p\right) \tag{4}$$

These 3-dimensional convolution layers were stacked with max-pooling and fully-connected layers. Kernels were initialized with Gaussian distribution and parameters were tuned with standard stochastic gradient descent and cross-entropy loss minimization.

2.2. Image Thresholding to Detect Acute Hemorrhage

The CNN extracts features from the input images, and most of the features are associated with edges, shape, and curves present in the input images. It can also be seen by visualizing the different layers of CNN that the first layer mostly picks up edges present in the images, while the second layer picks up curves, and the third layer picks up shapes. The human eye is similar in that it also observes an image for its constituent edges, curves, and shapes, as suggested by the presence of ocular dominance and orientation columns in the primary visual cortex of the mammalian brain. In detecting anomalies on medical images, a region of interest (ROI) may be subtle or not apparent to visual inspection by either human or CNN. Just as thresholding aids a human radiologist to emphasize possibly abnormal areas, we propose a threshold operator to accentuate individual sharp edges, such that this improves the likelihood of a CNN detecting a feature, and, therefore, performing a correct classification.

To improve the discriminatory ability of our model, we propose a novel thresholding method to detect acute hemorrhage. We build on the work of Zhang et al. [23] who used spatial histograms to detect cars in images. There is a range of pixels intensities which is common in both normal and abnormal CT scans. Therefore, we can discard these intensities before feeding the image to the CNN without any loss of information. This is similar to the dimensionality reduction process. In our proposed method, we generate the average pixel intensity histograms of 2 classes of CT scans (such as Normal and SAH), to overlap and search for an optimal pixel intensity threshold that accentuates their difference. Pixel intensities below this threshold are then disregarded. This process creates sharp edges and shapes around a ROI which helps CNN in feature extraction. An added benefit is that a CNN will require less time for training.

The intuition underlying this approach is two-fold. First, we observed that normal CT brain scans, even across different patients are largely homogeneous, which makes a histogram representation of the class meaningful. Abnormal scans can be thought of as the addition of extraneous blood signal to normal scans. Second, we also observed that radiologists adjust image contrast levels when reading CT scans, to accentuate subtle amounts of blood, and to downplay the appearance of normal surrounding tissue. Our method attempts to model this behavior in human visual cognition. Using our method improves the overall performance of the classifier across all classes of acute brain hemorrhage.

2.3. Dataset and Pre-Processing

The dataset consists of non-contrast CT brain images from the National Neuroscience Institute at Tan Tock Seng Hospital, Singapore. After institutional review board approval, a search of electronic discharge summaries with the diagnoses matching head injury or intracranial hemorrhage was performed. The resultant list of patient identifiers was used to query and retrieve relevant scans from the hospitals' Picture archiving and communication system (PACS) servers. Each scan was anonymized and manually checked by the authors to ensure ground truth. The final dataset consisted of 399 unique patient volumetric CT brain scans representing approximately 12,000 images, summarized by the different classes in Table 1. These scans had varying numbers of image slices and slice thickness, owing to variability in CT scanner model and scanning protocols. We prepare the data for 5 five-fold cross-validation. Training, validation, and test CT scan images were augmented eight-fold by flipping along the vertical axis, and rotation up to 45 degrees.

Table 1. Number of original unique patient computed tomography (CT) scans.

Normal	Subarachnoid Hemorrhage (SAH)	Intraparenchymal Hemorrhage (IPH)	Subdural Hemorrhage (ASDH)	Brain Polytrauma Hemorrhage (BPH)
130	141	61	32	35

2.4. Network Architecture

Table 2 shows the model architecture used in our 3D CNN. We experimented with various model architectures including VGGNet and GoogLeNet, to optimize for the trade-off between classification accuracy and computational time. We aimed to have a model with a straightforward design for easy trouble-shooting and to facilitate real-world implementation. Like Dou et al. [20], we were concerned with the impact of processing large files of volumetric brain images on computation time. However, while they opted for a two-part ensembled screening and discrimination stages, we opted for a single throughput architecture for simplicity and performance. Figure 2 is a pictorial diagram of our proposed 3D CNN. After the necessary preprocessing steps, input volumetric data of 3D CT scans are passed through a pre-defined threshold operator as discussed in Section 2.2 and becomes the input to the 3D CNN.

Table 2. Model architecture of the 3-dimensional convolutional neural networks (3D CNN) used in this work.

Layer	Kernel Size	Stride	Output Size (Width × Length × Depth × Filters)
Input	-	-	$50 \times 50 \times 28$
Convolution 1	$3 \times 3 \times 3$	1	$50 \times 50 \times 28 \times 32$
Pooling 1	$2 \times 2 \times 2$	2	$25 \times 25 \times 14 \times 32$
Convolution 2	$3 \times 3 \times 3$	1	$25 \times 25 \times 14 \times 64$
Pooling 2	$2 \times 2 \times 2$	2	$13 \times 13 \times 7 \times 64$
Convolution 3	$3 \times 3 \times 3$	1	$13 \times 13 \times 7 \times 128$
Pooling 3	$2 \times 2 \times 2$	2	$7 \times 7 \times 4 \times 128$
Fully Connected 1	-	-	$25{,}088 \times 1024$
Fully Connected 2	-	-	1024×2

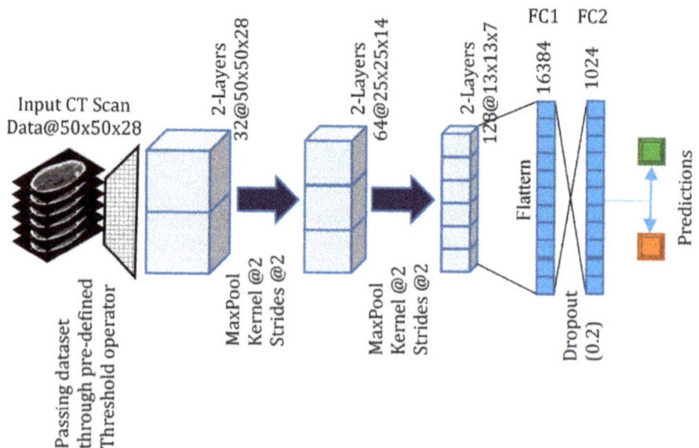

Figure 2. Proposed architecture for binary and multi-class classification of CT scans. The features are visualized using 3D deconvolution visualization methods at each pooling layer.

2.5. Training and Implementation

Training was performed on a computer with two Intel Xeon E5-2630 CPU processors, four NVIDIA GTX 1080 Ti Graphical processing units, and 128 GB of DDR4 Random access memory. The project was implemented using the Python programming language and the Google Tensorflow library. We used the rectified linear unit (ReLu) as the activation function, the Adam optimizer, and cross-entropy as the loss function. We used the grid search approach for optimizing the learning rate, dropout, and kernel size of the convolution and pooling layers. We varied learning rates from (0.1, 0.001, 0.0001), and dropout from (0, 0.1, 0.2, 0.3, 0.4, 0.5, 0.6) after each convolution layer and fully connected layer. The kernel sizes of convolution layers were varied from ($1 \times 1 \times 1$, $2 \times 2 \times 2$, $3 \times 3 \times 3$, $5 \times 5 \times 5$), and pooling kernel sizes ranged from ($1 \times 1 \times 1$, $2 \times 2 \times 2$). Eventually, we set the learning rate at 0.001 and set the dropout after the fully connected layer at 0.2. We employed a convolutional kernel size of $3 \times 3 \times 3$ at each layer, with a $2 \times 2 \times 2$ pooling kernel after each convolutional layer. We set $\beta1 = 0.001$, $\beta2 = 0.999$; $\varepsilon = 10^{-8}$.

Metrics

To evaluate our network performance, we measured the Sensitivity (S), Precision (P) and F_1 scores across each classification task. TP, FP, FN refer to true positive, false positive, and false negative, respectively. The F_1 score is defined as the harmonic average of Sensitivity and Precision and is a measure of a test's accuracy. An F_1 score of 1 indicates perfect Sensitivity and Precision, while a score of 0 indicates the opposite.

$$S = \frac{TP}{TP+FN}, \quad S = \frac{TP}{TP+FP} \tag{5}$$

$$F_1 = \frac{2}{\frac{1}{S}+\frac{1}{P}} \tag{6}$$

The confusion matrix for this 4 Class classification problem is tabulated in Table 3. Table 4 summarizes the Sensitivity, Precision, and F_1 scores, re-casting the multi-class problem into a 2-class problem (Normal versus a specific class) to calculate these metrics.

Table 3. Multi-Class Classification for Normal and Abnormal CT Scans.

		Actual			
		Normal	**SAH**	**IPH**	**ASDH**
Predicted	**Normal**	129	30	4	0
	SAH	7	100	35	3
	IPH	15	31	32	0
	ASDH	1	7	1	29

Table 4. 2-Class Classification Results (Normal versus a specific Abnormal Class).

Task	Sensitivity	Precision	F1 Score	AUC
Normal versus SAH (no thresholding)	0.947	0.818	0.878	0.900
Normal versus SAH (with thresholding)	1.000	0.864	0.927	0.950
Normal versus IPH (no thresholding)	0.819	0.881	0.849	0.958
Normal versus IPH (with thresholding)	0.944	0.919	0.932	0.989
Normal versus ASDH (no thresholding)	0.750	0.666	0.706	0.953
Normal versus ASDH (with thresholding)	0.938	0.968	**0.952**	0.999
Normal versus BPH (no thresholding)	0.925	0.881	**0.902**	0.989
Normal versus BPH (with thresholding)	0.850	1.00	0.919	0.990

Highest F1 scores are in bold.

3. Results

We performed experiments involving binary classification (Normal versus SAH, IPH, ASDH, BPH) and multi-class classification. In the latter, BPH was left out as BPH contains features of Normal, SAH, IPH, and ASDH. For the binary classification experiments, we implemented the experiments with and without the thresholding method described in Section 2.2.

For the multi-class classification experiments, the model discriminated between four classes with an overall F1 score of 0.684. Table 3 presents the confusion matrix for the multi-class classification. The actual number for each class represents the augmented test set after the original test set was augmented eight-fold. The respective F1 scores for each class are Normal: 0.819, SAH: 0.639, IPH: 0.427, ASDH: 0.829. Thresholding was not implemented for the multi-class classification as it is optimized for binary classification. The model performed well for Normal and ASDH scans, but only moderately well for SAH. Interestingly, although the training dataset for ASDH was the smallest, the model was able to discriminate this the best. This may be due to the fact that ASDH images are visually grossly asymmetrical compared to the other classes. This is due to brain compression from the significant subdural hemorrhage (see Figure 1), and may be a strongly activated discriminatory feature. Surprisingly, a significant number of IPH scans were misinterpreted as SAH, and we hypothesize that this may be due to subtle SAH traces that may appear on some of the IPH scans.

Table 4 summarizes the results from the 2-class classification experiments. Overall, our model was able to discriminate between normal and each of the classes with satisfactory results, matching or exceeding previously published results for similar work (Table 5). We demonstrate that for every class, the implementation of the thresholding technique improves all the evaluation metrics. The largest increase was seen in the ASDH class, where the F1 score increased from 0.706 to 0.952, despite a small training set. Various model architectures, input image, and filter sizes were modified to optimize for accuracy and computational time. The original image size of the CT scans was $512 \times 512 \times 28$ pixels (at 5 mm slice thickness), and the input size to our model was $50 \times 50 \times 28$ pixels. Further decreasing the input size resulted in deteriorating model performance. We posit that the thresholding technique improves the signal-to-noise ratio for each input image by downplaying the extraneous image features, thereby accentuating the presence of acute hemorrhage. We also found that the thresholding technique decreased training time for the respective models significantly. For example, in the Normal versus SAH classification task, the model training time was decreased from over 10 h to 1 h 32 min. 3D CNNs are computationally intensive to train, and any decrease in computational cost and time is advantageous in clinical deployment.

Table 5. Comparison of results involving brain hemorrhage detection in volumetric brain scans.

Task	Reference	Sensitivity Precision F1	Task	Reference
Detecting cerebral micro-bleeds on MRI brain scans	Dou et al. [20]	93.2 %	44.3 %	-
Brain hemorrhage classification on CT scan	Grewal et al. [24]	88.6 %	81.3 %	0.85
Brain hemorrhage classification on CT scan	Jnawali et al. [19]	77.0 %	87.0 %	0.83
Brain hemorrhage classification on CT scan	Our best performing method (no thresholding)	92.5 %	88.1 %	0.90
Brain hemorrhage classification on CT scan	Our best performing method (with thresholding)	93.8 %	96.8 %	0.95

Dou et al. expressed their evaluation metric as Sensitivity, Precision, and False positives per subject.

4. Discussion

Acute brain hemorrhage is a common neurosurgical emergency which can result in severe patient morbidity and mortality. It is the result of myriad causes, including trauma, hypertension, cerebral aneurysm rupture, and the treatment of each is different. Depending on the clinical condition, the patient may require close observation in a high dependency or intensive care setting, or immediate neurosurgical operation. It is imperative to minimize the time from diagnosis to treatment, to give the patient the best chance of recovery.

We propose an automated 3D CNN to classify volumetric CT brain data into various hemorrhagic variants, to assist doctors in expediting patient treatment. We trained and tested our model on 399 CT brain images from our hospital to classify CT brain scans into normal, SAH, IPH, ASDH and BPH. These classes were chosen as the neurosurgical treatment for each class is different.

We also proposed and implemented a novel pixel thresholding method to detect acute hemorrhage on CT brain scans. This method improved classifier performance on our dataset and can be conceivably exported for use in other datasets for other anatomical regions, where acute hemorrhage detection is required. Potentially, aside from detecting acute blood, this method can also be generalized for the detection of other abnormalities such as tumors. Figure 3 demonstrates an image slice of a CT brain with acute subdural hemorrhage. The hemorrhage is the white crescent on the left of the image, which is putting pressure on the grey areas of brain, and pushing it to the right of the image. To visualize the activations in the 3D CNN better, we used the deconvolution technique described by Yosinski et al. [25] to visualize this single image. The top row of the image (boxes B and C) represent convolution layer 1 and pooling layer 1, respectively. The bottom row (boxes D and E) represent the same layers with thresholding applied. The difference with thresholding is that the edges of the target hemorrhagic lesion appear more distinct and sharper, which may account for why thresholding improves classification accuracy.

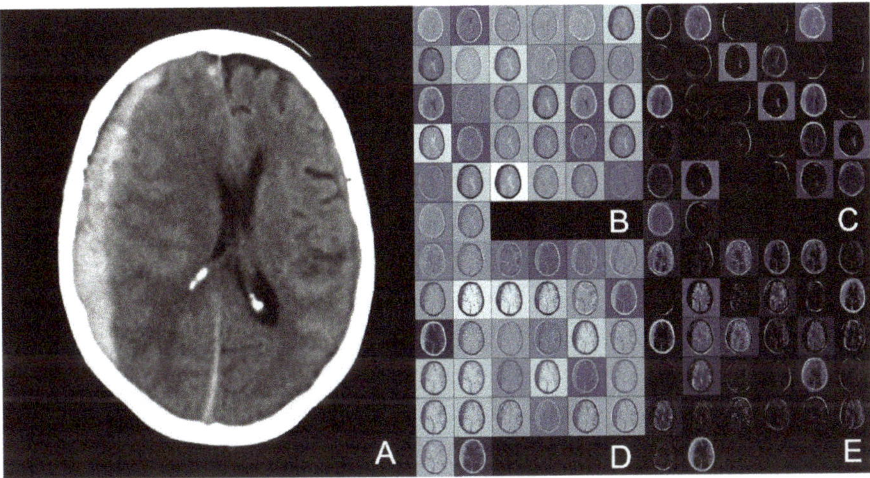

Figure 3. To demonstrate the effect of thresholding, a single slice of an ASDH CT Brain scan is shown, with the corresponding activations of convolution and pooling layers. **A**, original image. **B**, 1st convolution layer. **C**, 1st pooling layer. **D**, 1st convolution layer with thresholding applied. **E**, 1st pooling layer with thresholding applied. D and E appear sharper than B and C, demonstrating how thresholding can accentuate abnormal areas, and improve classifier performance.

In this paper, we demonstrate state-of-the-art 3D CNN classifier performance for different classes of acute brain hemorrhage. In addition, the application of our thresholding method for acute hemorrhage enables our model to achieve further additional improvement in classification. In the 4-class classification task, our model achieved an overall F1 score of 0.684. To our knowledge, there has been no published work involving multi-class classification of different classes of brain hemorrhage for comparison. In the 2-class classification task, the best performance was achieved in differentiating normal from ASDH scans, with a F1 score of 0.952 using our proposed thresholding method. The largest improvement with thresholding was also seen in the ASDH class, as the initial F1 score was only 0.706.

There has been a long history in the attempt to analyze hemorrhage in volumetric CT and MRI data. Before the widespread use of CNNs in image analysis, methods used included Hopfield networks [26], support vector machines [27], and segmentation masking with logistic classifiers [28]. Although these simple classifiers performed well, they were often applied to single image slices with obvious pathology, that were often manually chosen, which, therefore, represents an unrealistic problem scenario. Hybrid 2D CNN methods exemplified by Roth et al. [14] were a bridge to the 3D CNN training of networks. Fully 3D CNN model local and contextual spatial dependencies and extract features in all three dimensions of image voxels. Kamnitsas et al. [16] exploited the dense inference technique and small kernels, to segment lesional areas on brain MRI scans. Of note, they used a dual 11-layer 3D CNN pathway to process images at multiple scales, and Conditional Random Fields to decrease their false positive rate.

There are two points worth noting in our network architecture design. First, we deliberately kept our network architecture simple with relatively few layers, and did not leverage on other techniques, such as ensembling and transfer-learning. This was for both practical and theoretical considerations. The application of full 3D CNN to volumetric medical images is nascent, and by purposely keeping the model architecture straightforward, we are able to assess the effect size of various hyperparameter-tuning quickly, experiment with various architectures efficiently, and to troubleshoot network errors expediently. Running 3D CNN is computationally intensive, and a simpler network mitigates lengthy run-times.

From a more theoretical perspective, a straightforward architecture also allows us to grasp a sense of the baseline performance of a 3D CNN, and to describe our experimentations with a clear mathematical description. We believe that a firm theoretical framework will help in directing further areas for exploration, rather than blindly tuning hyperparameters, or implementing network boosting ensembling architectures without understanding.

The second point is that with a straightforward network architecture design, we were able to demonstrate the clear improvement that our thresholding method brought for detecting acute hemorrhage. We arrived at this thresholding method by observing radiologists as they scrolled through various patient CT scans. We noted that in situations where areas of acute brain hemorrhage were subtle, the radiologists would increase and decrease the image contrast to accentuate the appearance of the abnormality, which in this case was blood. We did the same thing while studying our dataset, and, therefore, explored how this optimization of human visual cognition or analysis could be implemented algorithmically. We took inspiration from the work of Zhang et al. [23] who used pixel intensity 'spatial histograms' for object detection within an image. Our underlying assumption was similar, that objects of a certain class would have similar patterns of pixel intensities. Where we differ is that while Zhang et al. were concerned with the spatial location of the object, we are concerned more with the actual presence of a particular pattern of pixel intensity, and the point of divergence from the pattern denoting a normal scan. This is because in CT brain scans depicting a hemorrhage, the blood even within the same class of brain hemorrhage can appear in several different areas of the brain.

The proposed architecture in this work has important clinical significance. The different abnormal diagnoses studied in this work all exert a significant epidemiological and social-economic toll on healthcare systems and societies. SAH, IPH, ASDH, and BPH are all neurosurgical emergencies that require immediate but different treatments to maximize the likelihood of patient survival and to achieve a good long-term functional outcome. Our work has a role in helping clinicians minimize the time between diagnosis and treatment, especially in hospitals that may not have a radiologist after hours, or in remote rural settings where no clinicians are available. Even in large tertiary care hospitals with 24 h-radiologists, our proposed architecture framework can assist radiologists by triaging important abnormal scans from the large numbers of normal scans that are read sequentially as patients are scanned.

One limitation of this work is the relatively small and un-balanced dataset that we worked with, which is a common issue in medical image analysis, with a bias towards normal samples. At 399 images, our dataset size is comparable to those used in many other works [20,24] but smaller than the almost 40,000 CT scans used by Jnawali et al. [19]. Despite this, data augmentation techniques have resulted in performance comparable to Jnawali et al.'s much larger dataset. This may be due to the fact that medical data is relatively homogeneous in appearance, compared to natural image processing tasks. It would be interesting to study what is the optimal dataset size for processing specific volumetric image classification tasks, in CT, MRI, 3D ultrasound images for different lesions. We addressed concerns of overfitting with known mitigating techniques, such as dropout, which were applied to the early layers in our network.

There are many avenues for further exploration in volumetric medical image analysis. CNNs and 3D CNNs have been the dominant network architecture in image analysis, but unsupervised learning methods for medical image analysis are emerging for 3D object generation [29,30], and they have been largely unexplored in the context of 3D medical image analysis. Generative architectures, such as variational autoencoders and generative adversarial networks, have not been applied to volumetric medical data, and these techniques may potentially mitigate the need for large well-labelled datasets. Specific to our work, we intend to explore if adding a screening stage to a 3D CNN, or multi-scale receptive fields can improve performance, as some authors have demonstrated [16,20] for MRI brain scans. The addition of a memory or attention-based component to model long term dependencies in 3D medical image analysis [24,31,32] is also interesting for further investigation, as there is evidence for a strong biological correlate with the mammalian visual system [33].

5. Conclusions

This work presented the implementation of a 3D CNN to classify and diagnose volumetric CT brain data. Normal CT brains and a variety of abnormal scans constituting different types of brain hemorrhage were classified by our 3D CNN. We also implement a novel optimization method to detect acute hemorrhage on CT scans. The proposed 3D CNN can automatically detect important normal and abnormal features of cerebral anatomy without handcrafting, or significant data pre-processing. Computational costs were also modest, which will add to straightforward implementation. Results from classification experiments demonstrated that the 3D CNN outperforms previously published methods in detecting abnormal brain scans with hemorrhage, with higher sensitivity and recall. Our 3D CNN can be applied to other volumetric medical data and can be used to expedite and improve patient care.

Author Contributions: Conceptualization, L.W., J.R., T.L., J.K.; software S.S. and Y.B.; method J.K. and S.S.; validation, S.S., J.K. and Y.B.; writing J.K.; supervision L.W., J.R. and T.L.

Funding: This research was funded by the National Neuroscience Institute-Nanyang Technological University Neurotechnology Fellowship Grant.

Conflicts of Interest: The authors declare no conflict of interest.

References

1. Caceres, J.A.; Goldstein, J.N. Intracranial hemorrhage. *Emerg. Med. Clin. North Am.* **2012**, *30*, 771. [CrossRef]
2. Warren, S.M.; Walter, P. A Logical Calculus of the Ideas Immanent in Nervous Activity. 1943. *Bull. Math. Biol.* **1990**, *52*, 99–115.
3. Kunihiko, F.; Sei, M. Neocognitron: A self-organizing neural network model for a mechanism of visual pattern recognition. *Compet. Coop. Neural Nets* **1982**, *36*, 267–285.
4. Yann, L.; Bernhard, B.; John, S.D.; Donnie, H.; Richard, E.H.; Wayne, H.; Lawrence, D.J. Backpropagation applied to handwritten zip code recognition. *Neural Comput.* **1989**, *1*, 541–551.
5. Alex, K.; Ilya, S.; Geoffrey, E.H. Imagenet Classification with Deep Convolutional Neural Networks. In *Advances in Neural Information Processing Systems*; MIT Press Ltd.: Cambridge, MA, USA, 2012; pp. 1097–1105.
6. Pranav, R.; Jeremy, I.; Kaylie, Z.; brandon, Y.; Hershel, M.; Tony, D.; Daisy, D.; Aarti, B.; Curtis, L.; Katie, S.; et al. CheXNet: Radiologist-Level Pneumonia Detection on Chest X-Rays with Deep Learning. *arXiv* **2017**, arXiv:1711.05225.
7. Varun, G.; Lily, P.; Marc, C.; Martin, C.S.; Derek, W.; Arunachalam, N.; Subhashini, V.; Kasumi, W.; Tom, M.; Jorge, C.; et al. Development and validation of a deep learning algorithm for detection of diabetic retinopathy in retinal fundus photographs. *JAMA* **2016**, *316*, 2402–2410.
8. Andre, E.; Brett, K.; Roberto, A.N.; Justin, K.; Susan, M.S.; Helen, M.B.; Sebastian, T. Dermatologist-level classification of skin cancer with deep neural networks. *Nature* **2017**, *542*, 115–118.
9. Babak, E.B.; Mitko, V.; Paul, J.V.D.; Bram, V.G.; Nico, K.; Geert, L.; Jeroen, A.V.D.L.; Meyke, H.; Quirine, F.M.; Maschenka, B.; et al. Diagnostic assessment of deep learning algorithms for detection of lymph node metastases in women with breast cancer. *JAMA* **2017**, *318*, 2199–2210.
10. Barnes, S.R.; Haacke, E.M.; Ayaz, M.; Boikov, A.S.; Kirsch, W.; Kido, D. Semiautomated detection of cerebral microbleeds in magnetic resonance images. *Magn. Reson. Imaging* **2011**, *29*, 844–852. [CrossRef] [PubMed]
11. Chan, T. Computer aided detection of small acute intracranial hemorrhage on computer tomography of brain. *Comput. Med. Imaging Gr.* **2007**, *31*, 285–298. [CrossRef]
12. Keshavamurthy, K.N.; Leary, O.P.; Merck, L.H.; Kimia, B.; Collins, S.; Wright, D.W.; Allen, J.W.; Brock, J.F.; Merck, D. Machine Learning Algorithm for Automatic Detection of CT-Identifiable Hyperdense Lesions Associated with Traumatic Brain Injury. In *Medical Imaging 2017: Computer-Aided Diagnosis*; SPIE: Bellingham, DA, USA, 2017; Volume 10134, p. 101342G.
13. Prasoon, A.; Petersen, K.; Igel, C.; Lauze, F.; Dam, E.; Nielsen, M. Deep Feature Learning for Knee Cartilage Segmentation Using a Triplanar Convolutional Neural Network. In Proceedings of the International Conference on Medical Image Computing and Computer-Assisted Intervention, Nagoya, Japan, 22–26 September 2013; pp. 246–253.

14. Roth, H.R.; Lu, L.; Seff, A.; Cherry, K.M.; Hoffman, J.; Wang, S.; Liu, J.; Turkbey, E.; Summers, R.M. A New 2.5 D Representation for Lymph Node Detection Using Random Sets of Deep Convolutional Neural Network Observations. In Proceedings of the International Conference on Medical Image Computing and Computer-Assisted Intervention, Boston, MA, USA, 14–18 September 2014; pp. 520–527.

15. Shin, H.C.; Roth, H.R.; Gao, M.; Lu, L.; Xu, Z.; Nogues, I.; Yao, J.; Mollura, D.; Summers, R.M. Deep convolutional neural networks for computer-aided detection: CNN architectures, dataset characteristics and transfer learning. *IEEE Trans. Med. Imaging* **2016**, *35*, 1285–1298. [CrossRef]

16. Kamnitsas, K.; Ledig, C.; Newcombe, V.F.; Simpson, J.P.; Kane, A.D.; Menon, D.K.; Rueckert, D.; Glocker, B. Efficient multi-scale 3D CNN with fully connected CRF for accurate brain lesion segmentation. *Med. Image Anal.* **2017**, *36*, 61–78. [CrossRef]

17. Moeskops, P.; Viergever, M.A.; Mendrik, A.M.; de Vries, L.S.; Benders, M.J.; Išgum, I. Automatic segmentation of MR brain images with a convolutional neural network. *IEEE Trans. Med. Imaging* **2016**, *35*, 1252–1261. [CrossRef]

18. Havaei, M.; Davy, A.; Warde-Farley, D.; Biard, A.; Courville, A.; Bengio, Y.; Pal, C.; Jodoin, P.M.; Larochelle, H. Brain tumor segmentation with deep neural networks. *Med. Image Anal.* **2017**, *35*, 18–31. [CrossRef]

19. Jnawali, K.; Arbabshirani, M.R.; Rao, N.; Patel, A.A. Deep 3D Convolution Neural Network for CT Brain Hemorrhage Classification. In *Medical Imaging 2018: Computer-Aided Diagnosis*; SPIE: Bellingham, WA, USA, 2018; Volume 10575, p. 105751C.

20. Dou, Q.; Chen, H.; Yu, L.; Zhao, L.; Qin, J.; Wang, D.; Mok, V.C.; Shi, L.; Heng, P.A. Automatic detection of cerebral microbleeds from MR images via 3D convolutional neural networks. *IEEE Trans. Med. Imaging* **2016**, *35*, 1182–1195. [CrossRef]

21. Bouvrie, J. Notes on Convolutional Neural Networks. Unpublished Work. 2006. Available online: http://cogprints.org/5869/1/cnn_tutorial.pdf (accessed on 1 May 2019).

22. Goodfellow, I.; Bengio, Y.; Courville, A.; Bengio, Y. *Deep Learning*; MIT Press: Cambridge, MA, USA, 2016; Volume 1.

23. Zhang, H.; Gao, W.; Chen, X.; Zhao, D. Object detection using spatial histogram features. *Image Vis. Comput.* **2006**, *24*, 327–341. [CrossRef]

24. Grewal, M.; Srivastava, M.M.; Kumar, P.; Varadarajan, S. RADNET: Radiologist Level Accuracy Using Deep Learning for Hemorrhage Detection in CT Scans. *arXiv* **2017**, arXiv:1710.04934.

25. Yosinski, J.; Clune, J.; Nguyen, A.; Fuchs, T.; Lipson, H. Understanding neural networks Through Deep Visualization. *arXiv* **2015**, arXiv:1506.06579.

26. Cheng, K.S.; Lin, J.S.; Mao, C.W. The application of competitive Hopfield neural network to medical image segmentation. *IEEE Trans. Med. Imaging* **1996**, *15*, 560–567. [CrossRef]

27. Liu, R.; Tan, C.L.; Leong, T.Y.; Lee, C.K.; Pang, B.C.; Lim, C.T.; Tian, Q.; Tang, S.; Zhang, Z. Hemorrhage Slices Detection in Brain CT Images. ICPR 2008. In Proceedings of the 19th International Conference on Pattern Recognition, Tampa, FL, USA, 8–11 December 2008; pp. 1–4.

28. Al-Ayyoub, M.; Alawad, D.; Al-Darabsah, K.; Aljarrah, I. Automatic detection and classification of brain hemorrhages. *WSEAS Trans. Comput.* **2013**, *12*, 395–405.

29. Brock, A.; Lim, T.; Ritchie, J.M.; Weston, N. Generative and discriminative voxel modeling with convolutional neural networks. *arXiv* **2016**, arXiv:1608.04236.

30. Wu, J.; Zhang, C.; Xue, T.; Freeman, B.; Tenenbaum, J. Learning a Probabilistic Latent Space of Object Shapes Via 3d Generative-Adversarial Modeling. In *Advances in Neural Information Processing Systems*; MIT Press Ltd.: Cambridge, MA, USA, 2016; pp. 82–90.

31. Ypsilantis, P.P.; Montana, G. Recurrent convolutional networks for pulmonary nodule detection in CT imaging. *arXiv* **2016**, arXiv:1609.09143.

32. Chen, J.; Yang, L.; Zhang, Y.; Alber, M.; Chen, D.Z. Combining fully convolutional and recurrent neural networks for 3d biomedical image segmentation. In *Advances in Neural Information Processing Systems*; MIT Press Ltd.: Cambridge, MA, USA, 2016; pp. 3036–3044.

33. Brady, T.F.; Konkle, T.; Alvarez, G.A.; Oliva, A. Visual long-term memory has a massive storage capacity for object details. *Proc. Natl. Acad. Sci. USA* **2008**, *105*, 14325–14329. [CrossRef]

Article

Rapid Multi-Sensor Feature Fusion Based on Non-Stationary Kernel JADE for the Small-Amplitude Hunting Monitoring of High-Speed Trains

Jing Ning [1,2,*], Mingkuan Fang [1], Wei Ran [1], Chunjun Chen [1] and Yanping Li [1]

[1] School of Mechanical Engineering, Southwest Jiaotong University, Chengdu 610031, Sichuan, China; 13971658422@163.com (M.F.); ranwei@swjtu.edu.cn (W.R.); cjchen@swjtu.edu.cn (C.C.); signal2020_swj@163.com (Y.L.)
[2] Technology and Equipment of Rail Transit Operation and Maintenance Key Laboratory of Sichuan Province, Chengdu 610031, Sichuan, China
* Correspondence: ningjing@swjtu.cn; Tel./Fax: +86-28-87600690

Received: 29 March 2020; Accepted: 16 June 2020; Published: 18 June 2020

Abstract: Joint Approximate Diagonalization of Eigen-matrices (JADE) cannot deal with non-stationary data. Therefore, in this paper, a method called Non-stationary Kernel JADE (NKJADE) is proposed, which can extract non-stationary features and fuse multi-sensor features precisely and rapidly. In this method, the non-stationarity of the data is considered and the data from multi-sensor are used to fuse the features efficiently. The method is compared with EEMD-SVD-LTSA and EEMD-JADE using the bearing fault data of CWRU, and the validity of the method is verified. Considering that the vibration signals of high-speed trains are typically non-stationary, it is necessary to utilize a rapid feature fusion method to identify the evolutionary trends of hunting motions quickly before the phenomenon is fully manifested. In this paper, the proposed method is applied to identify the evolutionary trend of hunting motions quickly and accurately. Results verify that the accuracy of this method is much higher than that of the EEMD-JADE and EEMD-SVD-LTSA methods. This method can also be used to fuse multi-sensor features of non-stationary data rapidly.

Keywords: high-speed trains; hunting; non-stationary; feature fusion; multi-sensor fusion

---

1. Introduction

Hunting motion is a self-excited vibration that is a serious obstacle to the safety of high-speed trains [1]. Monitoring systems are designed to detect hunting only after it has developed to a specific degree. Besides, in most cases, the recognition result is only obtained using a single observation. Therefore, the accuracy and real-time performance of monitoring systems need to be further improved [2]. With the increasing performance requirements of high-speed trains, it is important to establish an accurate and rapid feature extraction method for hunting detection through multi-characterizations before the phenomenon has developed to any significant degree.

The structure of high-speed trains is very complicated and their working conditions are very poor, resulting in non-stationary vibration signals [3]. In the existing feature extraction research, a variety of data extraction algorithms are utilized. These algorithms can be roughly divided into two categories: those designed for stationary data and those for non-stationary data. Feature extraction methods for stationary data include Singular Value Decomposition (SVD), Linear Discriminant Analysis (LDA) [4,5], Principal Component Analysis (PCA) [6], Locality Preserving Projection (LPP) [7], and so on. However, it is hard to extract features of non-stationary data using these methods. For non-stationary data, manifold learning is a good feature extraction method [8–11]; however, it incurs a high computational cost and has extremely long calculation times, which means that the diagnostic information cannot be fed back to the system in time. Therefore, a rapid yet precise method to extract the non-stationary signal in practical engineering applications is needed. Cardoso [12,13] proposed a method named Joint Approximate Diagonalization of Eigen-matrices (JADE) in the field of blind source separation, which is used to quickly separate multiple features. Because the method is simple and effective, it is also widely used in pattern recognition. Liu [14] improved the JADE method and achieved a good feature fusion performance by simplifying the calculations, and then developed a method based on kernel JADE to identify the rolling bearing fault [15]. The method based on JADE was adopted to diagnose the faulty parts of the rolling bearing in [16], and a similar method was applied to predict coaxial bearing performance degradation in [17]. However, in this application, the non-stationary and rapid character of the signal is not considered.

The structure of high-speed trains is complex. The vibration signals from high-speed trains are affected by many factors, such as the propagation path and the measuring point location, which lead the signal to couple with different characteristic information. Pattern recognition based on single sensor vibration signals has limitations, which cannot completely render the evolutionary characteristics of high-speed train vibrations [18]. Therefore, it is necessary to utilize a multi-sensor fusion method to extract the features of the train's operating state.

Furthermore, the evaluation parameters of lateral stability of railway passenger trains in different countries are different. Lateral force on the rail and on the wheel axis, lateral acceleration of the bogie frame, acceleration of the axle-box, and lateral acceleration of the vehicle body can be used as evaluation parameters, respectively [19–22]. However, only one parameter is used in each standard, and this parameter has different limits. For example, the influence of frequency is not considered when lateral acceleration of the bogie frame is used. In fact, the acceleration of the axle-box is a particularly important index in hunting monitoring. Therefore, in this paper, we try to fuse the signal from the bogie frame and axle box to extract the hunting motion features. To monitor the state of hunting of high-speed trains, four basic states are classified: normal, small convergence, small divergence, and hunting. Our aim is to identify the small divergence state before hunting occurs. Thus, in this paper, the monitoring of small amplitude hunting enables us to rapidly distinguish between these four basic states online [23].

Moreover, for real-time classification, it is extremely important to ensure that once the small amplitude appears to diverge, the entire calculation process can be completed immediately. Therefore, the time of the whole calculation process must be short enough for real-time classification.

Considering the above problems, a rapid multi-sensor feature fusion method based on the Non-stationary Kernel JADE method is proposed. The JADE method is a fast and accurate feature fusion algorithm, but it is generally used in stable environments. In order to use the algorithm in a non-stationary environment, the whole time series is divided into M time periods [24] and the kernel function is introduced. Then, M kernel matrices are obtained by the M time periods, which are jointly decomposed. After this, Jacobian rotation is used to obtain the unitary matrix by diagonalizing multiple kernel matrices simultaneously to extract the non-stationary fusion features. In addition, in order to visualize the data features, the extracted fusion features are expressed in three dimensions. The between-class indicator and within-class indicators [25] are also employed to describe the clustering performance of the features quantitatively.

In this paper, a multi-sensor data feature extraction framework is provided, in which a rapid feature fusion method using Non-stationary Kernel JADE (NKJADE) is proposed. This framework consists of the following series of steps. First, the Ensemble Empirical Mode Decomposition (EEMD) method is utilized to decompose the preprocessed signals to Intrinsic Mode Functions (IMFs). Then, the energy matrices are obtained using the IMFs and the fusion features are obtained through NKJADE, followed by inputting the extracted features to an LSSVM [26,27] for training and recognition.

Data from Case Western Reserve University (CWRU) are used to verify the performance of this method against that of the SVD and JADE methods. In this paper, the proposed method is also applied to identify the evolutionary trend of the hunting motion quickly and accurately. The results verify that the accuracy of this method is much higher than that of the JADE method.

The remainder of this paper is organized as follows: In Section 2, the theoretical backgrounds of Ensemble Empirical Mode Decomposition (EEMD) [28] and a separability metric are introduced. The proposed method of non-stationary Kernel JADE is also described. The framework of the proposed method is introduced in Section 3. In Section 4, data from Case Western Reserve University (CWRU) are used to test the proposed method, and the operational data of multiple states of high-speed trains are used to verify the accuracy and rapidity of the proposed method. Finally, the conclusion is presented in Section 5.

2. Theoretical Background

2.1. EEMD

The basic idea of the EEMD method is to use the sifting process to decompose the signal into several intrinsic mode functions, when Gaussian white noise is added. For the original signal, the specific steps to use EEMD to decompose it into IMFs are as follows:

(1)   Obtain the overall signal by adding Gaussian white noise to the original signal:

$$a'(t) = a(t) + \omega(t) \tag{1}$$

(2)   The overall signal is decomposed to obtain the IMF components of each order, where i represents the i-th component, r is the residual term, and n is the number of IMFs:

$$a'(t) = \sum_i^n c_i + r \tag{2}$$

(3)   Repeat step (1) and (2), each time adding different white noise sequences of the same amplitude:

$$a'_j(t) = \sum_{i=1}^n c_{ij} + r_j \tag{3}$$

In Equation (3), c_{ij} is the i-th IMF component of the decomposition to which white noise at for the j-th time, while r_j is the residual value of the decomposition.

(4)   Using the zero-mean principle of the Gaussian white noise frequency, the effect of white noise can be eliminated, and the IMF component corresponding to the original signal can be expressed as:

$$c_i(t) = \frac{1}{N} \sum_{j=1}^N c_{ij}(t) \tag{4}$$

where n represents the number of times the white noise is added and c_i represents the i-th IMF component obtained by the EEMD decomposition of the original signal.

2.2. IMF Energy Matrix

For the IMF component $c'_i = [c'_i(1), \ldots \ldots, c'_i(M)]$, the energy feature can be expressed as:

$$e_i = \sum_{j=1}^M [c'_i(j)]^2 \tag{5}$$

All IMF components of a vibration signal form a feature energy vector $e = [e_1, e_2, \ldots\ldots e_n]$, where n is the dimension of vector e and represents the number of IMF components.

2.3. Proposed Non-Stationary Kernel JADE

Since the original JADE algorithm is based on stationary signal analysis, considering the non-stationary nature of high-speed train signal, the kernel JADE method [15] is applied to a non-stationary environment.

The main idea of the kernel is to map the input matrix into the nonlinear space Φ Suppose $X = \{x_1, x_2, \ldots, x_m\}$; then, the processing can be defined as follows:

$$\{x_1, x_2, \ldots, x_m\} \rightarrow \{\Phi(x_1), \Phi(x_2), \ldots, \Phi(x_m)\} \tag{6}$$

During implementation, we need to calculate the inner product of two eigenvectors which have been mapped into a nonlinear space using a kernel function, and a kernel matrix will be calculated using Equation (7):

$$K_{ij} = k(x_i, x_j) = <\Phi(x_i), \Phi(x_j)> \tag{7}$$

where x_i, x_j are vectors. The commonly used kernel function includes the following [29]:

$$k(x_i, x_j) = \exp(-\frac{\|x_i - x_j\|^2}{2\delta^2}) \tag{8}$$

$$k(x_i, x_j) = (\alpha x_i^T x_j + c)^d \tag{9}$$

$$k(x_i, x_j) = \tanh(\alpha x_i^T x_j + c) \tag{10}$$

Similar to KPCA [30], the centered kernel matrix K can be calculated through Equation (11):

$$K_{ij} = K_{ij} - \frac{1}{M}\sum_{r=1}^{M} K_{ir} - \frac{1}{M}\sum_{r=1}^{M} K_{rj} + \frac{1}{M^2}\sum_{r,s=1}^{M} K_{rs} \tag{11}$$

The whole time series is divided into M segment intervals $T_1, \ldots\ldots, T_M$, and consequently M covariance matrices $\mathbf{S}_{T_1}, \ldots\ldots, \mathbf{S}_{T_M}$ can be generated. These M covariance matrices are jointly diagonalized to find a unitary matrix $\mathbf{U}$ which can diagonalize M covariance matrices simultaneously; then, the energy features can be extracted.

Step 1: For M segment intervals $T_1, \ldots\ldots, T_M$, the covariance matrices of signal $x(t)$ can be expressed as:

$$Cov_{T_k} = \mathbf{S}_{T_k} = \frac{1}{|T_m|}\sum_{t \in T_m} [k(s_t, s_t)] \tag{12}$$

where $s_t = x_t - E(x_t)$, $E(x_t)$ denotes the mean of x_t.

Step 2: The most common method to diagonalize matrices $\mathbf{S}_{T_1}, \ldots\ldots, \mathbf{S}_{T_M}$ is to diagonalize the first matrix, and then transform the remaining $M-1$ matrices into diagonalization. $\mathbf{W}$ is the diagonalized matrix of the covariance matrix $\mathbf{S}_{T_1}$:

$$\mathbf{W} = \mathbf{S}_{T_1}^{-1/2} = \mathbf{V}^H \mathbf{\Lambda}^{-1/2} \mathbf{V} \tag{13}$$

In Equation (13), $\mathbf{V}$ is the eigen-matrix of $\mathbf{S}_{T_1}$, while $\mathbf{\Lambda}$ is the eigenvector of $\mathbf{S}_{T_1}$.
For the remaining $M-1$ matrices, the diagonalization matrix can be defined respectively:

$$\mathbf{S}_{T_m}^* = \mathbf{S}_{T_1}^{-1/2}\mathbf{S}_{T_m}(\mathbf{S}_{T_1}^{-1/2})^H, m = 2, \ldots, M \tag{14}$$

Step 3: The approximate joint diagonalization problem is equivalent to finding an orthogonal matrix **U** that minimizes:

$$\sum_{m=2}^{M} \|\mathrm{off}(\mathbf{US}_{T_m}^* \mathbf{U}^H)\|^2 = \sum_{m=2}^{M} \sum_{b \neq d} (\mathbf{US}_{T_m}^* \mathbf{U}^H)_{bd}^2 \tag{15}$$

where $\mathrm{off}(\mathbf{US}_{T_m}^* \mathbf{U}^H)$ has the same off-diagonal elements as $\mathbf{US}_{T_m}^* \mathbf{U}^H$, and the diagonal element is zero, while b and d represent the b-th row and d-th column of the matrix, respectively.

Since the sum of squares remains the same when multiplied by an orthogonal matrix, the problem is equivalent to maximizing the sum of squares of the diagonal elements:

$$\sum_{m=2}^{M} \|\mathrm{diag}(\mathbf{US}_{T_m}^* \mathbf{U}^H)\|^2 = \sum_{m=2}^{M} \sum_{b=1}^{p} (\mathbf{US}_{T_m}^* \mathbf{U}^H)_{bb}^2 \tag{16}$$

where p represents the dimension to which the feature is extracted.

Step 4: Givens rotation is used to transform the set of matrices to a more diagonal form, two rows and two columns at a time. The Givens rotation matrix is given by:

$$G(i,j,\theta) = \begin{pmatrix} 1 & \cdots & 0 & \cdots & 0 & \cdots & 0 \\ \vdots & \ddots & \vdots & & \vdots & & \vdots \\ 0 & \cdots & \cos(\theta) & \cdots & -\sin(\theta) & \cdots & 0 \\ \vdots & & \vdots & \ddots & \vdots & & \vdots \\ 0 & \cdots & \sin(\theta) & \cdots & \cos(\theta) & \cdots & 0 \\ \vdots & & \vdots & & \vdots & \ddots & \vdots \\ 0 & \cdots & 0 & \cdots & 0 & \cdots & 1 \end{pmatrix} \tag{17}$$

where

$$\theta = \frac{1}{2}\mathrm{arccot}\left[\frac{(\mathbf{S}_{T_m}^*)_{22} - (\mathbf{S}_{T_m}^*)_{11}}{2(\mathbf{S}_{T_m}^*)_{12}}\right] \tag{18}$$

The initial value for the orthogonal matrix U is the identity matrix I. Matrices $\mathbf{S}_{T_2}, \ldots, \mathbf{S}_{T_m}$ are then updated using:

$$\begin{aligned} \mathbf{S}_{T_m}^* &\leftarrow \mathbf{G}(1,2,\theta)\mathbf{S}_{T_m}^* \mathbf{G}(1,2,\theta), m = 2,\ldots,M \\ \mathbf{U} &\leftarrow \mathbf{U}\mathbf{G}(1,2,\theta) \end{aligned} \tag{19}$$

When the values of all non-diagonal elements are less than a given threshold ε, the iteration is completed, and the joint approximation diagonalization is achieved. The unitary matrix $\hat{\mathbf{U}}$ is obtained so that multiple matrices are diagonalized jointly.

Step 5: The transform matrix A can be calculated as $\mathbf{A} = \hat{\mathbf{U}}\mathbf{W}^{\#}$, where superscript # denotes the pseudo-inverse.

As such, the original signal $s(t)$ can be expressed as:

$$s(t) = \mathbf{A} \cdot \mathbf{K} \tag{20}$$

Since the NKJADE method can extract the nonlinear relationships hidden in the high-dimensional feature space, the fusion features can be estimated through the joint feature decomposition using multiple inputs. The fusion features can express the nonlinear and non-stationary relationships hidden in the inputs well, so they can represent the characteristic relationship of different states, and then distinguish different feature states quickly and accurately.

2.4. Separability Evaluation

To illustrate the merit of our proposed algorithm, the separability J is utilized to demonstrate the algorithm's ability to form distinct classes. The capability of the feature extraction in pattern classification can be described quantitatively using between-class scatter S_b, within-class scatter S_w [31], and separability [32]. Assuming that the data have a total of C classifications, the vector of the i-th classification is:

$$x^i = \left(x_1^i, x_2^i, \ldots, x_{n_i}^i\right) \tag{21}$$

where n_i is the number of i-th classification.

Between-class scatter S_b and within-class scatter S_w can be calculated as follows:

$$S_b = \sum_{i=1}^{C} p_i(m_i - m)(m_i - m)^T \tag{22}$$

$$S_w = \sum_{i=1}^{C}\left[\frac{p_i}{n_i}\sum_{k=1}^{n_i}(x_k^i - m_i)(x_k^i - m_i)^T\right] \tag{23}$$

where $p_i = n_i/\sum_{j=1}^{C} n_j$, $m_i = mean(x_{n_i}^i)$, $m = \sum_{i=1}^{C} p_i m_i$.

The between-class scatter S_b describes how far different classes are separated, and the within-class scatter S_w indicates how compactly each class of samples is distributed. Based on between-class scatter and within-class scatter, separable evaluation J is introduced to describe the clustering ability of different methods. Separable evaluation J could be calculated as follows:

$$J = \text{trace}(S_b/S_w) \tag{24}$$

where function *trace* refers to the sum of diagonal elements.

3. Methodology

For the different characteristics of vibration signals, a multi-sensor data feature extraction framework is provided, in which a rapid feature fusion method using NKJADE is proposed. The framework of the feature extraction method is shown in Figure 1. The method can quickly and accurately extract different features of information contained in non-stationary vibration signals. The main steps of this framework are as follows:

(1) The m position sensor data and l classes are selected, and a total of m sample matrices are obtained.
(2) The EEMD algorithm is used to separately decompose the sample matrices of m positions, then n IMF components of each sample signal can be obtained.
(3) By extracting the energy features of the n components of IMF obtained using decomposition, an IMF energy matrix is obtained from each position. In this paper, a total of m energy matrices of IMF were obtained.
(4) For the obtained energy matrices of IMF of m positions, a rapid feature fusion method using NKJADE is proposed, so the dimensionality is reduced to three for a better-observed performance.
(5) The LSSVM is trained on and used to test the fusion features to verify the accuracy of the method.

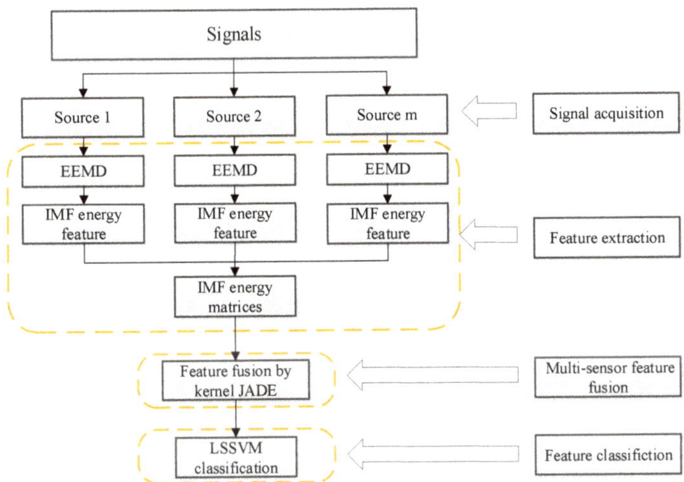

Figure 1. Method framework.

4. Experiment Results and Analysis

In order to verify the valid of the method, we applied it to bearing fault identification (Case I) and small amplitude hunting monitoring of high-speed trains (Case II). Some traditional approaches, such as Ensemble Empirical Mode Decomposition Joint Approximate Diagonalization of Eigen-matrices (EEMD-JADE) and Ensemble Empirical Mode Decomposition Singular Valuable Decomposition Learning Technology Systems Architecture (EEMD-SVD-LTSA), were compared with the proposed method.

4.1. Case I—Case Western Reserve University Data

4.1.1. Data Description

The bearing test data for normal and faulty bearings were from the Case Western Reserve University (CWRU). The test signal contained normal state, ball fault, and inner and outer race faults; for the latter three, the fault diameters were 0.07, 0.14 and 0.21 inches, respectively.

The data used in this paper were collected from the drive end. The sampling frequency was 12 KHZ and the speed of the shaft was 1725 r/min, corresponding to 400 points collected per revolution. In order to reduce the influence of equipment fluctuations, each sample contained 800 points.

As shown in Table 1, two datasets were selected. Dataset A had the same fault location (inner race fault), but the fault size was different. Dataset B had different fault locations, but the fault size was the same, and the outer race fault was at the location of 3 o'clock. Each of the datasets was divided into three categories.

Table 1. Description of dataset.

Datasets	Class	Number of Training Samples	Number of Testing Samples	Label
Dataset A	Inner race, 0.07″	70	30	1
	Inner race, 0.14″	70	30	2
	Inner race, 0.21″	70	30	3
Dataset B	Inner race, 0.07″	70	30	1
	Ball, 0.07″	70	30	4
	Outer Race, 0.07″	70	30	5

4.1.2. Signal Processing Results

The scatter plots obtained by applying the EEMD-SVD-LTSA, EEMD-JADE method and EEMD-NKJADE methods on dataset A are shown in Figure 2a–c, while the corresponding scatter plots for dataset B are shown in Figure 2d–f. The results of selecting parameters for kernel are shown in Figure 3. The results of the application of the three algorithms in dataset A are shown in Table 2, while the results for dataset B are shown in Table 3.

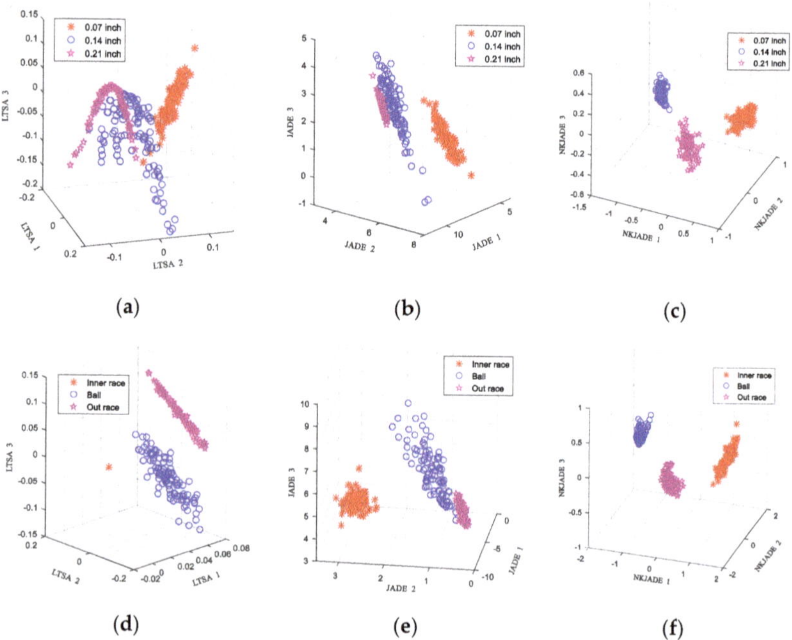

Figure 2. Scatter plots of faults. (**a**) EEMD-SVD-LTSA on dataset A; (**b**) EEMD-JADE on dataset A; (**c**) EEMD-NKJADE on dataset A; (**d**) EEMD-SVD-LTSA on dataset B; (**e**) EEMD-JADE on dataset B; (**f**) EEMD-NKJADE on dataset B.

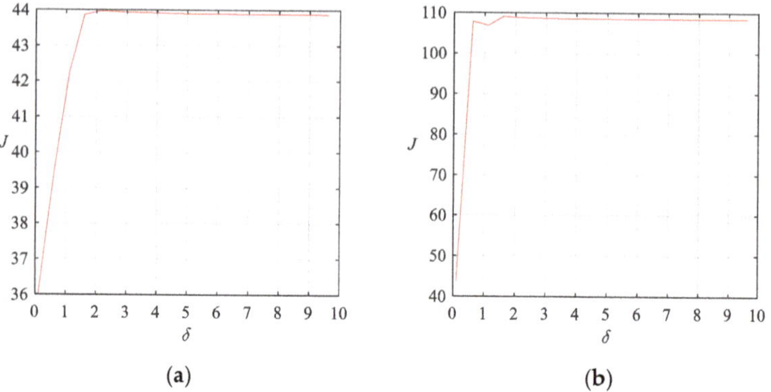

Figure 3. The results of selecting parameters for the Gaussian kernel. (**a**) Parameter selection results of Gaussian kernel of EEMD-NKJADE method on dataset A; (**b**) Parameter selection results of Gaussian kernel of EEMD-NKJADE method on dataset B.

Table 2. Results from dataset A.

Method	EEMD-SVD-LTSA	EEMD-JADE	EEMD-NKJADE
Accuracy (%)	98.33	98.89	100
J	36.26	39.24	43.91
Time (s)	1.7593	0.7820	1.0070

Table 3. Results from dataset B.

Method	EEMD-SVD-LTSA	EEMD-JADE	EEMD-NKJADE
Accuracy (%)	100	98.89	100
J	1.195×10^{24}	47.96	87.96
Time (s)	1.5513	0.7190	0.9650

4.1.3. Discussion

We found that the generalization ability of the Gaussian kernel is better than that of the other kernels. Therefore, we focused on the parameters of the Gaussian kernel.

According to the results of selecting parameters in [15], we tried to set the range of σ to (1, 10). The parameter was incremented step by step (parameter step is 0.5, shown in X-axis), and the separability J was calculated (shown in Y-axis). The larger J is, the better the classification result will be. Therefore, for the parameter selection of EEMD method, the optimal parameter of the Gaussian kernel on dataset A and dataset B is 2.

From Figure 2 and Table 2, we see that all three methods could extract the features with satisfactory results. The accuracy of the three methods was nearly 100% in all cases. The result may be attributed to the CWRU bearing fault data having great differences, which are quite easy to classify.

However, the separability J in Table 2 is different, which is consistent with (a)–(c) in Figure 2. In this respect, the classification ability of EEMD- NKJADE algorithm is obviously better than other algorithms. In addition, the results of the running time in Table 2 show that the time required for EEMD-JADE calculation was relatively short (the PC configuration was as follows: Intel Core i5-4460, 12GB of memory, NVIDIA GeForce GT720).

In order to further verify the robustness of the algorithm, we also tested the results of the algorithm on dataset B.

In Table 3, the accuracy of the three methods in dataset B was almost the same as that in dataset A. However, the separability of dataset B was the highest after being processed using the EEMD-SVD-LTSA algorithm, while in dataset A, the separability achieved using EEMD-SVD-LTSA was the lowest, which indicates that the selection of dataset had a greater impact on EEMD-SVD-LTSA. Compared with EEMD-SVD-LTSA, the results from the EEMD-NKJADE and EEMD-JADE algorithms were less affected by the different dataset, and therefore seem to be more robust. Therefore, the method of EEMD-NKJADE offers superior performance with respect to the classification effect and the robustness, but its calculation time is longer than EEMD-JADE.

4.2. Case II—Small Amplitude Hunting Monitoring of High-Speed Trains

4.2.1. Problem Description

The stability of hunting has always been a key problem in the study of vehicle lateral motion stability [33]. Small amplitude hunting is a sign of hunting instability. In China, hunting phenomena are considered to occur when the amplitude of lateral acceleration signals from the bogie frame reaches or exceeds 8–10 m/s² more than 5 times continuously after a 10 Hz low-pass filter [34]. In Figure 4, the lateral acceleration of the bogie frame signals will sometimes go through a normal operation/small amplitude hunting/normal operation periodic cycle, which is a gradually convergent process. The hunting amplitude in this situation is small and convergent. Sometimes, the signals will go through a normal operation/small amplitude hunting/critical hunting process, which is a gradually divergent process. The hunting amplitude in this situation starts small and then diverges.

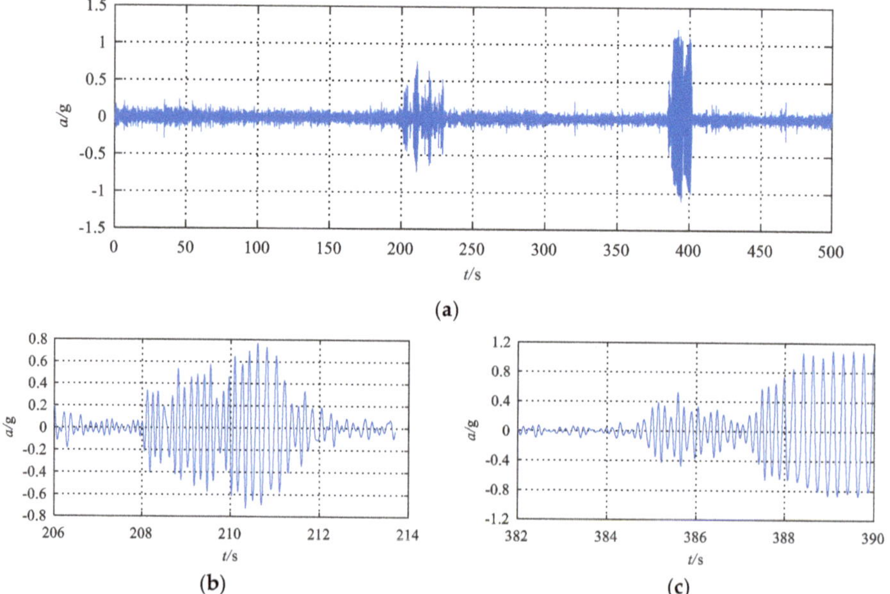

Figure 4. Lateral acceleration signal of the bogie frame when small-amplitude hunting motion occurs during an online test. (**a**) Bogie frame acceleration; (**b**) Small amplitude convergent hunting; (**c**) Small amplitude divergent hunting.

Therefore, it is necessary to extract the features of different states rapidly and accurately, especially the small amplitude divergent hunting states, to guarantee the system is alerted in time to ensure the safe operation of the train.

4.2.2. Data Acquisition

The data used in this paper were lateral acceleration signals of the bogie frame and axle box from an online tracking experiment. The CRTS II ballastless track and seamless rail were used on the whole line. The speed of the train was 320–350 km/h. The sampling frequency was 2500 Hz. All the data were acquired in accordance with China's Railway Passenger Traffic Safety Monitoring Standard [35]. Although in China the amplitude of lateral acceleration signals from the bogie frame is used as the testing parameter to monitor hunting motion, research has proven that many other testing parameters are also important for hunting monitoring. As such, in this paper, acceleration signals of the bogie frame and the axle box were used.

The installation locations of the accelerometers are shown in Figure 5, where two accelerometers are installed in the diagonal direction on the H-shaped bogie frame. The lateral accelerometers from the bogie are respectively denoted as S1 and S2. Also, considering that the vibration of the axle box is important for hunting [36], a sensor located on the axle box was used. In Figure 6, the lateral accelerometer on the axle box is denoted as S3. Figure 5 shows a photograph of the installation site.

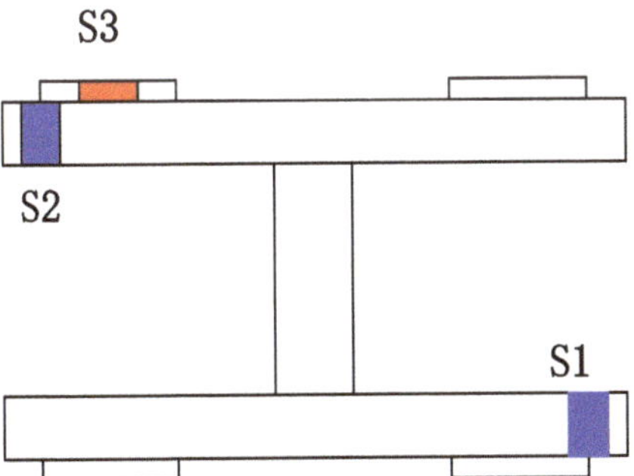

Figure 5. Installation locations of the accelerometers. S1, S2: Accelerometer on the bogie frame; S3: Accelerometer on the axle box.

Figure 6. Site installation drawing.

Considering the applications of other researchers, the sampling frequency of the original signal was set to 2500 Hz. However, for hunting, the characteristic frequency of the lateral acceleration of the frame is only 3–7 Hz. Therefore, in the preprocessing stage, a 250 Hz resampling method was applied [37], and a low-pass filter of 0–10 Hz was applied. Then, a zero-mean smoothing process was used for preprocessing to eliminate trend terms. In accordance with commonly followed practices, parameters such as the amplitude of the lateral acceleration signals from the bogie frame and axel box were used to obtain a synthetic assessment of the lateral stability of the high-speed train tested. In this paper, the filtered lateral acceleration signals were divided into four states: normal, small amplitude convergent hunting, small amplitude divergent hunting, and hunting. Ten groups of sample data were used for each of the four states and each sensor, yielding a total of 120 sample groups. The length of each sample was 500 points, corresponding to a sample time of 2 s.

Nonlinear factors [38,39] have been proven to affect the bifurcation evolution of small amplitude hunting and, according to [23], all the values of the Lyapunov exponent of the lateral acceleration are greater than 1, which means that the lateral acceleration signals from the bogie frame have non-stationary characteristics.

4.2.3. Feature Fusion

First, EEMD was applied on each signal. Then, 7 IMFs and 1 residue were obtained using the EEMD method. Figure 7 shows an EEMD decomposition view of the three signals at different points during the same time period.

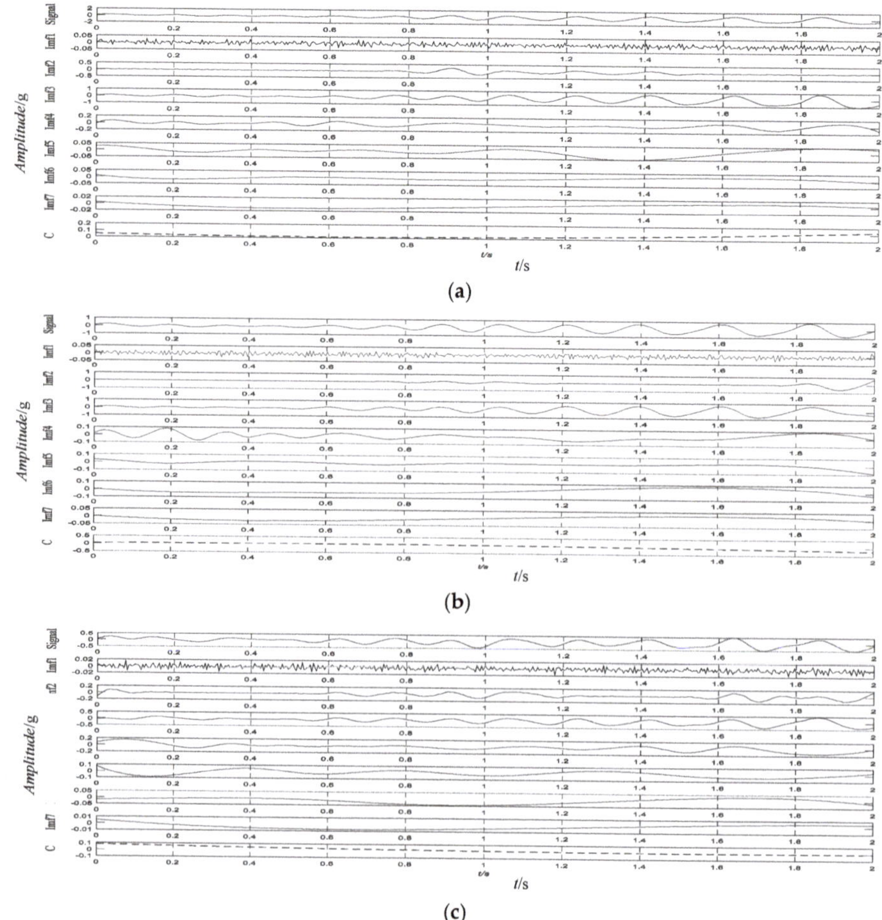

Figure 7. Ensemble Empirical Mode Decomposition (EEMD) diagram of signals at different positions. (**a**) S1 position (bogie frame); (**b**) S2 position (bogie frame); (**c**) S3 position (axle box).

The samples of the three sensors were processed, and the three corresponding energy matrices E1, E2, and E3, with a size of 40 × 8 were obtained. The NKJADE method was used to fuse the three high-dimensional energy matrices and the data dimensionality was reduced.

4.3. Result and Discussion

4.3.1. Single Sensor Classification Using NKJADE

A scatter plot of the features extracted using a single sensor and NKJADE is shown in Figure 8. The separability *J* and accuracy of the NKJADE method are shown in Table 4.

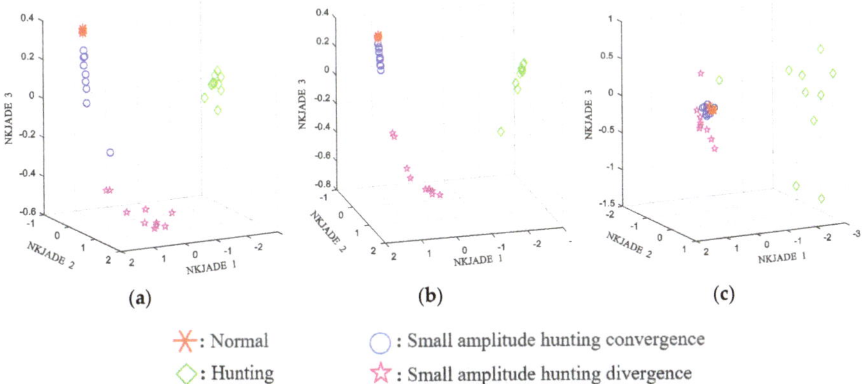

(**a**)   (**b**)   (**c**)

$*$: Normal $\bigcirc$: Small amplitude hunting convergence

$\diamondsuit$: Hunting $\star$: Small amplitude hunting divergence

Figure 8. Scatter plot of feature extraction from single sensor using Non-stationary Kernel JADE (NKJADE). (**a**) S1 (bogie frame); (**b**) S2 (bogie frame); (**c**) S3 (axle box).

Table 4. Separability J running time and accuracy of NKJADE method.

Sensors	Only S1	Only S2	Only S3	S1, S2 and S3
J	65.6	61.1	21.1	155.7
Accuracy (%)	97.23	96.54	29.85	100
Run Time (s)	0.0392	0.0415	0.0387	0.1293

From Table 4, the accuracy achieved using S1 data only was 97.23%, which was greater than that attained at the other sensor locations. However, when the three sensors were used together, the accuracy rate became 100%, and J reached a much greater value.

4.3.2. Multi-Sensor Fusion Using NKJADE

The EEMD-SVD-LTSA and EEMD-JADE methods were used to compare the identification accuracy and calculation speed of the feature fusion method in multi-sensor conditions. The results of the methods are shown in Figure 9. The separability J and accuracy obtained using the different sensor fusion methods with data from all three sensors are shown in Table 5.

The JADE and SVD-LTSA methods were compared with the proposed method in the multi-sensor feature fusion conditions. From Table 5, the accuracy rate of the SVD-LTSA method was 93.75%, which used the non-stationary method LTSA. The accuracy rate of the JADE method was only 30.12%, which used the stationary method. The accuracy rate of NKJADE (Gaussian kernel, $\delta = 0.6$) was 100%, while the separability J of NKJADE was greater than those of the other methods. Considering that data from high-speed trains have more typical no-stationary characteristics than the bearing fault data from CWRU, this result shows that the proposed method may more be suitable for non-stationary data. The separability J of the JADE method was only 26.67, which is even lower than that achieved using single sensor S2 or S3 data. The reason for this is probably that in the JADE method, the non-stationary condition is not considered.

Besides, because real-time processing is a significant factor for the small amplitude hunting monitoring of high-speed trains, the calculation time is a very important factor. If the calculation time is too long, the diagnostic information cannot be fed back to the system in time. Compared with the SVD-LTSA method, fusion features can be extracted very quickly using the proposed NKJADE method. The run time of the NKJADE method was nearly the same as that of the JADE method, and the separability of J of NKJADE was greater than JADE. This shows that NKJADE is a rapid multi-sensor feature fusion method based on non-stationary condition, which outperforms the SVD-LTSA and

JADE methods. It can be applied to the small amplitude hunting bifurcation evolution monitoring in high-speed trains.

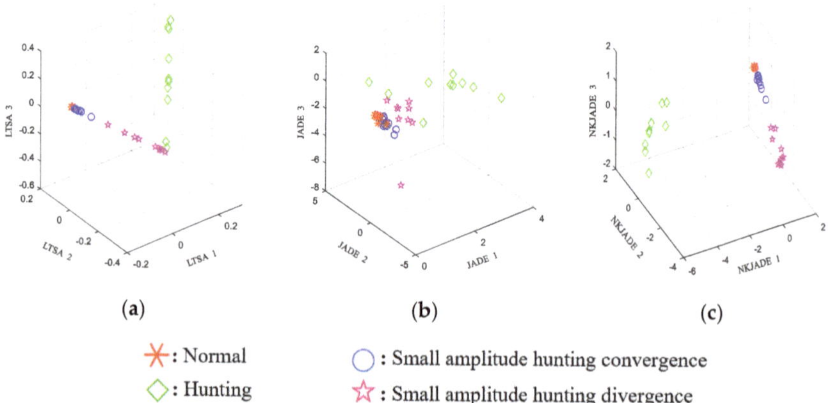

Figure 9. Scatter plot of feature extraction from multi-sensor data. (**a**) EEMD-SVD-LTSA; (**b**) EEMD-JADE; (**c**) EEMD-NKJADE.

Table 5. The separability *J*, running time, and accuracy using different sensor fusion methods.

Method	EEMD-SVD-LTSA	EEMD-JADE	EEMD-NKJADE
J	52	26.67	155.7
Accuracy (%)	93.75	30.12	100
Run time (s)	0.9972	0.1216	0.1298

As shown in Figure 10, the parameter is incremented step by step (parameter of step is 0.01, shown in X-axis), and the separability index *J* is calculated (shown in Y-axis). The larger *J* is, the better the classification result will be.

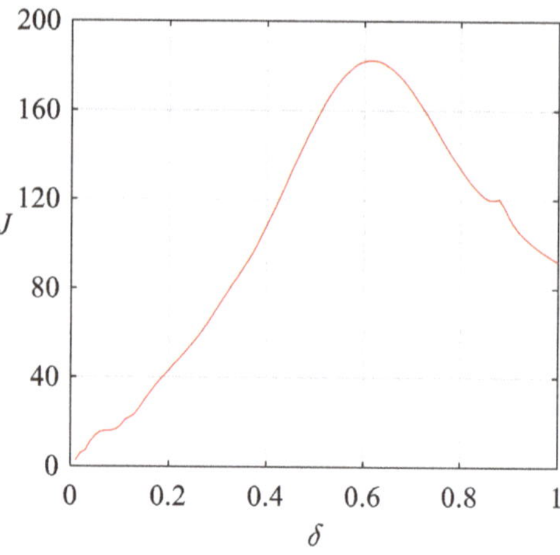

Figure 10. Parameter selection results of Gaussian kernel of EEMD-NKJADE method.

Because the raw data collected are mainly distributed in the range of −1–1 g (gravity). In Equation (8), σ is the width parameter of the kernel function, so we tried to set the range of σ to (0, 1). As shown in Figure 10, with the increase of δ, J increases at first and then decreases. It indicates that the range of parameter selection is reasonable. When $\delta = 0.6$, the classification result is the best.

5. Conclusions

In this paper, a rapid multi-sensor feature fusion method based on NKJADE is proposed, with which the features of multiple sensors can be extracted quickly and accurately. In order to use the algorithm in a non-stationary environment, the whole time series is divided into M time periods, and the kernel function is introduced. Then, Jacobian rotation is used to obtain the unitary matrix by diagonalizing multiple kernel matrices simultaneously to extract the non-stationary fusion features.

Our main findings are that:

1. The fusion features can be extracted quickly and efficiently using the proposed method NKJADE compared to the SVD-LTSA and the JADEs methods with bearing fault data from Case Western Reserve University.
2. The NKJADE method can extract the fusion features from non-stationary data effectively compared to the JADE method. In case I, the accuracy rate of the three methods (SVD-LTSA, JADE, and NKJADE) was nearly the same (100%), but in case II, the accuracy rate of the three methods was very different.
3. The NKJADE method can extract the multi-sensor fusion features effectively. The data from hunting monitoring of high-speed trains were used to verify the validity of the method in multi-sensor conditions.

Author Contributions: Data analysis and funding acquisition, J.N.; Writing—original draft preparation, M.F.; data collection: W.R.; experiment design, C.C.; detection equipment, Y.L. All authors have read and agreed to the published version of the manuscript.

Funding: This research was funded by the National Natural Science Foundations of China, grant number 51975486 and 51975487.

References

1. De Pater, A.D. The approximate determination of the hunting movement of a railway vehicle by aid of the method of krylov and bogoljubow. *Appl. Sci. Res.* **1961**, *10*, 205–228. [CrossRef]
2. Yao, J.; Sun, L.; Hou, F. Study on evaluation methods for lateral stability of high-speed trains. *J. China Railway Sci.* **2012**, *33*, 132–139.
3. Chu, F.; Peng, Z.; Feng, Z. *Modern Signal Processing Methods in Machinery Fault Diagnosis*; Science Press: Beijing, China, 2009.
4. Guo, K.; Zhu, Y.; San, Y. Analog circuit fault diagnosis using LDA and OAOSVM approach. *Adv. Mater. Res.* **2012**, *490*, 1130–1134. [CrossRef]
5. Park, W.; Lee, S.; Joo, W.; Song, J. A mixed algorithm of PCA and LDA for fault diagnosis of induction motor. In *Advanced Intelligent Computing Theories and Applications*; Springer: Berlin/Heidelberg, Germany, 2007; pp. 934–942.
6. Moura, E.; Souto, C.; Silva, A.; Irmao, M. Evaluation of principal component analysis and neural network performance for bearing fault diagnosis from vibration signal processed by RS and DF analyses. *Mech. Syst. Signal Process.* **2011**, *5*, 1765–1772. [CrossRef]
7. He, Q.; Ding, X.; Pan, Y. Machine fault classification based on local discriminant bases and locality preserving projections. *Math. Probl. Eng.* **2014**, *5*, 1–12. [CrossRef]

8.   He, Q. Vibration signal classification by wavelet packet energy flow manifold learning. *J. Sound Vib.* **2013**, *7*, 1881–1894. [CrossRef]

9.   Hettiarachchi, R.; Peters, J.F. Multi-manifold LLE learning in pattern recognition. *Pattern Recognit.* **2015**, *48*, 2947–2960. [CrossRef]

10. Bu, Y.; Chen, F.; Pan, J. Stellar spectral subclasses classification based on Isomap and SVM. *New Astron.* **2014**, *28*, 35–43. [CrossRef]

11. Sun, W.; Halevy, A.; Benedetto, J.J.; Czaja, W.; Li, W.; Liu, C.; Shi, B.; Wang, R. Nonlinear dimensionality reduction via the ENH-LTSA method for hyperspectral image classification. *IEEE J. Sel. Top. Appl. Earth Obs. Remote Sens.* **2014**, *7*, 375–388. [CrossRef]

12. Cardoso, J. High-order contrasts for independent component analysis. *Neural Comput.* **1999**, *1*, 158–191. [CrossRef]

13. Cao, S.; Ouyang, H. Multi-damage identification based on joint approximate diagonalization and robust distance measure. *J. Phys. Conf. Ser.* **2017**, *842*, 012022. [CrossRef]

14. Liu, F.; Liu, Y.; Chen, F.; He, B. Residual life prediction for ball bearings based on joint approximate diagonalization of eigen-matrices and extreme learning machine. *Proc. Inst. Mech. Eng. Part C J. Mech. Eng. Sci.* **2017**, *9*, 1699–1711. [CrossRef]

15. Liu, Y.; He, B.; Liu, F. Feature fusion using kernel joint approximate diagonalization of eigen-matrices for rolling bearing fault identification. *J. Sound Vib.* **2016**, *385*, 389–401. [CrossRef]

16. Wu, T.; Liu, C.C.; He, C. "Fault Diagnosis of Bearings Based on KJADE and VNWOA-LSSVM Algorithm". *Math. Probl. Eng.* **2019**, *14*, 1–19. [CrossRef]

17. Liu, F.; Li, L.; Liu, Y.; Cao, Z.; Yang, H.; Lu, S. HKF-SVR Optimized by Krill Herd Algorithm for Coaxial Bearings Performance Degradation Prediction. *Sensors* **2020**, *20*, 660. [CrossRef] [PubMed]

18. Ning, J.; Liu, Q.; Ouyang, H. A multi-sensor fusion framework for detecting small amplitude hunting of high-speed trains. *J. Vib. Control* **2018**, *17*, 3797–3808. [CrossRef]

19. Europeenne, British Standard Norme. *Testing ans Approval of Railway Vehicles from the Point of View of Their Dynamic Behaviour-Safety-Track Fatigue-Ride Quality*; UIC Code 518; International Union of Railways: Paris, France, 2005.

20. Czech Institute for Normalisation. *EN14363, B.S. Railway Applications-Testing for the Acceptance of Running Characteristics of Railway Vehicles-Testing of Running Behaviour and Stationary Tests*; Czech Institute for Normalisation: London, UK, 2016.

21. TSI; HSRST. Technical Specification for Interoperability Relating to the 'Rolling Stock' Sub-System of the Trans-European High-Speed Rail System. *Off. J. Eur. Union L* **2008**, *25*, 22–282.

22. Federal Railroad Administration. *Vehicle/Track Interaction Safety Standards, High-Speed and High Cant Deficiency Operations*; National Archives and Records Administration: College Park, MD, USA, 2013.

23. Ning, J. Feature recognition of small amplitude hunting signals based on the MPE-LTSA in high-speed trains. *Measurement* **2019**, *131*, 452–460. [CrossRef]

24. Miettinen, J.; Nordhausen, K.; Taskinen, S. Blind source separation based on joint diagonalization in R: The packages JADE and BSSasymp. *J. Stat. Softw.* **2017**, *76*, 1–31. [CrossRef]

25. Ding, X.; He, Q.; Luo, N. A fusion feature and its improvement based on locality preserving projections for rolling element bearing fault classification. *J. Sound Vib.* **2015**, *335*, 367–383. [CrossRef]

26. Ye, Y.; Ning, J. Small Amplitude Hunting Instability of High-speed Train Diagnosis Method Based on Modified Ensemble Empirical Mode Decomposition, Shannon Entropy and Least Square Support Vector Machine. In Proceedings of the 2016 4th International Conference on Mechanical Materials and Manufacturing Engineering, Wuhan, China, 15–16 October 2016; Atlantis Press: Paris, France, 2016.

27. Luo, B.; Wang, H.; Liu, H.; Li, B.; Peng, F. Early fault detection of machine tools based on deep learning and dynamic identification. *IEEE Trans. Ind. Electron.* **2018**, *66*, 509–518. [CrossRef]

28. Wang, J.; Gao, R. Integration of EEMD and ICA for wind turbine gearbox diagnosis. *Wind Energy* **2014**, *5*, 757–773. [CrossRef]

29. Cheng, C.; Peng, Z.; Dong, X.; Zhang, W.; Meng, G. A novel damage detection approach by using Volterra kernel functions based analysis. *J. Frankl. Inst.* **2015**, *8*, 3098–3112. [CrossRef]

30. Hoffmann, H. Kernel PCA for novelty detection. *Pattern Recognit.* **2007**, *3*, 863–874. [CrossRef]

31. He, Q. Time-frequency manifold for nonlinear feature extraction in machinery fault diagnosis. *Mech. Syst. Signal Process.* **2013**, *35*, 200–218. [CrossRef]

32. Zhang, S. *Research on Methods of Machinery Condition Recognition Based on Manifold Learning*; Southwest Jiaotong University: Chengdu, China, 2014.

33. Dong, H. *Study on Stability and Bifurcation Types of Railway Vehicles*; Southwest Jiaotong University: Chengdu, China, 2014.

34. *TB/T3188-2007, Technical Specification for Railway Car Safety Monitor and Diagnosis*; China Academy of Railway Sciences Locomotive and Car and Research Institute: Beijing, China, 2007.

35. *TB 10761, Ministry of Railways of the People's Republic of China, Technical Regulations for Dynamic Acceptance for High-Speed Railways Construction*; China Academy of Railway Sciences Locomotive and Car and Research Institute: Beijing, China, 2013.

36. Ning, J.; Lin, J.; Zhang, B. Time-frequency processing of track irregularities in high-speed train. *Mech. Syst. Signal Process.* **2016**, *66*, 339–348. [CrossRef]

37. Madiena, C. Color and vector flow imaging in Parallel Ultrasound with Sub-Nyquist sampling. *IEEE Trans. Ultrason. Ferroelectr. Freq. Control* **2018**, *5*, 795–802. [CrossRef]

38. Bruni, S.; Vinolas, J.; Berg, M. Modelling of suspension components in a railvehicle dynamics context. *Veh. Syst. Dyn.* **2011**, *49*, 1021–1072. [CrossRef]

39. Wen, Z.F.; Jin, X.S. Effects of lateral deformations of wheelset/track on creepforces of wheel/rail. *J. Mech. Strength* **2002**, *24*, 383–387.

Article

A Cognitive Method for Automatically Retrieving Complex Information on a Large Scale

Yongyue Wang [1], Beitong Yao [1], Tianbo Wang [2], Chunhe Xia [1,3] and Xianghui Zhao [4,*

[1]   School of Computer Science and Engineering, Beijing Key Laboratory of Network Technology, Beihang University, Beijing 100191, China; buaawyy@buaa.edu.cn (Y.W.); yao6921@buaa.edu.cn (B.Y.); xch@buaa.edu.cn (C.X.)
[2]   School of Cyber Science and Technology, Beihang University, Beijing 100191, China; wangtb@buaa.edu.cn
[3]   School of Computer Science and Information Technology, Guangxi Normal University, Guilin 541004, China
[4]   China Information Technology Security Evaluation Center, Beijing 100085, China
*   Correspondence: zhaoxianghui@126.com

Received: 19 April 2020; Accepted: 26 May 2020; Published: 28 May 2020

Abstract: Modern retrieval systems tend to deteriorate because of their large output of useless and even misleading information, especially for complex search requests on a large scale. Complex information retrieval (IR) tasks requiring multi-hop reasoning need to fuse multiple scattered text across two or more documents. However, there are two challenges for multi-hop retrieval. To be specific, the first challenge is that since some important supporting facts have little lexical or semantic relationship with the retrieval query, the retriever often omits them; the second challenge is that once a retriever chooses misinformation related to the query as the entities of cognitive graphs, the retriever will fail. In this study, in order to improve the performance of retrievers in complex tasks, an intelligent sensor technique was proposed based on a sub-scope with cognitive reasoning (2SCR-IR), a novel method of retrieving reasoning paths over the cognitive graph to provide users with verified multi-hop reasoning chains. Inspired by the users' process of step-by-step searching online, 2SCR-IR includes a dynamic fusion layer that starts from the entities mentioned in the given query, explores the cognitive graph dynamically built from the query and contexts, gradually finds relevant supporting entities mentioned in the given documents, and verifies the rationality of the retrieval facts. Our experimental results show that 2SCR-IR achieves competitive results on the HotpotQA full wiki and distractor settings, and outperforms the previous state-of-the-art methods by a more than two points absolute gain on the full wiki setting.

Keywords: information retriever sensor; multi-hop reasoning; evidence chains; complex search request

---

1. Introduction

Most previous work about retrievers focus on searching the relative lexicon from a single paragraph, and some easy search queries may be satisfied reluctantly using a single paragraph or document by the traditional retrieval methods of search engines, such as matching lexical or semantic similarity or relatedness. However, a very challenging retrieval task for complex search requests, called multi-hop reasoning retrieval, requires combining evidence from multiple sources, as shown in Figure 1. Single-hop retrieval methods are almost incompetent for intricate search tasks. The objective of a multi-hop retriever is to obtain and fuse scattered information from the internet step by step [1,2]. Figure 1 provides an example of a search request requiring multi-hop reasoning. In response to the search request, only when one first infers from the first context in the search request can the Internet information corresponding to the query be inferred with any confidence from the second context.

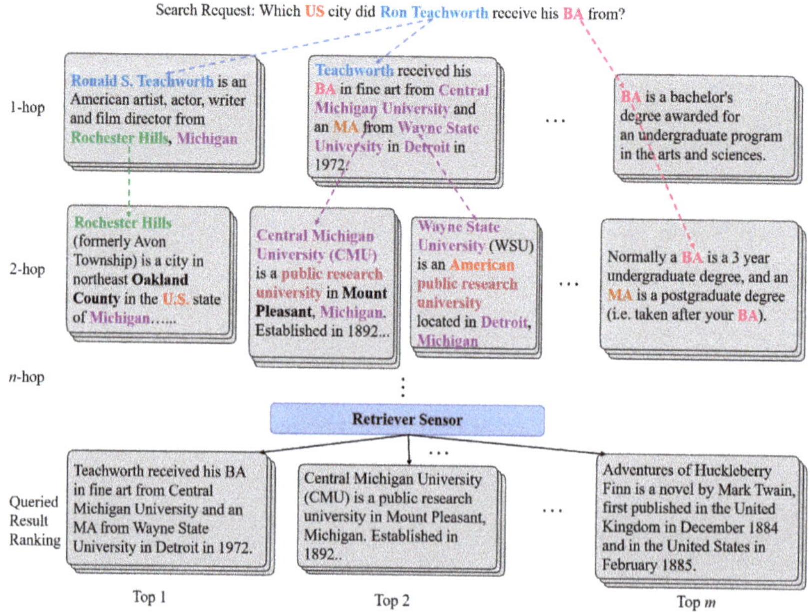

Figure 1. An example of a multi-hop retrieval task. The method in this paper mimics the human reasoning process while searching for information online. The retrieval sensor will finally provide users with reasoning chains, which will help obtain structured retrieval information.

The multi-hop IR is of great difficulty, because it requires reading and aggregating information over multiple pieces of textual evidence. Although the Reading Comprehension model is applied to candidates of the paragraph for prediction (e.g., QFE [3], MINIMAL [4]), the aggregation of evidence from sources for supporting fact prediction is beset with critical difficulty. One of the challenges is that since some pivotal supporting facts have little lexical or semantic relationship with the original retrieval query, the retriever often omits these crucial facts, as illustrated in Figure 2. The reason for this challenge is that multi-hop IR is open domain retrieval on a large scale, where search requests are given without any accompanying contexts. In this case, one is required to locate relevant contexts to queries from a large internet source before extracting the most relevant information. Retrieving a fixed list of documents independently does not capture the relationships between evidence documents through bridge entities required for multi-hop reasoning. They often fail to retrieve the required evidence for the multi-hop query. The recent open domain IR methods [5] do not capture lexical information in entities, but are encountered with challenges for entity-centric problems. The complex cognitive process in humans seems to provide more inspiration for this retrieval process. Recently, some researchers began to mimic the human reasoning process to search for complex information. Although Ding's Cognitive Graph [6] takes the entity relations, this method cannot find the reasoning path automatically. Multi-Step Reasoner introduces a multi-step reasoning method [7], but it only repeats the retrieval process for a fixed number, resulting in ineffectively integrated information. Qi et al. propose GoldEn IR [8], which cannot use the information of relations between documents during the iterative process. Most state-of-the-art approaches [2,9–11] for the open-domain retrieval leverage non-parameterized models to retrieve a fixed set of documents, where an information span is extracted by a neural model. However, they often fail to retrieve the required evidence needed by multi-hop reasoning. Some research on the reasoning process do not separate the information from the cognitive graph. If helpful information for users is not selected on the cognitive graph, the retriever may fail. In addition, the explosion of online textual content with unknown verification raises a vital concern about misinformation, such as fake news, rumors, and online "water army" opinions [12–14]. While some misinformation may have semantic relations with the retrieval query, they cannot be efficaciously identified by some methods [15–17] with ideal performance. If the scattered information does not have stronger interactions with neighbors on the cognitive graph, a retriever is apt to select this misinformation in comparison with the whole nodes. While RE^3 and Multi-Passage [11,18] introduce an extract-then-select framework to re-rank candidates to improve these interactions, they overlook the information between different evidence candidates.

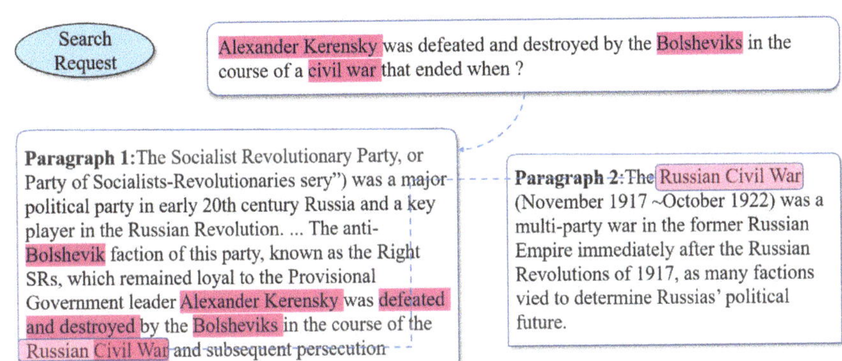

Figure 2. A complex example of an open-domain retrieval task from HotpotQA. Paragraph 2 is difficult to be retrieved using traditional retrievers due to little lexical overlap with the given query.

In this paper, we study the problem of multi-hop retrieval methods, which requires multi-hop reasoning among evidence scattered around multiple raw documents on a large scale. Despite the given query utterance and a set of accompanying documents, not all of them are relevant. The information can be obtained by selecting two or more pieces of evidence from documents and reasoning among them. The models are expected to search the most useful information for search request in open domains. We propose a sub-scope cognitive reasoning information retrieval (2SCR-IR) sensor, a novel method to address the above-mentioned difficulties for multi-hop retrieving, as shown in Figure 3 (Nodes are fundamental entities, with the color denoting the identical entity in paragraphs. The blue edges are the relationship between the different entities in the same paragraph, while the red ones are the relationship between the same entities in different paragraphs. One search request and three paragraphs are given. Our 2SCR-IR sensor conducts multi-step reasoning over the supporting evidence by constructing a cognitive graph from the supporting information, automatically adjusting the sub-scope to select a subgraph, propagating information through the graph, verifying the reliability of the information, and finally producing supporting evidence chains. The circles, denoting the sub-scope, are chosen by 2SCR-IR in each step). For the first challenge, the 2SCR-IR sensor constructs a self-adjusting sub-scope cognitive graph based on entities referred to in the search request and web documents, which is iterated to accomplish multi-hop reasoning. In each step, the 2SCR-IR sensor is produced and reasons on a dynamically adjusted graph, with unrelated entities left out and reasoning information exclusively preserved in a scope prediction process. To solve the second difficult problem, we use an information fusion process in the 2SCR-IR sensor to eliminate misinformation. The process of iteratively expanding with clues in sub-scope can discover the weak paragraphs relative to the query in our framework, which can also play a pivotal role in filtering out misinformation. For the purpose of further verifying the authenticity of retrieved information, we introduce pageview verification to our sensor. Overall, our work incorporates a four-fold contribution:

1. We proposed a 2SCR-IR sensor, a novel means for the multi-hop text-based retrieval tasks.
2. We proposed a novel sub-scope module in the 2SCR-IR sensor capable of settling the problem that resulted from weak correlations between the queries and useful documents.

3. To increase the retrieved information's credibility, we introduce the fusion module and pageview information to our sensor to filter misinformation.

4. Most IR methods using the "black box" merely provide users with retrieval results, generating utter ignorance of users about the logical relationship of the information they list. We proposed an explicit way to explain the reasoning process of the retriever. As a result, our 2SCR-IR sensor can produce reasoning chains, which can provide users with a range of structured documents or paragraphs. Compared with scattered and unstructured text, the reasoning chains can help users raise the efficiency of searching for complex information.

2. Related Works

2.1. Search Engine and Information Retrieval

The search engine is a system that assists users in retrieving information that they wish to obtain after submitting searching queries. One search engine, in response to a retrieval query, typically compares keywords of the query with an index generated from a sea of web sources, such as text files, image files, video files, and other content items. Based on this comparison, it will provide users with the most relevant content items [19]. In our paper, only textual information retrieved from the Internet is taken into account. With the rapid development of deep learning, researchers have an increasing interest in bringing neural networks into IR tasks. Although IR tasks frequently use sentence-similarity, as discussed by Guo et al. [20], they pay more attention to relevance-matching, in which the match of specific content plays a significant part. Some researchers [21] have confirmed that IR is more about retrieving sentences with the identical semantic meaning, while traditional methods based on relevance matching show a serious deficiency in semantic understanding. Therefore, we used natural language reasoning techniques, instead of relevance-focused IR methods, to solve this issue. In many cases, the paragraph containing the information corresponding to the searching query has great lexical overlap with the query, adding to its difficulty for the retrieval of common search engines from a large open scope. For instance, the accuracy of a BM25 retriever for finding all supporting evidence for a query diminishes from 57.3% to 25.9% on the "easy" and "hard" subsets, respectively [2].

(**a**)

Figure 3. *Cont.*

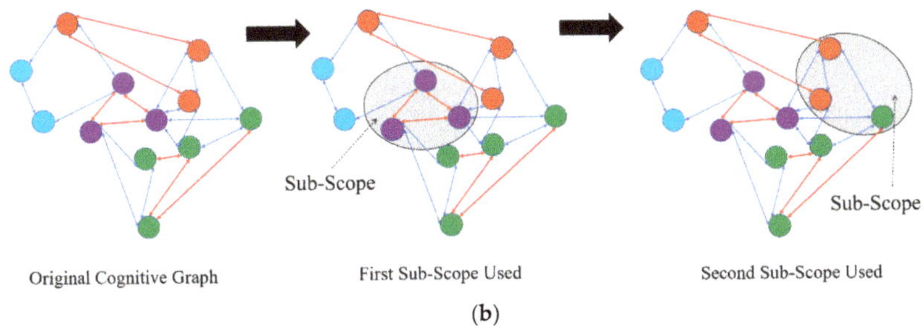

Figure 3. An example of a multi-hop text-based information retrieval task. (**a**). Search request and corresponding information source. (**b**). Reasoning process.

2.2. Single-Hop and Multi-Hop QA

Relying on the complexity in underlying reasoning, information retrieval tasks can be categorized into single-hop and multi-hop ones. The former only needs one piece of evidence extracted from the underlying information, e.g., "which city is the capital of China". On the contrary, a multi-hop retrieval task requires recognizing multiple relevant evidence and reasoning about them, e.g., "what is the capital city of the largest province in China". Many IR techniques that are able to reply the single-hop searching query [22] are hardly introduced in multi-hop tasks, since single evidence can only partially satisfy its searching query.

2.3. Open-Domain Retrieval Task

Open-Domain retrieval task refers to the setting where the search scope of the supporting evidence is extremely large and complex. Recent open domain IR systems based on deep learning methods follow a two-step process, namely selecting potentially relevant documents from a large corpus and extracting useful sources from these selected documents. As the first to successfully apply this framework to Chen et al. [23], IR [24], introduced new benchmarks, [11,25] improved this framework by introducing feedback signals, and [26] proposed a method that can dynamically adjust the number of retrieved documents. However, these mainstream methods are weak in mining the information loosely related to the searching queries and fail to identify the misinformation.

2.4. Similarity Matching Based Method

Traditional frequency-based approaches, such as the "bag of words" model [27], have been employed to extract features in the search engine. However, these frequency-based techniques do not preserve the text sequence, bringing about the lack of understanding of the context's full meaning [28]. While other text-similarity-matching approaches, such as tree kernels [29] and high order n-grams [30], may have the perception of the word order and semantics, they cannot master the context meaning completely, thus heavily affecting the accuracy of recognition.

2.5. Multi-Hop Reasoning for Retrieval Task

Previous research on popular GNN frameworks has shown promising results in retrieval tasks requiring multi-hop reasoning. Coref-GRN aggregates information in disparate references from scattered paragraphs [31]. The different mentions of the same entity are combined with a graph recurrent neural network (GRN) to produce aggregated entity representations [32]. Based on Coref-GRN, MHQA-GRN [33] refines the graph construction procedure with more connections, which shows further improvements. Entity-GRN [34] proposes a method to distinguish different relations in the cognitive graphs through the convolutional neural network. Besides, Core-GRN [35] explores the cognitive construction problem. Nonetheless, it is rarely investigated by researchers about how to effectively reason on the constructed cognitive graphs, which is the primary problem studied in this paper.

Other frameworks, such as Memory Networks [36], deal with multi-hop reasoning. These frameworks develop representations for queries and retrieval sources, between which interactions are then made in a multi-hop manner. The IR-Net [37] generates an entity state and a relation state at each step, computing the similarity degree between all entities and relations given by the dataset. However, these frameworks perform reasoning on simple datasets with a limited number of entities and relations, which is quite different from our work with the large-scale retrieval and intricate search queries.

2.6. Pageview

The pageview is a collection of Web objects or resources representing a specific user event, such as clicking on a link or viewing a product page [38]. Besides, pageview frequency has been shown to help improve the performance of evidence retrieval results [39].

2.7. Peasoning Chains

A reasoning chain is a sequence that logically connect users' search requests to supporting facts [40]. Reasoning chains should be intuitively related: they should exhibit a logical structure [41], or some other kind of textual relation that would allow human readers to quickly obtain what they really need.

3. System Architecture

3.1. Overview

In this section, we describe our new graph-based reasoning method that learns to find supporting evidence as reasoning paths for complex retrieval tasks. Our inspiration is motivated by the human cognitive process for searching an open question online, and our method starts from selecting the critical entity in the retrieval query. In order to make better use of the current entity's surrounding information, our method is to connect the start entity to some related entities either searched in the neighborhood or connected by the same surface mentioned. The abovementioned process is repeated to construct reasoning chains.

The main part to implement the 2SCR-IR framework is comprised of two components in our proposed system (Figure 4): a system to construct the cognitive graph, on which the other system reasons explicitly. Our method uses a strong interconnection between evidence to prevent important documents from being omitted in the reasoning paths.

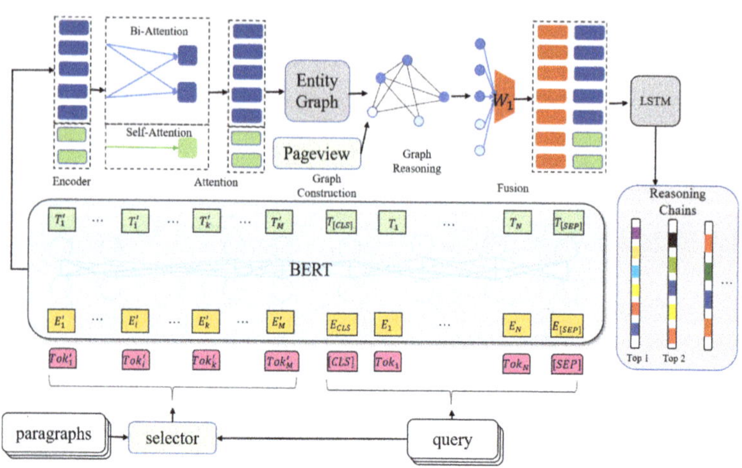

Figure 4. Model architecture of the proposed 2SCR-IR sensor.

We used Wikipedia to retrieve relevant information for an open-domain search request, in which each article falls into paragraphs, resulting in billions of paragraphs in total. Any paragraph p can be perceived as our retrieval target. Given a query q, our framework aims at deriving its relevant textual sources to make up evidence chains by retrieving and reasoning paths, each of which is represented with a sequence of paragraphs.

3.2. Gold Paragraph Selection

To ease the computational burden of following steps, we first applied paragraph selection to the given context. Documents with titles matching the whole query were our priority. The selector will calculate the relevance score of the query and each paragraph between 0 and 1. We can set a threshold parameter τ, and paragraphs with predicted scores greater than τ are chosen, which are concatenated together as the context C. Since not every piece of text is relevant to the query, we further search for paragraphs that contain entities appearing in the searching query. If the two previous ways fail, a BERT-based paragraph ranker [42] will be used to select the paragraph with the scores higher than τ. Q and C are further processed by upper layers. After the first-hop paragraphs are identified, the next step is to search for evidence within paragraphs leading to other relevant paragraphs. Unlike the dependence on entity linking, we use hyperlinks in the first-hop paragraphs to find second-hop paragraphs to avert introducing noise. After all the links are chosen, we add edges between the evidence with these links and hyperlinks. When the two-hop paragraph is selected, we can obtain some candidate paragraphs. In order to further lower noise, we use the paragraph ranking model based on the BERT encoder to select the top-N paragraphs in each step. Besides, a retrieval query q and each of the 10 paragraphs p_i are introduced into a BERT model [42]. Furthermore, a soft-max function is utilized to calculate the probability $P\,(q, p_i)$ of p_i. A paragraph p_i is chosen as a gold paragraph for the query q if $P\,(q, p_i)$ is larger than the setting τ. To collect most of the gold paragraphs for each retrieval query, the threshold value of 97.0% for recall and 69.0% for precision was set.

3.3. Encoder and Attention Module

We introduced a BERT model to acquire the representation of the query $Q = [q_1, q_2 \dots q_k] \in \mathbb{R}^{k \times d}$ and that of the context $C = [C_1, \dots C_l] \in \mathbb{R}^{l \times d}$, where k and l are the lengths of the searching query and the context, and d is the size of the BERT hidden layer.

To gain more semantic information, following [43], we use the bidirectional attention module to update the representation for each word and a weighted self-aligned vector to represent the query.

The representation of the context C is

$$C = [C; CQ; C \odot CQ; QC \odot CQ] \in \mathbb{R}^{l \times 4d_h} \tag{1}$$

where $\odot$ is elementwise multiplication, and d_h is the hidden dimension.

The representation of the searching query is

$$\varphi = softmax\left(w_\varphi h\right) \tag{2}$$

$$q = w_a tanh\left(W_a \sum_k \varphi_k h_k\right) \tag{3}$$

where h represents the query embedding; h_k denotes the k-th word in the query; and w_φ, w_a, and W_a are linear projection parameters.

3.4. Construction of Cognitive Graph

To perform the multi-hop reasoning, we first needed to construct a graph of the information source to cover all the Wikipedia paragraphs. The Stanford CoreNLP Toolkit [44] was employed in an attempt to generate semantic graphs from the source text. The cognitive graph was constructed with entities as nodes and edges, while the number of extracted entities was denoted as N. We constructed the entity graph based on Entity-GCN [34] and DFGN [16], signifying that all mentions of the candidates searched in the documents are used as nodes in this cognitive graph. We used hyperlinks in Wikipedia to develop the direct edges, while the undirected edges were defined in line with the positional properties of every pair of nodes. The edges in this graph were split into two categories:

1. The cross-document edge for every pair of nodes with the same entity located in different documents.
2. The within-document edge for every pair of nodes located in the same document, including sentence-level, paragraph-level, and context-level links.

It needs to be emphasized that we do not apply the co-reference resolution to pronouns because it will introduce complex and unnecessary links.

3.5. Multi-Step Reasoning

After context encoding, the 2SCR-IR sensor performs reasoning over the cognitive graph. With the embedding process of the query Q and the context C, there is a huge challenge: to identify support entities and the text span of potential evidence paragraphs, as well as capturing relationships between evidence documents with little lexical overlap or semantic relationship to the original query. Based on the Cognitive Graph [6] and the DFGN method [16] for the QA system, an advanced knowledge-fusion module is adopted to mimic humans' step-by-step thinking and reasoning behavior—starting from Q_0 as well as C_0 and finding one-step supporting entities. The module achieves the following:

1. passing information from tokens to entities (Doc2Graph flow);
2. propagating information on the entity graph;
3. passing information from the entity graph to tokens (Graph2Doc flow).

After obtaining the entity embeddings from the input context C_t, we applied a graph neural network to propagate the node information to its neighbors. Furthermore, a dynamic graph attention mechanism, which is a hierarchical and top-down process, was utilized to imitate humans' step-by-step exploring and reasoning behavior. It first attentively reads all knowledge graphs and then all triples in each graph for the final word generation. Yan et al. [45] show that the higher relevance given to the query can help the neighbor entity obtain more information from nearby. In order to calculate the degree of relation between the context and the query, 2SCR-IR calculates the attention between them. If one entity is more relative to the query, the entity is more pivotal. We followed [16] to multiply this relevance score:

$$\gamma_i^{(t)} = sigmoid\left(\frac{\tilde{q}^{(t-1)} V^{(t)} e_i^{(t-1)}}{\sqrt{d_2}}\right) \tag{4}$$

$$\tilde{E}^{(t-1)} = \gamma^{(t)} \odot e^{(t-1)} \tag{5}$$

where $V^{(t)}$ is a linear projection matrix; d_2 represents the dimension of each entity; and $\gamma_i^{(t)}$ is the relevance score of the entity e_i in step t. The entity graph is updated in a loop.

As a result, this step of the information propagation is confined to a dynamic scope on the cognitive graph. The next step is to disseminate information across the dynamic scope.

We introduce how to integrate the output of the entity into the context. To identify which sentences are more crucial to answer the question, we used a weighted self-aligned vector to represent the context. For each sentence s_i, we compute weight ε_i by

$$\varepsilon_i = softmax\left(w_\varphi s_i\right) \tag{6}$$

The context vector is computed by

$$C_V = w_a tanh\left(W_a \sum_i \varepsilon_i s_i\right) \tag{7}$$

where w_φ, w_a are linear projection parameters, the context vector $C_V \in \mathbb{R}^d$.

After the context vector is concatenated to each word in the context, the i-th word c_i in the context C is updated as follows:

$$c_i = [c_i; C_V] \tag{8}$$

After completing the contextual information, we used an LSTM layer to fuse the output of the entity graph and the context:

$$C = LSTM([C; E]) \tag{9}$$

The structure layer we used is the same as that used by Yang et al. [2]. This prediction process has three functions, including support sentences as well as the start and end position of the information needed, denoted as O_{sup}, O_{start}, O_{end}, respectively. We put three RNNs P_i layer by layer to solve the problem of output dependency. The information of the fusion block is sent to the first RNN P_0, and the output of this module is as follows:

$$\mathbf{O}_{sup} = P_0\left(\mathbf{C}^{(t)}\right) \tag{10}$$

$$\mathbf{O}_{start} = P_1\left(\left[\mathbf{C}^{(t)}, \mathbf{O}_{sup}\right]\right) \tag{11}$$

$$\mathbf{O}_{end} = P_2\left(\left[\mathbf{C}^{(t)}, \mathbf{O}_{sup}, \mathbf{O}_{end}\right]\right) \tag{12}$$

Each P_i outputs a logit $\mathbf{O} \in \mathbb{R}^{M \times d_2}$.

In order to optimize the combined effect of these three sections, we compute these three cross entropy losses:

$$L = \mu_{start}L_{start} + \mu_{end}L_{end} + \mu_{spport}L_{support} \tag{13}$$

Our method learns to extract reasoning paths within the constructed cognitive graph as evidence chains. Besides, evidence resources for a complex searching query do not necessarily have lexical overlaps. We adopted an automatically adjusting sub-scope to collect those entities with some sort of relation. Once one of the resources is retrieved, its entity mentions and the query often entail another resource.

3.6. Producing Plausible Evidence Chains

Finally, the 2SCR-IR sensor will provide users with reasoning chains, which first verifies every reasoning path, and then extracts a retrieval span from the most reasonable paths using a common method [42]. Our retriever model learns to predict plausible reasoning paths by capturing the paragraph interactions through the BERT (CLS) representations, after independently encoding the paragraphs along with the searching query. This paragraph interaction is crucial for multi-hop reasoning [46], especially when faced with open-domain information. However, as there is invariably some noise or misinformation, the best evidence chain is not always enough to fully cover all information to be retrieved. Hence, we re-computed the probability of the path with the increase in the pageview to lower the uncertainty and make our framework more robust. In the end, we select the top-m evidence chains for users.

3.7. Example

After N paragraphs and the query are put in, the selector module will filter out irrelevant paragraphs. Then, BERT is used as the encoder module to represent the filtered content in line with Formulas (1)–(3). After the process of knowledge representation, a classification layer is applied to the prediction of relevance scores between the paragraphs and the query. One paragraph will be signed as 1 if it contains supporting facts. Otherwise, it will be signed as 0. In the inference process, the paragraph will be chosen if the relevance score exceeds τ. In the process of reasoning, in order to make inferences on information, we calculated the degree of the relationship between the entities and the query in the light of Formulas (4) and (5) and built the entity graph. After we got the output of the entity graph, it was fused into the context. We used Formulas (6)–(9) to recognize the relevance of the search request to sentences. Next, Formulas (10)–(12) were utilized to predict the supporting sentences, as well as the start and end position corresponding to each search request. Moreover, Formula (13) was employed to optimize the combined effect. Finally, we will get supporting facts and construct reasoning chains in correspondence with the search request.

4. Experiments

4.1. Datasets

We evaluate our method on an open-domain Wikipedia-sourced dataset named HotpotQA [2], which is a human-annotated large-scale dataset that needs multi-hop reasoning and provides annotations to evaluate the prediction of supporting sentences. Almost 84% of queries require multi-reasoning. We used the full-wiki and the distractor setting of HotpotQA to conduct our experiments, with our primary target being the full wiki setting. Due to its open-domain scenario, we used the distractor setting to evaluate the performance of our method in a closed scenario where the evidence candidates are provided. For the sake of optimization, we used the Adam Optimizer [47] with an initial learning rate of 0.0005 and a mini-batch size of 32.

In addition, a good information retriever should be robust enough to prevent noise from disturbing themselves. IR can be armed with the well-structured and large-scale knowledge graph DBPedia [48] to offer some robust knowledge between concepts. However, the biggest disadvantage of this approach lies in that our cognitive model cannot be trained end-to-end and the errors may be cascaded. Inspired by [23], our retriever is trained to recognize relevant and irrelevant paragraphs at each step. We therefore mixed the noise information (negative examples) and generally right paragraphs together; to be more specific, we used two types of negative examples: Term Frequency–Inverse Document Frequency-based (TF–IDF-based) and hyperlink-based ones. When it comes to the single-hop retriever, merely the former type is used. Regarding the multi-hop retriever, we used both sorts, with the latter type carrying more weight to prevent our retriever from being distracted by reasoning paths without correct answer spans. Generally, the number of negative examples is set to 50.

4.2. Implementation Details

We performed our experiment in the Co-lab. In the section paragraph sector, the setting of the threshold τ is 0.1, with the length of the context restricted to 512. Besides, the number of entities in the cognitive graph was set to be 100. The input dimension of the BERT, the hidden dimension, and the batch size were set to be 800, 310, and 10, respectively. The total train epoch was set to be 30 and the first 5 epochs.

4.3. Baseline

We used Yang's model [2] proposed in the original HotpotQA paper as a baseline model that follows the retrieval–extraction framework of DrQA [23] and introduces the advanced techniques, such as self-attention and bi-attention.

4.4. Evaluation Metrics

First, we introduced two common metrics to evaluation metrics: the Exact Match (EM) and the F1 score [2]. For better a performance evaluation of the multi-hop reasoning concerning our sensor, we also introduced the supporting fact retrieval based on the common metrics. We adopted Supporting fact Prediction F1 (SPF1) and Supporting fact Prediction EM (SPEM) to evaluate the sentence-level supporting fact retrieval accuracy. It is worth noting that the paragraph-level retrieval accuracy matters for the multi-hop reasoning as well. Thus, we adopted Paragraph Recall (PR), which evaluates if at least one of the ground-truth paragraphs is included among the retrieved paragraphs. To evaluate whether both of the ground-truth paragraphs used for multi-hop reasoning are included among the retrieved paragraphs, we put Paragraph Exact Match (PEM) into use.

5. Results

5.1. Overall Results

Table 1 shows the performance of dissimilar IR models on the HotpotQA development set. From the table, we can see that the 2SCR-IR sensor outperforms all previous results under both the full wiki and distractor settings. Compared with the state-of-the-art model [15], 2SCR-IR achieves 1.1 F1 and 1 EM gains on the full wiki, as well as 1.2 F1 and 1.9 EM gains on the distractor wiki. The result shows that 2SCR-IR achieves improvement in predicting supporting facts both in the full wiki and distractor wiki settings.

Table 1. Primary results for HotpotQA development set results: SP results on the HotpotQA's full wiki and distractor settings. The highest value per column is marked in bold.

Models	Full-Wiki SP		Distractor SP	
	F1	EM	F1	EM
Baseline [2]	40.7	5.4	68.3	23.2
MUPPET [7]	47.7	16.5	76.1	45.6
Cognitive Graph [6]	57.2	24.6	60.3	38.2
GoldEn Retriever [8]	65.2	31.4	69.8	45.4
Semantic Retrieval [9]	72.3	41.2	73.2	50.7
DFGN [16]	73.2	43.7	85.1	58.3
QFE [3]	45.2	13.9	82.9	56.7
Asai [15]	76.1	49.2	85.1	59.3
Our method	**77.2**	**50.2**	**86.3**	**61.2**

After comparing our 2SCR-IR sensor with other competitive retrieval methods on SQuAD datasets, our model is found to outperform the current state-of-the-art model [49] by 2.6 F1 and 3 EM scores, as shown in Table 2 At the beginning of the paper, we predicted that the lower lexical overlap between the search query and contexts will pose a challenge to the methods using lexical-based retrievers in finding relevant articles. The experiment proved that our prediction is valid.

Table 2. SQuAD results: We report the F1 and EM scores on SQuAD, following previous work. The highest value per column is marked in bold.

Models	F1	EM
R^3 [11]	37.8	28.9
Multi-step Reasoner [41]	39.3	32.7
MINIMAL [4]	42.6	34.9
DENSPI-hybrid [5]	45.1	36.7
BERTserini [9]	45.9	38.4
MUPPET [7]	46.3	39.3
RE^3 [18]	51.2	42.3
multi-passage [49]	61.6	54.1
Our method	**64.2**	**57.1**

5.2. Performance of Evidence Chain Retrieval

Retrieval results in Table 3 display that our 2SCR-IR sensor generates the improvement of 1.3 PR, 7.2 PEM, and 7.8 EM, respectively. The conspicuous improvement from TF-IDF to Entity-centric IR demonstrates that exploring different granularity reasoning helps to retrieve the paragraphs with fewer term overlaps. Moreover, the comparison of our retriever with Entity-centric IR Retrieval shows the importance of explicitly retrieved reasoning paths in the cognitive graph, especially for the complex multi-hop searching query.

Table 3. Retrieval evaluation: Comparing our retrieval method with other methods across Paragraph Recall, Paragraph EM, and EM metrics. The highest value per column is marked in bold.

Methods	Paragraph Recall (PR)	Paragraph EM (PEM)	EM
TF-IDF [23]	66.7	9.8	18.3
Re-rank [50]	86.2	28.9	36.1
Entity-centric IR [17]	87.2	35.3	41.6
Cognitive Graph [6]	88.2	56.9	38.4
Semantic Retrieval [39]	92.9	65.7	46.3
Our method	**94.2**	**72.9**	**54.1**

5.3. Ablation Study

To evaluate the performance of the disparate components in our 2SCR-IR sensor, we performed ablation studies, where we simply use golden supporting facts as the input context. Table 2 shows the ablation results of the multi-hop retrieval performances in the development set of HotpotQA.

From Table 4, we can observe that the sub-scope and pageview parts can help our sensor obtain from 3 to 7% relative and 5 to 8% gains for F1 and EM, respectively. "-attention module" means discarding the information of bidirectional attention. Similarly, the performance drops a lot in each metric, which proves the capability of the bidirectional module to comprehend semantics.

Table 4. Ablation results on dev set.

Model	F1	EM
2SCR-IR	86.3	61.2
-sub-scope and pageview part	78.6	56.2
-attention module	82.7	53.9

At the beginning of the paper, we predicted that the pageview and the sub-scope can improve the ability to find useful supporting facts. In the experiment, our observation and proposed scheme are proved valid.

6. Case Study

Our case study is presented in Figure 5, which signifies the reasoning process with a 2SCR-IR sensor. Firstly, our model produces scope 1 as the first entity of reasoning by comparing the searching query with entities, in which "Ran Paul" and "hotel" are selected as the first entities of two reasoning chains, the information of which is then passed to their neighbors on the cognitive graph. Secondly, mentions of the same entity "Galt house" are detected by scope 2, serving as a bridge for propagating information across two paragraphs. Thirdly, these two reasoning chains are linked together by the bridge entity "Galt house". Both reasoning chains and supporting facts are provided to users, which will assist them in obtaining things they want to retrieve very promptly and efficiently. Finally, users can acquire "The Ran Paul presidential campaign 2016 event was held at a hotel which is on the Ohio River", nearly without any complex manual retrieval process.

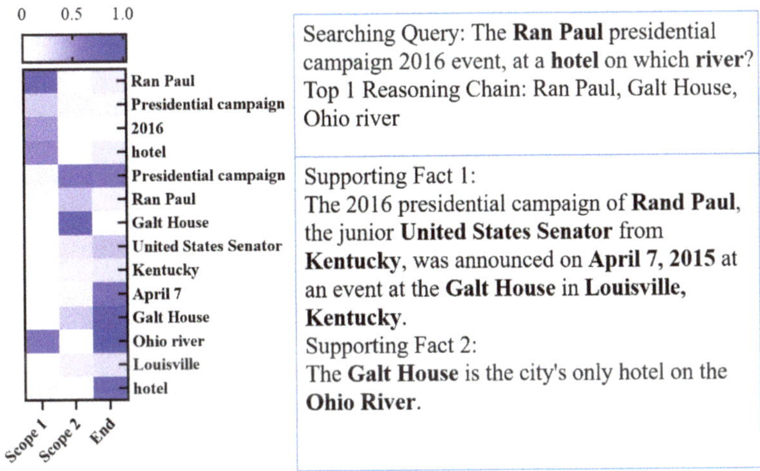

Figure 5. A case study of the development set. The number on the left side shows the importance scores of the predicted sub-scope. The text on the right side includes queries, predicted top-1 reasoning chains, and supporting facts.

Figure 6 shows another multi-hop reasoning example of 2SCR-IR from the HotpotQA development set. In this case, we need to search information about the actress who played *Corliss Archer* in *A Kiss for Corliss*. The information relevant to the search request may appear in multiple positions in texts, but they have little lexical or semantic relationship to the original retrieval query. It is generally difficult for common search engine system to know which entity mentions might eventually lead to texts containing what the user really needs. Our 2SCR-IR can provide users with a logically structured text. From the reasoning chain and three supporting facts, we can easily get the information: *The woman who portrayed Corliss Archer in the film A Kiss for Corliss held Chief of Protocol in the government*. As shown in Figure 7, our 2SCR-IR method has been successfully applied to a KLBNT Retrieval System. For complex problems, we can get the reasoning process and the logically structured documents.

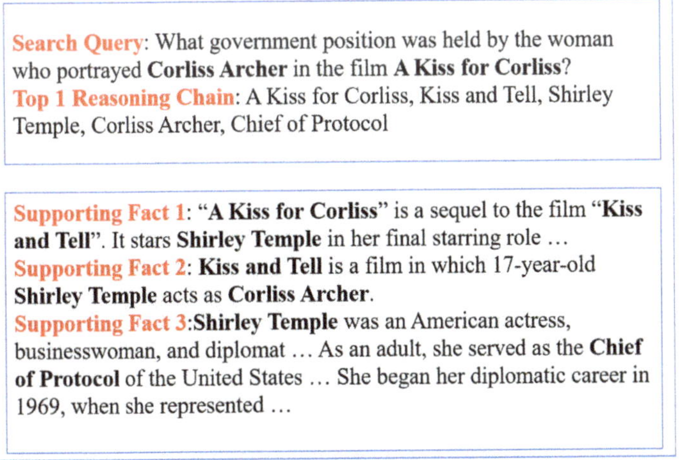

Figure 6. Another case study in the development set. Several documents are given as supporting facts for the search query.

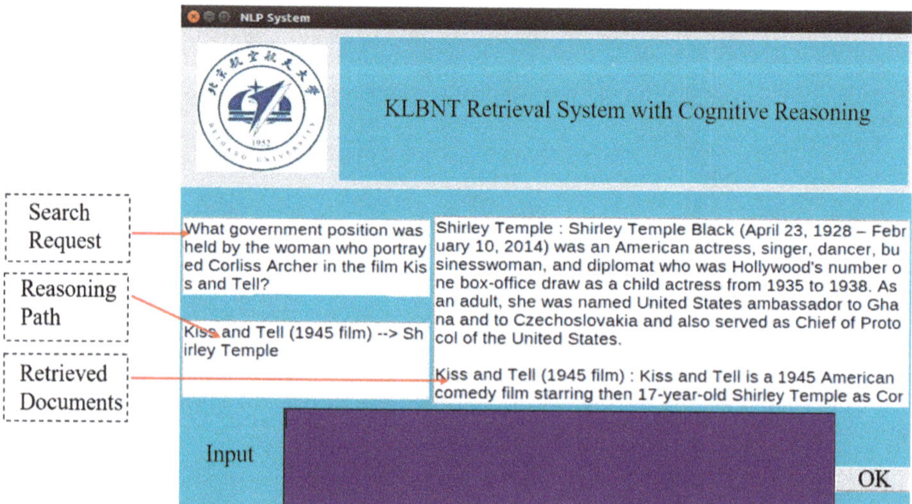

Figure 7. The KLBNT Retrieval System supported by 2RCR-IR technology.

7. Conclusions

We present a new framework 2SCR-IR sensor to tackle multi-hop retrieval problems on a large scale, which retrieves reasoning paths over the cognitive graph to provide users with useful explicit evidence chains. Our retriever model learns to sequentially retrieve evidence paragraphs to construct reasoning paths, which is subsequently re-ranked by the sensor that determines the final information presented as the one extracted from the best reasoning path. Our retriever obtains state-of-the-art results using the HotpotQA dataset, which shows the efficiency of our framework. The state-of-the-art performance on SQuAD is achieved, demonstrating the robustness of our method. Besides, our analysis shows that 2SCR-IR can produce reliable and explainable reasoning chains. In the future, we may incorporate new advances in building cognitive graphs from the web context to solve more difficult reasoning problems.

Author Contributions: X.Z. and C.X. planned and supervised the whole project; Y.W. developed the main theory and wrote the manuscript; B.Y. and T.W. contributed to doing the experiments and discussing the results. All authors have read and agreed to the published version of the manuscript.

Funding: This research was funded by the National Natural Science Foundation of China with the grant number of U1636208, NO 61862008 and NO. 61902013.

Conflicts of Interest: The authors declare no conflict of interest.

References

1. Rajpurkar, P.; Jia, R.; Liang, P. Know what you don't know: Unanswerable questions for SQuAD. *arXiv* **2018**, arXiv:1806.03822. Available online: https://arxiv.org/abs/1806.03822 (accessed on 27 May 2020).
2. Yang, Z.; Qi, P.; Zhang, S.; Bengio, Y.; Cohen, W.; Salakhutdinov, R.; Manning, C.D. Hotpotqa: A dataset for diverse, explainable multi-hop question answering. In Proceedings of the 2018 Conference on Empirical Methods in Natural Language Processing, Brussels, Belgium, 31 October–4 November 2018.
3. Nishida, K.; Nishida, K.; Nagata, M.; Otsuka, A.; Saito, I.; Asano, H.; Tomota, J. Answering while Summarizing: Multi-task Learning for Multi-hop QA with Evidence Extraction. In Proceedings of the 57th Annual Meeting of the Association for Computational Linguistics, Florence, Italy, 28 July–2 August 2019; pp. 2335–2345.
4. Min, S.; Zhong, V.; Xiong, C.; Socher, R. Efficient and Robust Question Answering from Minimal Context over Documents. In Proceedings of the 56th Annual Meeting of the Association for Computational Linguistics, Melbourne, Australia, 15–20 July 2018.

5. Seo, M.; Lee, J.; Kwiatkowski, T.; Parikh, A.P.; Farhadi, A.; Hajishirzi, H. Real-time open-domain question answering with dense-sparse phrase index. In Proceedings of the 57th Annual Meeting of the Association for Computational Linguistics, Florence, Italy, 28 July–2 August 2019.

6. Ding, M.; Zhou, C.; Chen, Q.; Yang, H.; Tang, J. Cognitive graph for multi-hop reading comprehension at scale. In Proceedings of the 57th Annual Meeting of the Association for Computational Linguistics, Florence, Italy, 28 July–2 August 2019.

7. Feldman, Y.; Elyaniv, R. Multi-Hop Paragraph Retrieval for Open-Domain Question Answering. In Proceedings of the 57th Annual Meeting of the Association for Computational Linguistics, Florence, Italy, 28 July–2 August 2019.

8. Qi, P.; Lin, X.; Mehr, L.; Wang, Z.; Manning, C.D. Answering Complex Open-domain Questions Through Iterative Query Generation. In Proceedings of the 2019 Conference on Empirical Methods in Natural Language Processing and the 9th International Joint Conference on Natural Language Processing (EMNLP-IJCNLP), Hong Kong, China, 3–7 November 2019; pp. 2590–2602.

9. Yang, W.; Xie, Y.; Lin, A.; Li, X.; Tan, L.; Xiong, K.; Li, M.; Lin, J. End-to-end open-domain question answering with bertserini. In Proceedings of the 2019 Conference of the North American Chapter of the Association for Computational Linguistics, Minneapolis, MA, USA, 2–7 June 2019.

10. Lee, J.; Yun, S.; Kim, H.; Ko, M.; Kang, J. Ranking paragraphs for improving answer recall in open-domain question answering. In Proceedings of the 2018 Conference on Empirical Methods in Natural Language Processing, Brussels, Belgium, 31 October–4 November 2018.

11. Wang, S.; Yu, M.; Guo, X.; Wang, Z.; Klinger, T.; Zhang, W.; Chang, S.; Tesauro, G.; Zhou, B.; Jiang, J. R3: Reinforced ranker-reader for open-domain question answering. In Proceedings of the Thirty-Second AAAI Conference on Artificial Intelligence, New Orleans, LA, USA, 2–7 February 2018.

12. Wang, J.; Jiang, C.; Zhang, H.; Ren, Y.; Chen, K.-C.; Hanzo, L. Thirty Years of Machine Learning: The Road to Pareto-Optimal Wireless Networks. *IEEE Commun. Surv. Tutor.* **2020**. [CrossRef]

13. Zhang, W.; Lu, J. An online water army detection method based on network hot events. In Proceedings of the 2018 10th International Conference on Measuring Technology and Mechatronics Automation (ICMTMA), Changsha, China, 10–11 February 2018; pp. 191–193.

14. Zannettou, S.; Sirivianos, M.; Blackburn, J.; Kourtellis, N. The web of false information: Rumors, fake news, hoaxes, clickbait, and various other shenanigans. *J. Data Inf. Qual.* **2019**, *11*, 1–37. [CrossRef]

15. Asai, A.; Hashimoto, K.; Hajishirzi, H.; Socher, R.; Xiong, C. Learning to Retrieve Reasoning Paths over Wikipedia Graph for Question Answering. In Proceedings of the International Conference on Learning Representations, Virtual Conference, Formerly Addis Ababa, Ethiopia, 26 April–1 May 2020.

16. Xiao, Y.; Qu, Y.; Qiu, L.; Zhou, H.; Li, L.; Zhang, W.; Yu, Y. Dynamically fused graph network for multi-hop reasoning. In Proceedings of the 57th Annual Meeting of the Association for Computational Linguistics, Florence, Italy, 28 July–2 August 2019.

17. Godbole, A.; Kavarthapu, D.; Das, R.; Gong, Z.; Singhal, A.; Zamani, H.; Yu, M.; Gao, T.; Guo, X.; Zaheer, M.; et al. Multi-step Entity-centric Information Retrieval for Multi-Hop Question Answering. *arXiv* **2019**, arXiv:1909.07598. Available online: https://arxiv.org/abs/1909.07598 (accessed on 27 May 2020).

18. Hu, M.; Peng, Y.; Huang, Z.; Li, D. Retrieve, Read, Rerank: Towards End-to-End Multi-Document Reading Comprehension. In Proceedings of the 57th Annual Meeting of the Association for Computational Linguistics, Florence, Italy, 28 July–2 August 2019; pp. 2285–2295.

19. Wallenberg, M.; Arngren, T. Search Engine for Textual Content and Non-Textual Content. U.S. Patent No. 20160085860A1, 15 October 2019.

20. Guo, J.; Fan, Y.; Ai, Q.; Croft, W.B. A deep relevance matching model for ad-hoc retrieval. In Proceedings of the 25th ACM International on Conference on Information and Knowledge Management, Indianapolis, IN, USA, 24–28 October 2016; pp. 55–64.

21. Yoneda, T.; Mitchell, J.; Welbl, J.; Stenetorp, P.; Riedel, S. Ucl machine reading group: Four factor framework for fact finding (hexaf). In Proceedings of the First Workshop on Fact. Extraction and VERification (FEVER), Brussels, Belgium, 1 November 2018; pp. 97–102.

22. Rajpurkar, P.; Zhang, J.; Lopyrev, K.; Liang, P. Squad: 100,000+ questions for machine comprehension of text. In Proceedings of the 2016 Conference on Empirical Methods in Natural Language Processing, Austin, TX, USA, 1–5 November 2016.

23. Chen, D.; Fisch, A.; Weston, J.; Bordes, A. Reading wikipedia to answer open-domain questions. In Proceedings of the 55th Annual Meeting of the Association for Computational Linguistics, Vancouver, BC, Canada, 30 July–4 August 2017.

24. Dhingra, B.; Mazaitis, K.; Cohen, W.W. Quasar: Datasets for question answering by search and reading. *arXiv* **2017**, arXiv:1707.03904. Available online: https://arxiv.org/abs/1707.03904 (accessed on 27 May 2020).

25. Wang, S.; Yu, M.; Jiang, J.; Zhang, W.; Guo, X.; Chang, S.; Wang, Z.; Klinger, T.; Tesauro, G.; Campbell, M. Evidence aggregation for answer re-ranking in open-domain question answering. *arXiv* **2017**, arXiv:1711.05116. Available online: https://arxiv.org/abs/1711.05116 (accessed on 27 May 2020).

26. Kratzwald, B.; Feuerriegel, S. Adaptive document retrieval for deep question answering. In Proceedings of the 2018 Conference on Empirical Methods in Natural Language Processing, Brussels, Belgium, 31 October–4 November 2018.

27. Harris, Z.S. Distributional structure. *World* **1954**, *10*, 146–162.

28. Le, Q.; Mikolov, T. Distributed representations of sentences and documents. In Proceedings of the International conference on machine learning, Beijing, China, 21–26 June 2014; pp. 1188–1196.

29. Collins, M.; Duffy, N. New ranking algorithms for parsing and tagging: Kernels over discrete structures, and the voted perceptron. In Proceedings of the 40th annual meeting on association for computational linguistics, Philadelphia, PA, USA, 7–12 July 2002; pp. 263–270.

30. Hirsimaki, T.; Pylkkonen, J.; Kurimo, M. Importance of high-order n-gram models in morph-based speech recognition. *IEEE Trans. Audio Speech Lang. Process.* **2009**, *17*, 724–732. [CrossRef]

31. Dhingra, B.; Jin, Q.; Yang, Z.; Cohen, W.; Salakhutdinov, R. Neural models for reasoning over multiple mentions using coreference. *arXiv* **2018**, arXiv:1804.05922. Available online: https://arxiv.org/abs/1804.05922 (accessed on 27 May 2020).

32. Song, L.; Zhang, Y.; Wang, Z.; Gildea, D. A graph-to-sequence model for amr-to-text generation. In Proceedings of the 56th Annual Meeting of the Association for Computational Linguistics, Melbourne, Australia, 15–20 July 2018.

33. Song, L.; Wang, Z.; Yu, M.; Zhang, Y.; Florian, R.; Gildea, D. Exploring graph-structured passage representation for multi-hop reading comprehension with graph neural networks. *arXiv* **2018**, arXiv:1809.02040. Available online: https://arxiv.org/abs/1809.02040 (accessed on 27 May 2020).

34. De Cao, N.; Aziz, W.; Titov, I. Question answering by reasoning across documents with graph convolutional networks. In Proceedings of the 2019 Conference of the North American Chapter of the Association for Computational Linguistics, Minneapolis, MA, USA, 2–7 June 2019.

35. Kipf, T.N.; Welling, M. Semi-supervised classification with graph convolutional networks. In Proceedings of the International Conference on Learning Representations, Toulon, France, 24–26 April 2017.

36. Sukhbaatar, S.; Weston, J.; Fergus, R. End-to-end memory networks. In Proceedings of the 28th International Conference on Neural Information Processing Systems, Montreal, QC, Canada, 8–13 December 2015; pp. 2440–2448.

37. Zhou, M.; Huang, M.; Zhu, X. An interpretable reasoning network for multi-relation question answering. *arXiv* **2018**, arXiv:1801.04726. Available online: https://arxiv.org/abs/1801.04726 (accessed on 27 May 2020).

38. Singh, B.; Singh, H.K. Web Data Mining research: A survey. In Proceedings of the IEEE International Conference on Computational Intelligence & Computing Research, Kanyakumari, India, 15–18 December 2011.

39. Nie, Y.; Chen, H.; Bansal, M. Combining fact extraction and verification with neural semantic matching networks. In Proceedings of the AAAI Conference on Artificial Intelligence, Honolulu, HI, USA, 27 January–1 February 2019; Volume 33, pp. 6859–6866.

40. Chen, J.; Lin, S.-T.; Durrett, G. Multi-hop Question Answering via Reasoning Chains. *arXiv* **2019**, arXiv:1910.02610.

41. Das, R.; Dhuliawala, S.; Zaheer, M.; McCallum, A. Multi-step Retriever-Reader Interaction for Scalable Open-domain Question Answering. In Proceedings of the Seventh International Conference on Learning Representations, New Orleans, LA, USA, 6–9 May 2019.

42. Devlin, J.; Chang, M.-W.; Lee, K.; Toutanova, K. Bert: Pre-training of deep bidirectional transformers for language understanding. *arXiv* **2018**, arXiv:1810.04805. Available online: https://arxiv.org/abs/1810.04805 (accessed on 27 May 2020).

43. Seo, M.; Kembhavi, A.; Farhadi, A.; Hajishirzi, H. Bidirectional attention flow for machine comprehension. *arXiv* **2016**, arXiv:1611.01603. Available online: https://arxiv.org/abs/1611.01603 (accessed on 27 May 2020).

44. Manning, C.D.; Surdeanu, M.; Bauer, J.; Finkel, J.R.; Bethard, S.; McClosky, D. The Stanford CoreNLP natural language processing toolkit. In Proceedings of the 52nd annual meeting of the association for computational linguistics: System demonstrations, Baltimore, MD, USA, 22–27 June 2014; pp. 55–60.

45. Yan, D.; Li, K.; Ye, J. Correlation analysis of short text based on network model. *Physica A* **2019**, *531*, 121728. [CrossRef]

46. Wang, H.; Yu, M.; Guo, X.; Das, R.; Xiong, W.; Gao, T. Do Multi-hop Readers Dream of Reasoning Chains? *arXiv* **2019**, arXiv:1910.14520. Available online: https://arxiv.org/abs/1910.14520 (accessed on 27 May 2020).

47. Kingma, D.P.; Ba, J. Adam: A method for stochastic optimization. In Proceedings of the 3rd International Conference on Learning Representations, ICLR 2015, San Diego, CA, USA, 7–9 May 2015.

48. Auer, S.; Bizer, C.; Kobilarov, G.; Lehmann, J.; Cyganiak, R.; Ives, Z. DBpedia: A nucleus for a web of open data. In *The Semantic Web*; Aberer, K., Choi, K.-S., Noy, N., Allemang, D., Lee, K.-I., Nixon, L., Golbeck, J., Mika, P., Maynard, D., Mizoguchi, R., et al., Eds.; Springer: Heidelberg, Germany, 2007; pp. 722–735.

49. Wang, Z.; Ng, P.; Ma, X.; Nallapati, R.; Xiang, B. Multi-passage BERT: A Globally Normalized BERT Model for Open-domain Question Answering. In Proceedings of the international joint conference on natural language processing, Hong Kong, China, 3–7 November 2019; pp. 5877–5881.

50. Nogueira, R.; Cho, K. Passage Re-ranking with BERT. *arXiv* **2019**, arXiv:1901.04085. Available online: https://arxiv.org/abs/1901.04085 (accessed on 27 May 2020).

Article

Multi-View Visual Question Answering with Active Viewpoint Selection

Yue Qiu [1,2,*] **Yutaka Satoh** [1,2]**, Ryota Suzuki** [2]**, Kenji Iwata** [2] **and Hirokatsu Kataoka** [2]

[1]   Graduate School of Science and Technology, University of Tsukuba, Tsukuba 305-8577, Japan; yu.satou@aist.go.jp

[2]   National Institute of Advanced Industrial Science and Technology (AIST), Tsukuba 305-8560, Japan; ryota.suzuki@aist.go.jp (R.S.); kenji.iwata@aist.go.jp (K.I.); hirokatsu.kataoka@aist.go.jp (H.K.)

*   Correspondence: s1830151@s.tsukuba.ac.jp

Received: 31 March 2020; Accepted: 14 April 2020; Published: 17 April 2020

Abstract: This paper proposes a framework that allows the observation of a scene iteratively to answer a given question about the scene. Conventional visual question answering (VQA) methods are designed to answer given questions based on single-view images. However, in real-world applications, such as human–robot interaction (HRI), in which camera angles and occluded scenes must be considered, answering questions based on single-view images might be difficult. Since HRI applications make it possible to observe a scene from multiple viewpoints, it is reasonable to discuss the VQA task in multi-view settings. In addition, because it is usually challenging to observe a scene from arbitrary viewpoints, we designed a framework that allows the observation of a scene actively until the necessary scene information to answer a given question is obtained. The proposed framework achieves comparable performance to a state-of-the-art method in question answering and simultaneously decreases the number of required observation viewpoints by a significant margin. Additionally, we found our framework plausibly learned to choose better viewpoints for answering questions, lowering the required number of camera movements. Moreover, we built a multi-view VQA dataset based on real images. The proposed framework shows high accuracy (94.01%) for the unseen real image dataset.

Keywords: visual question answering; three-dimensional (3D) vision; reinforcement learning; deep learning; human–robot interaction

---

1. Introduction

Recent developments in deep neural networks have resulted in significant technological advancements and have broadened the applicability of human–robot interaction (HRI). Vision and language tasks, such as visual question answering (VQA) [1–6], and visual dialog [7,8], can be extremely useful in HRI applications. In a VQA system, the input is an image along with a text question that the system needs to answer by recognizing the image, interpreting the natural language in the question, and determining the relationships between them. VQA tasks play an essential role in various real-world applications. For example, in HRI applications, VQA can be used to connect robot perceptions with human operators through a question-answering process. In video surveillance systems, the question–answer process can serve as an interface to help avoid the manual checking of each video frame, significantly reducing labor costs.

Although VQA methods can be useful in real-world applications, there are numerous problems related to their implementation in real-world environments. For example, conventional VQA methods answer a given question based on a single image. However, in real-world environments, because it is challenging to take photographs continuously from optimal viewpoints, objects can be greatly occluded and thus answering questions based on single-view images could be difficult. Considering that multi-view observations are possible in HRI applications, this study discusses VQA under multi-view settings. Qiu et al. [9] proposed a multi-view VQA framework that uses perimeter viewpoint observation for answering questions. However, perimeter viewpoints may be difficult to set up in real-world environments due to environmental constraints, making their method difficult to implement. In addition, observing each scene from perimeter viewpoints is relatively inefficient, especially for applications that require real-time processing. Moreover, the authors did not evalutate their method under real-world image settings.

Here, we propose a framework, shown in Figure 1, in which a scene is actively observed. Answers to questions are obtained based on previous observations of the scene. The overall framework consists of three modules, namely a scene representation network (SRN) that integrates multi-view images into a compact scene representation, a viewpoint selection network that selects the next observation viewpoint (or ends the observation) based on the input question and the previously observed scene, and a VQA network that predicts the answer based on the observed scene and question. We built a computer graphics (CG) multi-view VQA dataset with 12 viewpoints. For this dataset, the proposed framework achieved accuracy comparable to that of a state-of-the-art method [9] (97.11% (Ours) vs. 97.37%) that answers questions based on the perimeter observation of scenes from fixed directions, while using an average of just 2.98 viewpoints as a contrast with 12 viewpoints of the previous method. In addition, we found that our framework learned to choose viewpoints efficiently for answering questions by ending observation while the observed scene contains all objects needed to answer the question, or additionally observing the scene from viewpoints with large spacing to enhance accessible scene information, thereby lowering the camera movement cost. Furthermore, to evaluate the effectiveness of the proposed method in realistic settings, we also created a real image dataset. In experiments conducted with this dataset, the proposed framework outperformed the existing method by a significant margin (+11.39%).

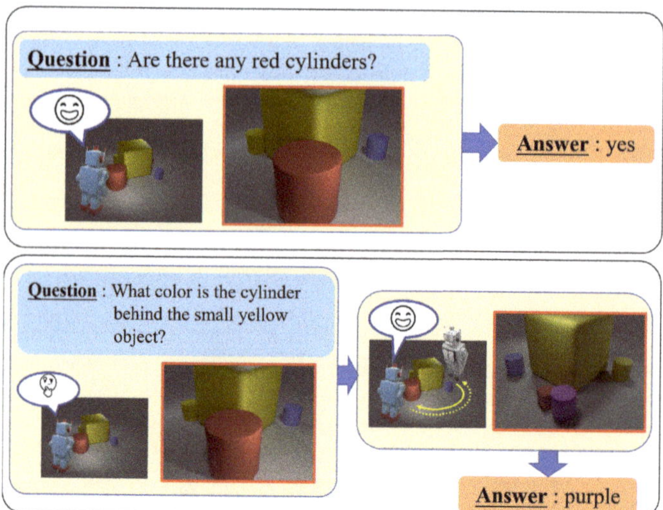

Figure 1. Illustration of multi-view VQA with active viewpoint selection. Our framework selects the observation viewpoint (highlighted in red bounding boxes) to answer questions based on the observed scene information.

The contributions of our work are three-fold:

- we discuss the VQA task under a multi-view and interactive setting, which is more representative of a real-world environment compared to traditional single-view VQA settings. We also built a dataset for this purpose.

- we propose a framework that actively observes scenes and answers questions based on previously observed scene information. Our framework achieves high accuracy for question answering with high efficiency in terms of the number of observation viewpoints. In addition, our framework can be applied in applications that have restricted observation viewpoints.
- we conduct experiments on a multi-view VQA dataset that consists of real images. This dataset can be used to evaluate the generalization ability of VQA methods. The proposed framework shows high performance for this dataset, which indicates the suitability of our framework for realistic settings.

2. Related Work

2.1. Visual Question Answering

In recent years, a variety of VQA methods have been proposed. Holistic VQA methods, such as feature-wise linear modulation (FiLM) [10], combine image features with question features by concatenation [1], bilinear pooling [6], or the Hadamard product [3], and then predict answers from these combined multi-modal features.

A series of VQA datasets have been proposed. The VQA 2.0 dataset [2], one of the most popular datasets, contains real images sampled from the MS COCO dataset [11], and question–answer pairs collected using crowdsourcing (Amazon Mechanical Turk).Johnson et al. [12] pointed out that the VQA2.0 dataset has high dataset distribution biases and proposed the Compositional Language and Elementary Visual Reasoning (CLEVR) [12] dataset, in which images and questions are generated by pre-defined programs. The CLEVR dataset has been widely adopted for evaluating visual reasoning ability. Although significant progress related to VQA tasks has been made, there is relatively little discussion about multi-view VQA. Qiu et al. [9] proposed a multi-view VQA framework. However, the authors used a limited perimeter viewpoint setting, and their study did not conduct real-world environment experiments. Here, we propose a framework with high efficiency in terms of the number of observation viewpoints and discuss its application in real-world image settings.

2.2. Learned Scene Representation

Traditional approaches based on three-dimensional (3D) convolutional neural networks (CNNs) learn a 3D representation from 3D data, such as point cloud data [13,14], mesh [15], and voxel data [16]. These methods can extract representations that contain the underlying 3D structure of the input data. However, due to the discrete nature of 3D data formats, such as point clouds, the resolution of the learned representation is limited. In addition, learning from direct 3D data usually requires a massive amount of training data, high memory cost, and a long execution time.

In contrast, a series of continuous 3D scene representations have recently been proposed. Generative query network [17] is a network based on a conditional variational autoencoder [18] that learns a meaningful 3D representation described by the parameters of a scene representation network (SRN) from a massive amount of 3D scene data annotated with only camera viewpoint information. Deep signed distance function [19] is an auto decoder-based structure that learns continuous signed distance functions to map 3D coordinates into the signed distance of a group of objects. Scene representation networks [20] proposed by Sitzmann et al. predict 3D range maps along with a scene representation of the input scene and thus exhibit robustness to unknown camera poses.

A learned continuous scene representation contains the underlying 3D structure and object information of the input scene and usually requires a relatively small amount of annotation signals and relatively short execution costs. Therefore, in this work, we use a continuous SRN to extract scene information and use the representation to answer questions.

2.3. Deep Q-Learning Networks

Q-learning [21] is a type of value-based reinforcement learning framework that holds a Q-table of the values of taking actions a in environment states s. However, for a large number of environment states s or continuous state spaces, the time and space consumption of Q-learning can become rather high. Instead of holding a Q-table in memory, deep q-learning networks (DQNs) [22] use neural networks to predict action values from environment states s. Conventional DQN methods that involve image information learn policies from raw image inputs [23]. In the present work, instead of raw images, we use scene representations that contain underlying 3D structure and object information of scenes and encoded question information as input and then adopt a DQN to select observation viewpoints from the previously observed scene and question information.

2.4. Embodied Question Answering

The work most similar to ours is the embodied question answering (EQA) task defined by Das et al. [24]. The authors define an EQA task for which an agent is embodied within a 3D environment. The agent attempts to answer given questions by navigating and gathering needed image information using egocentric perceptions of the environment. The authors proposed an EQA baseline method that splits the EQA task into a recurrent neural network-based hierarchical planner-controller navigation policy for navigating to the target region in the environment and a VQA model that combines the weighted sum of the image features of five images and the question features to give a final answer. Yu et al. [25] proposed a multi-target EQA task, in which every question involves two targets. In [26], the authors extended the original CG EQA dataset setting to photorealistic environments and suggested the use of point cloud data.

Most EQA methods focus on the vision-language grounded navigation process and finding the shortest paths in a given environment. They usually use relatively simple VQA methods, such as in [24], where the authors simply concatenate five images into a VQA model for answering questions. In the above study, relatively little attention was given to the kind of information necessary to answer questions, and no evidence was given that the framework answers questions based on a 3D understanding of the given scene. In contrast, our work focuses on exploring necessary scene information to answer questions by selecting observation viewpoints. Therefore, our approach can avoid unnecessary observation and reduce camera movement costs. In addition, our framework is based on an integrated 3D understanding of the given scene observed from multiple viewpoints. Therefore, our framework can associate 3D information with viewpoints and determine the minimum necessary scene information required to answer questions via the selection of observation viewpoints.

3. Approach

In real-world HRI applications, it can be challenging to obtain photographs from perimeter viewpoints. In addition, it is efficient to end the observation when the input scene information is sufficient to answer the question. Based on the above, we propose a framework that actively observes the environment and decides when to answer the question based on previously observed scene information.

As shown in Figure 2, the proposed framework consists of three modules, namely an SRN, a viewpoint selection network, and a VQA network.

The inputs of the overall framework are the default viewpoint v_0, the image x_0 of the scene observed from v_0, and the question q. v_0 and x_0 are first processed by the SRN to obtain the original scene representation s_0. The question is processed by an embedding layer and a two-layered long short-term memory (LSTM) [27] network to obtain q'. Then, the viewpoint selection network predicts the next observation viewpoint (or selects the end action) based on s_0 and q'. If viewpoint v_t is chosen at time step t, the agent obtains the image x_t from viewpoint v_t of the scene (e.g., for a robot, a scene image from v_t is taken). Next, the SRN updates the scene based on x_t and v_t. If the end action is chosen, the VQA network predicts an answer based on the q' and the integrated scene representation at that time. In the following sections, we discuss these three networks in greater detail.

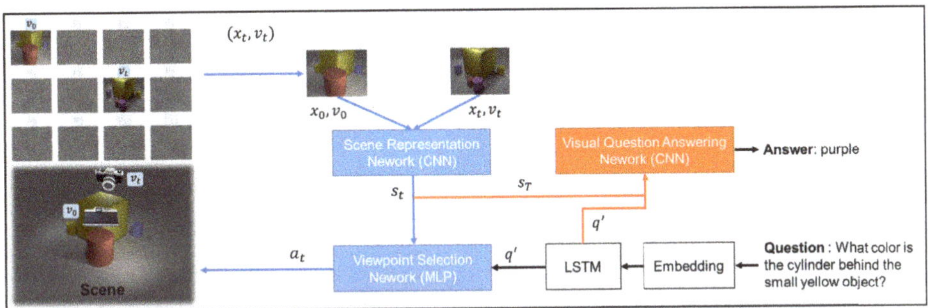

Figure 2. Overall framework. The viewpoint selection network (multi-layer perceptron (MLP)-structured) chooses the observation viewpoint (or ends the observation) iteratively based on the scene s_t and question q' (blue arrow flow). When the end action is chosen, the scene representation up to that time step s_T and the encoded question q' will be processed by the VQA network to provide an answer (orange arrow flow).

3.1. Scene Representation

We use the SRN proposed by Eslami et al. [17] to obtain integrated scene representations s_t from viewpoints $\{v_0, v_1, ..., v_t\}$ and images $\{x_0, x_1, ..., x_t\}$. For scene i observed from K viewpoints, the observation o_i is defined as follows:

$$o_i = \{(x_i^k, v_i^k)\}_{k=0,...,K-1} \tag{1}$$

The scene representation $s = f_{SRN}(o_i)$ and $g(x^m|v^m, s)$ is jointly trained for image rendering from arbitrary viewpoint m to maximize the likelihood between the predicted x^m and the ground truth images. f_{SRN} integrates multi-view information into a compact scene representation. We use the above framework to train the SRN.

3.2. Viewpoint Selection

For VQA in real-world environments, it is necessary to choose an observation based on the question and previously observed visual information. For example, for the question " is there a red ball?", if a red ball has previously been observed, the question can be answered instantly. Additionally, for highly occluded scenes, it may be necessary to make observations from a variety of viewpoints. Therefore, we propose a DQN-based viewpoint selection network f_{VS} to actively choose actions.

More specifically, assuming that the scene can be observed from K viewpoints, we define an action set $A = \{a_i\}_{i=0,...,K}$, where $a_0, ..., a_{K-1}$ denote the viewpoint selection actions and a_K represents the end observation action.

The viewpoint selection network f_{VS} predicts a K+one-dimensional, vector that represents the obtainable reward value of each action (after it is executed) from the input of the previously observed scene s_t and question q'. Assuming that the j-th action a_t is chosen at time step t, we denote the predicted reward value r_t^{eval} of action a_t under the environment state s_t as follows:

$$r_t^{eval} = \{f_{VS}(s_t, q')\}_j \tag{2}$$

The real reward value of action a_t can be formulated as Equation (3), where r_t^{env} denotes the reward obtained from the environment, and γ is the discount factor of future rewards:

$$r_t^{real} = r_t^{env} + \gamma argmax_{a_{t+1}} f_{VS}(a_{t+1}|s_{t+1}, q') \tag{3}$$

The overall objective of viewpoint selection is to minimize the distance between r_t^{eval} and r_t^{real}.

We designed the reward r^{env} based on the correctness of the question answering and the numbers of selected viewpoints. For each added new observation viewpoint or viewpoint that is chosen repeatedly, we assign penalties p_{sp} and p_{rt} (hyperparameters). We denote the VQA loss as $loss_{vqa}$ (normalized to $[-1, 1]$). We designed r^{env} for three action types, as shown below.

$$r^{env} = \begin{cases} p_{sp} + p_{rt} - loss_{vqa} & (i) \\ p_{sp} + p_{rt} & (ii) \\ -loss_{vqa} & (iii) \end{cases} \tag{4}$$

For the final viewpoint selection action, the reward is (i); for the other viewpoint selection actions, the reward is (ii); for the end action, the reward is (iii).

3.3. Visual Question Answering

VQA predicts an answer based on integrated scene information s and the processed question q'. We denote the VQA network to be trained as f_{VQA}. The answer *ans* can be predicted by the following:

$$ans = argmax_{ans} f_{VQA}(ans|s, q') \tag{5}$$

The network is optimized by minimizing the cross-entropy loss between the predicted *ans* and the ground truth answer. In this study, we used the state-of-the-art FiLM method [10] as the VQA network. However, it is noteworthy that the VQA framework can be arbitrary.

The proposed framework could not deal with ill-structured or non-English questions that are not included in the training dataset. However, the framework could be expanded for these situations by further integrating sentence structure checking or translation modules.

4. Experiments

We conducted experiments using settings with CG and real image datasets, respectively. In the following sections, we discuss the experimental details for these two settings.

4.1. Implementation Details

Our overall framework is composed of three modules, namely an SRN, a viewpoint selection network, and a VQA network. Here, we introduce the implementation details of these three modules. The implementation setting is used for both CG and real image experiments. We used PyTorch [28] as the implementation framework. rtwoWe also show the detailed network structure in Figure 3.

SRN: we adopted a 12-layered tower-structured SRN proposed in [17] to extract scene representations. The SRN transforms the input of $64 \times 64 \times 3$-dimensional images and $1 \times 1 \times 7$-dimensional camera information (a vector represents camera position and rotation) into a $16 \times 16 \times 256$-dimensional scene representation. We used Adam optimizer [29] and an initial learning rate of 5×10^{-4} for training. In all experiments, we trained the SRN network for 140 epochs.

Viewpoint Selection Network: to extract question information, we first adopted an embedding layer (torch.nn.Embedding defined in PyTorch), which transforms the one-hot vector of each word into a 300-dimensional vector. Next, we adopted a two-layered LSTM with hidden dimensions of 1024 to encode each question sentence into a 1024-dimensional vector. Alternative structures, such as the gated recurrent unit [30] or transformer [31], can also be used for question feature extraction. Next, we combined the encoded question with the scene representation by concatenation, which resulted in a $1024 + 16 \times 16 \times 256$-dimensional feature. Next, we processed the concatenated feature via three fully connected layers to predict a K+one-dimensional, action reward value vector (K indicates the viewpoint number). We adopted a tangent activation function before the final fully connected layer. We set the hyperparameter γ of the viewpoint selection network to 0.8 and ϵ-greedy to choose actions. The initial ϵ was set to 0.5 and multiplied by 1.04 every three epochs. The penalties for viewpoint action and repeated action were set to –0.25 and –0.05, respectively. This network was jointly trained with the VQA network.

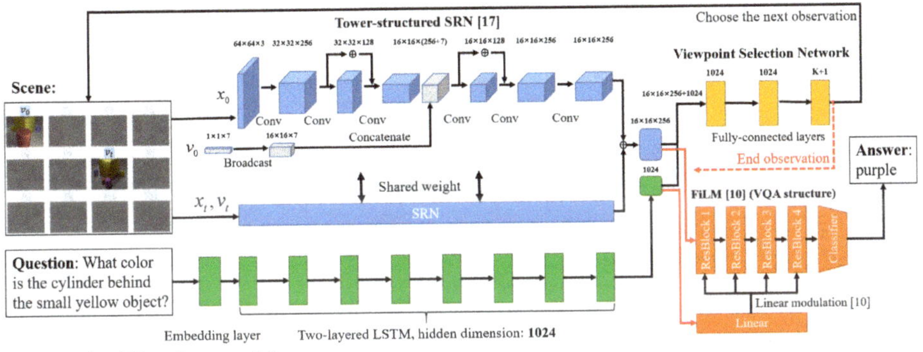

Figure 3. Detailed network structure. Our overall framework is composed of a tower-structured SRN [17] (shown in blue), a viewpoint selection network (shown in yellow), and a VQA framework (shown in orange), which is modified from FiLM [10]. The viewpoint selection network and VQA share the question feature extraction structure (shown in green). Based on the previous observation, the viewpoint selection network chooses the next observation or ends observation (red dotted arrow flow). While the end observation is chosen, the scene representation to that time step, and the question feature (red arrow flows) will be processed by the VQA framework to give an answer prediction:

VQA: we modified the FiLM implementation code [32] for our VQA module. We removed the image feature extraction component used in [32] and used integrated scene representation as the input image feature. The question extraction component is shared with the viewpoint selection network during training. We jointly trained the modified FiLM network with the viewpoint selection network with an initial learning rate of 3×10^{-4} and Adam optimizer for 40 epochs in all experiments.

4.2. Experiments with CG Images

Dataset Setting: we modified the CLEVR dataset generation program by placing multiple virtual cameras in each scene and then generating a multi-view CLEVR dataset (Multi-view-CLEVR_12views_CG). The camera setup, which is widely used in multi-view object recognition tasks (e.g., [33,34]), is shown in Figure 4 (left). The cameras are elevated 30 degrees above the scene and placed at intervals of 30 degrees around the center of the scene. We place a relatively large object in the middle of the scene and three to five other objects around the center object. The setting was designed to make it difficult to observe all objects from a single viewpoint, thereby providing a simple testbed for viewpoint selection efficiency. We created CG objects and set the illumination, lighting, background of scenes, and virtual camera positions through a 3D creation software blender [35] and Python-blender source code based on the the CLEVR dataset generation program [36]. We set two types of questions: exist (query the existence of an object) and query color (query an object's color). Then, by considering the existence of relationship words (left of, right of, behind, in front of), the questions can be further separated into spatial and non-spatial types. Example images and questions of a scene from Multi-view-CLEVR_12views_CG dataset are shown in Figure 4 (right) and the detailed statistics of the dataset are shown in Table 1.

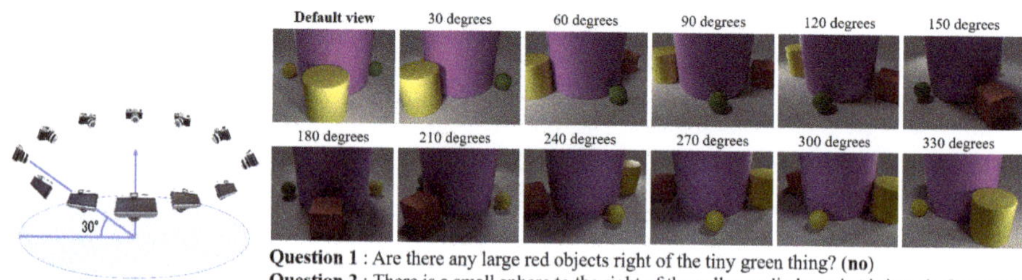

Question 1 : Are there any large red objects right of the tiny green thing? (**no**)
Question 2 : There is a small sphere to the right of the yellow cylinder; what is its color? (**green**)
Question 3 : Are any big red cylinders visible? (**no**)
Question 4 : What color is the big cube? (**red**)

Figure 4. Virtual camera setup (left) and example images and questions from Multi-view-CLEVR_12views_CG dataset (right). The default view determines the spatial relationships of objects in each scene.

Table 1. Statistics for datasets used in this study.

Datasets	No. of Scenes	No. of Q–A Pairs	Spatial		Non-Spatial	
	(Train/Test)	(Train/Test)	Exist	Query Color	Exist	Query Color
Multi-view-CLEVR_12views_CG (**CG**)	(40,000 / 3,000)	(157,232 / 11,801)	43,000	42,993	43,000	40,040
Multi-view-CLEVR_4views_CG (**CG**)	(40,000 / 3000)	(158,860 / 11,928)	43,000	42,997	43,000	41,791
Multi-view-CLEVR_4views_3DMFP (**CG**)	(10,000 / 0)	(39,694 / 0)	10,000	9994	10,000	9,700
Multi-view-CLEVR_4views_Real (**Real**)	(0 / 500)	(0 / 1974)	500	500	500	474

Quantitative Results: the overall and per-question accuracies are shown in Table 2. Our framework (SRN_FiLM_VS) showed a slight drop in overall accuracy compared with that for SRN_FiLM proposed in [9]. However, our framework used many fewer viewpoints on average (2.98 vs. 12). This indicates that the proposed framework learned to obtain the required information for answering questions, lowering the camera movement cost. This ability, which is especially crucial for applications with restricted observation viewpoints, makes our framework more efficient because it decreases the required number of camera (or robot) movements.

Table 2. Evaluation results for Multi-view-CLEVR_12views_**CG**.

Method	Overall Accuracy	Spatial		Non-Spatial		Average no. of Used Viewpoints
		Exist	Query Color	Exist	Query Color	
SRN_FiLM	97.37%	94.20%	98.20%	99.03%	98.11%	12
SRN_FiLM_VS	97.11%	95.27%	96.90%	99.03%	97.25%	2.98

Qualitative Results: the results for four example questions are shown in Figure 5. For Question 1, the queried object "tiny cube" appears in the default view, so our model ended viewpoint selection instantly and answered the question. For Questions 2 and 3, the queried objects do not appear in the default view, so our model selected an additional viewpoint for observation. These results show that our model learned how to make observations to obtain the necessary scene information for answering questions. For Question 4, our model used three additional viewpoints and gave an incorrect answer. In this case, the queried object was a purple cylinder. The shadow cast by the center object made it difficult for the model to recognize the color. It is believed that using higher-resolution input images could improve performance in such situations.

Effect of Viewpoint Selection: we conducted an experiment using the input of randomly sampled viewpoints (SRN_FiLM_Random) and equally sampled viewpoints (SRN_FiLM_Equal), in which all the viewpoints from Multi-view-CLEVR_12views_CG were evenly sampled. The results are shown in Table 3. We first evaluated the accuracy of answering questions based on a randomly picked viewpoint (SRN_FiLM_Random (1 view)). For this setup, an overall accuracy of 66.71% was achieved, which indicates that our dataset is challenging with single-view images. Moreover, because our viewpoint selection network predicts answers based on an average of 2.98 viewpoints, we sampled three viewpoints for equity. The results show that our viewpoint selection network greatly outperforms the randomly and equally sampled settings, indicating that the proposed viewpoint selection network makes the utilization of viewpoint information more effective through constructing scene representations needed for answering questions from the partial observation of scenes.

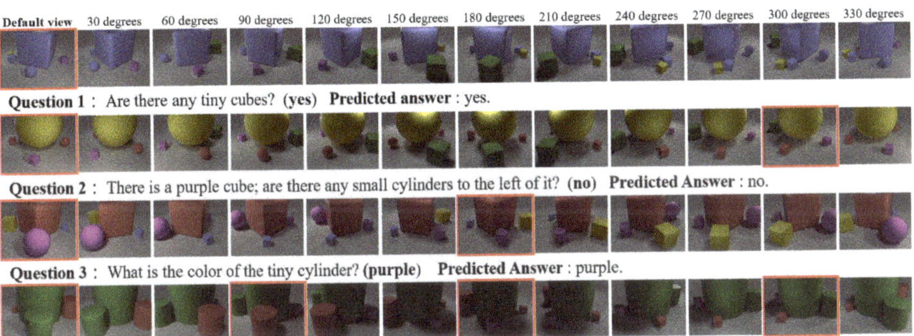

Figure 5. Example results of SRN_FiLM_VS for the Multi-view-CLEVR_12views_**CG** dataset. The default view and selected view images are highlighted in red bounding boxes. The incorrect answers are shown in red.

Table 3. Effect of viewpoint selection on accuracy.

Method (No. of Used Viewpoints)	Overall Accuracy
SRN_FiLM_Random (1 view)	66.71%
SRN_FiLM_Random (3 views)	81.39%
SRN_FiLM_Equal (3 views)	82.90%
SRN_FiLM_VS (2.98 views)	97.11%

Runtime information: the test process was conducted on an Intel(R) Core(TM) i9-7920X CPU @ 2.90 GHz machine with a graphic processing unit of Nvidia TITAN V. The runtime is 0.043 s per scene on average with a single GPU and single process.

4.3. Experiments with Real Images

4.3.1. Training on CG Images and Testing on the Real Images Dataset

Real-world environment applications, such as HRI applications, require the ability to sense and recognize real-world scenes. Thus, the ability to transfer the learned scene representation and understanding obtained in a simulated environment to a real-world scene is important for AI-based frameworks. To evaluate the transfer ability of our framework, we built a multi-view dataset with real images of objects similar to those in the CLEVR dataset. We then evaluated the performance of the model that learned in the simulated environment (multi-view CLEVR dataset) on the real image dataset.

Dataset Setting: to evaluate the model performance for real images, we built a dataset (Multi-CLEVR_4views_Real) that consisted of photographed real images. We printed out and colored models with attributes similar to those of the objects described in the previous section using a 3D printer. We then placed these models on a scene table of a laboratory with normal fluorescent lighting. We used a camera (Sony Alpha 7 III Mirrorless Single Lens Digital Camera [37]) to take photographs of scenes from four viewpoints. The camera was elevated 30 degrees above the table and placed at 90-degree intervals around it with a distance of 1.1 m from the scene center.

An average of seven photographs were taken for each viewpoint, with the camera slightly moved and rotated. In total, we collected 100 scene photographs, which were augmented to obtain 500 scene photographs by sampling images taken randomly from each viewpoint. The question–answer pair settings were the same as that those used in the previous section. The statistics are shown in Table 1. Several scene examples are shown in Figure 6 (middle).

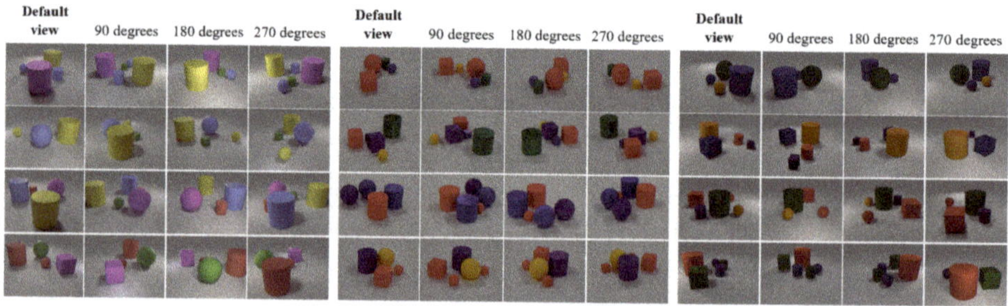

Figure 6. Example images from Multi-view-CLEVR_4views_**CG** (**left**) which is entirely built from simulated scenes, Multi-view-CLEVR_4views_**Real** (**middle**), which are created from photographing real table scenes composed of real object models, and a Multi-view-CLEVR_4views_**3DMFP** (**right**) dataset that is built from importing scanned real object models into simulated scenes and aimed for fine-tuning.

Quantitative Results: we first trained and tested the SRN_FiLM and our framework on Multi-view-CLEVR_4views_CG (Figure 6 (left)) for 40 epochs (the statistics are shown in Table 1). The test split results are shown in Table 4. Our framework achieved performance comparable to that of SRN_FiLM while using fewer viewpoints. We then evaluated the performance for the Multi-view-CLEVR_4views_Real, the results of which are shown in Table 5. Compared with the test results for CG images (Table 4), those for the real images dropped by nearly 30% for both methods without fine-tuning (first and fifth rows of Table 5).

Table 4. Evaluation results for Multi-view-CLEVR_4views_**CG**.

Method	Accuracy	No. of Used Viewpoints
SRN_FiLM	97.67%	4
SRN_FiLM_VS	97.64%	**2.02**

Table 5. Evaluation results for Multi-view-CLEVR_4views_**Real**.

Method	Fine-Tuning		Accuracy
	SRN	FiLM	
SRN_FiLM	-	-	67.88%
SRN_FiLM	-	✓	76.14%
SRN_FiLM	✓	-	77.30%
SRN_FiLM	✓	✓	82.62%
SRN_FiLM_VS	-	-	66.82%
SRN_FiLM_VS	-	✓	79.99%
SRN_FiLM_VS	✓	-	91.56%
SRN_FiLM_VS	✓	✓	94.01%

4.3.2. Fine-Tuning on the Semi-CG Dataset

This reduction in performance is likely caused by the domain gap between CG and real images, such as color distribution, lighting, and object texture differences. To obtain high performance for real images with models that were directly trained using simulated images, a common approach is to fine-tune the models with real images. However, in our data setting, each scene is composed of objects with random attributes and locations and is observed from four viewpoints. Therefore, photographing a real scene under our dataset setting requires manually arranging objects and moving cameras, which is both labor- and time-consuming.

To solve the problem mentioned above, we introduce a method for creating a fine-tuning dataset that scans real objects and imports the obtained models into a simulation environment. In detail, the dataset construction process includes three steps. First, we placed the real objects used for building the Multi-view-CLEVR_4views_Real in a 3D photography studio (3D MFP, Ortery [38]) for scanning. The 3D MPF photographs object models from multiple viewpoints. Next, we used the software 3DSOM Pro V5 [39] to generate textured 3D models, which is accomplished by aligning and integrating multiview images obtained by the 3D MFP. Finally, we imported the obtained 3D object models into the CLEVR image generation process and created a dataset, which we denoted as Multi-view-CLEVR_4views_3DMFP, using the same scene and camera setup used for Multi-view-CLEVR_4views_Real. We use the Multi-view-CLEVR_4views_3DMFP dataset for fine-tuning (the statistics of which are shown in Table 1, example images are shown in Figure 6 (right)).

Next, we conducted ablation experiments on the fine-tuned network parts (freezing the VQA model, freezing the SRN model, and then fine-tuning both models). The results are shown in Table 5. We found that fine-tuning the models improved their performance by a large margin, especially when both the SRN and VQA models were fine-tuned (up to +27.19%). After fine-tuning, the proposed model achieved an accuracy of 94.01% on the unseen real image dataset, outperforming SRN_FiLM by 11.39% (fourth and eighth rows of Table 5). This performance increase is attributed to the proposed SRN_FiLM_VS model being relatively less affected by noise in images from unnecessary viewpoints, which enables the model to adapt faster to unfamiliar scenes. Furthering discussion is left for future work. It is noteworthy that using the fine-tuning dataset Multi-view-CLEVR_4views_3DMFP helps significantly improve the performance of both models. This result shows that, by introducing the "semi-CG" dataset consisting of the scanned real object models and simulation environments for the fine-tuning process, real-world applications have the potential to improve the performance of models trained from pure simulation data.

Qualitative Results: Here, we discuss the qualitative results shown in Figure 7. The SRN_FiLM method answered questions based on all four viewpoints; in contrast, our framework selected the observation viewpoints (highlighted in red bounding boxes). For Questions 1, 2, and 3, the proposed model made reasonable viewpoint selections and answered the questions correctly. For Question 4, the spatial reasoning was difficult, and both models failed to give the correct answer. Narrowing the gap between the CG and real images might improve performance.

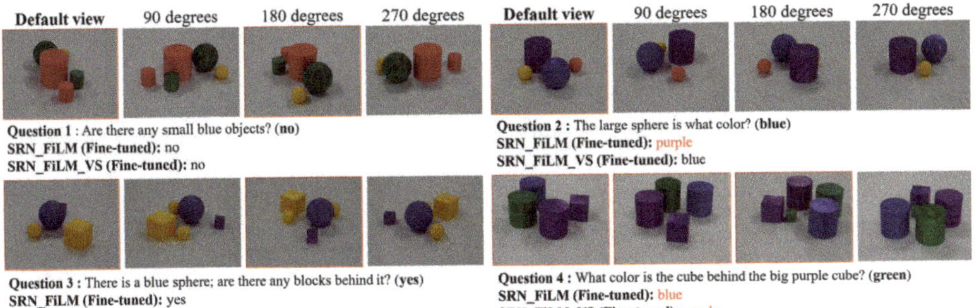

Figure 7. Example results for Multi-view-CLEVR_4views_**Real** dataset. The default view and selected view images are highlighted in red bounding boxes. Incorrect answers are shown in red.

5. Conclusions

In this study, we proposed a multi-view VQA framework that actively chooses observation viewpoints to answer questions. VQA is defined as the task of answering a text question based on an image that is essential in various HRI systems. Existing VQA methods answer questions based on single-view images. However, in real-world applications, single-view image information can be insufficient for answering questions, and observation viewpoints are usually limited. Moreover, the ability to observe scenes efficiently from optimal viewpoints is crucial for real-world applications with time restrictions. To resolve these issues, we propose a framework that makes iterative observations under a multi-view VQA setting. The proposed framework iteratively selects additional observation viewpoints and updates scene information until the scene information is sufficient to answer questions. the proposed framework achieves performance comparable to that of a state-of-the-art method on a VQA dataset with CG images and, in the meantime, greatly reduces the number of observation viewpoints (a reduction of 12 to 2.98). in addition, the proposed method outperforms the existing method by a significant margin (+11.39% on overall accuracy) on a dataset with real images, which is closer to the real-world setting, making our method more promising to be applied to real-world applications. However, directly applying a model trained on CG images to real images results in a performance gap (a drop of -30.82% on accuracy). In the future, we will consider using various methods to narrow this gap.

Author Contributions: Y.Q. has proposed and implemented the approaches and conducted the experiments. She also wrote the paper, together with Y.S., R.S., K.I., and H.K. All authors have read and approved the final version of the manuscript.

Funding: This research received no external funding.

Acknowledgments: The authors would like to acknowledge the assistance and comments of Hikaru Ishitsuka and Tomomi Satoh.

Conflicts of Interest: The authors declare no conflict of interest.

Abbreviations

The following abbreviations are used in this manuscript:

VQA	Visual Question Answering
HRI	Human–Robot Interaction
3D	Three-Dimensional
SRN	Scene Representation Network
CG	Computer Graphics
FiLM	Feature-wise Linear Modulation
CLEVR	Compositional Language and Elementary Visual Reasoning
CNNs	Convolutional Neural Networks
DQNs	Deep Q-learning Networks
MLP	Multi-Layer Perceptron
EQA	Embodied Question Answering
LSTM	Long Short-Term Memory
3DMFP	3D photography studio, 3D MFP, Ortery

References

1. Antol, S.; Agrawal, A.; Lu, J.; Mitchell, M.; Batra, D.; Zitnick, C.L.; Parikh, D. Vqa: Visual question answering. In Proceedings of the IEEE International Conference on Computer Vision, Las Condes, Chile, 11–18 December 2015; pp. 2425–2433.
2. Goyal, Y.; Khot, T.; Summers-Stay, D.; Batra, D.; Parikh, D. Making the V in VQA matter: Elevating the role of image understanding in Visual Question Answering. In Proceedings of the IEEE Conference on Computer Vision and Pattern Recognition, Honolulu, HI, USA, 21–26 July 2017; pp. 6904–6913.

3. Anderson, P.; He, X.; Buehler, C.; Teney, D.; Johnson, M.; Gould, S.; Zhang, L. Bottom-up and top-down attention for image captioning and visual question answering. In Proceedings of the IEEE Conference on Computer Vision and Pattern Recognition, Salt Lake City, UT, USA,18–24 June 2018; pp. 6077–6086.

4. Fukui, A.; Park, D.H.; Yang, D.; Rohrbach, A.; Darrell, T.; Rohrbach, M. Multimodal compact bilinear pooling for visual question answering and visual grounding. *arXiv* **2016**, arXiv:1606.01847.

5. Hudson, D.A.; Manning, C.D. Compositional attention networks for machine reasoning. *arXiv* **2018**, arXiv:1803.03067.

6. Kim, J.H.; Jun, J.; Zhang, B.T. Bilinear attention networks. *arXiv* **2018**, arXiv:1805.07932v2.

7. Das, A.; Kottur, S.; Gupta, K.; Singh, A.; Yadav, D.; Moura, J.M.F.; Parikh, D.; Batra, D. Visual dialog. In Proceedings of the IEEE Conference on Computer Vision and Pattern Recognition, Honolulu, HI, USA, 21–26 July 2017; pp. 326–335.

8. Das, A.; Kottur, S.; Moura, J.M.F.; Lee, S.; Batra, D. Learning cooperative visual dialog agents with deep reinforcement learning. In Proceedings of the IEEE international conference on computer vision, Venice, Italy, 22–29 October 2017; pp. 2951–2960.

9. Qiu, Y.; Satoh, Y.; Suzuki, R.; Kataoka, H. Incorporating 3D Information into Visual Question Answering. In Proceedings of the IEEE International Conference on 3D Vision (3DV), Quebec City, QB, Canada, 16–19 September 2019; pp. 756–765.

10. Perez, E.; Strub, F.; De Vries, H.; Dumoulin, V.; Courville, A. Film: Visual reasoning with a general conditioning layer. In Proceedings of the Thirty-Second AAAI Conference on Artificial Intelligence, New Orleans, LA, USA, 2–7 February 2018.

11. Lin, T.Y.; Maire, M.; Belongie, S.; Bourdev, L.; Girshick, R.; Hays, J.; Perona, P.; Ramanan, D.C.; Zitnick, L.; Dollar, P. Microsoft Coco: Common Objects in Context. In Proceedings of the European Conference on Computer Vision, Zurich, Switzerland, 6–12 September 2014; Springer: Cham, Swityerland, 2014; pp. 740–755.

12. Johnson, J.; Hariharan, B.; van der Maaten, L.; Li, F.-F.C.; Zitnick, L.; Girshick, R. Clevr: A diagnostic dataset for compositional language and elementary visual reasoning. In Proceedings of the IEEE Conference on Computer Vision and Pattern Recognition, Honolulu, HI, USA, 21–26 July 2017; pp. 2901–2910.

13. Su, H.; Jampani, V.; Sun, D.; Maji, S.; Kalogerakis, E.; Yang, M.-H.; Kautz, J. Splatnet: Sparse lattice networks for point cloud processing. In Proceedings of the IEEE Conference on Computer Vision and Pattern Recognition, Salt Lake City, UT, USA, 18–23 June 2018; pp. 2530–2539.

14. Qi, C.R.; Su, H.; Mo, K.; Guibas, L.J. Pointnet: Deep learning on point sets for 3d classification and segmentation. In Proceedings of the IEEE Conference on Computer Vision and Pattern Recognition, Honolulu, HI, USA, 21–26 July 2017; pp. 652–660.

15. Gkioxari, G.; Malik, J.; Johnson, J. Mesh r-cnn. In Proceedings of the IEEE International Conference on Computer Vision, Seoul, Korea, 27–28 October 2019; pp. 9785–9795.

16. Liu, Z.; Tang, H.; Lin, Y.; Han, S. Point-Voxel CNN for efficient 3D deep learning. Advances in Neural Information Processing Systems. In lProceedings of the Conference on Neural Information Processing Systems, Vancouver, BC, Canada, 8–14 December 2019; pp. 963–973.

17. Eslami, S.M.A.; Rezende, D.J.; Besse, F.; Viola, F.; Morcos, A.S.; Garnelo, M.; Ruderman, A.; Rusu, A.A.; Danihelka, I.; Gregor; K.; et al. Neural scene representation and rendering. *Science* **2018**, *360*, 1204–1210. [CrossRef] [PubMed]

18. Kingma, D.P.; Mohamed, S.; Rezende, D.J.; Mohamed, S.; Welling, M. Semi-supervised learning with deep generative models. *arXiv* **2014**, arXiv:1406.5298.

19. Park, J.J.; Florence, P.; Straub, J.; Newcombe, R.; Lovegrove, S. Deepsdf: Learning continuous signed distance functions for shape representation. In Proceedings of the IEEE Conference on Computer Vision and Pattern Recognition, Long Beach, CA, USA, 16–20 January 2019; pp. 165–174.

20. Sitzmann, V.; Zollhöfer, M.; Wetzstein, G. Scene representation networks: Continuous 3D-structure-aware neural scene representations. *arXiv* **2019**, arXiv:1906.01618 2019.

21. Watkins, C.J.C.H.; Dayan, P. Q-learning. *Mach. Learn.* **1992**, *8*, 279–292. [CrossRef]

22. Mnih, V.; Kavukcuoglu, K.; Silver, D.; Graves, A.; Antonoglou, I.; Wierstra, D.; Riedmiller, M. Playing atari with deep reinforcement learning. *arXiv* **2013**, arXiv:1312.5602.

23. Levine, S.; Finn, C.; Darrell, T.; Abbeel, P. End-to-end training of deep visuomotor policies. *J. Mach. Learn. Res.* **2016**, *17*, 1334–1373.

24. Das, A.; Datta, S.; Gkioxari, G.; Lee, S.; Parikh, D.; Batra, D. Embodied question answering. In Proceedings of the IEEE Conference on Computer Vision and Pattern Recognition, Salt Lake City, UT, USA, 18–23 June 2018; pp. 2054–2063.

25. Yu, L.; Chen, X.; Gkioxari, G.; Bansal, M.; Berg, T.L.; Batra, D. Multi-target embodied question answering. In Proceedings of the IEEE Conference on Computer Vision and Pattern Recognition, Long Beach, CA, USA, 16–20 June 2019; pp. 6309–6318.

26. Wijmans, E.; Datta, S.; Maksymets, O.; Das, A.; Gkioxari, G.; Lee, S.; Essa, I.; Parikh, D.; Batra, D. Embodied question answering in photorealistic environments with point cloud perception. In Proceedings of the IEEE Conference on Computer Vision and Pattern Recognition, Long Beach, CA, USA, 16–20 Junuary 2019; pp. 6659–6668.

27. Gers, F.A.; Schmidhuber, J.; Cummins, F. Learning to forget: Continual prediction with LSTM. *Neural Comput.* **2000**, *12*, 2451–2471. [CrossRef] [PubMed]

28. Paszke, A.; Gross, S.; Massa, F.; Lerer, A.; Bradbury, J.; Chanan, G.; Killeen, T.; Lin, Z.; Gimelshein, N.; Antigaet, L.; et al. PyTorch: An imperative style, high-performance deep learning library. *arXiv* **2019**, arXiv:1912.01703.

29. Kingma, D.P.; Ba, J. Adam: A method for stochastic optimization. *arXiv* **2014**, arXiv:1412.6980.

30. Cho, K.; Van Merriënboer, B.; Gulcehre, C.; Bahdanau, D.; Bougares, F.; Schwenk, H.; Bengio, Y. Learning phrase representations using RNN encoder-decoder for statistical machine translation. *arXiv* **2014**, arXiv:1406.1078.

31. Vaswani, A.; Shazeer, N.; Parmar, N.; Uszkoreit, J.; Jones, L.; Gomez, A.N.; Kaiser, L.; Polosukhinl, I. Attention is all you need. *arXiv* **2017**, arXiv:1706.03762.

32. FiLM Implementation Code. Available online: https://github.com/ethanjperez/film (accessed on 8 April 2020).

33. Su, H.; Maji, S.; Kalogerakis, E.; Learned-Miller, E. Multi-view convolutional neural networks for 3D shape recognition. In Proceedings of the IEEE International Conference on Computer Vision, Santiago, Chile, 7–13 December 2015; pp. 945–953.

34. Kanezaki, A.; Matsushita, Y.; Nishida, Y. Rotationnet: Joint object categorization and pose estimation using multiviews from unsupervised viewpoints. In Proceedings of the IEEE Conference on Computer Vision and Pattern Recognition, Salt Lake City, UT, USA, 18–23 January 2018; pp. 5010–5019.

35. Software Blender Site. Available online: https://www.blender.org/ (accessed on 8 April 2020).

36. CLEVR Dataset Generation Implementation Code. Available online: https://github.com/facebookresearch/clevr-dataset-gen (accessed on 8 April 2020).

37. Sony Alpha 7 III Mirrorless Single Lens Digital Camera Site. Available online: https://www.sony.jp/ichigan/products/ILCE-7RM3/ (accessed on 8 April 2020).

38. 3D-MFP Site. Available online: https://www.ortery.jp/photography-equipment/3d-photography-ja/3d-mfp/ (accessed on 8 April 2020).

39. 3DSOM Pro V5 Site. Available online: http://www.3dsom.com/new-release-v5-markerless-workflow/ (accessed on 8 April 2020).

MDPI

St. Alban-Anlage 66

4052 Basel

Switzerland

Tel. +41 61 683 77 34

Fax +41 61 302 89 18

www.mdpi.com

Sensors Editorial Office

E-mail: sensors@mdpi.com

www.mdpi.com/journal/sensors